T0192206

Experimental Vibration Analysis for Civil Structures

Testing, Sensing, Monitoring, and Control

Taylor and Francis Series in Resilience and Sustainability in Civil, Mechanical, Aerospace and Manufacturing Engineering Systems

Series Editor:
Mohammad Noori
Cal Poly San Luis Obispo

PUBLISHED TITLES

Resilience of Critical Infrastructure Systems
Emerging Developments and Future Challenges
Zhishen Wu, Xilin Lu, and Mohammad Noori

Experimental Vibration Analysis for Civil Structures
Testing, Sensing, Monitoring, and Control
Jian Zhang, Zhishen Wu, Mohammad Noori, and Yong Li

For more information about this series, please visit: www.routledge.com

Experimental Vibration Analysis for Civil Structures
Testing, Sensing, Monitoring, and Control

Edited by
Jian Zhang, Zhishen Wu, Mohammad Noori, and Yong Li

CRC Press is an imprint of the
Taylor & Francis Group, an **informa** business

MATLAB® is a trademark of The MathWorks, Inc. and is used with permission. The MathWorks does not warrant the accuracy of the text or exercises in this book. This book's use or discussion of MATLAB® software or related products does not constitute endorsement or sponsorship by The MathWorks of a particular pedagogical approach or particular use of the MATLAB® software

First edition published 2021
by CRC Press
6000 Broken Sound Parkway NW, Suite 300, Boca Raton, FL 33487-2742

and by CRC Press
2 Park Square, Milton Park, Abingdon, Oxon, OX14 4RN

© 2021 Taylor & Francis Group, LLC

CRC Press is an imprint of Taylor & Francis Group, LLC

ISBN: 978-0-367-54746-2 (hbk)
ISBN: 978-1-003-09056-4 (ebk)

Typeset in Times
by codeMantra

Contents

Preface

This book is a collection of selected papers presented at the International Conference on Experimental Vibration Analysis for Civil Engineering Structures (EVACES), held in Nanjing, China in September 2019.

EVACES is a premier venue where recent progress in the field of civil infrastructure systems, especially the latest developments in experimental vibration analysis research in civil engineering structures, is presented and discussed by experts from the global community. Over the past decade, due to emergence of new technologies and computational tools, the field of civil engineering and infrastructure systems has witnessed a significant advancement in the aforementioned areas. EVACES provides a forum for the global research community to share and exchange the latest achievements in these areas. After the first seven successful EVACES conferences which took place in Bordeaux, France (2005), Porto, Portugal (2007), Wroclaw, Poland (2009), Varenna, Italy (2011), Ouro Preto, Brazil (2013), Dubendorf, Switzerland (2015), and San Diego, United States of America (2017), EVACES 2019 was organized by the Southeast University (SEU) at Nanjing, held from September 5–8, 2019, on the main campus of SEU. The main topics of EVACES 2019 included (1) damage identification and structural health monitoring, (2) testing, sensing, and modeling, (3) vibration isolation and control, (4) system and model identification, (5) coupled dynamical systems (including human-structure, vehicle structure, and soil-structure interaction), and (6) application of big data and artificial intelligence techniques. The key objective of this meeting was to present and discuss the latest developments in the aforementioned areas in order to better understand how to assess the state of health of structural, geo-structural, and complex infrastructure systems and how to predict their remaining useful life using vibration data collected from these systems when subjected to operational and extreme loads. It is noteworthy to indicate this was the first time that the importance of advanced data analysis (e.g., system and damage identification and artificial intelligence) methods in experimental vibration analysis of critical infrastructure, as an important and emerging area, was addressed to support decision making related to maintenance and inspection, retrofit, upgrade, and rehabilitation of these systems especially in case of emergency response.

This book contains selected papers prepared by the leading researchers that were presented at EVACES 2019. These contributions can provide strategic roadmaps for research in the area of experimental vibration analysis for civil engineering structures and for building more resilient and sustainable civil infrastructure systems.

We express our sincere thanks to the members of the Scientific Committee and, in particular, to all the authors and participants, for their essential and valuable contributions.

Workshop Co-Chairs and the Organizing Committee:

Jian Zhang, Southeast University, China (Co-Chair)
Zhishen Wu, Southeast University, China (Co-Chair)
A. Emin Aktan, Drexel University, USA (Co-Chair)
James Brownjohn, University of Exeter, UK (Co-Chair)
Gang Wu, Southeast University, China (Organizing Committee)
Jingquan Wang, Southeast University, China (Organizing Committee)
Joel P. Conte, University of California San Diego, USA (Organizing Committee)
Yong Li, Southeast University (Secretary General)
Zheng Liu, University of British Columbia, Canada (Organizing Committee)
Maosen Cao, Hehai University, China (Organizing Committee)
Yufeng Zhang, Jiangsu Transportation Institute (JSTI), China (Organizing Committee)

Yan Xu, Southeast University, China (Organizing Committee)
Mohammad Noori, Cal Poly, San Luis Obispo, USA (Organizing Committee)

MATLAB® is a registered trademark of The MathWorks, Inc. For product information, please contact:
The MathWorks, Inc.
3 Apple Hill Drive
Natick, MA 01760-2098 USA
Tel: 508-647-7000
Fax: 508-647-7001
E-mail: info@mathworks.com
Web: www.mathworks.com

Editors

Professor Jian Zhang is a professor and vice dean of the School of Civil Engineering, SEU, China. He received his PhD from Kyoto University, Japan, and worked at the University of California at San Diego and Drexel University, USA. In the area of structural health monitoring, he has published 4 books and over 50 first/corresponding-author Science Citation Index journal papers. His research results have been applied on over 20 long-span bridges including the Sutong Yangzi-River Bridge and the Second Humen Bridge. He is the editor board member of *Journal of Computer-Aided Civil and Infrastructure Engineering* and *Journal of Structural Control and Monitoring*. He serves as co-chair of the EVACES-8 conference and as the International Society for Structural Health Monitoring of Intelligent Infrastructure (ISHMII council member. He was awarded the first prize of Science and Technology Award of Jiangsu Province (twice), the first prize of Science and Technology of China Highway Society, and the second prize of National Award for Technological Invention.

Professor Zhishen Wu is a distinguished professor at Ibaraki University, Japan and at SEU, China. His research expertise includes structural/concrete/maintenance engineering and advanced composite materials. He is the author or co-author of over 600 refereed papers including over 200 journal articles and has given 50 keynote or invited papers. He also holds 50 patents. Dr Wu was awarded the JSCE Research Prize by Japan Society of Civil Engineering in 1990, the JSCM Technology Award by the Japan Society for Composite Materials in 2005 and 2009, SHM Person of the Year Award by *International Journal of Structural Health Monitoring*, and the National Prize for Progress in Science and Technology (2nd) of China in 2012. He is also a member of Japan National Academy of Engineering. He chairs China Chemical Fibers Association Committee on Basalt Fibers and is the president of the ISHMII. He is also an elected follow of American Society of Civil Engineering (ASCE), Japan Society of Civil Engineers (JSCS), ISHMII, and the International Institute of Fiber Reinforced Polymers (FRP) in Construction. He also serves as an editor, associate editor, and editorial board member for more than ten international journals. He is the founding chief editor of *International Journal of Sustainable Materials and Structural Systems*. Professor Wu has supervised a large number of national research projects in China and was the founder of the International of Institute for Urban Systems Engineering, at SEU.

Professor Mohammad Noori is a professor of mechanical engineering at Cal Poly, San Luis Obispo, a Fellow of the American Society of Mechanical Engineering, and a recipient of the Japan Society for Promotion of Science Fellowship. Dr Noori's work in nonlinear random vibrations, seismic isolation, and the application of artificial intelligence methods for structural health monitoring is widely cited. He has authored over 250 refereed papers, including over 100 journal articles and 6 scientific books, and has edited 25 technical and special journal volumes. Noori has supervised over 90 graduate students and post-doc scholars, and has presented over 100 keynote, plenary, and invited talks. He is the founding executive editor of an international journal and has served on the editorial boards of over 10 other journals and as a member of numerous scientific and advisory boards. He has been a Distinguished Visiting Professor at several highly ranked global universities, and directed the Sensors Program at the National Science Foundation in 2014. He has been a founding director or co-founder of three industry-university research centers and held Chair professorships at two major universities. He served as the dean of engineering at Cal Poly for five years, has also served as the Chair of the national committee of mechanical engineering department heads, and was one of 7 co-founders of the National Institute of Aerospace, in partnership with NASA Langley Research Center. Noori also serves as the Chief Technical Advisor for several scientific organizations and industries.

Dr. Yong Li is an assistant professor in the School of Civil Engineering, SEU, China. He received his undergraduate degree in 2012 and his PhD in 2017, both degrees in aerospace engineering and applied mechanics, at Tongji University, in Shanghai, China. His research focused on the coupling relationship between mechanics and chemical reaction. He has published 1 book and over 20 journal papers in leading journals. He is also a reviewer for several journals in the fields of mechanics and physics. Dr Li is the recipient of an Outstanding Young Investigator Award supported by the National Natural Science Foundation of China. He is also engaged in several other national key research and development programs as an investigator. He served as secretary general of the International Conference on EVACES-8.

Contributors

J. Abell
Faculty of Engineering and Applied Sciences
Universidad de los Andes
Santiago, Chile

Zulkifl Ahmed
School of Resource and Civil Engineering
Northeastern University
Shenyang, China

D. Ai
School of Civil Engineering and Mechanics
Huazhong University of Science and
    Technology
Wuhan, China

J. S. Ali
Department of Civil Engineering
Aliah University
Kolkata, India

Mohammad Almutairi
School of Civil Engineering
University of Leeds
Leeds, United Kingdom
and
Sustainability and Infrastructure Programme,
    Energy and Building Research Center
Kuwait Institute for Scientific Research
Safat, Kuwait

Wael A. Altabey
International Institute for Urban Systems
    Engineering
Southeast University
Nanjing, Jiangsu, China.
and
Faculty of Engineering, Department of
    Mechanical Engineering
Alexandria University
Alexandria, Egypt

D. Anastasopoulos
Department of Civil Engineering
University of Leuven (KU Leuven)
Leuven, Belgium

R. Astroza
Faculty of Engineering and Applied Sciences
Universidad de los Andes
Santiago, Chile

Onur Avci
Department of Civil Engineering
Qatar University
Doha, Qatar

Bai Wen
Institute of Engineering Mechanics
China Earthquake Administration
Beijing, China
and
Key Laboratory of Earthquake Engineering
    and Engineering Vibration of China
    Earthquake Administration
Harbin, China

A. Bakhtiar
Department of Engineering
Sepehr Institute of Higher Education
Isfahan, Iran

D. Bandyopadhyay
Department of Construction Engineering
Jadavpur University
Kolkata, India

Baozhuang Zhang
State Key Laboratory of Green Building in
    Western China
Xi'an University of Architecture &
    Technology
Xi'an, China

N. Barrientos
Faculty of Engineering and Applied Sciences
Universidad de los Andes
Santiago, Chile

Ivan Bartoli
Department of Civil, Architectural and
    Environmental Engineering
Drexel University
Philadelphia, Pennsylvania

Farhad Behnamfar
Department of Civil Engineering
Isfahan University of Technology
Isfahan, Iran

Binbin Li
ZJU-UIUC Institute
Zhejiang University
Haining, China

B. Briseghella
Sustainable and Innovative Bridge Engineering
    Research Center (SIBERC)
College of Civil Engineering, Fuzhou
    University
Fuzhou, P. R. China

P. Chansukho
Field of Structural Engineering
Asian Institute of Technology
Pathumthani, Thailand

Chaoyang Tang
School of Civil Engineering
Fuzhou University
Fuzhou, China
and
Fujian Highway Administrative Bureau
Fuzhou, China

Z. Chen
School of Civil Engineering
Tianjin University
Tianjin, China

Zhiwei Chen
Department of Civil Engineering
Xiamen University
Xiamen, China

Chi Peihong
Institute of Earthquake Protection and Disaster
    Mitigation
Lanzhou University of Technology
Lanzhou, China

Chiu Jen Ku
Department of Civil and Environmental
    Engineering
Shantou University
Guangdong, China

Chuan-Zhi Dong
Department of Civil, Environmental, and
    Construction Engineering
University of Central Florida
Orlando, Florida

Chunyu Qian
China Jikan Research Institute of Engineering
    Investigations and Design, Co., Ltd
Xi'an, Shaanxi 710043, China

Dai Junwu
Institute of Engineering Mechanics
China Earthquake Administration
Beijing, China
and
Key Laboratory of Earthquake Engineering
    and Engineering Vibration of China
    Earthquake Administration
Harbin, China

Ji Dang
Department of Civil and Environmental
    Engineering
University of Saitama
Saitama, Japan

De-Hui Tang
Harbin Institute of Technology (Shenzhen)
School of Civil Engineering
Shenzhen, China

G. De Roeck
Department of Civil Engineering
University of Leuven (KU Leuven)
Leuven, Belgium
and
Structural Mechanics Division
University of Leuven (KU Leuven)
Leuven, Belgium
and
Sustainable and Innovative Bridge Engineering
    Research Center (SIBERC)
College of Civil Engineering, Fuzhou
    University
Fuzhou, P. R. China

Contributors xvii

Y. Ding
Key Laboratory of Coast Civil Structures
  Safety of Ministry of Education
Tianjin University
Tianjin, China

and

Department of Civil Engineering
Zhejiang University
Hangzhou, China

Dong Furui
School of Resource and Civil Engineering
Northeastern University
Shenyang, China

Du Yongfeng
Institute of Earthquake Protection and
  Disaster Mitigation
Lanzhou University of Technology
Lanzhou, China

A. Emin Aktan
Department of Civil, Architectural and
  Environmental Engineering
Drexel University
Philadelphia, Pennsylvania

Fang Dengjia
Institute of Earthquake Protection and
  Disaster Mitigation
Lanzhou University of Technology
Lanzhou, China

S. E. Fang
School of Civil Engineering
Fuzhou University
Fuzhou, Fujian Province, China
and
National and Local United Research Center
  for Seismic and Disaster Informatization of
  Civil Engineering
Fuzhou University
Fuzhou 350108, China

Feng Gu
Department of Civil and Environmental
  Engineering
Shantou University
Guangdong, China

Feng Qianshuo
School of Civil Engineering
Central South University
Changsha, China

K. Gao
School of Civil Engineering and Mechanics
Huazhong University of Science and
  Technology
Wuhan, P. R. China

H. Ge
Department of Civil Engineering
Meijo University
Nagoya, Japan

M. R. Ghasemi
Faculty of Engineering, Department of
  Civil Engineering
University of Sistan and Baluchestan
Zahedan, Iran

R. Ghiasi
Faculty of Engineering, Department of
  Civil Engineering
University of Sistan and Baluchestan
Zahedan, Iran

C. S. Goit
Department of Civil and Environmental
  Engineering
Saitama University
Saitama, Japan

Kirk Grimmelsman
Intelligent Infrastructure Systems
Philadelphia, Pennsylvania

M. Guarini
Faculty of Engineering and Applied Sciences
Universidad de los Andes
Santiago, Chile

J. M. Gutierrez
Faculty of Engineering and Applied Sciences
Universidad de los Andes
Santiago, Chile

Jianping Han
Key Laboratory of Disaster Prevention
    and Mitigation in Civil Engineering of
    Gansu Province
Lanzhou University of Technology
Lanzhou, China

J. He
Key Laboratory of Wind and Bridge
    Engineering of Hunan Province
College of Civil Engineering, Hunan
    University
Changsha, China

L. He
Sustainable and Innovative Bridge Engineering
    Research Center (SIBERC)
College of Civil Engineering, Fuzhou
    University
Fuzhou, P. R. China

W. Y He
Department of Civil Engineering
Hefei University of Technology
Hefei, Anhui Province, China

X. H. He
School of Civil Engineering
Central South University
Changsha, China
and
National Engineering Laboratory for High
    Speed Railway Construction
Changsha, China
and
Joint International Research Laboratory of
    Key Technology for Rail Traffic Safety
Changsha, China

Zhiqiang Hou
The Fifth Construction Co., Ltd. of China
    TIESIJU Civil Engineering Group
Jiujiang, China

J. Y. Huang
School of Civil Engineering
Fuzhou University
Fuzhou, Fujian Province, China

F. Jaramillo
Department of Electrical Engineering
University of Chile
Santiago, Chile

Jiang Lizhong
School of Civil Engineering
Central South University
Changsha, China

Jiayan Lei
School of Architecture and Civil Engineering
Xiamen University
Xiamen, China

Jiyuan Shi
Department of Civil and Environmental
    Engineering
University of Saitama
Saitama, Japan

Jun Li
Centre for Infrastructure Monitoring and
    Protection, School of Civil and Mechanical
    Engineering
Curtin University
Bentley, Australia

Jun Teng
Harbin Institute of Technology (Shenzhen)
Shenzhen, China

Junhao Zheng
College of Civil Engineering
Fuzhou University
Fuzhou, Fujian, China

Junlong Zhu
Shanghai Baitong Project Management
    Consulting, Co., Ltd
Shanghai, China

Kang Liu
College of Civil Engineering
Fuzhou University
Fuzhou, Fujian, China

Muhammad Israr Khan
School of Resources and Civil Engineering
Northeastern University
Shenyang city, Liaoning Province, China

D. Li
School of Civil Engineering
Hefei University of Technology
Hefei, Anhui, China

H. Li
School of Civil Engineering and
    Mechanics
Huazhong University of Science and
    Technology
Wuhan, China

Li Junxing
Department of Civil Engineering
Harbin Institute of Technology
Harbin, China

Li Xu
College of Civil Engineering
Fuzhou University
Fuzhou, Fujian, China

Liao Yuchen
School of Civil Engineering
Southeast University
Nanjing, China

Lin Dinan
Fujian Academy of Bulding Research
Fujian Key Laboratory of Green Building
    Technology
Fuzhou, China

Linfeng Hu
School of Architecture and Civil
    Engineering
Xiamen University
Xiamen, China

H. Liu
School of Civil Engineering
Tianjin University
Tianjin, China

Liyuan Wang
College of Civil Engineering
Fuzhou University
Fuzhou, China

H. Luo
School of Civil Engineering and Mechanics
Huazhong University of Science and
    Technology
Wuhan, China

Luo Sihui
School of Civil Engineering
Central South University
Changsha, China

H. Mao
School of Ecological Environment and
    Urban Construction
Fujian University of Technology
Fuzhou, China

G. C. Marano
Sustainable and Innovative Bridge Engineering
    Research Center (SIBERC)
College of Civil Engineering, Fuzhou
    University
Fuzhou, P. R. China

Matteo Mazzotti
Department of Civil, Architectural and
    Environmental Engineering
Drexel University
Philadelphia, Pennsylvania

Mingxin Gao
Department of Bridge and Tunnel Engineering
Harbin Institute of Technology
Harbin, China

T. Nagayama
Department of Civil Engineering
The University of Tokyo
Tokyo, Japan

Mohammadreza Najar
Department of Civil Engineering
University of Isfahan
Isfahan, Iran

F. Necati Catbas
Department of Civil, Environmental, and
    Construction Engineering
University of Central Florida
Orlando, Florida

Z. H. Nie
College of Science & Engineering
Jinan University
Guangzhou, Guangdong, China
and
The Key Laboratory of Disaster Forecast and
    Control in Engineering
Guangdong, Guangzhou, China

Nikolaos Nikitas
School of Civil Engineering
University of Leeds
Leeds, United Kingdom

Mohammad Noori
International Institute for Urban Systems
    Engineering (IIUSE)
Southeast University
Nanjing, Jiangsu, China
and
Department of Mechanical Engineering
California Polytechnic State University
San Luis Obispo, California

M. Orchard
Department of Electrical Engineering
University of Chile
Santiago, Chile

Pengyu Wang
School of Resources and Civil Engineering
Northeastern University
Shenyang City, Liaoning Province, China

M. C. Qi
Key Laboratory of Wind and Bridge
    Engineering of Hunan Province
College of Civil Engineering, Hunan
    University
Changsha, China

Qifang Xie
State Key Laboratory of Green Building in
    Western China
Xi'an University of Architecture &
    Technology
Xi'an, China

G. Z. Qu
China Railway Construction Bridge Design and
    Research Branch
China Railway Siyuan Survey and Design
    Group Co., LTD.
Wuhan, Hubei, China

Qunjun Huang
College of Civil Engineering
Fuzhou University
Fuzhou, Fujian, China

W. X. Ren
Department of Civil Engineering
Hefei University of Technology
Hefei, Anhui Province, China

E. Reynders
Department of Civil Engineering
University of Leuven (KU Leuven)
Leuven, Belgium
and
Structural Mechanics Division
University of Leuven (KU Leuven)
Leuven, Belgium

Z. F. Shen
College of Science & Engineering
Jinan University
Guangzhou, Guangdong, China

S. Shi
College of Civil Engineering
Fuzhou University
Fuzhou, China

Y. Shi
Key Laboratory of Coast Civil Structures
    Safety of Ministry of Education
Tianjin University
Tianjin, China

Shiqi Fu
College of Civil Engineering
Huaqiao University
Xiamen, P. R. China

Shuhong Wang
School of Resource and Civil Engineering
Northeastern University
Shenyang, China

A. I. Silik
International Institute for Urban Systems
    Engineering (IIUSE)
Southeast University
Nanjing, China
and
Faculty of Engineering Sciences,
    Department of Civil Engineering
Nyala University
Nyala, Sudan

Siyang Ma
College of civil Engineering
Fuzhou University
Fuzhou, China

Soto, C.
Faculty of Engineering and Applied Sciences
Universidad de los Andes
Santiago, Chile

U. Starossek
Head of Institute, Structural Analysis
    Institute
Hamburg University of Technology
Hamburg, Germany

D. Su
Department of Civil Engineering
The University of Tokyo
Tokyo, Japan

X. T. Sun
School of Civil Engineering
Hefei University of Technology
Hefei, Anhui, China

R. Terrill
Structural Analysis Institute
Hamburg University of Technology
Hamburg, Germany

Tianyu Zheng
College of Civil Engineering
Fuzhou University
Fuzhou, China

Wang Fan
School of Civil Engineering
Central South University
Changsha, China

Luyao Wang
Department of Civil and Environmental
    Engineering
Saitama University
Saitama, Japan

T. Wang
International Institute for Urban Systems
    Engineering
Southeast University
Nanjing, Jiangsu, China

Xin Wang
Department of Architecture and Building
    Engineering
Tokyo University of Science
Tokyo, Japan

P. Warnitchai
Department of Structural Engineering
Asian Institute of Technology
Pathumthani, Thailand

Wei Shi
School of Architecture and Civil Engineering
Xiamen University
Xiamen, China

Weibiao Yang
CSCEC Strait Construction and Development
    Co., Ltd
Fuzhou, China

Wei-Hua Hu
School of Civil Engineering
Harbin Institute of Technology (Shenzhen)
Shenzhen, China

Wen Wang
School of Civil Engineering and Architecture
Guangxi University of Science and Technology
Guangxi, China
and
State Key Laboratory of Disaster Reduction in
    Civil Engineering
Tongji University
Shanghai, China

Wen Xiong
Department of Bridge Engineering,
    School of Transportation
Southeast University
Nanjing, China

S. Weng
School of Civil Engineering and
    Mechanics
Huazhong University of Science and
    Technology
Wuhan, P. R. China

Wenqian Chen
College of Civil Engineering
Fuzhou University
Fuzhou, Fujian, China

Z. Wu
International Institute for Urban Systems
    Engineering
Southeast University
Nanjing, Jiangsu, China

Y. Xia
Department of Civil and Environmental
    Engineering
The Hong Kong Polytechnic University
Hong Kong, P. R. China

Xiangang Lai
Department of Civil, Architectural and
    Environmental Engineering
Drexel University
Philadelphia, Pennsylvania

Y. K. Xie
College of Science & Engineering
Jinan University
Guangzhou, Guangdong, China
and
The Key Laboratory of Disaster Forecast and
    Control in Engineering
Guangdong, Guangzhou, China

Xuanneng Gao
College of Civil Engineering
Huaqiao University
Xiamen, P. R. China

K. Xue
Department of Civil Engineering
The University of Tokyo
Tokyo, Japan

J. F. Yan
China Railway Construction Bridge Design and
    Research Branch
China Railway Siyuan Survey and Design
    Group Co., LTD.
Wuhan, Hubei, China

Yan Jiang
School of Civil Engineering and Architecture
Guangxi University of Science and Technology
Guangxi, China

X. Yan
College of Civil Engineering
Fuzhou University
Fuzhou, China

Yan-An Gao
Faculty of Architecture and Civil Engineering
Huaiyin Institute of Technology
Huaian, P. R. China

J. Yang
School of Civil Engineering
Central South University
Changsha, China
and
National Engineering Laboratory for
    High Speed Railway Construction
Changsha, China
and
Joint International Research Laboratory of
    Key Technology for Rail Traffic Safety
Changsha, China

Yang Liu
Department of Bridge and Tunnel Engineering
Harbin Institute of Technology
Harbin, China

Yang Xu
College of Civil Engineering
Fuzhou University
Fuzhou, Fujian, China

Yan-Long Xie
ZJU-UIUC Institute
Zhejiang University
Haining, China

A. A. Yazdani
Department of civil engineering
University of Isfahan
Isfahan, Iran

K. Ye
Department of Civil Engineering
Hefei University of Technology
Hefei, China

X. W. Ye
Department of Civil Engineering
Zhejiang University
Hangzhou, China

P. C. Yin
China Railway Construction Bridge Design and
    Research Branch
China Railway Siyuan Survey and
    Design Group Co., LTD.
Wuhan, Hubei, China

Z. Ying
International Institute for Urban Systems
    Engineering
Southeast University
Nanjing, Jiangsu, China

Yongjian Chen
College of civil Engineering
Fuzhou University
Fuzhou, China

Yu Cheng
Department of Bridge Engineering,
    School of Transportation
Southeast University
Nanjing, China

X. S. Yu
China Railway Construction Bridge Design and
    Research Branch
China Railway Siyuan Survey and Design
    Group Co., LTD.
Wuhan, Hubei, China

V. Zabel
Institute of Structural Mechanics
Bauhaus-Universität Weimar
Weimar, Germany

C. Zhang
College of Civil Engineering
Fuzhou University
Fuzhou, China

Hongyu Zhang
Key Laboratory of Disaster Prevention and
    Mitigation in Civil Engineering of Gansu
    Province
Lanzhou University of Technology
Lanzhou, China

J. Zhang
Department of Civil Engineering
Hefei University of Technology
Hefei, China

Zhang Kun
School of Civil Engineering
Southeast University
Nanjing, China

Zhang Wenyuan
Department of Civil Engineering
Harbin Institute of Technology
Harbin, China

X. X. Zhang
Key laboratory of wind and bridge engineering
    of Hunan Province
College of Civil Engineering, Hunan University
Changsha, China

L. Zhou
College of Civil Engineering
Fuzhou University
Fuzhou, China

H. P. Zhu
School of Civil Engineering and Mechanics
Huazhong University of Science and
    Technology
Wuhan, P. R. China

Zhu Zhihui
School of Civil Engineering
Central South University
Changsha, China

Zhuhong Ouyang
Department of Civil and Environmental
    Engineering
Shantou University
Shantou, China

Zihao Wang
School of Architecture and Civil Engineering
Xiamen University
Xiamen, China

Zong Zhouhong
School of Civil Engineering
Southeast University
Nanjing, China

Y. F. Zou
School of Civil Engineering
Central South University
Changsha, China
and
National Engineering Laboratory for
    High Speed Railway Construction
Changsha, China
and
Joint International Research Laboratory of
    Key Technology for Rail Traffic Safety
Changsha, China

Rongzhi Zuo
Department of Civil and Environmental
    Engineering
University of Saitama
Saitama, Japan

1 Wavelet-Based Damage-Sensitive Features Extraction

A. I. Silik
Southeast University
Nyala University

Mohammad Noori
California Polytechnic State University

W. A. Altabey
Southeast University
Alexandria University

CONTENTS

1.1 INTRODUCTION

Signal-based damage detection techniques have been introduced to extract features to represent the signal characteristics which are used to answer the questions associated to damage levels [1,2]. Over the last decades, various methods have been developed to improve the feature extraction (FE) procedure [3–10]. Yoon et al. [11] and Staszewski [12] discussed the signal processing role for FE in composite materials and stated that the advances in damage detection attribute to the development of signal processing techniques such as denoising, FE, and optimal sensor location procedures. FE-based signal processing has been classified into time, frequency, and time-frequency methods [13]. It aims to transform data into a lower dimension by removing redundant information and preserving the useful elements [14]. In time domain, features are extracted from the signal using linear and nonlinear functions [15]. In frequency and time-frequency domain wavelet transform (wt) have been used for FE. Pittner and Kamarthi [16] explored the changes in wavelet coefficients for FE for pattern recognition tasks. Balafas and Kiremidjian [17] proposed damage features for earthquake damage estimation based on continuous wavelet transform (CWT), using ground motion and its structure responses. Patel et al. [18] used complex Gaussian wavelet to extract the discontinuity of the acceleration response of a reinforced concrete (RC) building. The results showed that wavelet weights are directly affected by the change in physical properties and can identify damage to a reasonable extent. The modified wavelet energy rate-based damage detection methods are proposed by several researchers [19–22]. The method based on Approximate Entropy [23] is proposed to assess the damaged condition from the vibration signals generated by excitation [24–26] on structures. It is able to capture the deterioration of beams with more sensitive characteristics. Although many works have been done using wt for damage detection [27,28], however, it cannot always identify the damage extent or its location. Most of them are not suited when the signal is nonstationary and is mainly restricted to small structures, simulated models, or laboratory tests. Moreover, some of them refer to specific load scenarios that are not common in real structures; their applicability is limited and the ground motion recording is neglected. Further, only one feature was considered to resemble the dynamic characteristic, and the base wavelet and resolution level are chosen in an arbitrary manner, which results in a negative effect in results. It's clear that there is a lack of a truly intelligent and comprehensive approach that can detect the slight damage in a large, complex system and in a noisy environment that may include a large number of relevant features, and not just a few dynamic features, to detect the location and the damage extent. Thus, an advanced intelligent algorithm for sensitive features extraction of large complex data is required. Other crucial issue in damage detection is how to select the correct features that describe the damage and improve the AI-algorithm accuracy. In this study, given the modulation characteristics of structural dynamic responses (SDR), the wavelet is used for damage-sensitive parameter (DSF) extraction which cannot be obtained either by short Fourier transform or Wigner-Ville distribution. Here, FE is used to describe how to find the best features which represent various damage conditions, whereas feature selection is used to define how to select features for damage detection based wavelet DSFs. The proposed algorithm is able to de-noise and extract the features that define the structure's true state. The algorithm's novelty is to identify and extract several key features that can determine the key dynamic characteristics of the structure and correlate to the damages.

1.2 DATA PREPROCESSING

First, the raw data are visualized as in Figure 1.1 to get an intuitive feeling for the data. Some transitory characteristics are quite evident in the time domain, while the frequency content is quite clear and the user can specify a range of frequencies for further analysis. Hence, it seems necessary to do some form of preprocessing prior to any statistical model development. Preprocessing involves normalization, trend isolation, and denoising. Noise presence may affect the detection of high-frequency (HF) components related to damage, so, signal de-noising is necessary, and in some situations, the frequency band of an unwanted signal lies within the same signal frequency range which is required to remove and preserve the useful information. Figure 1.1 visualizes the excitation in various domains. The upper right plot shows the discrete Fourier transform (DFT) of the signal. It indicates that the signal has frequencies between 0.0 and 10 Hz, but neither gives any indication of their evolution or the presence of the discontinuity. The lower left plot is the time-frequency plot of the CWT showing the localization in time and frequency to identify changes in the signal frequency as time evolves, which is a contour map.

1.2.1 FEATURE DETECTION AND EXTRACTION

FE is a key aspect of damage diagnosis using a statistical model for civil structures [15]. DSFs are used to track the structure change to detect the damage. However, it is hard to extract useful features from SDR just by observing them in the time domain, so, several techniques have been developed for this purpose. The specific technique choice is usually tied to the application under study and specific requirements. Due to the modulation characteristics of SDR, a wavelet-based method has been developed to analyze nonstationary data in time and frequency domain simultaneously and then extract the features. The input and output variations that reflect changes in the structure due to damage are considered. Based on wavelet weights, DSFs are derived through statistical processing to indicate the true structure state. Due to variations in the input and output, DSFs are normalized

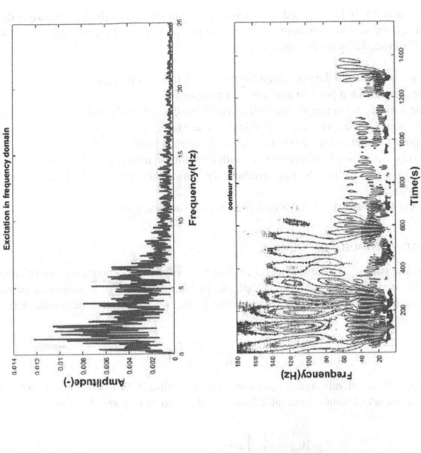

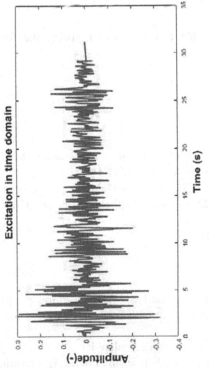

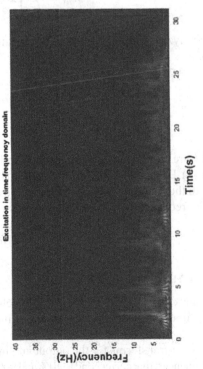

FIGURE 1.1   Excitation visualization in various domains.

to account for variations in loading and environmental conditions. To distinguish between the useful and poor features, the relationship between DSFs and the structure's physical parameters should be investigated. FE procedures are as follows:

1. Load and visualize the data obtained from sensor i, denoted by $x_i(t)$
2. Data are standardized prior to any subsequent analyses
3. Prepossess the signal to remove the artifacts such as all trends and environmental conditions
4. Choose a wavelet function and a set of scales to analyze.
5. Decompose the measured signal into various frequency bands
6. Study every frequency band component with a resolution matched to its scale.
7. Extract DSFs that define the key structure dynamic characteristics that correlated well with the damage.
8. Choose the features exposing the best classification accuracy of these features.

1.2.2 STATISTICAL MODEL FORMULATION

Once an analyzing wavelet $\psi(t)$ has been chosen, two basic parameters should be chosen and manipulations should be done so as to make the $\psi(t)$ more flexible. The $\psi(t)$ is stretched or squeezed through varying its dilation parameter s and moved through its translation parameter τ (i.e. along the localized time index τ).

$$\psi_{s,\tau}(t) = \frac{1}{\sqrt{s}}\psi\left(\frac{t-\tau}{s}\right) \tag{1.1}$$

Let $x(t)$ be the system acceleration response, where t denotes time. CWT of a function $x(t) \in$ L2 ($\Re$), where L2 ($\Re$) is the set of square integrable functions denoted as $W_{s,\tau}$ and defined as

$$W_{s,\tau} = \frac{1}{\sqrt{s}}\int_{\infty}^{-\infty}x(t)\cdot\psi^*\left(\frac{t-\tau}{s}\right)dt \tag{1.2}$$

The basic analyzed wavelet $\phi(t)$ is a square integrable function, and it meets the following relationship:

$$C_\Psi = \int_R \frac{|\psi(\omega)|^2}{|\omega|}d\omega < \infty \tag{1.3}$$

where s and τ are used to adjust the frequency and time location. $W_{s,\tau}$ shows how closely $\psi_{s,\tau}(t)$ correlated with $x(t)$. High scale gives low frequency (LF) and low scale offers HF. HF in a given scale is an indicator to the signal frequency content at that time [29]. At each scale, there is a corresponding characteristic frequency F_a that relates to s, time rate Δt, and wavelet center frequency F_c as

$$F_a = F_c/(s\cdot\Delta t) \tag{1.4}$$

1.2.3 SCALE SELECTION

Multiplier matrix is affected by the scale at which wt is evaluated. So, the procedures proposed by Balafas and Kiremidjian are followed to determine the scales at which wt can be evaluated.

1. The Fourier transforms for the output and input signals are calculated.
2. The frequencies that correspond to 10% and 90% cumulative Fourier amplitudes are determined.

3. The maximum and minimum frequencies determined above are taken as the outer bound frequencies.
4. The frequency boundaries are converted to wavelet scales to determine the scale bounds.
5. The scales at which the wavelet transform will be evaluated are linearly spaced between the scale bounds.

1.2.4 FEATURE DAMAGE INDEX IDENTIFICATION BASED WAVELET

$W_{s,\tau}$ is used to extract the feature vectors. To reduce its dimensionality, statistics over $W_{s,\tau}$ is used as

1. Maximum of the wavelet coefficients in each sub band.
2. Minimum of the wavelet coefficients in each sub band.
3. Mean of the wavelet coefficients in each sub band.
4. Standard deviation of the wavelet coefficients in each sub band.
5. Wavelet Entropy, Wavelet Energy, and Wavelet Power Spectrum
6. Residual Error between reconstructed signal and the actual measured signal

Wavelet Energy at a given scale can indicate damages such as loss of energy and how slowly the energy decays. Nair and Kiremidjian [15] stated that $W_{s,\tau}$ shift in a given scale is described by the energy as

$$E_a = \sum_{b=1}^{K} \left| W\ddot{x}(a,b) \right|^2 \tag{1.5}$$

For a multidegree system with mass (M), damping (C), and stiffness K, subjected to force $g(t)$, damping ratio in each mode is ξ. The transfer response function of the displacement at the k_{th} degrees of freedom is derived as

$$H_{kl}(s) = \sum_{r=1}^{N} \frac{\phi_{kr}\phi_{lr}}{\left(\omega_r^2 - s^2 + 2j\xi\omega_r s \right)} \tag{1.6}$$

And for acceleration at the k_{th} degrees of freedom as

$$\ddot{X}_k(s) = \sum_{r=1}^{N} \frac{-s^2\phi_{kr}\phi_{lr}G_l(s)}{\left(\omega_r^2 - s^2 + 2j\xi\omega_r s \right)} \tag{1.7}$$

$W_{s,\tau}$ of acceleration response $(\ddot{x}(t))$ with respect to the Morlet wavelet is derived as

$$W_M\ddot{x}_k(a,b) = \sqrt{\frac{a}{2\pi}} \int_{-\infty}^{\infty} \sum_{r=1}^{N} \frac{-s^2\phi_{kr}\phi_{lr}G_l(s)}{\left(\omega_r^2 - s^2 + 2j\xi\omega_r s \right)} \exp(jsb)\exp\left[-\frac{1}{2}(-as+\omega_0)^2 \right] ds \tag{1.8}$$

Instantons Energy is a feature that provides the distribution energy in each frequency band and is given as

$$\text{IE} = \log_{10} 1/N_j \cdot \sum_{s=1}^{N_s} \left(W_{s,\tau} \right)^2 \tag{1.9}$$

Teager Energy is a nonlinear operator used to track the signal energy fluctuations and is defined as

$$\text{TE} = \log_{10}\left(\left|\frac{1}{N_s}\sum_{s=1}^{N_s-1}\left(W_{s,\tau}\right)^2 - \left(W_{s,\tau}(t-1)\cdot W_{s,\tau}(t+1)\right)\right|\right) \tag{1.10}$$

Wavelet Entropy: Entropy change rate CR(t) can be used as a damage index to describe the structural state changes and defined as

$$\text{CR}(t) = \frac{s_{\text{wt}}(t) - s_{\text{wt}}(0)}{s_{\text{wt}}(0)} \tag{1.11}$$

where $s_{\text{wt}}(0)$ is undamaged entropy and $s_{\text{wt}}(t)$ for damaged. Approximate Entropy which is a measure of the irregularity and nonlinear time series complexity can be used as a damage feature. It is computed as

$$\text{ApEn} = \Phi_m - \Phi_m + 1 \tag{1.12}$$

where $\Phi_m = (N-m+1)^{-1}\sum_{i=1}^{N-m+1}\log(N_i)$, where N_i is the number within range points.

1.3 NUMERICAL SIMULATION AND RESULTS: DISCUSSION

To validate the effectiveness of the proposed method, a simulated two-story frame is done with lumped mass 100 kips/g at each floor, and same story stiffness 31.54 kips/in for all stories and the story height is 12 ft. The damping ratio for all natural modes is = 0.05. The structure is subjected to El Centro ground motion. The SDR of first floor is selected to illustrate the concept of the process. Figure 1.2 shows the response at first story and its filtered and decay versions representation in time and frequency domains. Figures 1.3 and 1.4 show the various features in wavelet domain based on amplitude, energy, and entropy. The magnitude gives the time-frequency behavior for the data. By looking at wt and the scales giving large $W_{s,\tau}$ the signal frequencies can be estimated. By visual correlation of time-frequency representation plots, it is possible to discriminate several time regions with various frequency behaviors, as well to define the instant of time separating those regions. An abrupt change in SDR means local maxima of wt modulus. Therefore, the ridges are recognized by seeking out the points in scalogram plot where the time-frequency transform coefficients take on local maximum values. The ridges can be observed as modal parameters which are based on finding the location of max $W_{s,\tau}$. In summary, from the wavelet results it is possible to extract the sensitive features which are used for further analysis.

1.4 CONCLUSIONS

In this study a sequence of DSFs for civil structures under arbitrary excitation has been proposed. DSFs are derived from the acceleration signals by using CWT and both the input ground motion and the output are taken into account. Here, various algorithms are identified and used in order to track the change of the structures and to extract the damage features. The procedures for extracting these features are completely data-driven; there are no assumptions about the structure and there is minimal user-input besides the input and output signals. These features could be able to distinguish not only a few dynamic characteristics for detecting the location and the damage extent but also the damage types in a structure. Further, it can identify the slightest damage in a large, complex system in a noisy environment subjected to arbitrary dynamic loads.a

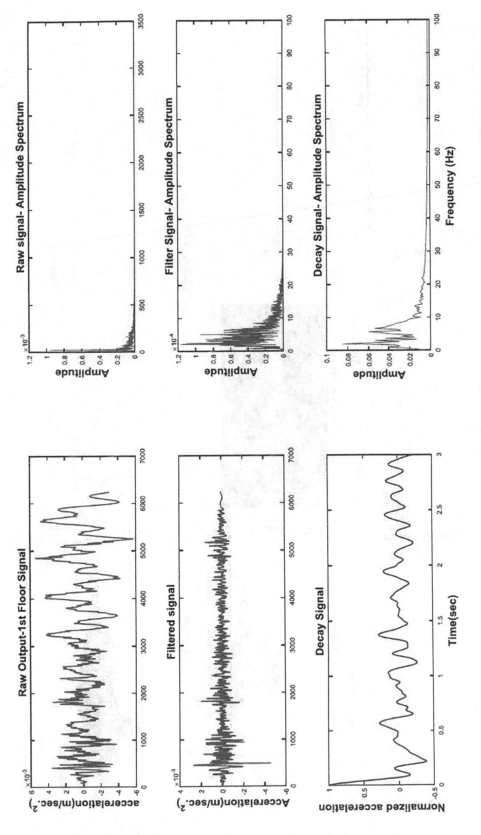

FIGURE 1.2 Single representations: (a) time domain and (b) frequency.

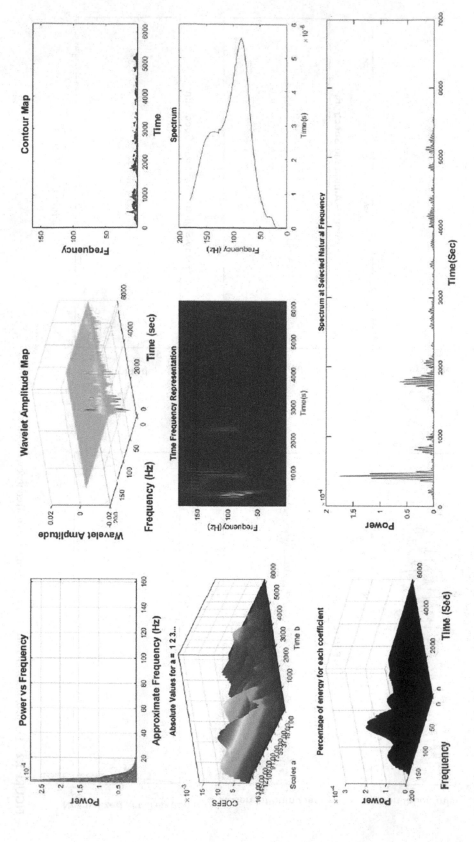

FIGURE 1.3 Visual representations of damage features.

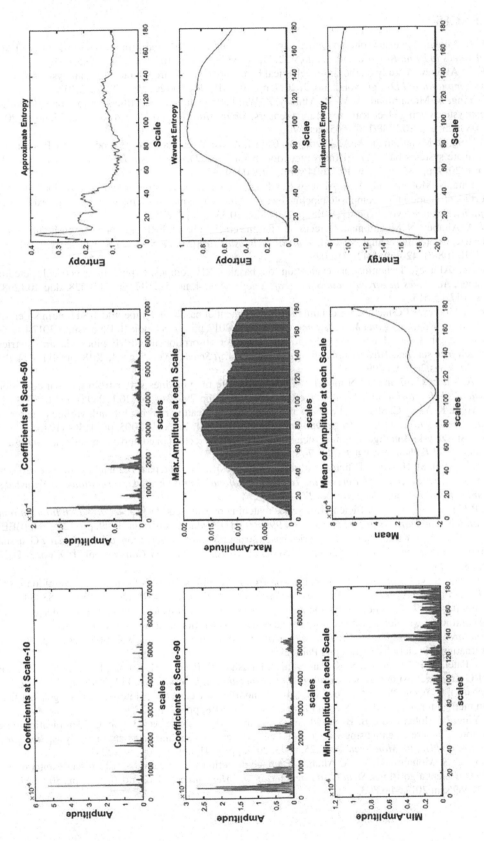

FIGURE 1.4 Additional visual representations of damage features.

REFERENCES

1. W. A. Altabey, An exact solution for mechanical behavior of BFRP nano-thin films embedded in NEMS, *Advances in Nano Research*, vol. 5, no. 4, 2017, pp. 337–357. doi: 10.12989/anr.2017.5.4.337.
2. W. A. Altabey, A study on thermo-mechanical behavior of MCD through bulge test analysis, *Advances in Computational Design*, vol. 2, no. 2, 2017, pp. 107–119. doi: 10.12989/acd.2017.2.2.107.
3. Z. Ying, N. Mohammad, A. W. A. Altabey, Z. Wu, Fatigue damage identification for composite pipeline systems using electrical capacitance sensors, *Smart Materials and Structures*, vol. 27, no. 8, 2018, p. 085023. doi: 10.1088/1361-665x/aacc99.
4. Z. Ying, N. Mohammad, W. A. Altabey, R. Ghiasi, Z. Wu, A fatigue damage model for FRP composite laminate systems based on stiffness reduction, *Structural Durability and Health Monitoring*, vol. 13, no. 1, 2019, pp. 85–103, doi: 10.32604/sdhm.2019.04695.
5. Z. Ying, N. Mohammad, W. A. Altabey, R. Ghiasi, Z. Wu, Deep learning-based damage, load and support identification for a composite pipeline by extracting modal macro strains from dynamic excitations, *Applied Sciences*, vol. 8, no. 12, 2018, p. 2564. doi: 10.3390/app8122564.
6. W. A. Altabey, N. Mohammad, Detection of fatigue crack in basalt FRP laminate composite pipe using electrical potential change method, *Journal of Physics Conference Series*, vol. 842, 2017, p. 012079. doi: 10.1088/1742-6596/842/1/012079.
7. W. A. Altabey, Delamination evaluation on basalt FRP composite pipe by electrical potential change, *Advances in Aircraft and Spacecraft Science*, vol. 4, no. 5, 2017, pp. 515–528. doi: 10.12989/aas.2017.4.5.515.
8. W. A. Altabey, EPC method for delamination assessment of basalt FRP pipe: Electrodes number effect, *Structural Monitoring and Maintenance*, vol. 4, no. 1, 2017, pp. 69–84. doi: 10.12989/smm.2017.4.1.069.
9. W. A. Altabey, N. Mohammad, Monitoring the water absorption in GFRE pipes via an electrical capacitance sensors, *Advances in Aircraft and Spacecraft Science*, vol. 5, no. 4, 2018, pp. 411–434. doi: 10.12989/aas.2018.5.4.499.
10. W. A. Altabey, FE and ANN model of ECS to simulate the pipelines suffer from internal corrosion, *Structural Monitoring and Maintenance*, vol. 3, no. 3, 2016, pp. 297–314. doi: 10.12989/smm.2016.3.3.297.
11. H. Yoon, K. Yang, C. Shahabi, Feature subset selection and feature ranking for multivariate time series, *IEEE Transactions on Knowledge and Data Engineering*, vol. 17, no. 9, 2005, pp. 1186–1198.
12. W. J. Staszewski, Intelligent signal processing for damage detection in composite materials, *Composites Science and Technology*, vol. 62, no. 7–8, 2002, pp. 941–950.
13. P. W. P. Man, M. H. Wong, Efficient and robust feature extraction and pattern matching of time series by a lattice structure. *In Proceedings of the Tenth International Conference on Information and Knowledge Management*, Atlanta, GA 2001, pp. 271–278, ACM.
14. K. P. Chan, A. W. C. Fu, Efficient time series matching by wavelets. *In Proceedings 15th International Conference on Data Engineering (Cat. No. 99CB36337)*, Sydney, Australia, 1999, pp. 126–133, IEEE.
15. K. K. Nair, A. S. Kiremidjian, Time series based structural damage detection algorithm using Gaussian mixtures modeling, *Journal of Dynamic Systems, Measurement, and Control*, vol. 129, no. 3, 2007, pp. 285–293.
16. S. Pittner, S. V. Kamarthi, Feature extraction from wavelet coefficients for pattern recognition tasks, *IEEE Transactions on Pattern Analysis and Machine Intelligence*, vol. 21, no. 1, 1999, pp. 83–88.
17. K. Balafas, A. S. Kiremidjian, Extraction of a series of novel damage sensitive features derived from the continuous wavelet transform of input and output acceleration measurements, *In Sensors and Smart Structures Technologies for Civil, Mechanical, and Aerospace Systems*, vol. 9061, 2014, p. 90611L. International Society for Optics and Photonics.
18. S. S. Patel, A. P. Chourasia, S. K. Panigrahi, J. Parashar, N. Parvez, M. Kumar, Damage identification of RC structures using wavelet transformation, *Procedia Engineering*, vol. 144, 2016, pp. 336–342.
19. M. Noori, H. Wang, W. A. Altabey, A. I. Silik, A modified wavelet energy rate-based damage identification method for steel bridges, *Scientia Iranica*, vol. 25, 2018, pp. 3210–3230.
20. Z. Ying, N. Mohammad, B. B. Seyed, W. A. Altabey, Mode shape based damage identification for a reinforced concrete beam using wavelet coefficient differences and multi-resolution analysis, *Structural Control and Health Monitoring*, vol. 25, no. 3, 2017, pp. 1–41. doi: 10.1002/stc.2041.
21. Z. Ying, N. Mohammad, W. A. Altabey, Damage detection for a beam under transient excitation via three different algorithms, *Structural Engineering and Mechanics*, vol. 63, no. 6, 2017, pp. 803–817. doi: 10.12989/sem.2017.64.6.803.

22. Z. Ying, N. Mohammad, W. A. Altabey, T. Awad, A comparison of three different methods for the identification of hysterically degrading structures using BWBN model, *Frontiers in Built Environment*, vol. 4, 2019, p. 80. doi: 10.3389/fbuil.2018.00080.

23. Z. K. Xie, G. H. Liu, Z. H. Zhang, Damage detection of concrete structure based on approximate entropy, *Applied Mechanics and Materials*, vol. 226, 2012, pp. 920–925. Trans Tech Publications.

24. W. A. Altabey, Free vibration of basalt fiber reinforced polymer (FRP) laminated variable thickness plates with intermediate elastic support using finite strip transition matrix (FSTM) method, *Vibroengineering*, vol. 19, no. 4, 2017, pp. 2873–2885. doi: 10.21595/jve.2017.18154.

25. W. A. Altabey, Prediction of natural frequency of basalt fiber reinforced polymer (FRP) laminated variable thickness plates with intermediate elastic support using artificial neural networks (ANNs) method, *Vibroengineering*, vol. 19, no. 5, 2017, pp. 3668–3678. doi: 10.21595/jve.2017.18209.

26. W. A. Altabey, High performance estimations of natural frequency of basalt FRP laminated plates with intermediate elastic support using response surfaces method, *Vibroengineering*, vol. 20, no. 2, 2018, pp. 1099–1107. doi: 10.21595/jve.2017.18456.

27. Z. Hou, A. Hera, M. Noori, Wavelet-Based Techniques for Structural Health Monitoring. In: *Health Assessment of Engineered Structures: Bridges, Buildings and Other Infrastructures*, ed., Achintya Haldar (University of Arizona), World Scientific, Singapore, 2013, pp. 179–202.

28. Z. Hou, M. Noori, R. S. Amand, Wavelet-based approach for structural damage detection, *Journal of Engineering Mechanics*, vol. 126, no. 7, 2000, pp. 677–683.

29. D. Hester, A. González, A wavelet-based damage detection algorithm based on bridge acceleration response to a vehicle, *Mechanical Systems and Signal Processing*, vol. 28, 2012, pp. 145–166.

2 Deep Learning for Automated Damage Detection
A Novel Algorithm in CNN Family for Faster and Accurate Damage Identification

Wael A. Altabey
Southeast University
Alexandria University

Mohammad Noori
California Polytechnic State University

CONTENTS

2.1   INTRODUCTION

For industrial and civil structures, structural health monitoring is more complicated, especially monitoring in a real time when the types of structures are complicated as well as when the measurement signal from sensors is corrupted with environmental noise. Therefore, structural health monitoring needs other techniques to identify structural damage. Deep Learning can solve this problem satisfactorily due to its superior adaptive learning of datasets [1–12].

Krizhevsky et al. [13] classify 1.2 million high-resolution images by using a large and deep convolutional neural network (CNN). Bouvrie [14] presented the derivation and implementation of CNNs. The detection and localization of damage based on real-time vibrations have been proposed using 1D CNNs [15–19]. A sparse coding algorithm was applied to a large number of untagged examples to train a feature extractor, and the features were then used to pass to a neural network classifier to distinguish various damage states from bridges [20]. A type of deep architecture CNN has been used to detect cracks in concrete without directly extracting the features of the defects, to overcome challenges such as lighting and shadow changes [21]. The damage in steel structures was indicated by using the differences in wavelet coefficients to analyze the responses of a healthy and loose connection structure [22,23]. The wavelets were used to detect strong earthquake response waves of framed structures indicating the occurrence of damage [24]. A new signal processing algorithm was proposed by the deep learning model, which consists of combined wavelets, neural networks, and Hilbert transform [25,26].

In addition, three types of deep neural network models, such as deep Boltzmann machines, deep belief networks, and stacked auto-encoders, were investigated to identify rolling bearing failure conditions [27]. Compared to a traditional support vector machine (SVM) and backpropagation neural network, planetary gearbox conditions could be effectively detected with the best diagnostic accuracy using deep CNNs [28]. The benefits of generalization and memorization ability for recommending systems were combined by using the wide and deep common learning by applying wide linear models and deep neural networks [29]. A multi-scale structural health monitoring system has been constructed to monitor health conditions and assess the usefulness of large-scale bridges using the Hadoop ecosystem (MS-SHM-Hadoop). A Bayesian network has been studied to assess the reliability of specific components as a function of ease of maintenance and inter-component correlations [30] and a hybrid response surface method [31]. Unlike buildings or bridges, for pipeline networks system, damage usually begins on the internal surface, especially when the internal temperature is higher than the external temperature. Micro damage, cracking of the matrix [32], delamination [33,34], Webpage [35], and fiber failure are the damage cases that result in ultimate failure of fiber reinforced polymers (FRP) composite structure, and damages may probably initiate and concentrate inside the pipes rather than outside the surface. Previous studies have shown that by acquiring images from structural damage areas (mostly outside surfaces of structures), it may be good to distinguish different health states of structures. Images captured from the external surface of pipelines may not be an appropriate way to identify damage. The specific purpose of current work is to identify and locate closed-circuit television (CCTV) image pipeline failures through deep learning to overcome the limitations of conventional computer vision techniques. Automatic detection of image faults and binary segmentation (mapping) with very high precision were established; the current disk format can have a high degree of accuracy in measuring the type and location of image errors, especially in mini image errors. The current approach provides pointers for others to apply in-depth learning techniques that address similar topics, such as identifying and locating civic infrastructure failures in building and property management. Practical implementation of the proposed approach in industry is expected to significantly reduce inspection time and resources and improve the efficiency of pipeline status assessment.

The present work generally relates to a new algorithm in Region-based CNN (R-CNN) family architecture and is called Faster Dual/Multi R-CNN (Faster D/M-R-CNN) since it provides a fast and accurate damage detection and classification for various structures and especially pipelines at real time. This method has the potential to be used as the fastest system for damage identification in various structures such as different infrastructures (bridges, high buildings, dams, pipelines, tanks, etc.), for traffic control system, and transportation systems. It can also be used for image analysis and processes occurring in smart cities and traffic control and transportation systems. Figure 2.1 shows the block diagram depicting the new algorithm (Faster D/M-R-CNN) for damage detection and image classification.

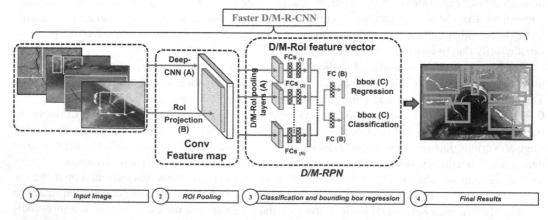

FIGURE 2.1 A block diagram depicting the new algorithm (Faster D/M-R-CNN) for damage detection and classification.

2.2 DEEP LEARNING BASED APPROACHES FOR IMAGE CLASSIFICATION AND OBJECT DETECTION

2.2.1 DEEP LEARNING FOR IMAGE CLASSIFICATION

The damage in structures has several important signs that indicate structural degradation and even the inception of catastrophic failure. Image-based damage detection has been attempted in research communities that bear the potential of replacing human-based inspection.

Deep learning has been widely developed and applied in various areas such as computer vision, speech recognition, and natural language processing through various deep learning architectures, among which CNNs are commonly applied.

As shown in Figure 2.2 a CNN model typically consists of feature extraction through a stack of layers on the input image such as convolution, activation and poling, and classification through fully connected (FC) layers for outputting the scores for each class.

Each layer is responsible for different functions and uses the result from the previous layer as the input.

For supervised computer vision tasks, CNNs (1) extract features from raw images, (2) feed the features forward using filters assigned with initial random weights and bias to predict the classes, (3) calculate the loss between predicted scores and the ground truth, and (4) apply backpropagation to adjust the filter weights and bias continuously to finally obtain an optimized model.

Compared with conventional approaches, CNNs require less image preprocessing, and image features are extracted through learning. Therefore, there is no requirement of expertise for manual design of complex feature extractors.

The region-based CNN (R-CNN) is one typical deep learning approach for object detection. As shown in Figure 2.3, region proposals are generated through an external method called selective search for the input image. Each warped region proposal image is forwarded into a CNN model to compute the features, which is then fed into an SVM classifier to calculate the classification scores. Bounding box (bbox) regression is then conducted for the classified image such that the location of each object

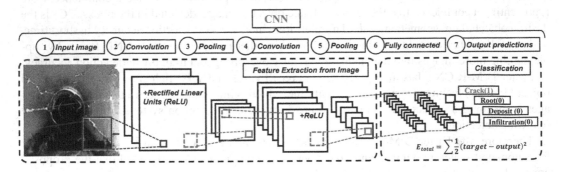

FIGURE 2.2 Architecture of a CNN model.

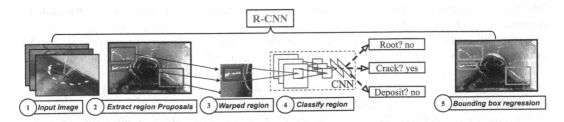

FIGURE 2.3 Example architecture of R-CNN.

can be predicted. One limitation of R-CNN is that the multi-staged training process which is time consuming requires a high computation cost. In addition, the detection speed is quite slow for each image as convolution, classification, and regression need to be implemented for each region's proposal.

Among the many methodologies, deep learning based damage detection is actively explored in recent years. However, how to automatically extract damage quickly and accurately at a pixel level, i.e. damage delineation (including both detection and segmentation) at real time, is a challenging issue.

This work proposes a new algorithm in R-CNN family-based architecture that is Faster D/M-R-CNN for real-time object (damage) detection and classification with high accuracy.

Let's quickly summarize the different algorithms in the R-CNN family (R-CNN, Fast R-CNN, and Faster R-CNN).

R-CNN extracts a bunch of regions from the given image using selective search, and then checks if any of these boxes contains an object. We first extract these regions, and for each region, CNN is used to extract specific features. Finally, these features are then used to detect objects. Unfortunately, R-CNN becomes rather slow due to the multiple steps involved in the process.

Fast R-CNN, on the other hand, passes the entire image to ConvNet which generates regions of interest (instead of passing the extracted regions from the image). Also, instead of using three different models (as we saw in R-CNN), it uses a single model which extracts features from the regions, classifies them into different classes, and returns the bboxs.

All these steps are done simultaneously, thus making it execute faster as compared to R-CNN. Fast R-CNN is, however, not fast enough when applied on a large dataset as it also uses selective search for extracting the regions.

Faster R-CNN fixes the problem of selective search by replacing it with Region Proposal Network (RPN). We first extract feature maps from the input image using ConvNet and then pass those maps through an RPN which returns object proposals. Finally, these maps are classified and the bboxs are predicted.

A block diagram depicting the new algorithm (Faster D/M-R-CNN) for object (damage) detection and classification is shown in Figure 2.1. As shown in the figure, apply Dual/Multi RPN (D/M-RPN) {3} on these feature maps {2}, get Dual/Multi object (damage) proposals for each candidate object (damage) in the image, and compare between these proposals to create a confidence score representing a confidence that the each candidate object (damage) detected in the bboxs {3C} is the desired object (damage) meticulously. This algorithm increases the detection accuracy in short time, without following the traditional method followed in other networks in the R-CNN family by supplementing more images into the database to reduce overfitting and to improve detection accuracy.

Faster D/M-R-CNN has high precision and recall value and very high speed for extracting all object (damage) features from images at real time, which is important for accurately detecting damages from images collected that increased previous damage detection systems ability to achieve real-time detection.

Therefore, the Faster D/M-R-CNN will be investigated and applied in this invented product for automated structure damage and object detection from image. We expected that the proposed approach will be demonstrated to be applicable for detecting structure damages and object detection with high accuracy and speed for extracting all damage features from images.

We expected this innovative product will increase the speed and accuracy of object (damage) detection and improves dataset size, training mode, and network hyper-parameters, which have influence on model performance.

Specifically, when an increase of dataset size and convolutional layers can improve the model speed and accuracy, it can achieve a mean average precision (mAP) of up to 98%–99% approximately.

The new designed network will lay the foundation for applying a new generation of deep learning techniques in structural damage detection systems as well as addressing deficiencies in the previous structural damage detection systems based on deep learning.

The comparison between the algorithms in R-CNN family and the new algorithm described herein is shown in Table 2.1.

TABLE 2.1

Comparison between the Algorithms in R-CNN Family and the New Algorithm (Faster D/M-R-CNN) Described Herein Using Damage Detection and Classification

Algorithm	Features	Prediction Time/Image	Limitations
CNN	Divides the image into multiple regions and then classifies each region into various classes	–	Needs a lot of regions to predict accurately and hence high computation time
R-CNN	Uses selective search to generate regions. Extracts around 2,000 regions from each image	40–50 seconds	High computation time as each region is passed to the CNN separately. Also, it uses three different models for making predictions
Faster R-CNN	Each image is passed only once to the CNN and feature maps are extracted. Selective search is used on these maps to generate predictions. Combines all the three models used in R-CNN together	2 seconds	Selective search is slow and hence computation time is still high
Faster R-CNN	Replaces the selective search method with RPN, which makes the algorithm much faster	0.2 seconds	Object proposal takes time and as there are different systems working one after the other, the performance of systems depends on how the previous system has performed
Faster D/M-R-CNN	Applied D/M-RPN to get Dual/Multi object (damage) proposals for each candidate object in the same image and compare between these proposals to result in the desired object, which makes the algorithm with the highest accuracy and the fastest	0.1–0.2 seconds	Needs to obtain an optimum number of RPN to investigate the target

2.3 METHODOLOGY

As shown in Figure 2.1, the steps followed by a Faster D/M-R-CNN algorithm to detect objects (damage) in various structure image are summarized below:

1. Take an input image {1} and pass it to the Deep-CNN {2A} which returns feature maps {2} for the image.
2. Apply D/M-RPN {3} on these feature maps {2} and get Dual/Multi object (damage) proposals for each candidate object (damage) in the image.
3. Compare between the Dual/Multi object (damage) proposals to create a confidence score representing a confidence that the object (damage) detected in the bboxs {3C} is the desired object (damage).
4. Pass these proposals to an FC {3B} layer in order to classify and predict the bboxs {3C} for the image.
5. Finally, receive the final image results {4} and compute a confidence score associated with it. The computing device can then provide a requestor with an output including object classification and/or confidence score.

The flowchart of the new algorithm process steps of the Faster D/M-R-CNN is introduced in Figure 2.4.

To ensure the accuracy of the object (damage) detection and classification network, the D/M-RPN {3} can be trained with training images. A discussion of the object (damage) detection and classification network training will be discussed in greater detail with regard to Figure 2.5.

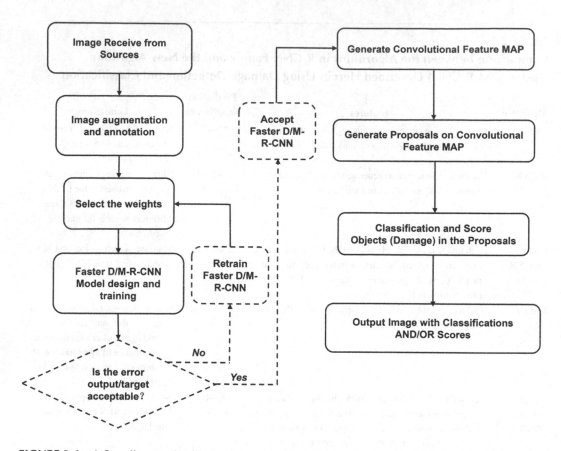

FIGURE 2.4 A flow diagram that illustrates a process flow for damage detection and classification network.

The Faster D/M-R-CNN training process consists of four steps. The first step is training a D/M-RPN {3} initialized from pre-trained Deep-CNN {2A}, in this case generating the Dual/Multi object (damage) proposals using the trained D/M-RPN {3}. The second step is training a Faster D/M-R-CNN initialized from the pre-trained Deep-CNN on the Dual/Multi object (damage) obtained by the first step. In the third step, the D/M-RPN {3} is trained again using the weights from step two as shown in Figure 2.4, without changing the Deep-CNN {2A} layers.

After generating new Dual/Multi object (damage) proposals from the trained D/M-RPN {3}, Faster D/M-R-CNN is trained again in step four with the parameters trained in the previous step.

The classification process is presented in Figure 2.6. A series of images from the image of a single target (i.e., as a sequence of temporally consecutive frames of the structures damage types) are fed to the D/M-CNN that is applied to feature extraction.

The two-step learning method is applied, i.e., the D/M-CNN is trained with the first $N-1$ layers viewed as feature maps, and these maps are used to train a Dual/Multi SVM (D/M-SVM) classifier. The output of the SVM from each CNN is compared with them and all damage features from images with high accuracy are collected and presented as P-Tensors as follows:

$$P_{(i,j)_k} = \begin{bmatrix} c_{(1,1)_1} & \cdots & c_{(1,j)_{nc}} \\ \vdots & \ddots & \vdots \\ c_{(i,1)_1} & \cdots & c_{(i,j)_{nc}} \end{bmatrix}, \ k = 1,\ldots, n, \quad (2.1)$$

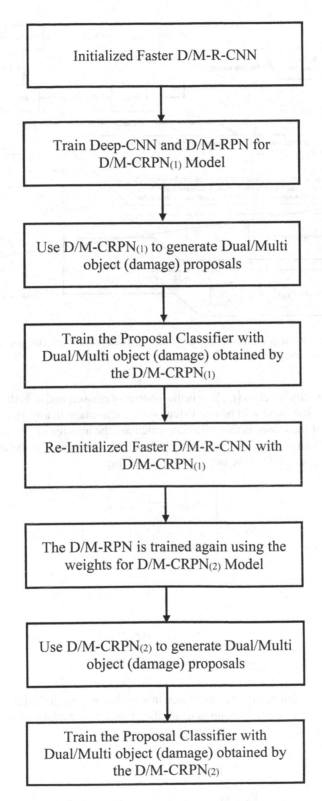

FIGURE 2.5 A flow diagram that illustrates a process flow of training one or more parameters of an object (damage) detection network.

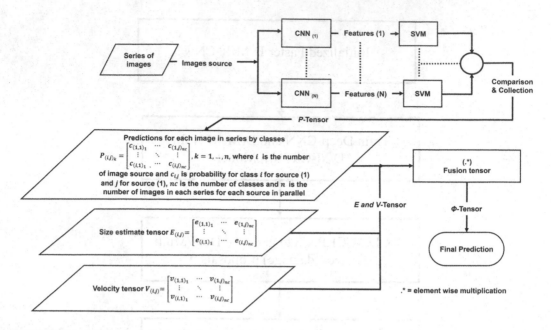

FIGURE 2.6 Data flow diagram depicting an example architecture of the process of the damage classification network described in Figure 2.1.

where $c_{i,j}$ is the probability for class (i,j), nc is the number of classes, and n is the number of images of training examples; thus there will be one P-Tensor for each image in any given image series.

The composition of the tensor is the following: calculate the average of the size estimates of the image series, check from the size-look-up table all the classes that contain the average size, e, and turn those elements to one and set the others to zero, yielding

$$E_{(i,j)} = \begin{bmatrix} e_{(1,1)_1} & \cdots & e_{(1,j)_{nc}} \\ \vdots & \ddots & \vdots \\ e_{(i,1)_1} & \cdots & e_{(i,j)_{nc}} \end{bmatrix} \quad (2.2)$$

With elements:

$$e_{(i,j)} = \begin{cases} 1 & \text{if } e \text{ fits class } i,j, \\ 0 & \text{otherwise.} \end{cases} \quad (2.3)$$

The velocity of the target damage type is composed in a similar way as the E-Tensor in Size Estimate (2.2), i.e., check from the velocity-look-up table all the classes that contain the provided velocity, v, and turn those elements to one and the others to zero.

$$V_{(i,j)} = \begin{bmatrix} v_{(1,1)_1} & \cdots & v_{(1,j)_{nc}} \\ \vdots & \ddots & \vdots \\ v_{(i,1)_1} & \cdots & v_{(i,j)_{nc}} \end{bmatrix} \quad (2.4)$$

With elements:

$$v_{(i,j)} = \begin{cases} 1 & \text{if } v \text{ fits class } i, j, \\ 0 & \text{otherwise.} \end{cases} \tag{2.5}$$

The final classification is achieved by a fusion between the parameters provided by the radar and the predictions from the image classifier. The combined P-Tensor for a series of images is

$$P_{(i,j)} = \sum_{k=1}^{n} P_{(i,j)_k} \tag{2.6}$$

where n is the number of images in each series and the fusion vector, Φ, is

$$\Phi_{(i,j)} = P_{(i,j)} \cdot {}^{*}V_{(i,j)} \cdot {}^{*}E_{(i,j)} \tag{2.7}$$

where (*) denotes element wise multiplication. The score, S, for final prediction is

$$S_{(i,j)} = \max_{m} \Phi_{(i,j)} \tag{2.8}$$

$$m = \arg\max_{m} \Phi_{(i,j)} \tag{2.9}$$

Some examples are given to explain Dual/Multi convolution and pooling processes in the Dual/Multi convolution operation, and we can see from Figure 2.7 that the input data is composed of $7 \times 7 \times 3$ dataset, where 7×7 represents width and height pixels, and 3 represents R, G, B color channels.

M/D-Filter $W0_{(i,j)}$ and M/D-Filter $W1_{(i,j)}$ are two different filter sets. The stride is two, indicating that the window extracts 3×3 local data, and strides two steps for each time. Zero-padding $= 1$. With the left window moving smoothly, the filter sets convolve by using different local data covered by the window. Respectively, two filter sets are used to calculate Dual/Multi convolution operation, and Dual/Multi of results separate into two groups as shown.

In the Dual/Multi CNN (D/M-CNN), D/M Filters (neurons with a group of fixed weights) are used to operate convolution for local input data. After calculating the data in each window, the data window moves smoothly with a specific stride, until all the convolution operation is finished. There are a few parameters that need to be figured out: (1) Depth: the number of neurons (filters), determining the depth, (2) Stride: the number of stride covering through the data, (3) Zero-padding: Supplement a few zeros to make the window more from the initial location to the end of the dataset.

Figure 2.8 is max pooling operation, which means taking the maximum value of the specific data window area. The other pooling method is average pooling in the Faster D/M-R-CNN algorithm, i.e., take the average value of the specific data window area.

The basic architecture of the connection of D/M-CNN layer and D/M-sub-sampling layer is depicted in Figure 2.9. $C_{(1,j)}$ is a D/M-CNN layer composed of six feature maps for each CNN layer. By doing convolutional operation, the feature of original signal can be enhanced and also reduce noise effects. Each neuron of feature map is connected with a 16×16 neighborhood of input image. The feature map size is 196×196. $C_{(1,j)}$ has 156 tuned parameters (each filter has 16×16 unit parameters and a bias parameter, six filters in total, so $(16 \times 16 + 1) \times 6 = 1{,}542$ parameters in total. One kernel is used between input and $C_{(1,j)}$, so $1{,}542 \times (196 \times 196) = 59{,}237{,}472$ connections in total.

$S_{(2,j)}$ is a D/M-sub-sampling layer. According to the local correlation principle of image, each sub-sampling can be applied to images, thus decreasing data processing ability and retaining

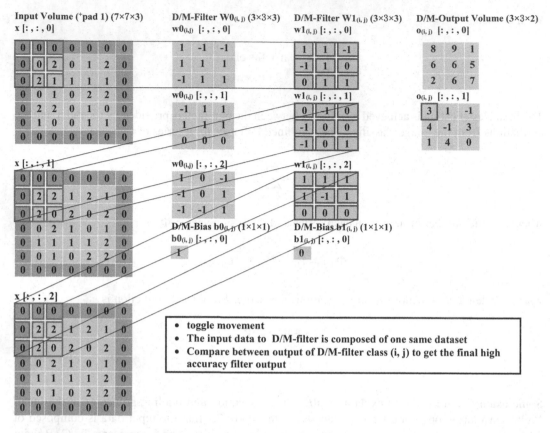

FIGURE 2.7 D/M-CNN operation in network.

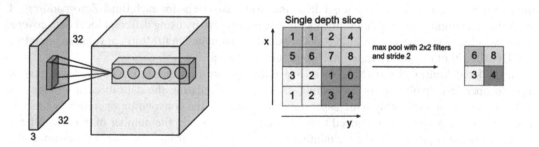

FIGURE 2.8 Max pooling operation in network.

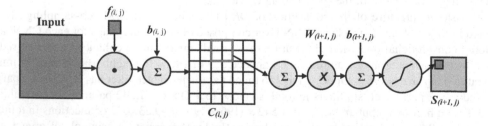

FIGURE 2.9 The connection of D/M-CNN layer and D/M-sub-sampling layer use in network.

useful information. Two 98 × 98 feature maps are used for each. Each unit of feature map is connected with an 8 × 8 neighborhood of $C_{(1,j)}$. Sixteen inputs of each unit of $S_{(2,j)}$ are added, multiplied by a tuned parameter with a tuned bias. The result can be calculated by sigmoid function. Tuned parameters and bias control the nonlinearity of sigmoid function. If these parameters are relatively small, the operation is as similar as linear operation. Each sub-sampling is as equivalent as fuzzy images by decreasing the pixels of images. If these parameters are relatively larger, each sub-sampling can be regarded as "or" or "and" operations with noise. An 8 × 8 receptive field is not overlapped for each unit; therefore the size of each feature map in $S_{(2,j)}$ is 1/4 of that of $C_{(1,j)}$. $S_{(2,j)}$ has $(1 + 1) \times 2 = 4$ tuned parameters and $(8 \times 8 + 1) \times 2 \times (98 \times 98) = 1{,}248{,}520$ connections for each.

2.4 CONCLUSIONS

The main goal of this research was to develop a vision-based damage detection that can automatically detect and identify, in real time, damages in pipelines as well as address shortcomings in damage identification in other fields in civil and transportation engineering such as infrastructure, traffic control systems and transportation systems. The new algorithm in R-CNN family has been designed to be able to detect the damage in images capture from structures at real time and to increase the damage detection accuracy. The new algorithm is more accurate and faster, thus it is called Faster D/M-R-CNN architecture. In this new algorithm, the D/M-RPN is applied on these feature maps to get Dual/Multi damage scenarios for each candidate damage in the image, and compare these scenarios to create a confidence score representing a confidence that the damage detected in the bbox is the desired damage identification. This algorithm increases the detection accuracy at short time, without following the traditional method followed in the other networks by supplementing more images into the database to reduce overfitting to improve the detection accuracy.

ACKNOWLEDGEMENTS

Involvement and contributions to this research project was made possible, for the second co-author, in part due to a release time funding from Donald E. Bently Center for Engineering Innovation, Mechanical Engineering, Cal Poly, San Luis Obispo. Herein this support is acknowledged.

REFERENCES

1. Zhao, Y.; Noori, M.; Altabey, W. A.; Ghiasi, R.; and Wu, Z.; Deep learning-based damage, load and support identification for a composite pipeline by extracting modal macro strains from dynamic excitations, *Applied Sciences*, 8(12), 2018, 2564. doi: 10.3390/app8122564.
2. Ying, Z.; Mohammad, N.; Seyed, B.B.; and Altabey, W.A.; Mode shape based damage identification for a reinforced concrete beam using wavelet coefficient differences and multi-resolution analysis, *Journal of Structural Control and Health Monitoring*, 25(3), 2017, 1–41. doi: 10.1002/stc.2041.
3. Ying, Z.; Mohammad, N.; and Altabey, W. A.; Damage detection for a beam under transient excitation via three different algorithms, *Journal of Structural Engineering and Mechanics*, 63(6), 2017, 803–817. doi: 10.12989/sem.2017.64.6.803.
4. Zhao, Y., Noori, M.; Altabey, W. A.; and Awad, T. A.; Comparison of three different methods for the identification of hysterically degrading structures using BWBN model. *Frontiers in Built Environment*, 4, 2019, 80. doi: 10.3389/fbuil.2018.00080.
5. Altabey, W. A.; An exact solution for mechanical behavior of BFRP nano-thin films embedded in NEMS, *Journal of Advances in Nano Research*, 5(4), 2017, 337–357. doi: 10.12989/anr.2017.5.4.337.
6. Altabey, W. A.; A study on Thermo-Mechanical Behavior of MCD through bulge test analysis, *Advances in Computational Design*, 2(2), 2017, 107–119. doi: 10.12989/acd.2017.2.2.107.
7. Ying, Z.; Mohammad, N.; Altabey, W. A.; and Zhishen, W.; Fatigue damage identification for composite pipeline systems using electrical capacitance sensors, *Journal of Smart Materials and Structures*, 27(8), 2018, 085023. doi: 10.1088/1361-665x/aacc99.

8. Zhao, Y., Noori, M.; Altabey, W. A.; Ramin, G.; and Zhishen, W.; A fatigue damage model for FRP composite laminate systems based on stiffness reduction, *Structural Durability and Health Monitoring*, 13(1), 2019, 85–103. doi: 10.32604/sdhm.2019.04695.

9. Altabey, W. A.; FE and ANN model of ECS to simulate the pipelines suffer from internal corrosion, *Journal of Structural Monitoring and Maintenance*, 3(3), 2016, 297–314. doi: 10.12989/smm.2016.3.3.297.

10. Altabey, W. A.; Detecting and predicting the crude oil type inside composite pipes using ECS and ANN, *Journal of Structural Monitoring and Maintenance*, 3(4), 2016, 377–393. doi: 10.12989/smm.2016.3.4.377.

11. Altabey, W. A.; The thermal effect on electrical capacitance sensor for two-phase flow measurements, *Journal of Structural Monitoring and Maintenance*, 3(4), 2016, 335–347. doi: 10.12989/smm.2016.3.4.335.

12. Kost, A.; Altabey, W. A.; Noori M.; and Awad, T.; Applying neural networks for tire pressure monitoring systems, *Journal of Structural Durability and Health Monitoring*, 13(3), 2019, 247–266. doi: 10.32604/sdhm.2019.07025.

13. Krizhevsky, A.; Sutskever, I.; and Hinton, G. E.; Imagenet classification with deep convolutional neural networks, *Journal of Advances in Neural Information Processing Systems*, 2012, 1097–1105. doi: 10.1145/3065386.

14. Bouvrie, J.; Notes on convolutional neural networks, 2006.

15. Altabey, W. A.; Free vibration of basalt fiber reinforced polymer (FRP) laminated variable thickness plates with intermediate elastic support using finite strip transition matrix (FSTM) method, *Journal of Vibroengineering*, 19(4), 2017, 2873–2885. doi: 10.21595/jve.2017.18154.

16. Altabey, W. A.; Prediction of natural frequency of basalt fiber reinforced polymer (FRP) laminated variable thickness plates with intermediate elastic support using artificial neural networks (ANNs) method, *Journal of Vibroengineering*, 19(5), 2017, 3668–3678. doi: 10.21595/jve.2017.18209.

17. Altabey, W. A.; High performance estimations of natural frequency of basalt FRP laminated plates with intermediate elastic support using response surfaces method, *Journal of Vibroengineering*, 20(2), 2018, 1099–1107. doi: 10.21595/jve.2017.18456.

18. Al-tabey, W. A.; Vibration analysis of laminated composite variable thickness plate using finite strip transition matrix technique, In *MATLAB Verifications MATLAB- Particular for Engineer*, ed., Kelly Bennett (InTech), 21, 2014, 583–620. doi: 10.5772/57384.

19. Abdeljaber, O.; Avci, O.; Kiranyaz, S.; Gabbouj, M.; and Inman, D. J.; Real-time vibration-based structural damage detection using one-dimensional convolutional neural networks, *Journal of Sound and Vibration*, 388, 2017, 154–170. doi: 10.1016/j.jsv.2016.10.043.

20. Guo, J.; Xie, X.; Bie, R.; and Sun, L.; Structural health monitoring by using a sparse coding-based deep learning algorithm with wireless sensor networks, *Journal of Personal and Ubiquitous Computing*, 18(8), 2014, 1977–1987. doi: 10.1007/s0077.

21. Cha, Y. J.; Choi, W.; and Büyüköztürk, O.; Deep learning-based crack damage detection using convolutional neural networks, *Journal of Computer: Aided Civil and Infrastructure Engineering*, 32(5), 2017, 361–378. doi: 10.1111/mice.12263.

22. Pnevmatikos, N. G.; Blachowski, B.; Hatzigeorgiou, G. D.; and Swiercz, A.; Wavelet analysis based damage localization in steel frames with bolted connections. *Journal of Smart Structures and Systems*, 18(6), 2016, 1189–202. doi: 10.12989/sss.2016.18.6.1189.

23. Noori, M.; Haifegn, W.; Altabey, W. A.; and Ahmad, I. H. S.; A modified wavelet energy rate based damage identification method for steel bridges, *International Journal of Science and Technology, Scientia Iranica*, 25(6), 2018, 3210–3230. doi: 10.24200/sci.2018.20736.

24. Pnevmatikos, N. G. and Hatzigeorgiou, G. D.; Damage detection of framed structures subjected to earthquake excitation using discrete wavelet analysis, *Journal of Bulletin of Earthquake Engineering*, 15(1), 2017, 227–248. doi: 10.1007/s10518-016-9962-z.

25. Kanarachos, S.; Christopoulos, S. R.; Chroneos, A.; and Fitzpatrick, M. E.; Detecting anomalies in time series data via a deep learning algorithm combining wavelets, neural networks and Hilbert transform. *Expert Systems with Applications*, 85, 2017, 292–304. doi: 10.1016/j.eswa.2017.04.028.

26. Ghiasi, R.; Ghasemi, M. R.; Noori, M.; and Altabey, W.; A non-parametric approach toward structural health monitoring for processing big data collected from the sensor network, *12th International Workshop on Structural Health Monitoring*, 2019, Stanford, CA.

27. Chen, Z.; Deng, S.; Chen, X.; Li C.; Sanchez, R. V.; and Qin, H.; Deep neural networks-based rolling bearing fault diagnosis, *Journal of Microelectronics Reliability*, 75, 2017, 327–333. doi: 10.1016/j.microrel.2017.03.006.

28. Jing, L.; Wang, T.; Zhao, M.; and Wang, P.; An adaptive multi-sensor data fusion method based on deep convolutional neural networks for fault diagnosis of planetary gearbox, *Journal of Sensors*, 17(2), 2017, 414. doi: 10.3390/s17020414.

29. Cheng, H. T.; Koc, L.; Harmsen, J.; Shaked, T.; Chandra, T.; Aradhye, H.; Anderson, G.; Corrado, G.; Chai, W.; Ispir, M.; and Anil, R.; Wide and deep learning for recommender systems, *In Proceedings of the 1st Workshop on Deep Learning for Recommender Systems*, Boston, MA, ACM, 2016 September 15, pp. 7–10.

30. Liang, Y.; Wu, D.; Liu, G.; Li, Y.; Gao, C.; Ma, Z. J.; and Wu, W.; Big data-enabled multiscale serviceability analysis for aging bridges, *Journal of Digital Communications and Networks*, 2(3), 2016, 97–107. doi: 10.1016/j.dcan.2016.05.002.

31. Ying, Z.; Mohammad, N.; Altabey, W. A.; and Naiwei, L.; Reliability evaluation of a laminate composite plate under distributed pressure using a hybrid response surface method, *International Journal of Reliability, Quality and Safety Engineering*, 24(3), 2017, 1750013. doi: 10.1142/S0218539317500139.

32. Altabey, W. A. and Mohammad, N.; Detection of fatigue crack in basalt FRP laminate composite pipe using electrical potential change method, *Journal of Physics: Conference Series*, 842, 2017, 012079. doi: 10.1088/1742-6596/842/1/012079.

33. Altabey, W. A.; Delamination evaluation on basalt FRP composite pipe by electrical potential change, *Journal of Advances in Aircraft and Spacecraft Science*, 4(5), 2017, 515–528. doi: 10.12989/aas.2017.4.5.515.

34. Altabey, W. A.; EPC method for delamination assessment of basalt FRP pipe: Electrodes number effect, *Journal of Structural Monitoring and Maintenance*, 4(1), 2017, 69–84. doi: 10.12989/smm.2017.4.1.069.

35. Altabey, W. A. and Mohammad, N.; Monitoring the water absorption in GFRE pipes via an electrical capacitance sensors, *Journal of Advances in Aircraft and Spacecraft Science*, 5(4), 2018, 411–434. doi: 10.12989/aas.2018.5.4.499.

3 Seismic Protection of Cultural Relics Using Three-Dimensional Base-Isolation System

Bai Wen and Dai Junwu
Key Laboratory of Earthquake Engineering and Engineering
Vibration of China Earthquake Administration

CONTENTS

3.1 INTRODUCTION

Earthquakes have caused damage to relics many times. For example, more than 3,000 pieces of relics are damaged during the 2008 Wenchuan earthquake [1,2]. Researchers have been working on the behavior of art objects since many decades [3,4].

Many applications have been used to protect relics against earthquakes. Most of these applications are used in Japan, United States, Italy, and China [5]. Base isolation, as a mature and effective method, is now becoming more and more popular in the seismic protection of freestanding items [6–9]. Most base-isolation researches focus on lateral isolation. The influence of vertical components is not well considered. However, past earthquakes show that at epicentral areas, the amplitude of vertical motion is sometimes stronger than that of lateral motion. The impact of vertical motion thus might not be ignored.

In order to evaluate the influence of vertical motion necessary for vertical isolation, a porcelain vase, as a typical fragile relic, is modeled and finite element analyses under lateral 2-dimensional (2D) isolation and 3-dimensional (3D) isolation are conducted. Response history analyses under 3 different ground motions are given. An incremental dynamic analysis (IDA) method is used. Sliding and overturn are the two indexes used to evaluate the effectiveness of isolation.

The PGA ratio between the vertical and lateral direction of each ground is adjusted to consider the impact of vertical motion.

The ground motions used in this paper and the corresponding finite element models are given in Sections 3.2 and 3.3, respectively. Dynamic responses are compared in Section 3.4. Finally, the conclusions are given in Section 3.5.

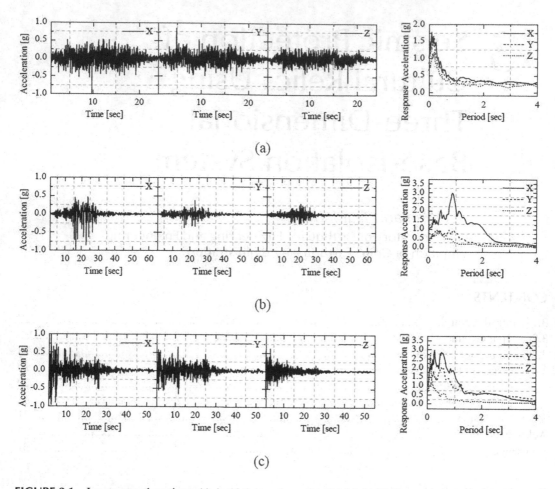

(a)

(b)

(c)

FIGURE 3.1 Input ground motions. (a) Artificial motions, (b) Chichi, (c) El Centro.

3.2 INPUT GROUND MOTIONS

Three sets of 3D ground motions, including two natural ground motions, such as El Centro and Chichi, and one artificial ground motions are used. Acceleration time histories of these three sets of ground motions and the related acceleration spectra at 5% damping ratio are plotted in Figure 3.1.

3.3 FINITE ELEMENT MODELS

All analyses are conducted using finite element method. LS DYNA (R8.1.0) with single precision at win 10 × 64 platform is used to build these finite element models. Figure 3.2 shows these models. Figure 3.2a is the vase without control. Figure 3.2b is the vase with lateral 2D isolation. Figure 3.2c is the vase with 3D isolation. The height of the vase is 1,200 mm and the related vase bottom diameter is 200 mm. The size of the support is 2,000 mm × 2,000 mm × 400 mm (Height). Element Shell 163 is used to build the vase and support. Base isolation is achieved using element Combi 165. As for the parameters, the static friction factor between the support and the vase is set as 0.2, and the corresponding dynamic friction factor is set as 0.18. The period at both perpendicular and lateral direction of both 2D and 3D isolation devices is set as 2 seconds, and the related damping ratios are all set as 20%. The vertical period of the 3D isolation device is set as 1.5 seconds and the related damping ratio is set as 10%.

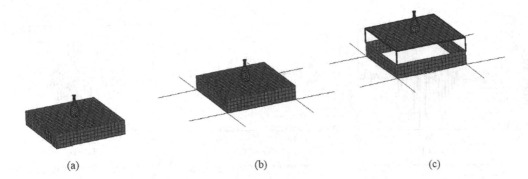

FIGURE 3.2   Vase without control (a), and with 2D (b) and 3D control (c).

3.4   RESPONSE HISTORY ANALYSES

3.4.1   Overturn

Three indexes, including overturning PGA, maximum sliding displacement, and sliding at the same PGA are used to compare the response differences among the three schemes. As for Overturn, the critical PGA values are compared. The isolation scheme with higher critical PGA is considered as a better isolation scheme. Maximum sliding displacements before the vase overturns are listed. The sliding responses at the same PGA levels are compared. The smaller the responses are, the more effective the isolation scheme is. The IDA method is used to conduct the response history analyses, and the PGA increases from 0.05 g until the vase overturns. The PGA discrimination is 0.05 g. The overturn results of these different schemes are given in Table 3.1. Taking the artificial ground motion as an example, the vase overturns at PGA 0.40 g under non-control and overturns at 1.10 and 1.00 g under lateral isolation and 3D isolation, respectively.

Time histories and related Fourier spectra of the isolated support under Chichi ground motion at PGA 0.4 g with the three schemes are plotted in Figure 3.3.

It should be mentioned that during the simulation. Model hourglass happened while using the default shell formulation and it affects the results. In order to solve the issue, fully integrated Belytschko-Tsay is used for shell formulation.

It can be seen from Figure 3.3 that after isolation, the support responses reduce. Since the parameters of 2D and 3D isolation at both lateral directions are the same, the lateral isolation effect of 2D and 3D is almost the same. As for vertical isolation, the 3D isolation shows that the effect and the vertical responses of both non-control and 2D are exactly the same. It can be concluded that both 2D and 3D isolation works, and the related responses at corresponding directions reduce. However, it can be seen from Table 3.1 that the 3D isolation does not show better performance than 2D isolation on vase overturn lateral isolation. The reduction of the acceleration responses does not directly reflect to an increase of the overturn PGA.

TABLE 3.1

Critical PGAs for Different Isolation Schemes under Different Ground Motions

	Non-Control (g)	Lateral 2D Isolation (g)	3D Isolation (g)
Artificial	0.40	1.10	1.00
Chichi	0.40	0.35	0.40
El Centro	0.35	0.50	0.45

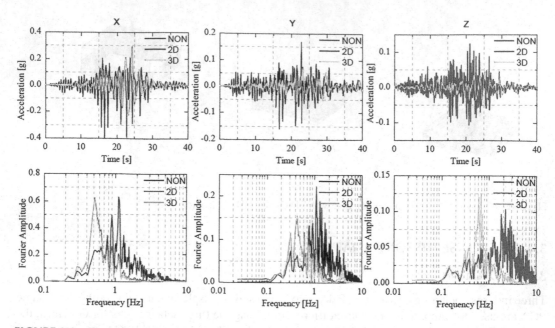

FIGURE 3.3 Time histories and Fourier spectra of the isolated support under Chichi at PGA 0.4 g.

3.4.2 SLIDING

The sliding distances of the vase under the same PGA and ground motion with the three schedules are compared, and the results are listed in Table 3.2. Meanwhile, the maximum sliding displacements of the vase at critical PGA are listed in Table 3.3. The sliding displacement histories of the vase under these cases are plotted in Figures 3.4 and 3.5.

From the above tables and figures it can be seen that isolation can reduce vase sliding. Hence, 3D isolation shows a better effect than 2D isolation. As for the artificial ground motion, isolation significantly increases the critical PGA. However, the sliding distances also increase. The support should be designed big enough to accommodate the sliding in case the vase does not overturn but drop off from the support edge.

TABLE 3.2
The Vase Maximum Sliding distances at X Direction at PGA 0.35 g Level

	Non-Control (mm)	Lateral 2D Isolation (mm)	3D Isolation (mm)
Artificial	5.2	2.3	1.3
Chichi	81.7	48.0	43.4
El Centro	15.3	2.3	1.1

TABLE 3.3
The Vase Maximum Sliding Displacements at X Direction at Critical PGA

	Non-Control (g)	Lateral 2D Isolation (g)	3D Isolation (g)
Artificial	29.7	207.3	151.7
Chichi	143.4	48.0	83.1
El Centro	15.3	31.1	18.4

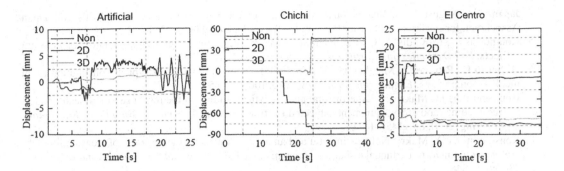

FIGURE 3.4 The vase sliding displacement histories at X directions at PGA 0.35 g level.

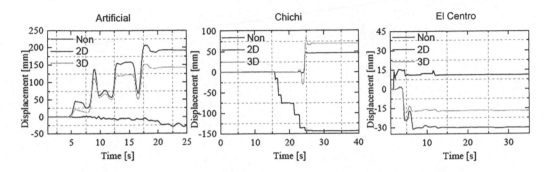

FIGURE 3.5 The vase sliding displacement histories at X directions at critical PGA.

3.5 CONCLUSIONS

Based on the simulation conducted in this paper. It can be concluded that isolation can reduce sliding and 3D isolation performs better than 2D lateral isolation. Isolation may increase the critical PGA of vase overturn. The extent of increase varies as the ground motion changes. The 3D isolation does not show better performance than 2D isolation at vase overturn. In future research, more ground motions reflecting different characteristics should be used. More parameters like different friction coefficients, isolation periods, and damping ratios should be considered. Also, a shake table test might be conducted to validate the conclusions.

ACKNOWLEDGMENT

This work is supported by Heilongjiang Provincial Natural Science Foundation of China (LH2019E094), Program for Innovative Research Team in China Earthquake Administration.

REFERENCES

1. Qian Z., Wei-ming Y. Analysis of damage to free-standing cultural relics caused by the Wenchuan earthquake. *Sciences of Conservation and Archaeology*, 2010, 22(3):36–43 (In Chinese).
2. Quan W., Xian-dan C. Experiences from damages of movable cultural relics by Wenchuan earthquake. *Sichuan Cultural Relics*, 2008, 4:10–13 (In Chinese).
3. Agbabian M., Ginell W., Masri S., Nigbor, R. Evaluation of seismic measures for art objects. *In Proceedings of the 4th U.S. National Conference on Earthquake Engineering*, Palm Springs, CA, 20–24 May 1990. Earthquake Engineering Research Institute (EERI), Oakland, CA, 1990, pp. 3–12.
4. Agbabian M., Masri S., Nigbor R., Ginell W. Seismic damage mitigation concepts for art objects in museums. *In Proceedings of the 9th World Conference on Earthquake Engineering*, Tokyo-Kyoto,

Japan, 2–9 August 1988. International Association of Earthquake Engineering (IAEE) and Japan Association for Earthquake Disaster Prevention, Tokyo, Japan, 1988, pp. 235–240.

5. Koumousis V., Moysidis A. On the dynamic behavior of nonlinear lightweight isolator for museum artifacts. *Soil Dynamics and Earthquake Engineering*, 2019, 117:251–262.

6. Di Sarno L., Petrone C., Magliulo G., Manfredi G. Dynamic properties of typical consultation room medical components. *Engineering Structures*, 2015, 100:442–454.

7. Konstantinidis D., Makris N. Experimental and analytical studies on the response of 1/4-scale models of freestanding laboratory equipment subjected to strong earthquake shaking. *Bulletin of Earthquake Engineering*, 2010, 8(6):1457–1477.

8. Konstantinidis D., Makris N. Experimental and analytical studies on the response of freestanding laboratory equipment to earthquake shaking. *Earthquake Engineering and Structural Dynamics*, 2009, 38(6):827–848.

9. Neurohr T., McClure G. Shake table testing of museum display cases. *Canadian Journal of Civil Engineering*, 2008, 35(12):1353–1364.

4 Combined Actuator-Shake Table Test with Optimized Input Energy

Farhad Behnamfar
Isfahan University of Technology

Mohammadreza Najar
Isfahan University of Technology

CONTENTS

4.1 INTRODUCTION

Various methods for earthquake engineering tests have been developed in the recent decades, including the pseudo-static, pseudo-dynamic, shake table, and hybrid experimental methods. In the pseudo-static testing the specimen is deformed up to predefined values by the forces generated by hydraulic actuators. The applied forces and the produced deformations change slowly, and therefore the inertial and damping forces that are specific to the dynamic response are not accounted for in this type of testing. The main idea behind this method is to evaluate the hysteresis behavior of different structural members instead of structural systems [1]. In the pseudo-dynamic testing again the rate of loading on the model is slow but the inertial forces are simulated with the aid of a computer controlling the progress of the test. This method is usually used for testing structural systems that can be as large as a full-scale building. In this method the lateral displacements of stories are analytically calculated using time integration methods at each time step. Then the actuators push at different levels to their calculated lateral displacements, and the actuator forces required to produce the above displacements are directly recorded by the load cells. The story shear at each level is determined as the sum of its above lateral forces. The calculated shear forces are utilized in the nonlinear dynamic equations of the system, using which the lateral displacements for the next time step are calculated by the computer controlling the testing procedure. Since there is no need to apply forces at high speeds in this method, the experimental apparatus is simpler and cheaper; therefore larger and even full-scale models can be tested. However, the effects of inertial and viscous damping forces produced by rapid motions of system during earthquake is only taken into account analytically and cannot be generated in the physical part of the test [1]. Shake tables are known for their ability to simulate the dynamic loading precisely, but they are the most expensive testing machines that are limited in number and size around the world [1].

The hybrid testing is meant to overcome the limitations of the previous experimental methods. In this procedure the system is divided into physical and analytical parts. The physical part, with a seismic behavior difficult to precisely simulate, is to be tested using hydraulic actuators while the analytical portion is analyzed in a parallel computer. The forces/displacements at the interface between the experimental and analytical parts are computed in the computer and applied by the actuators to the experimental part. If the forces are applied, the displacements will be recorded and input to the computer and vice versa. These boundary responses are used in the next time step of analysis. Efforts are being taken to raise the method to a level that can be visualized as a real-time testing method like the shake table test. This desire is not completely fulfilled yet because of technological limitations in flow capacity of the pumps, the accumulation system, and the size of the hydraulic actuators. Moreover, the rate of testing is limited by the dynamic performance of the actuators characterized by their maximum velocity and frequency response function [1].

This study proposes a combined type of seismic testing, called the combined testing method, by using force actuators and a shake table concurrently. The initial idea was first suggested by Kausel [2]. This paper presents the outcome of the efforts by the authors to extend and apply the method to various cases. A number of alternatives to specify the share of each testing apparatus from the ground motion are examined first. Then an optimization procedure is presented, which results in a minimized power input of the test.

4.2   EQUATIONS OF MOTION OF THE EXPERIMENTAL SDF MODEL

The governing equations for different testing methods in earthquake engineering are described here for a single degree of freedom (SDF) model.

The dynamic equation of equilibrium for an SDF oscillator with possible nonlinear behavior can be shown as follows:

$$m\ddot{u} + f(y, \dot{y}) = 0 \tag{4.1}$$

In Eq. (4.1), m shows the mass and u is the lateral displacement of the oscillator, and the dots represent the derivation with respect to time. Also,

$$u = y + u_g \tag{4.2}$$

in which y is the relative displacement respective to the base and u_g is the ground (base) displacement. As is seen in Eq. (4.1), the base shear is shown by $f(y, \dot{y})$ as a function of relative displacement and velocity of the oscillator. This function can be linear or nonlinear based on the level of shaking. For a linear system, the well-known relation $f(y, \dot{y}) = ky + c\dot{y}$ exists, in which k and c represent the lateral stiffness and the damping of the system.

In the pseudo-static testing, the force in the hydraulic actuator is $f(y, \dot{y})$ in which by lengthening the time of the experiment and lowering the rate of change of actuator force, dependence of f on $\dot{y}$ is minimized. On the other hand, in the pseudo-dynamic testing Eq. (4.1) is used at each time step to calculate the relative displacement y to be applied by the actuator to the model. Then the force in the actuator $f(y, \dot{y})$ to make this displacement is recorded and used in Eq. (4.1) again to calculate the relative displacement for the next time step.

For a system vibrating on a shake table, the terms $m\ddot{u}$ or $f(y, \dot{y})$ in Eq. (4.1) will show the force introduced by the shake table to the oscillator, and vice versa. Equation (4.1) can be rewritten in an effective force format by resorting to relative displacements, as shown in Eq. (4.3):

$$m\ddot{y} + f(y,\dot{y}) = -m\ddot{u}_g \tag{4.3}$$

In Eq. (4.3), the right side exhibits the effective force needed to be applied to the model if the base is kept unmoved. This is the basic idea behind the effective force method of testing [2]. The combined testing method introduced in this paper is a combination of testing with shake table and the effective force method. Here, the ground acceleration is divided into two parts as follows:

$$\ddot{u}_g = \ddot{u}_{g1} + \ddot{u}_{g2} \tag{4.4}$$

The term $\ddot{u}_{g1}$ is the acceleration of the shaking table and $\ddot{u}_{g2}$ refers to the share of the actuators. Substituting Eq. (4.4) in Eq. (4.1) results in

$$m(\ddot{y} + \ddot{u}_{g1} + \ddot{u}_{g2}) + f(y,\dot{y}) = 0 \tag{4.5}$$

Comparing Eqs. (4.5) and (4.1) results in

$$u = y + u_{g1} + u_{g2} \tag{4.6}$$

Equation (4.5) can be further written as Eq. (4.7)

$$m\ddot{u}_1 + f(y,\dot{y}) = -m\ddot{u}_{g2} \tag{4.7}$$

in which

$$u_1 = y + u_{g1} \tag{4.8}$$

Equation (4.7) governs the response of the oscillator in the combined testing method. In Eq. (4.7), the term $m\ddot{u}_{g2}$ represents the effective force to be applied by the force actuators to the system.

4.3   THE TESTING POWER

Apparently, the design basis of the combined testing method is the criterion used for dividing the ground motion between the two apparatuses in order to have the minimum testing power. The electric power needed for implementing a dynamic experiment is defined here as the product of the external force applied to the system by the testing facility and the velocity at which this force is applied. This is calculated in the following.

The external force provided by, or on, the shake table is simply the axial force in the hydraulic actuators driving the table itself as part of the shake table mechanism (not to be mistaken by the actuators forcing the oscillator by $-m\ddot{u}_{g2}$). Equilibrium of forces at the base is more clearly seen in Eq. (4.1), for which Eq. (4.7) is only a rearrangement. Therefore, from Eqs. (4.1) or (4.7), the force driving the shaking table is $P_1 = m(\ddot{y} + \ddot{u}_{g1} + \ddot{u}_{g2}) + m_b\ddot{u}_{g1} = m\ddot{u} + m_b\ddot{u}_{g1}$ that is applied at a velocity of $\dot{u}_{g1}$. m_b is mass of the shaking table. As shown in Eq. (4.7), the force produced by the actuators is $P_2 = -m\ddot{u}_{g2}$ applied at the velocity of the oscillator, $\dot{u}_1$, in order not to interfere with the oscillation. The testing power is computed by the addition of powers consumed by each facility in the test. However, as force and velocity are vector quantities, the instantaneous power can be a negative number at certain instants. Since the actuator driving the shake table and that applying the effective force directly on the oscillator are different, the total power needed at each time instant is the sum of the absolute values of individual powers (otherwise the total power can be zero at some instants while both apparatuses are working that is obviously incorrect), or

$$W = |W_1| + |W_2|$$ (4.9)

where W_1 is the power needed to drive the shaking table and W_2 is that of the force actuators, as

$$W_1 = \left(m\ddot{u} + m_b \ddot{u}_{g1}\right)\dot{u}_{g1}, \quad W_2 = -m\ddot{u}_{g2}\dot{u}_1$$ (4.10)

Then, using Eqs. (4.6) and (4.8):

$$|W_1| = W_1: \quad \text{if } \dot{u}_{g1}\left[\ddot{y} + \ddot{u}_{g2} + (1+\beta)\ddot{u}_{g1}\right] \geq 0$$ (4.11a)

$$|W_1| = -W_1: \quad \text{if } \dot{u}_{g1}\left[\ddot{y} + \ddot{u}_{g2} + (1+\beta)\ddot{u}_{g1}\right] < 0$$ (4.11b)

$$|W_2| = W_2: \quad \text{if } \ddot{u}_{g2}\left(\dot{y} + \dot{u}_{g1}\right) < 0$$ (4.11c)

$$|W_2| = -W_2: \quad \text{if } \ddot{u}_{g2}\left(\dot{y} + \dot{u}_{g1}\right) \geq 0$$ (4.11d)

Therefore, W has to be calculated in four cases. For the first case, Eqs. (4.11a) and (4.11c) are substituted in Eq. (4.9), which results in

$$\frac{W}{m} = \left(\ddot{u} + \beta\ddot{u}_{g1}\right)\dot{u}_{g1} - \ddot{u}_{g2}\dot{u}_1$$ (4.12)

where $\beta = m_b/m$. Substituting Eq. (4.8) in Eq. (4.12) results in

$$\frac{W}{m} = \left(\ddot{u} - \ddot{u}_{g2} + \beta\ddot{u}_{g1}\right)\dot{u}_{g1} - \ddot{u}_{g2}\dot{y}$$ (4.13)

Using Eq. (4.6) with Eq. (4.13):

$$\frac{W}{m} = \left(\ddot{y} + (1+\beta)\ddot{u}_{g1}\right)\dot{u}_{g1} - \ddot{u}_{g2}\dot{y} = \dot{u}_g\ddot{u}_g\frac{\dot{u}_{g1}}{\dot{u}_g}\frac{\ddot{u}_{g1}}{\ddot{u}_g}(1+\beta) + \ddot{y}\dot{u}_g\frac{\dot{u}_{g1}}{\dot{u}_g} - \dot{y}\ddot{u}_g\frac{\ddot{u}_{g2}}{\ddot{u}_g}$$ (4.14)

The parameter α is introduced as the fraction of the base acceleration associated with the force actuators as the following:

$$\alpha = \frac{\ddot{u}_{g2}}{\ddot{u}_g} \Rightarrow \frac{\ddot{u}_{g1}}{\ddot{u}_g} = 1 - \alpha$$ (4.15)

Note that the same relation also exists between velocities. Substitution of Eq. (4.15) in Eq. (4.14) results in

$$\frac{W}{m} = (1-\alpha)^2(1+\beta)\dot{u}_g\ddot{u}_g + (1-\alpha)\dot{u}_g\ddot{y} - \alpha\dot{y}\ddot{u}_g: \quad \dot{u}_{g1}\left[\ddot{y} + \ddot{u}_{g2} + (1+\beta)\ddot{u}_{g1}\right] \geq 0, \quad \ddot{u}_{g2}\left(\dot{y} + \dot{u}_{g1}\right) < 0$$ (4.16a)

Calculation of $\dfrac{W}{m}$ for the other three cases mentioned in Eq. (4.11) can be done in a similar way. The results are

$$\frac{W}{m} = \left[(1-\alpha)^2(1+\beta) + 2\alpha(1-\alpha)\right]\dot{u}_g\ddot{u}_g + (1-\alpha)\dot{u}_g\ddot{y} + \alpha\dot{y}\ddot{u}_g :$$

$$\dot{u}_{g1}\left[\ddot{y} + \ddot{u}_{g2} + (1+\beta)\ddot{u}_{g1}\right] \geq 0 \;\; \ddot{u}_{g2}\left(\dot{y} + \dot{u}_{g1}\right) \geq 0 \tag{4.16b}$$

$$\frac{W}{m} = -\left[(1-\alpha)^2(1+\beta) + 2\alpha(1-\alpha)\right]\dot{u}_g\ddot{u}_g - (1-\alpha)\dot{u}_g\ddot{y} - \alpha\dot{y}\ddot{u}_g :$$

$$\dot{u}_{g1}\left[\ddot{y} + \ddot{u}_{g2} + (1+\beta)\ddot{u}_{g1}\right] < 0, \ddot{u}_{g2}\left(\dot{y} + \dot{u}_{g1}\right) < 0 \tag{4.16c}$$

$$\frac{W}{m} = -(1-\alpha)^2(1+\beta)\dot{u}_g\ddot{u}_g - (1-\alpha)\dot{u}_g\ddot{y} + \alpha\dot{y}\ddot{u}_g : \;\; \dot{u}_{g1}\left[\ddot{y} + \ddot{u}_{g2} + (1+\beta)\ddot{u}_{g1}\right] < 0, \;\; \ddot{u}_{g2}\left(\dot{y} + \dot{u}_{g1}\right) \geq 0$$

$$\tag{4.16d}$$

Obviously, it is desired to make the testing power need a minimum value. This is the key point in calculating the value of the division factor α. Calculation of α can be done in different ways, which is the subject of discussion in the next section.

4.4 PROCEDURES FOR DIVIDING THE GROUND MOTION BETWEEN THE SHAKE TABLE AND ACTUATORS

In this section three different methods are proposed for dividing the ground motion between the force actuators and the shake table in the combined testing method. These methods are compared numerically in the next section. The methods are dividing in time domain, dividing in frequency domain, and the dynamic optimization, which are described in the following.

4.4.1 DIVIDING IN TIME DOMAIN

In this method, the dividing factor α is varied between zero and unity. This identifies the values of $\ddot{u}_{g1}$ and $\ddot{u}_{g2}$ in Eq. (4.15). Then the testing power is calculated from Eq. (4.16a–d). Comparing the results, the value of α for which the power is a minimum value is the desired dividing factor of the ground motion.

4.4.2 DIVIDING IN FREQUENCY DOMAIN

Generally, the shake tables are designed to work at lower frequencies compared with force actuators. Therefore it will be to the benefit of the experiment if the record is broken up into a pair of low-frequency and high-frequency components, in order to input the former to the shake table and the latter to the actuators. This is done conveniently through a direct and an inverse fast Fourier transform consequently. The appropriate selection of a cutoff frequency ω_c between the two components is the main challenge in this method. The simplest way is calculating the maximum required power for different values of ω_c and finding the ω_c at which the maximum power reaches its smallest value.

4.4.3 DYNAMIC OPTIMIZATION

The most important restriction in the method of dividing in time domain is the assumption that the dividing factor α is constant with respect to time. This means that α takes on a constant value all along a ground motion record in power calculations. It can be stipulated that if the value of the dividing factor of the ground motion is "tuned" with the dynamic characteristics of the record and the structure at each time step, i.e. if it changes value with time, the maximum required testing power can be even smaller. This is the basic idea of the dynamic optimization procedure.

To find α and the corresponding (minimum) W in this case, the derivative of W with respect to α in Eq. (4.16a–d) is calculated to be a null value. Then the desired value of α is determined to be

$$\alpha = 1 + \frac{1}{2(\beta+1)}\left(\frac{\ddot{y}}{\ddot{u}_g} + \frac{\dot{y}}{\dot{u}_g}\right): \ \dot{u}_{g1}\ddot{u}_{g2}\left[\ddot{y}+\ddot{u}_{g2}+(1+\beta)\ddot{u}_{g1}\right]\left(\dot{y}+\dot{u}_{g1}\right) < 0 \qquad (4.17a)$$

$$\alpha = \frac{1}{\beta-1}\left[\beta + \frac{1}{2}\left(\frac{\ddot{y}}{\ddot{u}_g} - \frac{\dot{y}}{\dot{u}_g}\right)\right]: \text{Otherwise} \qquad (4.17b)$$

Note that Eq. (4.17a) corresponds to the cases of Eqs. (4.16a) and (4.16d) and Eq. (4.17b) to the cases of Eqs. (4.16b) and (4.16c). Substituting the value of α from Eq. (4.17a,b) in Eq. (4.16a–d), results in the minimum value of the testing power, W_{min}, as follows:

$$\frac{W_{min}}{m} = -\left[\frac{1}{4(1+\beta)}\left(\frac{\ddot{y}}{\ddot{u}_g} + \frac{\dot{y}}{\dot{u}_g}\right)^2 + \frac{\dot{y}}{\dot{u}_g}\right]\dot{u}_g\ddot{u}_g: \ \dot{u}_{g1}\left[\ddot{y}+\ddot{u}_{g2}+(1+\beta)\ddot{u}_{g1}\right] \geq 0, \ \ddot{u}_{g2}\left(\dot{y}+\dot{u}_{g1}\right) < 0 \quad (4.18a)$$

$$\frac{W_{min}}{m} = \left\{\frac{1}{1-\beta}\left[\beta + \frac{1}{2}\left(\frac{\ddot{y}}{\ddot{u}_g} - \frac{\dot{y}}{\dot{u}_g}\right)\right]^2 + \frac{\ddot{y}}{\ddot{u}_g} + 1 + \beta\right\}\dot{u}_g\ddot{u}_g: \ \dot{u}_{g1}\left[\ddot{y}+\ddot{u}_{g2}+(1+\beta)\ddot{u}_{g1}\right] \geq 0, \ \ddot{u}_{g2}\left(\dot{y}+\dot{u}_{g1}\right) \geq 0$$

$$(4.18b)$$

The above equations correspond to the cases of Eqs. (4.16a) and (4.16b), respectively. For the cases of Eqs. (4.16c) and (4.16d), Eqs. (4.18a) and (4.18b) should be used, respectively, with an opposite sign.

An interesting feature of Eqs. (4.17a,b) and (4.18a,b) is that in them, the values of α and W_{min} are functions of the response ratios $\frac{\dot{y}}{\dot{u}_g}$ and $\frac{\ddot{y}}{\ddot{u}_g}$, and not the responses themselves. Since the variation of response ratios with time in an earthquake is much smoother than the responses, it can be expected that α and W_{min} will not vary too much from a time step to the next during the experiment. Another important point is that $\dot{y}$ and $\ddot{y}$ can be calculated from Eq. (4.3) if the nonlinear characteristics of the system are known before the experiment with good accuracy. Otherwise, they are recorded during the test and are used in Eq. (4.17a,b) at each time step to determine the value of α for the next time step.

4.5 EXTENDING TO MULTI DEGREE OF FREEDOM (MDF) MODELS

The procedure developed above can be easily extended to multi degree of freedom (MDF) systems. It is obvious that Eqs. (4.1)–(4.8) are valid for each story of an MDF system. Then Eqs. (4.9)–(4.11) are also valid for the calculation of power, except that Eq. (4.10) should be modified as follows:

$$W_1 = \left[\sum_{j=1}^{n}(m\ddot{u})_j + m_b\ddot{u}_{g1}\right]\dot{u}_{g1}, \ W_2 = -\sum_{j=1}^{n}(m\dot{u}_1)_j\ddot{u}_{g2} \qquad (4.19)$$

in which j is the story number and n is the total number of stories.

4.6 NUMERICAL CALCULATIONS

Using the equations developed in Section 4.4, the required experimental power is calculated for SDF structural models differing in the natural frequency, f_n. The natural frequency varies in the range of 1–10 Hz with 1 Hz increments. The damping ratio is set at 0.05. Both elastic and elasto-plastic behaviors are considered for the SDF systems. For the elasto-plastic case, the yield ratio, i.e., ratio of the yield force to the weight, is assumed identically to be 0.2, which is a common value. Also, because the mass ratio of a shaking table to the structural model, β, usually changes between 0.5 and 2, four values of 0.5, 1, 1.5, and 2 are assumed for β. Since usually in seismic tests and analysis a suite of consistent ground motions is used, arbitrarily ten earthquake records, all recorded on the soil type C (intermediate soil, ASCE7–10 [3]) with epicentral distances of 20–50 km and magnitudes 6–7.5, are selected from Pacific Earthquake Engineering Research Center (PEER) NGA database [4]. Although relevant only for the elasto-plastic case, the ground motions are scaled to result in a ductility demand μ in the range of two to four for each SDF model, both to reduce the nonlinear response dispersion and to have a considerable nonlinear behavior.

Because of space limitation, details of calculations are presented only for the dynamic optimization method as described above. According to Eqs. (4.17a,b), at each time step the value of α is calculated for each ground motion, and for known β and μ values. This value is then used in Eq. (4.18) at the same step to calculate the required power. In this method the α values are calculated case by case for each model during each earthquake. For instance, variation of α with time is shown in Figure 4.1 for $\beta = 1$ and $f_n = 5$ Hz during Irpinia earthquake. Because the equations of this method have been derived on the basis of minimizing the test power rigorously, the absolute minimum testing power should belong to this method. To confirm this expectation, the maximum power needed during all earthquakes for each set of β and f_n is calculated for the time and frequency domain procedures with the optimum values of α and f_c being normalized to the corresponding values of the dynamic optimization procedure. The results are shown in Table 4.1. The minimum ratio of powers is 1.17. This clearly shows the success of the dynamic optimization method to result in the absolute minimum testing power. It is to be noted that a strictly exact value of α is not necessarily needed. Therefore, an acceptably accurate value of α can be calculated beforehand using an appropriate model of the structure and can be used to decompose the input motion of the experiment.

4.7 CONCLUSIONS

A combined method for implementing earthquake engineering experiments was described in this paper. In this method, a combination of shake table and actuators is used to simulate the effect of ground motion on the structural model. Different methodologies for dividing the ground motion record between the two testing facilities were presented. Both elastic and elasto-plastic behaviors were accounted for the structural model. Also, SDF and MDF systems were considered. Different shake table to oscillator mass ratios were considered in the practical range of interest. The first method was called dividing in the time domain in which the amplitude of the record was divided by a constant number to determine the shares of the shaking table and the actuators. After calculating for different cases, a spectrum was computed illustrating the optimum dividing factor versus the natural frequency for which the power demand was a minimum for the time domain procedure. The second method called dividing in the frequency domain was a method of decomposing the record into its low- and high-frequency components. The low-frequency component was input to the shake table and the high-frequency one to the actuators. A table showing the optimum range of the cut-off frequency, where the ground motion is split into its two components in the frequency domain, was calculated for different cases. The third method, meant to be the most efficient one, was the dynamic optimization method. In this method the derivative of the power formula with respect to the dividing factor of the ground motion was utilized such that the total power was a minimum at each instant. The dynamic optimization procedure resulted in the absolute minimum values of the testing power between the presented methods.

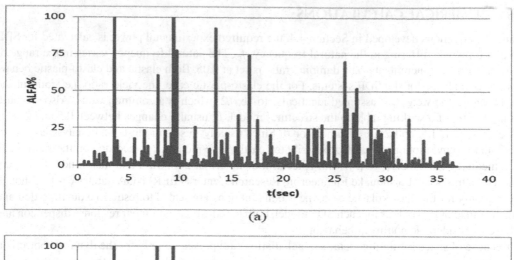

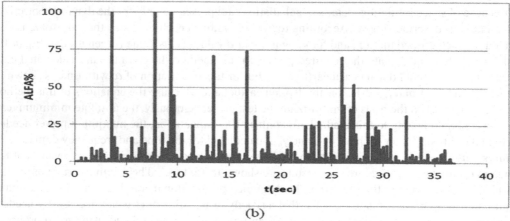

FIGURE 4.1 Variations of α of the dynamic optimization procedure with time $\beta = 1$ and $f_n = 5\,\text{Hz}$ during Irpinia earthquake. (a) Elastic behavior, (b) Elasto-plastic behavior, $\mu = 2$.

TABLE 4.1

Maximum Power Need of the Time and Frequency Domain Procedures Normalized to the Corresponding Values of the Dynamic Optimization Method

f_n (Hz)	β	Time Domain		Frequency Domain	
		$\mu = 1$	$\mu = 4$	$\mu = 1$	$\mu = 4$
1	0.5	1.88	2.63	2.24	2.82
	1.0	1.57	2.29	2.02	2.55
	1.5	1.34	1.89	1.63	1.76
	2.0	1.17	1.64	1.37	1.42
2	0.5	2.89	5.38	2.98	5.51
	1.0	2.29	4.16	2.41	4.37
	1.5	1.91	3.39	2.17	3.52
	2.0	1.63	2.84	1.85	3.04

(Continued)

TABLE 4.1 (*Continued*)
Maximum Power Need of the Time and Frequency Domain Procedures Normalized to the Corresponding Values of the Dynamic Optimization Method

		Time Domain		Frequency Domain	
f_n (Hz)	β	$\mu = 1$	$\mu = 4$	$\mu = 1$	$\mu = 4$
3	0.5	3.73	5.07	3.86	5.49
	1.0	3.03	3.92	3.12	4.21
	1.5	2.55	3.61	2.63	3.83
	2.0	2.19	3.11	2.31	3.45
4	0.5	4.15	4.73	4.52	4.88
	1.0	3.42	3.84	3.87	4.04
	1.5	2.91	3.26	3.19	3.44
	2.0	2.52	2.77	2.88	3.08
5	0.5	4.20	5.45	4.19	5.18
	1.0	3.47	4.29	3.34	4.42
	1.5	2.96	3.67	2.76	3.76
	2.0	2.56	3.09	2.43	3.29
6	0.5	4.28	6.58	4.28	6.72
	1.0	3.56	5.33	3.61	5.51
	1.5	3.03	4.25	3.06	4.47
	2.0	2.38	3.62	2.41	3.83
7	0.5	4.61	4.70	5.17	5.66
	1.0	3.79	3.86	3.98	4.58
	1.5	3.20	3.22	3.65	4.25
	2.0	2.79	2.79	3.07	2.89
8	0.5	4.56	5.54	4.74	5.69
	1.0	3.88	4.48	3.96	5.05
	1.5	3.20	4.27	3.29	4.32
	2.0	2.77	3.56	2.80	3.63
9	0.5	4.76	4.99	4.71	4.76
	1.0	3.91	4.01	3.89	3.97
	1.5	3.32	3.42	3.31	3.32
	2.0	2.87	2.97	2.85	2.88
10	0.5	4.77	4.91	4.83	4.95
	1.0	3.93	4.05	3.96	4.11
	1.5	3.34	3.43	3.33	3.47
	2.0	2.88	2.94	2.92	2.96

REFERENCES

1. Dimig, J. et al. 1999. Effective force testing: A method of seismic simulation for structural testing, *Journal of Structural Engineering*, ASCE, 125(9), pp. 1028–1037.
2. Kausel, E. 1998. New seismic testing method. I: Fundamental concepts, *Journal of Engineering Mechanics*, ASCE, 124(5), 565–570.
3. American Society of Civil Engineers (ASCE) 7–10, 2010. Minimum Design Loads for Buildings and Other Structures, Reston, VA.
4. http://peer.berkeley.edu/nga/search.html.

5 Design Spectra for Structures Subjected to Passing Underground Trains

F. Behnamfar
Isfahan University of Technology

A. A. Yazdani
Isfahan University of Technology

A. Bakhtiar
Sepehr Institute of Higher Education

CONTENTS

5.1 INTRODUCTION

Currently, the use of underground trains, or subways, are quite a common approach for urban transportation in large cities. The point that is a prime concern in structural design of such projects is how to evaluate precisely the induced vibrations due to passage of trains. In order to take remedial actions whenever needed, the maximum response of nearby structures has to be compared with the comfort thresholds set by the corresponding design codes.

Many researchers have worked out methods for the calculation of maximum structural responses under train vibrations. Eitzenberger (2008) reviewed the studies on initiation, transmission, and propagation of vibration in dynamical systems. It was concluded that a major need persisted for good estimations of train vibrations in structures. Bian et al. (2008) calculated the train vibrations using a two-dimensional (2D) finite element (FE) model perpendicular to the train path. Verbraken et al. (2012) compared the numerical and empirical methods of computing a transfer function for vibration analysis and derived a combined approach. Nicolosi et al. (2012) developed an integrated method for numerical analysis of train induced vibrations. Li (2011) made two assumptions for deriving the wheel loads. First, it was taken to be a single degree of freedom (SDF) moving mass, and second, a two degree of freedom (2DF) mass, damper, and stiffener moving load system. Woodward et al. (2013) utilized a dynamic equation governing one-fourth of a train oscillating in vertical direction. In his model, stiffness and damping of the primary and secondary suspension system were included. Huang and Chrismer (2013) and Connolly et al. (2014) also proposed a certain mass-damper-stiffener system for calculation of the dynamic train loading. Krylov (2014) presented a pseudo-static method for the derivation of wheel loads using the distance between the transverse beams and the train velocity. Presthus (2002) compared the efficiency of different air springs and studied a new one developed in Sweden. Sayyaadi and Shokouhi (2009) studied a full train system having 70 degrees of freedom. It consisted of four axles, four wagon, and two bogies on an air spring suspension system. They compared the existing models for the air spring and presented an optimum system for the same purpose. In all, in the above studies it is concluded that resonance is more probable in softer soils and for train velocities of 100 km/h and more. In modeling of the soil-tunnel system, the soil has been modeled with its several layers when an actual site has been under study. Otherwise, in a general parametric study, the researchers usually modeled the soil with one or two layers at most, if any. In a previous study, Behnamfar and Nikbakht (2013) calculated the maximum displacement, velocity, and acceleration of points on the ground surface due to the point loads of underground trains moving at different velocities. They presented their results as response spectra, i.e., graphs of maximum response of SDF systems against their natural periods and damping ratios. As an important limitation, they disregarded dynamic effects of train loading due to roughness of the railroad and wheels. Such effects have been reported by other researchers to be important for computing the loading model (Bian et al. 2008). Moreover, they only studied the movement of high-speed trains. They concluded that the maximum responses were sensitive to the train velocity, the soil type, the depth of tunnel, and the distance on the ground surface from the axis of tunnel. It is the only research presenting response spectra under train induced vibrations.

In the current paper, the work of Behnamfar and Nikbakht (2013) is extended by accounting for the dynamic effects of train loading on the railroad. For this purpose, three mostly in-use types of trains, three train velocities, three tunnel depths, three soil types, and three damping ratios for the SDF system are considered to cover all practical ranges. This study results in 243 graphs of spectral acceleration under train induced vibrations.

In the following sections, the trains and their dynamic wheel loads are represented, and the three-dimensional (3D) soil-tunnel model is identified and analyzed under the mentioned loads.

5.2 SPECIFICATIONS OF THE TRAINS

5.2.1 THE SELECTED TRAINS

Various trains are in operation around the world for underground city transportation. To make a general parametric study possible, it was decided to take three generic trains as shown in Table 5.1.

5.2.2 THE SUSPENSION SYSTEM

The suspension system is a vital component of train located at the base of the wagons over the wheels. As it appears, it is used for reducing the vibrations transferred to wagon and the riding passengers to provide comfort. Its correct modeling is very important in calculation of the loads applied to the wheels and rails for further computations. Today, three types of suspension systems are available. They are called the friction wedge, leaf spring, and air spring systems.

Opposite to the first two types, the air spring has a wide use in the suspension system of commutative trains of over and underground transportation. The suspension system is installed

TABLE 5.1

Characteristics of the Subway Trains in This Study

Train's Types	T_1	T_2	T_3
Number of wagons	4	6	9
Weight of wagons (tons)	40	48	30
Wheel spacing (multiplier of 0.2 m)	2&2.6	2&2.6	1.6&2
Bogie spacing (m)	15	13	10
Number of cars	4	6	9
Wheels per car	48	72	108

Layout of the T_1 train is shown in Figure 5.1.

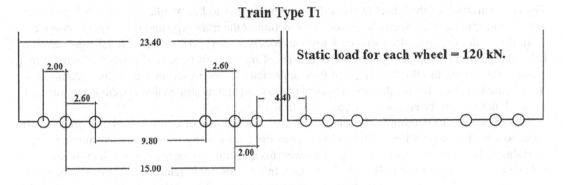

FIGURE 5.1 Loading configurations for the T_1 train.

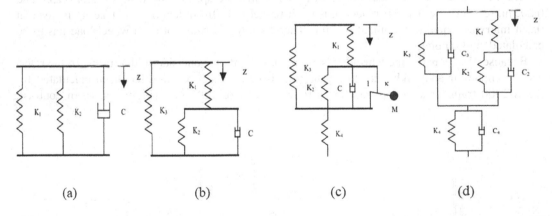

FIGURE 5.2 Dynamic models of the air. (a) Simple; (b) nishimura (Presthus 2002); (c) vampire; (d) simpack (Wu et al. 2014).

between the wheels and bogie as the main suspension and between the bogie and wagon as the secondary suspension system. The main system consists of a combination of dampers and spiral springs and the secondary system of damper and air springs.

5.2.2.1 Dynamical Model of the Secondary Suspension System

A number of models have been proposed in the literature for the air springs of the secondary suspension system. The more famous ones are shown in Figure 5.2.

The Nishimura model is used in this study. The parameters of this model include three spring coefficients K_1, K_2, and K_3 and the damping coefficient C, which according to (Sayyaadi and Shokouhi 2009, Presthus 2002) turn out to be $K_1 = 1{,}111.9$ kN/m, $K_2 = 694.95$ kN/m, $K_3 = 0$ kN/m, and $C = 52.62$ kN s/m, and the damping coefficient of each of the above eight dampers is taken to be 61.375 kN s/m.

5.2.2.2 Dynamical Model of the Main Suspension System

The main system consists of two vertical and horizontal springs and a damper at the connection of the bogie to each wheel.

According to (Sayyaadi and Shokouhi 2009), the damping coefficient is taken to be 40.8 kN s/m, and the horizontal and vertical spring coefficients are 366 and 683 kN/m, respectively.

5.3   THE LOADING PATTERN

The time variation of the wheel loads at different points has to be calculated before being applied to the soil-tunnel model. Such a variation is a function of the train type and the suspension system.

In this study the specialty software Universal Mechanism (UM), version 7.6.2.2, was selected. Like many other mechanical programs, equations of motion can be solved using the implicit and explicit algorithms in UM. The implicit Park approach has been used in this study because of its stability and the fact that the dynamical model of the train and its suspension system is not too complicated and it is preferred not to use too small time steps.

It is necessary to determine the time lapse between the output quantities of the program. In other words, it is imperative to find the answer to the question that values of each moving wheel load should be given at what instances. To answer this question it is necessary to select the distance between the nodal points of rails in the FE model. In this study, it is selected uniformly to be 0.2 m. Therefore, the value of the time steps for computations turns out to be $\Delta t = 0.009, 0.007$, and 0.005 for the train velocities of 80, 100, and 140 km/h, respectively.

Length, width, and thickness of the traverses, which are spaced at 0.6 m, are 2.5, 0.235, and 0.205 m, respectively. The FE model of the soil-tunnel is 120 m in length. The loading is given at 0.2 m intervals, equal to the nodal spacing. Therefore, time history of each wheel load has to be calculated at 600 points.

Because of symmetry, the wheel loads are recorded only for one side of the trains. In the reference (Behnamfar and Nikbakht, 2013), the (quasi-static) wheel reaction has been calculated for the X-2000 train for a smooth rail and a train velocity of 120 km/h (Figure 5.3a). In contrast,

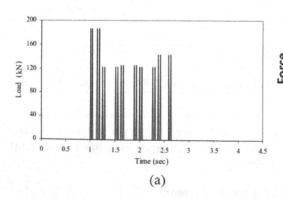
(a)

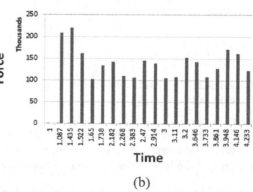
(b)

FIGURE 5.3   The wheel load. (a) Calculated in reference (Behnamfar and Nikbakht, 2013). (b) Calculated in this study.

Figure 5.3b shows the wheel load calculated in this study. Considering that the suspension system may dynamically cause an increase or decrease in load, a good similarity is observed.

5.4 THE 3D MODEL OF THE SOIL-TUNNEL SYSTEM

The soil-tunnel system is modeled three dimensionally using Abaqus, version 6.14.1, and analyzed using its implicit approach. The system consists of the soil with its natural/artificial boundaries, the rails and traverses, the ballast, and the tunnel lining. The model has a vertical plane of symmetry. Therefore, only half of the actual system is built in the software. Suitable boundary conditions of direct symmetry are applied at the plane of symmetry.

Dimensions of the soil body are $120 \times 45 \times 50$ m (length $\times$ width $\times$ thickness). These values have been optimized in the reference (Behnamfar and Nikbakht, 2013) such that the ground surface response at the middle of the model becomes insensitive to larger dimensions. The output of the computations will be the vertical response on the ground surface at the middle of the model.

5.4.1 MODELING OF THE COMPONENTS

Transverse section of the model is shown in Figure 5.4.

The steel rails take the wheel loads and transfer them to the traverses through elastic shock absorbers. Its section area and moment of inertia are 0.007587 m^2 and $3.38\text{E-}05$ m^4 (Behnamfar and Nikbakht 2013). They are modeled using beam elements. The elastic shock absorbers are placed between the rails and traverses in the model as spring and dampers spaced at 0.6 m, like traverses. The spring and damping coefficients of these components are taken to be 1.534×10^7 N/m and 1,350 N s/m (Behnamfar and Nikbakht 2013).

The traverses, modeled as beam elements, are assumed to be concrete beams with the mentioned dimensions. They are assumed to be resting on a concrete medium limited at bottom to the tunnel with a maximum thickness of 0.65 m (Behnamfar and Nikbakht 2013, Woodward et al. 2013).

The concrete tunnel lining is modeled using shell elements. Its diameter and thickness are 6 and 0.3 m. The distance from the ground surface to the top of the tunnel is defined as the tunnel depth in this study. It is taken to be 5, 10, and 15 m in different cases.

Three types of soils including stiff, intermediate, and soft are used for the medium around the tunnel. They are distinguished with their shear wave velocities to be 400, 200, and 100 m/s, respectively, and with their densities to be 19, 18, and 17 kN/m^3, respectively. Because of the low amplitude

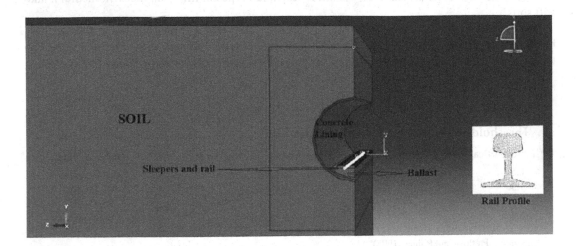

FIGURE 5.4   Transverse section of the soil-tunnel.

of vibrations due to the passage of trains, all of the material models are taken to be linearly elastic. The mesh size is 0.1 m for the rails, $0.2 \times 0.25 \times 0.25\,m^3$ for the tunnel lining, $0.2 \times 0.24 \times 0.25\,m^3$ for concrete medium under traverses, and $2 \times 1 \times 0.5\,m^3$ on average for the soil medium. If the use of five elements is taken to be sufficient to model a certain wave length, for the mentioned size of the soil elements, the smallest reliable period will be around 0.01 s for the average shear wave velocity.

5.4.2 THE OUTPUT

The time history of acceleration response is calculated at the middle point on the ground surface. Using Seismosignal version 4.3.0, it is converted to acceleration design spectra for SDF systems having damping ratios of 0%, 2%, and 5%. For the low amplitudes of train induced vibrations, a 2% structural damping seems to be more appropriate. The other two values are taken to make an interpolation possible and to study the effect of damping.

5.5 THE CALCULATED SPECTRA

5.5.1 INTRODUCTION

The vertical acceleration response spectra for the SDF system resting on the middle point of the ground surface have been calculated using the mentioned approach and are presented in this section. They are compared with the acceptable values shown in Table 5.2.

According to Table 5.2, the value 0.055 g is selected as the maximum acceptable acceleration for human comfort. This is shown as a horizontal line on the following graphs. Of course, what is to be compared with this value is the total spectral acceleration at each story of a multistory building, calculated using modal analysis based on the following response spectra.

5.5.2 THE NUMERICAL RESULTS

Because of space limitation, only results for the tunnel depth of 5 m and velocity of 140 km/h are illustrated here. The whole set of results can be found elsewhere (Yazdani 2017). The results of the other depths are only referred to comparatively. The results for the depth of 5 m are presented in three groups, categorized based on the soil type. In each group nine sets of graphs are shown for three trains comprising three sets of graphs for each train. Each set of graphs represents a certain train velocity.

Figure 5.5 shows the acceleration spectra for the three types of trains on the stiff, medium, and soft soils.

For the stiff soil, the shear wave velocity is 400 m/s. The vertical dimension of the soil medium has been selected such that it acts like a half space for the waves initiated at the rails and sensed at the ground surface. Therefore, the medium has only one Rayleigh mode independent from frequency

TABLE 5.2
Thresholds of Vibrations for Human Comfort

Peak Vertical Velocity (mm/s)	Peak Vertical Acceleration (mm/s²)	Human Feeling
0.50	34.00	Felt somewhat
1.30	100.00	Clearly felt
6.80	550.00	Annoying
13.80	1,800.00	Intolerable

Source: Pretlove and Rainer (1995).

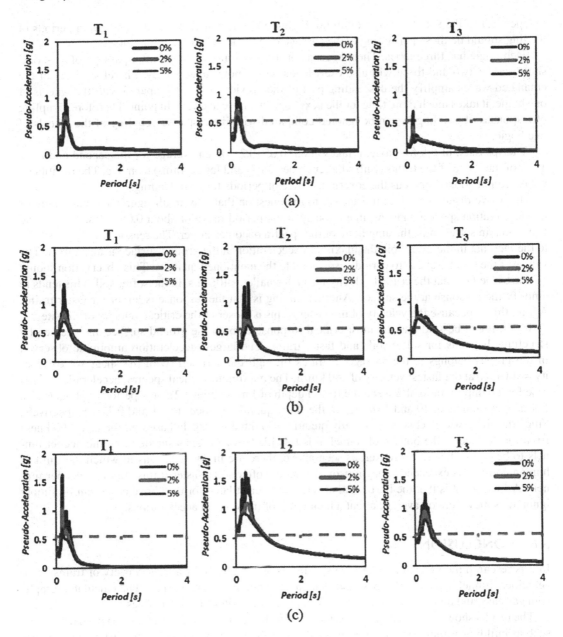

FIGURE 5.5 Acceleration spectra for the trains with tunnel depth = 5 m and V = 140/h. (a) Trains T_1, T_2, T_3 on the stiff soil, (b) trains T_1, T_2, T_3 on the medium soil, and (c) trains T_1, T_2, T_3 on the soft soil.

and the Rayleigh wave velocity is close to the shear wave velocity. The train type and velocity are the variables of the graphs.

The train velocity varies from 80 km/h, or 22.2 m/s to 140 km/h, or 38.9 m/s. All of the mentioned velocities are far from the critical (Rayleigh wave) velocity of the site. Moreover, the fundamental period of the site for vertical motions is well known to be four times the total depth (50 m) divided by the P-wave velocity. It turns out to be about 0.4 s. No amplification is seen at this period in the graphs. Therefore, if there is variation in the response under different train velocities, it should be primarily due to the change of the loading period. The letter is defined as the distance between the first and successive wheels of a wagon divided by the train velocity. Obviously, this quantity is independent of the soil type. By examining Figure 5.2, it is revealed that the governing periods of loading should be those pertaining to the passage of a bogie over a certain point.

Each bogie has three axles. Then, because of the proximity of axles, after passage of axle one, shortly axles two and three will also exert loads on the point under study. Therefore, each of the emanated waves amplify the descending part of the previous wave. Compared with the wheels of one bogie, it takes much more time for the next bogie to arrive at a certain point. Therefore, coupling of responses of two bogies can be ignored compared with the response coupling of the wheels of one bogie.

After passage of the first axle, it takes about 0.05–0.2 s on an average for the second and third axles of the bogie of the trains (with smaller values for T_3 and faster trains) to arrive. The mentioned values can be looked upon as the governing range of periods for train loading.

The above discussion gives the answer to the question that why in all figures the maximum of the acceleration spectra happens unanimously at the period range of about 0.05–0.2 s. The second fact is that in softer soils, the amplitude of the spectra becomes larger. The reason simply goes back to the fact that in the same soil, the maximum acceleration at the ground surface is also larger. The values of the spectra at a zero period correspond to the mentioned quantity. This observation should be due to the fact that the critical velocity is much smaller, <100 m/s, in the softer soil. This tends to amplify the response at the surface. Another finding is that the response is larger for faster trains. Again, this is because the velocity of the faster trains is closer to the critical velocity of the site.

In all, it can be said that the most critical response of both the ground surface and the nearby structures happens for softer soils and faster trains. The largest acceleration amplitude observed in the graphs belongs to the passage of trains through the soft soil (with the shear wave velocity = 100 m/s) at the fastest velocity of 140 km/h. The maximum vertical spectral acceleration at this case for a damping ratio of 2% and the tunnel depth of 5 m is about 1.2 g, at a period of about 0.15 s. For deeper tunnels of 10 and 15 m depth, the same quantity reduces to 1.1 and 0.75 g, respectively. Since the shear wave velocity is actually quantity weighted averagely based on the layer thickness from the surface to the bottom of tunnel, it is less likely for the sites surrounding the deeper tunnels to be so much soft. Moreover, as seen above, there are many occasions in which the limit of human comfort is exceeded. They mostly belong to softer soils, faster trains, and less deeper tunnels. A final point is that the horizontal accelerations calculated for the above cases can be simply ignored as they were on average about a hundredth of the vertical accelerations.

5.6 CONCLUSIONS

In this research the vertical acceleration spectra were calculated for a large variety of trains, train velocities, tunnel depths, soil types, and structural damping, as 243 graphs. The calculated graphs come as easy and quick tools for comfort evaluation at the sites of subway lines.

The results showed that in many cases the comfort limit can be exceeded for short period systems, such as building equipment. The most critical case belonged to softer soils and faster trains in less deeper tunnels.

REFERENCES

Behnamfar, F. and Nikbakht, R. 2013. Structural response spectra under passing underground trains, *In 3rd International Conference on Recent Advances in Railway Engineering*, Iran University of Science and Technology, Tehran, Iran, pp. 1–7.

Bian, X., Chen, Y. and Hu, T. 2008. Numerical simulation of high-speed train induced ground vibrations using 2.5 D finite element approach. *Science in China Series G: Physics, Mechanics and Astronomy*, 51(6), 632.

Connolly, D. P., Kouroussis, G., Giannopoulos, A., Verlinden, O., Woodward, P. K. and Forde, M. C. 2014. Assessment of railway vibrations using an efficient scoping model. *Soil Dynamics and Earthquake Engineering*, 58, 37–47.

Eitzenberger, A. 2008. *Train-Induced Vibrations in Tunnels: A Review.* Luleå tekniska universitet (Ed), Luleå, Swede.

Huang, H. and Chrismer, S. 2013. Dynamic train-track interaction model to study track performance under critical speed (No. 13–4397).

Krylov, V. V. 2014. Focusing of ground vibrations generated by high-speed trains, in: P. Sas (Ed.) *International Conference on Noise and Vibration Engineering*, 15–17 September 2014, KU Leuven, Leuven, Belgium, pp. 2007–2016.

Li, K. 2011. Study on the dynamic response of track structure under moving loads with variable speeds, in: G. D. Roeck (Ed.) *8th International Conference on Structural Dynamics, EURODYN*, 4–6 July 2011, Leuven, Belgium, pp. 846–851.

Nicolosi, V., D'Apuzzo, M. and Bogazzi, E. 2012. A unified approach for the prediction of vibration induced by underground metro. *Procedia-Social and Behavioral Sciences*, 53, 62–71.

Presthus, M. 2002. Derivation of air spring model parameters for train simulation. M.S. Thesis, Department of Applied Physics and Mechanical Engineering Division of Fluid Mechanics, Lulea University of Technology, p. 74.

Pretlove, J. H. and Rainer, A. J. 1995. Human response to vibrations, in: H. E. A. Bachman (Ed.) *Vibration Problems in Structures*, Birkhäuser, Basel, p. 237.

Sayyaadi, H. and Shokouhi, N. 2009. A new model in rail–vehicles dynamics considering nonlinear suspension components behavior. *International Journal of Mechanical Sciences*, 51(3), 222–232.

Verbraken, H., Lombaert, G. and Degrande, G. 2012. Experimental and numerical prediction of railway induced vibration. *Journal of Zhejiang University Science A*, 13(11), 802–813.

Woodward, P., Laghrouche, O. and El-Kacimi, A. 2013. The development and mitigation of ground mach cones for high speed railways. *In ICOVP 2013-International Conference on Vibration Problems, Lisbon, Portugal.*

Yazdani, A. A. 2017. Spectral analysis of structures under vibration due to underground passing trains, in: Faculty of Engineering, Islamic Azad University, Isfahan (Khorasgan) Branch, p. 142.

6 Frequency-Domain Fast Maximum Likelihood Estimation of Complex Modes

Binbin Li and Yan-Long Xie
Zhejiang University

CONTENTS

6.1 INTRODUCTION

Operational modal analysis (OMA) (Reynders, 2012; Brincker and Ventura, 2015; Au, 2017) aims at identifying structural modal parameters (e.g., modal frequencies, damping ratios, and mode shapes) by using only the measured structural response. The identified modal parameters play an important role in structural design verification and retrofits (Brownjohn et al., 2013), vibration control (Spencer and Nagarajaiah, 2003), and health monitoring (Brownjohn, 2007; Ntotsios et al., 2008). Since OMA does not require specific knowledge of the input force, it has gained popularity in the dynamic testing of civil engineering structures, where artificial excitation is costly or in many cases impractical.

In OMA, the input force is unknown and usually modeled as a wideband stochastic process. As a result, the identification process is more sophisticated compared with the input-output identification, and proper means must be devised to extract modal information from the stochastic response data in the absence of loading information. During the past decades, a large variety of methods have been developed in literature. Early methods such as random decrement (Cole, 1973) and peak-picking (Bendat and Piersol, 1980) are simple to use but they can only provide a rough estimate of the modal parameters. During the 1990s, two important techniques became available: the stochastic subspace identification (Van Overschee and De Moor, 1996) and the frequency-domain decomposition (Brincker, Zhang and Andersen, 2000). They are able to estimate modal parameters with good accuracy in most conditions. For a more thorough overview of OMA methods, the reader is referred to (Reynders, 2012; Brincker and Ventura, 2015; Au, 2017).

Recently, a fast algorithm (Au, 2012) for Bayesian fast Fourier transform (FFT) method has been developed for OMA. It can identify well-separated modes at the speed of seconds and closely spaced modes in less than 1 min with the identification error quantified, making it an ideal tool for practical applications. However, this method assumes a proportional damping, which may not capture the true dynamic characteristics of structures, e.g., when damper is introduced to suppress the vibration. This paper aims at developing its counterpart for the general non-proportional damping but with the principle of maximum likelihood (ML). In fact, the ML estimator should coincide with the Bayes estimator when the data is long enough, and especially the non-informative prior is used in the Bayesian inference. The potential gain of an ML estimator is that the identification uncertainty can be more efficiently computed. Unlike the fast algorithm (Au, 2012), which requires profound mathematical skill to understand and adept coding ability to program, we rely on the expectation-maximization (EM) algorithm to develop the ML estimation, yielding an easy-to-understand and simple-to-program approach.

6.2 PROBLEM FORMULATION

6.2.1 THE DETERMINISTIC MODEL

When a linear time-invariant dynamical system can be described or approximated by an n_d degrees of freedom (DoFs) numerical model with general viscous damping, its vibration behavior is governed by the following continuous-time state space model (SSM)

$$\dot{x}(t) = \mathbf{A}_c x(t) + \mathbf{B}_c \mathbf{u}(t) \tag{6.1}$$

where

$$x(t) = \begin{bmatrix} \mathbf{v}(t) \\ \dot{\mathbf{v}}(t) \end{bmatrix} \quad \mathbf{A}_c = \begin{bmatrix} \mathbf{0}_{n_d} & \mathbf{I}_{n_d} \\ -\mathbf{M}^{-1}\mathbf{K} & -\mathbf{M}^{-1}\mathbf{C} \end{bmatrix} \quad \mathbf{B}_c = \begin{bmatrix} \mathbf{0}_{n_d \times n_i} \\ \mathbf{M}^{-1}\mathbf{P} \end{bmatrix} \tag{6.2}$$

vectors $x(t) \in \mathbb{R}^{n_s}$ and $\dot{x}(t) \in \mathbb{R}^{n_s}$ are, respectively, the state variable and its time derivative; $\mathbf{v}(t) \in \mathbb{R}^{n_d}$ and $\dot{\mathbf{v}}(t) \in \mathbb{R}^{n_d}$ are the structural displacement and velocity response ($n = 2n_d$ is the model order). Matrices $\mathbf{A}_c \in \mathbb{R}^{n_s \times n_s}$ and $\mathbf{B}_c \in \mathbb{R}^{n_s \times n_i}$ are the state matrix and the input matrix, respectively. $\mathbf{0}_{n_d}$ and $\mathbf{I}_{n_d}$ represent n_d-by-n_d zero and identity matrices.

Considering the acceleration measurement, one has the observation equation

$$\mathbf{y}(t) = \mathbf{C}_c x(t) + \mathbf{D}_c \mathbf{u}(t) \tag{6.3}$$

where

$$\mathbf{C}_c = \begin{bmatrix} S_v - \mathbf{S}_{\ddot{v}}\mathbf{M}^{-1}\mathbf{K} & S_{\dot{v}} - \mathbf{S}_{\ddot{v}}\mathbf{M}^{-1}\mathbf{C} \end{bmatrix} \quad \mathbf{D}_c = \mathbf{S}_{\ddot{v}}\mathbf{M}^{-1} \tag{6.4}$$

$\mathbf{S}_{\ddot{v}} \in \mathbb{R}^{n \times n_d}$ are the selection matrices that define the location where measurements are made. Matrices $\mathbf{C}_c \in \mathbb{R}^{n \times n_s}$ and $\mathbf{D}_c \in \mathbb{R}^{n \times n_i}$ are the output matrix and the direct transmission matrix, respectively. In particular, one can show $\mathbf{D}_c = \mathbf{C}_c \mathbf{A}_c^{-1}\mathbf{B}_c$ (Peeters, 2000) when only the acceleration is measured.

Substituting the eigenvalue decomposition $\mathbf{A}_c = \mathbf{\Psi}_c \mathbf{\Lambda}_c \mathbf{\Psi}_c^{-1}$ into Eqs. (6.1) and (6.3), one can obtain the modal-form SSM as

$$\begin{cases} \dot{x}_m(t) = \mathbf{\Lambda}_c x_m(t) + \mathbf{L}_c^T \mathbf{u}(t) \\ \mathbf{y}(t) = \mathbf{\Phi}_c \mathbf{\Lambda}_c x_m(t) + \mathbf{\Phi}_c \mathbf{L}_c^T \mathbf{u}(t) \end{cases} \tag{6.5}$$

where we have defined the state vector $x_m(t) = \Psi_c^{-1} x(t)$ in the modal space. The modal input matrix $L_c^T = \Psi_c^{-1} B_c$ (".^{T}" denotes the non-conjugate transpose) contains the modal participation vectors l_j as its columns, and the modal output matrix $\Phi_c = C_c \Psi_c^{-1} \Lambda_c^{-1}$ contains the observed mode shapes ϕ_j as its columns.

Taking Fourier transform on both sides of Eq. (6.5) and consider a frequency band $[f_l, f_u]$ that includes m closely spaced modes, one can show that

$$y(f) \approx \sum_{j=1}^{m} \frac{\phi_j l_j^T \mathcal{U}(f)}{1 + i(2\pi f)^{-1} \lambda_j} = \Phi h(f) L^T u(f) \tag{6.6}$$

where $y(f)$ and $\mathcal{U}(f)$ are the Fourier transform of $y(f)$ and $u(f)$, respectively; i is the imaginary unit; $h(f) = \mathrm{diag}\left(\left[1 + i(2\pi f)^{-1} \lambda_1\right]^{-1}, \ldots, \left[1 + i(2\pi f)^{-1} \lambda_m\right]^{-1}\right)$ is the modal frequency response function (FRF) matrix.

Suppose we only have the structural response $y_{1:N} \triangleq \{y_1, \ldots, y_N\}$ sampled with an interval Δt sec. We then need to transform Eq. (6.6) into its corresponding discrete form

$$y_k \approx \Phi h_k L^T \mathcal{U}_k; h_k = \mathrm{diag}\left(\left[1 + i(2\pi f_k)^{-1} \lambda_1\right]^{-1}, \ldots, \left[1 + i(2\pi f_k)^{-1} \lambda_m\right]^{-1}\right) \tag{6.7}$$

where y_k can be obtained based on the scaled discrete Fourier transform (DFT) as

$$y_k = \sqrt{\Delta t / N} \sum_{j=0}^{N-1} y_j e^{-2\pi i jk/N} \tag{6.8}$$

and it corresponds to the frequency $f_k = k/N\Delta t$ (Hz), where $k \leq N_q = \mathrm{int}[N/2]$ ($\mathrm{int}[\cdot]$ denotes the integer part) and N_q is the index at the Nyquist frequency.

6.2.2 THE PROBABILISTIC MODEL

In the dynamic testing of large-scale or in-operation structures, artificial excitation is costly or in many cases impractical. The only feasible way is to intrigue the vibration by natural excitations, e.g., wind, traffic, and ground tremor. In this situation, the input force is unknown, and the conventional treatment is to model it as a wideband stochastic process. Corresponding to the deterministic model in Eq. (6.7), we construct the following probabilistic model:

$$y_k = \Phi h_k p_k + \varepsilon_k \tag{6.9}$$

where we have defined the DFT of modal force $p_k = L^T \mathcal{U}_k$, which is unknown and modeled statistically in our situation. Assuming zero-mean stationary modal excitation and long data (the number of DFT points within the frequency band), $N_f \gg 1$ p_k is then (circularly symmetric) complex Gaussian distributed and independent at different frequencies (Brillinger, 1981). Its covariance (i.e., power spectral density (PSD) of the process) is assumed to be a constant matrix $S \in \mathbb{C}^{m \times m}$ in the selected band, i.e., $p_k \sim \mathcal{CN}(0, S)$. The prediction error ε_k incorporates the modal truncation error and the measurement error. It is also assumed to be complex Gaussian distributed with zero mean and covariance matrix $\mathbf{E} \in \mathbb{C}^{n \times n}$ within the selected band. If we define the modal response $x_k = \mathbf{h}_k \boldsymbol{\eta}_k$, Eq. (6.9) then becomes a standard linear model

$$\boldsymbol{y}_k = \boldsymbol{\Phi} x_k + \boldsymbol{\varepsilon}_k \tag{6.10}$$

Since $\boldsymbol{\eta}_k \sim \mathcal{CN}(0, \mathbf{S})$, it is easy to see that $x_k \sim \mathcal{CN}(0, \mathbf{H}_k)$ where $\mathbf{H}_k = \mathbf{h}_k \mathbf{S} \mathbf{h}_k^*$. In addition, assuming the statistical independence between x_k and $\boldsymbol{\varepsilon}_k$, one can show the conditional distribution $\boldsymbol{y}_k \mid x_k, \boldsymbol{\theta} \sim \mathcal{CN}(\boldsymbol{\Phi} x_k, \mathbf{E})$ because of $\boldsymbol{\varepsilon}_k \sim \mathcal{CN}(0, \mathbf{E})$, where $\boldsymbol{\theta} = \{\boldsymbol{\Phi}, \boldsymbol{\Lambda}, \mathbf{S}, \mathbf{E}\}$ incorporates all the unknown parameters $(\boldsymbol{\Lambda} = \operatorname{diag}(\lambda_1, \ldots, \lambda_m))$. Therefore, we have the joint probability density function (PDF) of $\boldsymbol{y}_k$ and x_k given $\boldsymbol{\theta}$ as

$$f(\boldsymbol{y}_k, x_k \mid \boldsymbol{\theta}) = \frac{1}{\pi^{n+m} |\mathbf{E}| |\mathbf{H}_k|} \exp\left[-(\boldsymbol{y}_k - \boldsymbol{\Phi} x_k)^* \mathbf{E}^{-1} (\boldsymbol{y}_k - \boldsymbol{\Phi} x_k) - x_k^* \mathbf{H}_k^{-1} x_k \right] \tag{6.11}$$

where the joint covariance matrix Σ_k can be written as

$$\Sigma_k = \begin{bmatrix} \boldsymbol{\Phi} \mathbf{H}_k \boldsymbol{\Phi}^* + \mathbf{E} & \boldsymbol{\Phi} \mathbf{H}_k \\ \mathbf{H}_k \boldsymbol{\Phi}^* & \mathbf{H}_k \end{bmatrix} \tag{6.12}$$

Since the joint distribution of $\boldsymbol{y}_k$ and x_k given $\boldsymbol{\theta}$ is zero-mean multivariate complex Gaussian, one can easily derive the desired marginal and conditional distributions by manipulating the covariance matrix Σ_k. First, the conditional distribution of x_k given $\boldsymbol{y}_k$ and $\boldsymbol{\theta}$ is again a multivariant complex Gaussian distributed with the following mean and covariance matrix

$$\boldsymbol{\mu}_{xk} = \mathbf{P}_k^{-1} \boldsymbol{\Phi}^* \mathbf{E}^{-1} \boldsymbol{y}_k ; \mathbf{C}_{xk} = \mathbf{P}_k^{-1} \tag{6.13}$$

where we have defined $\mathbf{P}_k = \mathbf{H}_k^{-1} + \boldsymbol{\Phi}^* \mathbf{E}^{-1} \boldsymbol{\Phi}$. The reason we provide this distribution is that it is a key quantity in applying EM algorithm to maximize the likelihood function as introduced later. Second, the marginal distribution of $\boldsymbol{y}_k$ is simply a zero-mean multivariate Gaussian with the covariance matrix being $\boldsymbol{\Phi} \mathbf{H}_k \boldsymbol{\Phi}^* + \mathbf{E}$. Because of the statistical independence among $x_k, x_j, \boldsymbol{\varepsilon}_k$, and $\boldsymbol{\varepsilon}_j$ for $k \neq j$, the joint distribution of $\boldsymbol{y}_{1:N_f}$ is still a multivariate complex Gaussian with the PDF.

$$f(\boldsymbol{y}_{1:N_f} \mid \boldsymbol{\theta}) = \frac{1}{\pi^{n N_f} \prod_{k=1}^{N_f} |\boldsymbol{\Phi} \mathbf{H}_k \boldsymbol{\Phi}^* + \mathbf{E}|} \exp\left(-\sum_{k=1}^{N_f} \boldsymbol{y}_k^* \left[\boldsymbol{\Phi} \mathbf{H}_k \boldsymbol{\Phi}^* + \mathbf{E} \right]^{-1} \boldsymbol{y}_k \right) \tag{6.14}$$

6.3 ML ESTIMATION

In this paper, we choose the ML method to identify modal parameters for its theoretical asymptotic properties. Under mild assumptions on data ML method is consistent and efficient, i.e., asymptotically unbiased, convergent as sample size increases and its variance asymptotically achieves the smallest possible value of any unbiased estimator (Keener, 2010).

Given the measurements $\boldsymbol{y}_{1:N_f}$ and their joint PDF shown in Eq. (6.14), the log likelihood function (LLF) unknown parameter $\boldsymbol{\theta}$ is given as

$$L\left(\boldsymbol{\theta}\mid\boldsymbol{y}_{1:N_f}\right)=-nN_f\ln\pi-\sum_{k=1}^{N_f}\ln\left|\boldsymbol{\Phi}\mathbf{H}_k\boldsymbol{\Phi}^*+\mathbf{E}\right|-\sum_{k=1}^{N_f}\boldsymbol{y}_k^*\left[\boldsymbol{\Phi}\mathbf{H}_k\boldsymbol{\Phi}^*+\mathbf{E}\right]^{-1}\boldsymbol{y}_k$$

$$=-nN_f\ln\pi-N_f\ln|\mathbf{E}|-\sum_{k=1}^{N_f}\ln|\mathbf{H}_k|-\sum_{k=1}^{N_f}\ln|\mathbf{P}_k|-\sum_{k=1}^{N_f}\boldsymbol{y}_k^*\mathbf{E}^{-1}\boldsymbol{y}_k \qquad (6.15)$$

$$-\sum_{k=1}^{N_f}\boldsymbol{y}_k^*\mathbf{E}^{-1}\boldsymbol{\Phi}\mathbf{P}_k\boldsymbol{\Phi}^*\mathbf{E}^{-1}\boldsymbol{y}_k$$

which is simply the logarithm of Eq. (6.14), and we have simplified it in the second line using matrix inversion lemma.

Then, the ML estimator of modal parameters is given by

$$\boldsymbol{\theta}_{ML}=\arg\max_{\boldsymbol{\theta}}L\left(\boldsymbol{\theta}\mid\boldsymbol{y}_{1:N_f}\right) \qquad (6.16)$$

Note that the problem is not identifiable, i.e., $\boldsymbol{\theta}_{ML}$ is not unique. It can be seen that, for an arbitrary non-zero complex diagonal matrix $\boldsymbol{\Gamma}$, two different parameters $\{\boldsymbol{\Phi},\boldsymbol{\Lambda},\mathbf{S},\mathbf{E}\}$ and $\{\boldsymbol{\Phi}\boldsymbol{\Gamma},\boldsymbol{\Lambda},\boldsymbol{\Gamma}^{-1}\mathbf{S}\boldsymbol{\Gamma}^{-*},\mathbf{E}\}$ yield the same likelihood. In order to make the problem globally identifiable, we prescribe the element $\phi_{ij}=1$ for $j=1,2,\ldots,m$ and i is the index of a selected DoF.

6.4 EM ALGORITHM

The EM algorithm (Dempster, Laird and Rubin, 1977) is widely used to infer the statistical model that can be formulated as a latent variable model (Li and Der Kiureghian, 2017). In our model, the measurement $\boldsymbol{y}_k$ is the observed variable and the modal response x_k can be regarded as the latent variable. When applying the EM algorithm to our model, the closed-form update of mode shapes $\boldsymbol{\Phi}$ is not available due to the constraints on some of its elements. To resolve this difficulty, the parameter-expanded EM (PX-EM) is employed here.

The PX-EM algorithm is essentially an EM algorithm, but it performs inference on a larger full model with the expanded parameter set $\{\tilde{\boldsymbol{\theta}},\boldsymbol{\gamma}\}$. If there exists a reduction function $\boldsymbol{\theta}=R(\boldsymbol{\theta},\boldsymbol{\gamma})$ and $\boldsymbol{\theta}=R(\boldsymbol{\theta},\boldsymbol{\gamma}_0)$ for some fixed value $\boldsymbol{\gamma}_0$, $\boldsymbol{\theta}_{ML}$ can be computed via PX-E step. Compute

$$Q\left(\boldsymbol{\theta},\boldsymbol{\gamma}\mid\boldsymbol{\theta}^{(t)},\boldsymbol{\gamma}_0\right)=E_{x_{1:N_f}}\left[\mathcal{L}\left(\tilde{\boldsymbol{\theta}},\boldsymbol{\gamma}\middle|\boldsymbol{y}_{1:N_f},x_{1:N_f}\right)\middle|\boldsymbol{y}_{1:N_f},\boldsymbol{\theta}^{(t)},\boldsymbol{\gamma}_0\right] \qquad (6.17)$$

PX-M step. Find

$$\left\{\tilde{\boldsymbol{\theta}}^{(t+1)},\boldsymbol{\gamma}^{(t+1)}\right\}=\arg\max_{\tilde{\boldsymbol{\theta}},\boldsymbol{\gamma}}Q\left(\tilde{\boldsymbol{\theta}},\boldsymbol{\gamma}\mid\boldsymbol{\theta}^{(t)},\boldsymbol{\gamma}_0\right) \qquad (6.18)$$

and update $\theta^{(t+1)} = R\left(\tilde{\theta}^{(t+1)}, \gamma^{(t+1)}\right)$. Since it is the ordinary EM applied to the parameter-expanded complete-data model, PX-EM shares with EM its simplicity and stability. Liu et al. (Liu, Rubin and Wu, 1998) established theoretical results to show that PX-EM can converge no slower than EM.

The PX-EM algorithm works on an augmented model

$$\mathbf{y}_k = \breve{\Phi}\gamma\tilde{\mathbf{h}}_k \mathbf{x}_k + \varepsilon_k \qquad (6.19)$$

by introducing the auxiliary parameter $\gamma = \mathrm{diag}(\gamma_1, \gamma_2, ..., \gamma_m)$ $(\gamma_i \in \mathbb{C})$, where $\mathbf{x}_k \sim \mathcal{CN}\left(0, \tilde{\mathbf{S}}\right)$, $\varepsilon_k \sim \mathcal{CN}\left(0, \tilde{\mathbf{E}}\right)$, and $\tilde{\phi}_{ij} = 1$ for some specified i and j. By comparing the original and the augmented models, it is easy to find the following many-to-one mapping

$$\theta = \{\Phi, \Lambda, \mathbf{S}, \mathbf{E}\} = R\left(\tilde{\theta}, \gamma\right) = \left\{\breve{\Phi}, \tilde{\Lambda}, \gamma\tilde{\mathbf{S}}\gamma^*, \tilde{\mathbf{E}}\right\} \qquad (6.20)$$

Moreover, the original model corresponds to the augmented model when the value of γ is an $(m \times m)$ identity matrix, i.e., $\gamma_0 = \mathbf{I}_m$.

For the augmented model, we can derive the Q-function following similar steps for the original model to obtain

$$Q\left(\tilde{\theta}, \gamma \mid \theta^{(t)}, \gamma_0\right) = -(m+n)N_f \log\pi - Q_1\left(\breve{\Phi}, \tilde{\mathbf{E}} \mid \theta^{(t)}, \gamma_0\right) - Q_2\left(\tilde{\Lambda}, \tilde{\mathbf{S}} \mid \theta^{(t)}, \gamma_0\right) \qquad (6.21)$$

where

$$Q_1\left(\breve{\Phi}, \tilde{\mathbf{E}} \mid \theta^{(t)}, \gamma_0\right) = N_f \log\left|\tilde{\mathbf{E}}\right| + tr\left[\tilde{\mathbf{E}}^{-1}\left(\begin{array}{c} \displaystyle\sum_{k=1}^{N_f} \mathbf{y}_k \mathbf{y}_k^* + \breve{\Phi}\sum_{k=1}^{N_f} \mathbf{w}_{2k}\breve{\Phi}^* \\ -\breve{\Phi}\displaystyle\sum_{k=1}^{N_f} \mathbf{w}_{1k}\mathbf{y}_k^* - \sum_{k=1}^{N_f} \mathbf{y}_k\mathbf{w}_{1k}^*\breve{\Phi}^* \end{array}\right)\right] \qquad (6.22)$$

$$Q_2\left(\tilde{\Lambda}, \tilde{\mathbf{S}} \mid \theta^{(t)}, \gamma_0\right) = N_f \log\left|\tilde{\mathbf{S}}\right| + \sum_{k=1}^{N_f} \log\left|\tilde{\mathbf{h}}_k\tilde{\mathbf{h}}_k^*\right| + tr\left[\tilde{\mathbf{S}}^{-1}\sum_{k=1}^{N_f} \tilde{\mathbf{h}}_k^{-1}\mathbf{w}_{2k}\tilde{\mathbf{h}}_k^{-*}\right] \qquad (6.23)$$

$\mathbf{w}_{1k} = E\left[\mathbf{x}_k \mid \mathbf{y}_k, \theta^{(t)}, \gamma_0\right]$ and $\mathbf{w}_{2k} = E\left[\mathbf{x}_k\mathbf{x}_k^* \mid \mathbf{y}_k, \theta^{(t)}, \gamma_0\right]$, which are the first and second conditional moments of $\mathbf{x}_k$ given $\mathbf{y}_k$, $\theta^{(t)}$, and γ_0; and they can be simply calculated from Eqs. (6.24) and (6.25). Note that they are evaluated at parameter values in the previous iteration, they are constant in the M-step. Since $\breve{\Phi}$ and γ in Q_1 (and hence Q) always appear together, we have combined them into $\breve{\Phi} = \tilde{\Phi}\gamma$ which is free from norm constraints. Note that $Q_1\left(\breve{\Phi}, \tilde{\mathbf{E}} \mid \theta^{(t)}, \gamma_0\right)$ only depends on mode shapes $\breve{\Phi}$ and the covariance matrix of measurement error $\tilde{\mathbf{E}}$, while $Q_2\left(\tilde{\Lambda}, \tilde{\mathbf{S}} \mid \theta^{(t)}, \gamma_0\right)$ involves only the remaining unknown parameters. This implies that the unknown parameters can be optimized in different groups.

Once the Q-function is derived, we can then proceed to the PX-M step, i.e., to maximize $Q\left(\tilde{\theta}, \gamma \mid \theta^{(t)}, \gamma_0\right)$ w.r.t. $\tilde{\theta}$ and γ. We can derive the analytical update for $\breve{\Phi}$ and $\tilde{\mathbf{E}}$ from Q_1 in Eq. (6.36) and $\tilde{\mathbf{S}}$ from Q_2 in Eq. (6.37) as

$$\breve{\Phi} = \left(\sum_{k=1}^{N_f} \boldsymbol{y}_k \mathbf{w}_{1k}^* \right) \left(\sum_{k=1}^{N_f} \mathbf{w}_{2k} \right)^{-1} \qquad (6.24)$$

$$\tilde{\mathbf{E}} = \frac{1}{N_f} \left(\sum_{k=1}^{N_f} \boldsymbol{y}_k \boldsymbol{y}_k^* - \breve{\Phi} \sum_{k=1}^{N_f} \mathbf{w}_{2k} \breve{\Phi}^* \right) \qquad (6.25)$$

$$\tilde{\mathbf{S}} = \frac{1}{N_f} \sum_{k=1}^{N_f} \mathbf{h}_k^{-1} \mathbf{w}_{2k} \mathbf{h}_k^{-*} \qquad (6.26)$$

After $\breve{\Phi}$ has been updated, $\tilde{\Phi}$ and γ can be readily recovered by noting that $\tilde{\Phi} = \breve{\Phi} \gamma^{-1}$ and the diagonal elements of γ are simply $\breve{\Phi}_{ij}$, where the indices i and j satisfies $\tilde{\phi}_{ij} = 1$. For $\tilde{\Lambda}$, an analytical update has not been found (as is typical), and one has to rely on numerical optimization. Since the dimension (m) is not high, the MATLAB function "fminsearch" (The MathWorks, Inc, 2016) can provide an efficient solution using the simplex search method (Lagarias et al., 1998)

6.5 FIELD TEST

In order to validate the performance of the proposed PX-EM algorithm for the ML estimation of modal parameters, a field test study is presented here. We consider the modal analysis of the Guangzhou New TV Tower (GNTVT), which is a structural health monitoring benchmark building for high-rise structures (Ni et al., 2009). In the benchmark study, 24-hour field measurements of the structural acceleration time histories and the corresponding ambient conditions (temperature and wind) were provided. Twenty uniaxial accelerometers were installed along the tower, to measure the structural dynamic response. Figure 6.1 shows two typical time series measured from 2:00 to 3:00 on January 20, 2010, as well as the corresponding PSD and spectral acceleration, velocity (SV) spectrum. The selected frequency bands are marked in the SV spectra, including one three-mode band, one two-mode band, and seven single-mode band.

Before the analysis, we first divide each one-hour data into non-overlapping time windows at 20-minute intervals, which is about 1,300 cycles of fundamental mode, and found to be long enough for a good estimation. Though longer durations could reduce the posterior uncertainty of identified parameters, it also increases the risk of violating the stationary assumption. Regardless of whether this is explicitly recognized or not, this risk applies to many OMA methods as the assumption is quite common. The EM algorithm was applied to all 72 data sets. The identification results are plotted in Figure 6.2, with solid line representing the ML estimator and shaded area covering two standard deviations. The frequencies change slightly with time, while the damping ratios have a larger variation but still are in the same order of magnitude. The negative values of damping ratios in the plot is immaterial, merely due to the Gaussian distribution approximation and the large coefficient of variation (c.o.v).

6.6 CONCLUSION

This paper has presented an ML estimation of complex modes using EM algorithm. Regarding the modal response as a latent variable, the frequency-domain OMA model is formulated as a latent variable model. In order to free up the norm constraints on mode shapes, the PX-EM is introduced, allowing the closed-form analytical update of all parameters except the complex eigenvalues, i.e., frequency and damping ratio. The proposed EM algorithm has been verified with field test data, which shows that it is ready to be applied in practical test.

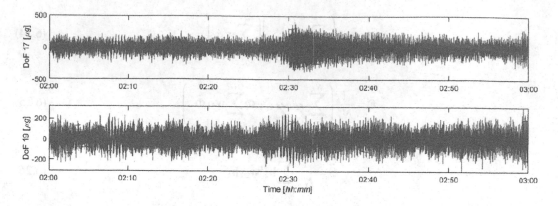

a)   Measured time history at two locations

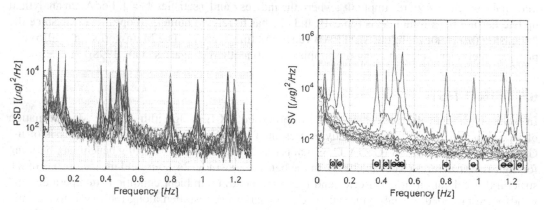

b) PSD and SV spectrum (Bracket: frequency bands used in the Bayesian FFT method)

FIGURE 6.1   Measured data, GNTVT (January 20, 2010). (a) Measured time history at two locations and (b) PSD and SV spectrum (Bracket: frequency bands used in the Bayesian FFT method).

ACKNOWLEDGEMENTS

The faculty research start-up fund from the ZJU-UIUC Institute at the Zhejiang University is gratefully acknowledged.

REFERENCES

Au, S. K. (2017) *Operational Modal Analysis: Modeling, Bayesian Inference, Uncertainty Laws*. Singapore: Springer.

Au, S. K. (2012) 'Fast Bayesian ambient modal identification in the frequency domain, Part II: Posterior uncertainty', *Mechanical Systems and Signal Processing*. Elsevier, 26(1), pp. 76–90.

Bendat, J. S. and Piersol, A. G. (1980) *Engineering Applications of Correlation and Spectral Analysis*, 1st edn. New York: John Wiley & Sons, Ltd.

Billinger, D.R. (2001), *Time Series: Data Analysis and Theory*, SIAM.

Brincker, R. and Ventura, C. (2015) *Introduction to Operational Modal Analysis*. London: Wiley.

Brincker, R., Zhang, L. and Andersen, P. (2000) 'Modal identification from ambient responses using frequency domain decomposition', *18th International Modal Analysis Conference (IMAC)*, San Antonio, Texas, USA, pp. 625–630.

Brownjohn, J. et al. (2013) 'Vibration monitoring and condition assessment of the University of Sheffield Arts Tower during retrofit', *Journal of Civil Structural Health Monitoring*, 3(3), pp. 153–168.

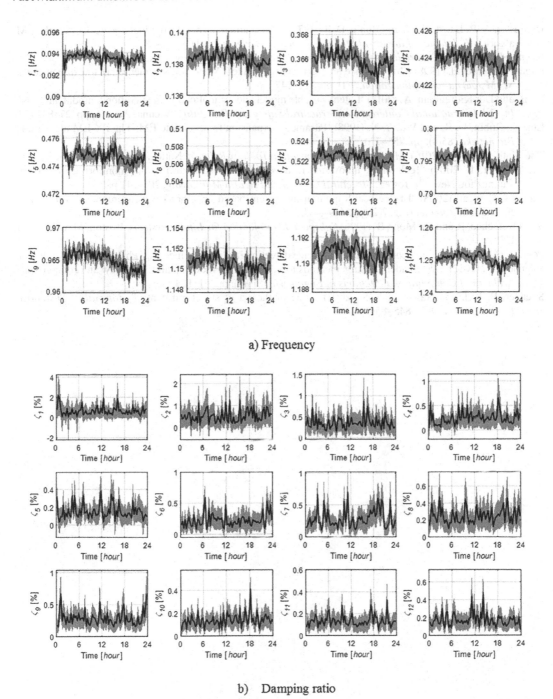

a) Frequency

b) Damping ratio

FIGURE 6.2 Identified modal parameters for different time GNTVTs: (a) Frequency and (b) damping ratio. Solid line: most probable value (MPV); shaded area: ±two standard deviations.

Brownjohn, J. M. W. (2007) 'Structural health monitoring of civil infrastructure', *Philosophical Transactions of the Royal Society A: Mathematical, Physical and Engineering Sciences*, 365(1851), pp. 589–622.

Cole, A. (1973) *On-line Failure Detection and Damping Measurement of Aerospace Structures by Random Decrement Signatures*. Mountain View, CA: Nielsen Engineering and Research, Inc.

Dempster, A. P., Laird, N. M. and Rubin, D. B. (1977) 'Maximum likelihood from incomplete data via the EM algorithm', *Journal of the Royal Statistical Society: Series B*, 39(1), pp. 1–38.

Keener, R. W. (2010) *Theoretical Statistics : Topics for a Core Course*. New York: Springer.

Lagarias, J. C. et al. (1998) 'Convergence properties of the Nelder--Mead simplex method in low dimensions', *SIAM Journal on Optimization*, 9(1), pp. 112–147.

Li, B. and Der Kiureghian, A. (2017) 'Latent variable model and its application in operational modal analysis', *In 12th International Conference on Structural Safety and Reliability*, Vienna, Austria, pp. 2748–2757.

Liu, C., Rubin, D. B. and Wu, Y. N. (1998) 'Parameter expansion to accelerate EM: The PX-EM algorithm', *Biometrika*, 85(4), pp. 755–770.

The MathWorks, Inc (2016). *MATLAB and Optimization Toolbox Release*. Natick, MA: The MathWorks, Inc.

Ni, Y. Q. et al. (2009) 'Technology innovation in developing the structural health monitoring system for Guangzhou New TV Tower', *Structural Control and Health Monitoring*, 2014, 11 p.

Ntotsios, E. et al. (2008) 'Bridge health monitoring system based on vibration measurements', *Bulletin of Earthquake Engineering*, 7(2), pp. 469–483.

Van Overschee, P. and De Moor, B. (1996) *Subspace Identification for Linear Systems*. Boston, MA: Springer US.

Peeters, B. (2000) System identification and damage detection in Civil Engineering. PhD thesis, K.U. Leuven.

Reynders, E. (2012) 'System identification methods for (operational) modal analysis: Review and comparison', *Archives of Computational Methods in Engineering*, 19(1), pp. 51–124.

Spencer, B. F. J. and Nagarajaiah, S. (2003) 'State of the art of structural control', *Journal of Structural Engineering*, 129(7), 845–856.

7 A Full Version of Vision-Based Structural Identification

Chuan-Zhi Dong and F. Necati Catbas
University of Central Florida

CONTENTS

7.1 INTRODUCTION

For safe operation, timely maintenance, and convenient management of civil structures and infrastructures, effective sensing technologies and analytical approaches are necessary to detect the structural changes and damages and to give reliable condition assessment and performance evaluation. In the last two decades structural health monitoring (SHM) has been widely explored and implemented on bridges all over the world. SHM systems can collect massive valuable information including structural input (loads and other external effects) and structural output (responses such as displacement, strain, and acceleration) and make diagnosis and prognosis to support the structural safety and decision making (Dong et al. 2018a,b).

In conventional SHM, various types of sensors are necessary to be installed on structures to monitor the structural input and output. During this procedure, a lot of time, cost, labor force, and cable wiring work are essential. In addition, bridge closure may be needed, and this would cause severe problems such as traffic jam of critical urban routes. With the development of camera with high resolution and speed and low cost, and the progress of computer vision, vision-based approaches are gathering increasing attention in the field of SHM due to the advantages such as non-contact, long distance, low cost, time saving, and ease of use. Currently most of the research on vision-based SHM focus on the structural output monitoring such as displacement and strain (Dong et al. 2015, Dong and Catbas 2019). However, if only response data are used for structural identification without knowing input force, the structural change and damage has to be large enough to induce significant change to the output responses. There are also very few researches focusing on identifying structural input using vision-based methods (Zaurin and Catbas 2010). By combining the vision-based structural input estimation and the conventional sensors for structural output monitoring, structural performance evaluation, condition assessment, and damage detection can be done. Although this strategy can solve the problems of response-only structural identification, the drawbacks of conventional SHM aforementioned still appear.

In this paper, the study of structural identification using input-output data will move forward from combining cameras and conventional sensors to a completely non-contact recognition system just using cameras. The input and output data are both obtained from portable cameras and computer vision techniques are employed to process the images and track the structural behaviors. The proposed recognition system will take unit influence line (UIL) as the target parameter for structural identification, and the proposed UIL extraction method can be extended to a fully non-contact damage detection approach.

7.2 METHODOLOGY

7.2.1 VISION-BASED STRUCTURAL INPUT ESTIMATION

To identify the vehicle location on bridge surface, in general there are four steps as shown in Figure 7.1. At first, the camera is calibrated to rectify the distortions such as projective distortion caused by camera pose and radial distortion caused by lenses. Then the objection detection algorithms are implemented to detect the category of vehicles and give the initial bounding boxes of detected vehicles, and they will be regarded as the tracking targets. The tracking targets can also be selected manually. In the third step, the visual tracking algorithms are implemented to track the detected or selected vehicles, and the vehicle location in each frame of the image sequence or video can be estimated. At last, the vehicle location in the image coordinates is transformed to the real-world coordinates to estimate the vehicle location on bridges with the camera calibration information.

The camera calibration is to find the relationship between the image and world coordinates and also to remove the distortion during projection. The projection from world coordinates to the image coordinates through camera coordinates can be expressed by the formula below:

$$s\begin{pmatrix} x \\ y \\ 1 \end{pmatrix} = \begin{bmatrix} f_x & \gamma & c_x \\ 0 & f_y & c_y \\ 0 & 0 & 1 \end{bmatrix} \begin{bmatrix} r_{11} & r_{12} & r_{13} & t_1 \\ r_{21} & r_{22} & r_{23} & t_2 \\ r_{31} & r_{32} & r_{33} & t_3 \end{bmatrix} \begin{pmatrix} X \\ Y \\ Z \\ 1 \end{pmatrix} \tag{7.1}$$

and it can be simplified as

$$s\mathbf{x} = \mathbf{K}[\mathbf{R} \mid \mathbf{t}]\mathbf{X} \tag{7.2}$$

where s is the scale factor, $\mathbf{x} = (x, y, 1)^T$ are image coordinates, $\mathbf{X} = (X, Y, Z, 1)^T$ are world coordinates, and $\mathbf{K}$ is the camera intrinsic parameters which represent the projective transformation from the three-dimensional (3D) real world to the two-dimensional (2D) image. In the intrinsic parameters, f_x and f_y are the focal lengths of the lens in horizontal and vertical directions, c_x and c_y are offsets of the optical axis in horizontal and vertical directions, and γ is the skew factor of the lens. $\mathbf{R}$ and $\mathbf{t}$ are camera extrinsic parameters which represents the rigid rotation and translation from the 3D real-world coordinates to the 3D camera coordinates, and r_{ij} ($i, j = 1, 2, 3$) and t_i ($i = 1, 2, 3$) are the elements of $\mathbf{R}$ and $\mathbf{T}$, respectively. In practical application, the calibration can be easily done with a chessboard and the MATLAB toolbox.

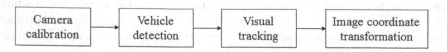

FIGURE 7.1 General procedure of vehicle tracking.

In this study, the vehicle detection can be done by the implementing the off-shelf visual detection algorithm such as Mask R-CNN (He et al. 2017) or can be selected manually. The visual tracking is done by implementing a discriminative correlation filter with channel and spatial reliability (CSRT) tracker (Lukezic et al. 2017).

7.2.2 Vision-Based Structural Output Estimation

The structural output estimation carried out in this study is vision-based displacement measurement. Usually there are four steps to estimate displacement from videos of structures using vision-based method (Ye et al. 2016a–d) as shown in Figure 7.2. First, camera calibration is done to calculate the geometric relationship between the image coordinates and the real-world coordinates. Second, measurement targets are selected from images as image subregions, and features in the subregions are extracted for object tracking. Third, visual tracking algorithms are employed by combining with the selected features to do object tracking, and the locations of measurement targets are updated in a consecutive image sequence. At the end, the displacement in image coordinates is calculated by comparing the location change of the measurement targets in each image and the final displacement in real-world coordinates is obtained by combining the camera calibration information and displacement in image coordinates.

The calibration procedure is the same with that presented in Section 2.1. In the step of feature extraction, image intensity, edges, or key points can be regarded as appropriate features. As shown in Figure 7.3, edges are selected as features, and template matching using normalized cross-correlation coefficient (NCC) of Canny edge is implemented to track the selected template of target in the

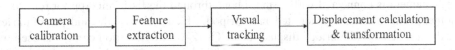

FIGURE 7.2 General procedure of vision-based displacement measurement.

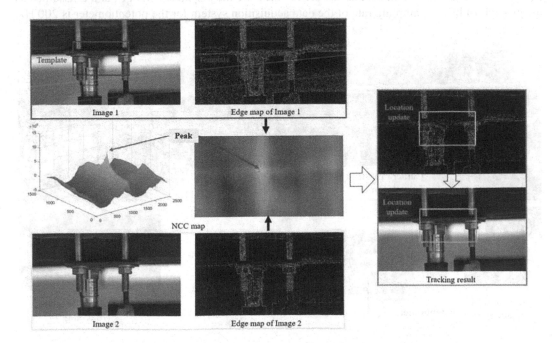

FIGURE 7.3 Template matching using normalize cross-correction coefficient of Canny edge.

consecutive image sequence. The location of target is obtained when the NCC between the edge map of the current image and that of the template achieves the maximum. For simplicity, scale ratio (Ye et al. 2015, Celik et al. 2018) can be used to do displacement calculation after visual tracking.

7.2.3 EXTRACT UIL FROM STRUCTURAL INPUT AND OUTPUT

Combining the displacement obtained from the vision-based method and the vehicle location information estimated using vehicle tracking, and also considering the weight distribution of the vehicle, the UIL is built. Since the displacement is the response under the moving load, it also includes the high vibration modes in the response signal. By applying the Fourier filter, the high vibration modes are removed and the final UIL is then extracted.

7.3 EXPERIMENTAL VERIFICATION

7.3.1 EXPERIMENTAL SETUP

The proposed framework is verified on the two-span bridge constructed in the University of Central Florida's Civil Infrastructure Technologies for Resilience and Safety (CITRS) Experimental Design and Monitoring (EDM) laboratory. As shown in Figure 7.4, the bridge consists of two 300-cm main continuous spans. The bridge deck includes a 3.18-mm steel sheet at 120 cm wide, which makes the deck 600 cm long by 120 cm wide. A toy truck (1.9 front weight) is used to model moving loads.

A fisheye camera with the resolution of $1,920 \times 1,080$ pixels, the frame rate of 30 frames per second (FPS), and a 170° wide angle lens is mounted on the tripod which is 2 m from the middle of the bridge. The camera is connected with a smartphone through the Ez iCam App for remote controlling. Another portable camera is mounted on the tripod, which is close to the midspan of the left span of the bridge to measure the bridge displacement. The Z-CAM E1 action camera is used with the resolution of $3,840 \times 2,160$ pixels with the speed of 30 FPS and a 75–300 mm zoom lens. The camera is also connected with a smartphone through the Z-CAM official application. A potentiometer (BEI Duncan 9615) is mounted under the deck to measure the displacement of P_1 and is assumed as the ground truth. The sampling rate of the data acquisition system for the potentiometer is 200 Hz,

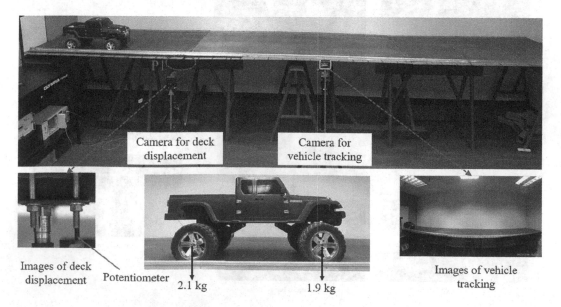

FIGURE 7.4 Experimental setup in laboratory.

which is then downsampled to 30 Hz during post-processing. During the experiments, the toy truck moves from one side of the bridge to the other while the potentiometer and the camera record the motion of P_1 (midspan of the left span) synchronously. The two cameras are synchronized by the audio signals from them using the audio cross correlation-based pattern matching.

7.3.2 RESULT ANALYSIS

Since the fisheye camera is used to track the vehicle, the images from the fisheye cameras have a big radial distortion as shown in Figure 7.4. The vehicle tracking is done after the image rectification with camera calibration information. Figure 7.5 shows the tracking results of the toy truck in the rectified images. During the loading process, even though the view and scale of the truck changes, the CSRT tracker can still successfully estimate the location of the toy truck in each image. Eventually the locations in the rectified images are converted to the location on the bridge deck. Figure 7.6 shows the displacement comparison between the proposed vision-based method and the potentiometer. It is easy to see that the result obtained from the proposed method is consistent with those obtained from the potentiometer. The NCC is calculated to evaluate the similarities between them. The NCC between the two methods is 99.1%, which shows a very high fit of goodness between the test methods.

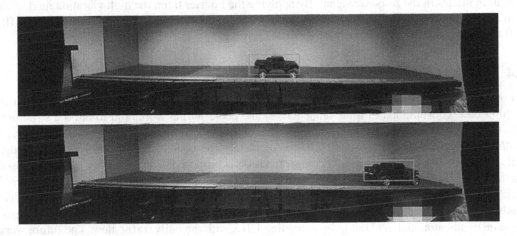

FIGURE 7.5 Vehicle tracking from the rectified images using fish camera.

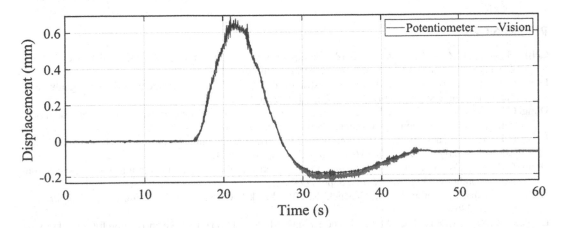

FIGURE 7.6 Displacement comparison between the proposed vision-based method and the potentiometer.

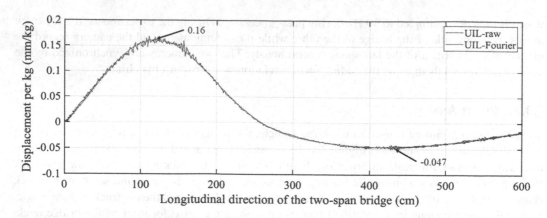

FIGURE 7.7 Extracted UIL using the proposed system.

In Figure 7.7, the black curve (UIL-raw) is the extracted UIL without any post-processing and filtering. As this bridge displacement is the response under the moving load, it also includes the high vibration modes in the response signal. By applying the Fourier filter, the high vibration modes are removed and the final UIL is shown as the red curve (UIL-Fourier). The maximum value of the UIL is 0.16 mm/kg and the minimum value is −0.047 mm/kg.

7.4   CONCLUSIONS

To overcome the inconveniences and disadvantages in the practices of the conventional SHM such as high cost, consumption of time, labor forces, and cable wiring work, and to build a structural identification framework with a normalized structural response indicator irrespective of the type and/or the loads for better decision making, a completely non-contact recognition system for bridge UIL using just portable cameras and computer vision is proposed. The feasibility of the proposed method is verified through a comparative study of a series of laboratory experiments. The proposed UIL recognition system also shows great probability to do damage detection by using statistical analysis of UILs, bridge load capacity evaluation by regarding UIL as a normalized structural performance indicator, and load rating by extracting UILs with the daily traffic flow. The future work will focus on the investigation of them and extending to more possible aspects of structural condition assessment at a global level.

REFERENCES

Celik, O., Dong, C.Z., and Catbas, F.N. 2018. A computer vision approach for the load time history estimation of lively individuals and crowds. *Comput. Struct.*, **200**, pp. 32–52.

Dong, C.Z. and Catbas, F.N. 2019. A non-target structural displacement measurement method using advanced feature matching strategy. *Adv. Struct. Eng.*, **13**(4), pp. 1–12.

Dong, C.Z., Celik, O., and Catbas, F.N. 2018a. Marker free monitoring of the grandstand structures and modal identification using computer vision methods. *Struct. Heal. Monit.*, **10**, pp. 1–19.

Dong, C.Z., Ye, X.W., and Jin, T. 2018b. Identification of structural dynamic characteristics based on machine vision technology. *Meas. J. Int. Meas. Confed.*, **126**, pp. 405–416.

Dong, C.Z., Ye, X.W., and Liu, T. 2015. Non-contact structural vibration monitoring under varying environmental conditions. *Vibroeng. Proc.*, **5**, pp. 217–222.

He, K., Gkioxari, G., Dollar, P., and Girshick, R. 2017. Mask R-CNN. *Proc. IEEE Int. Conf. Comput. Vis.*, pp. 2961–2969.

Lukezic, A., Vojir, T., Zajc, L.C., Matas, J., and Kristan, M. 2017. Discriminative correlation filter tracker with channel and spatial reliability. *Proc. IEEE Conf. Comput. Vis. Pattern Recognit.*, **2**, pp. 6309–6318.

Ye, X.W., Dong, C.Z., and Liu, T. 2016a. A review of machine vision-based structural health monitoring: Methodologies and applications. *J. Sensors*, **2016**, pp. 1–10.

Ye, X.W., Dong, C.Z., and Liu, T. 2016b. Image-based structural dynamic displacement measurement using different multi-object tracking algorithms. *Smart Struct. Syst.*, **17**(6), pp. 935–956.

Ye, X.W., Dong, C.Z., and Liu, T. 2016c. Force monitoring of steel cables using vision-based sensing technology: Methodology and experimental verification. *Smart Struct. Syst.*, **18**(3), pp. 585–599.

Ye, X.W., Yi, T.H., Dong, C.Z., and Liu, T. 2016d. Vision-based structural displacement measurement: System performance evaluation and influence factor analysis. *Meas. J. Int. Meas. Confed.*, **88**, pp. 372–384.

Ye, X.W., Yi, T.-H., Dong, C.Z., Liu, T., and Bai, H. 2015. Multi-point displacement monitoring of bridges using a vision-based approach. *Wind Struct. Int. J.*, **20**(2), pp. 315–326.

Zaurin, R. and Catbas, F. 2010. Integration of computer imaging and sensor data for structural health monitoring of bridges. *Smart Mater. Struct.*, **19**, pp. 1–15.

8 Damage Recognition of Wood Beam Based on Curvature Modal Technology

Chunyu Qian
China Jikan Research Institute of Engineering
Investigations and Design, Co., Ltd

Baozhuang Zhang
Xi'an University of Architecture & Technology

Junlong Zhu
Shanghai Baitong Project Management Consulting, Co., Ltd

Qifang Xie
Xi'an University of Architecture & Technology

CONTENTS

8.1   INTRODUCTION

The existing wooden historical buildings, because of the impacts of time and environment, have experienced different degrees of damages, which are shown in Figure 8.1. In view of the extremely high historical, artistic and scientific value of ancient wooden structures, it is imperative to strengthen the protection of wooden historical buildings. Health monitoring and damage detection are effective means to protect the wooden structure of ancient buildings. The general damage detection methods mainly include observation detection method, load test method onsite, non-destructive testing, and sampling destructive test under special circumstances [1]. Although these methods are easy to implement, the premise is that the approximate location of the structural damage needs to be known in advance, and the operation is time consuming, laborious, and expensive, and it is difficult to obtain damage information of the complicated concealed parts. In addition, the inspectors often judge the accuracy of the damage detection by subjective judgment, and fail to objectively and accurately evaluate the safety of the structural system.

It is, based on the urgent need for the protection of ancient wooden structures, urgent to have a damage detection technology with high detection accuracy, no damage to the original structure, and a reasonable damage determination theory, which can be used to improve the accuracy of wood detection and broaden the way of wood detection, and then it can be further better applied to engineering practice. In view of the fact that wood is an anisotropic material, steel structure and concrete are isotropic materials, which increases the difficulty in the identification of damage in ancient wooden structures. In recent years, domestic and foreign scholars have studied structural damage detection from different aspects. Liu [2] carried out studies to develop the numerical model of simply supported beam bridge and continuous beam bridge through the ANSYS program. The results show that the curvature mode is sensitive to local damage of the structure and is effective for damage identification. Li and Lu [3] clarified the theoretical basis and characteristics of curvature modal analysis (such as orthogonality and superposition), and derived the relevant formula. Xu and Wang [4] studied the wooden beams with different locations, size, and number of hole defects, and obtained the first-order displacement mode shape by modal test. Then the first-order curvature mode was obtained by the center difference and the curvature was observed, which was used to

FIGURE 8.1   Damage of local members of ancient wooden structures.

discriminate the structural damage, and the feasibility of using the curvature mode method for wood damage detection was discussed. Wang and Hu [5] carried out studies to use Xi'an Bell Tower as the prototype to simulate the damage of ancient wooden structures with finite element software under random excitation, and used wavelet packet to decompose the acceleration response signal of each node on the beam and transform it into wavelet packet energy. The difference in curvature was used to determine the damage of the structure. The research indicated that the index has certain anti-noise ability, which is sensitive to the damage of ancient wooden structures, and can accurately determine the damage location. Yang and Ishimaru [6] carried out studies to find that the mode shape and curvature mode are suitable for the preliminary positioning of wood knot defects. Hu and Afzal [7] developed a defect recognition algorithm and proved that it can initially identify the location and size of defects. Choi and Li [8] used experimental modal analysis methods to study the local defects in wood and explored the applicability of modal strain in quantitative detection of defects. Capecchi and Vestroni [9] and Cawley and Adams [10] proposed frequency methods, respectively, which are not sensitive to structural damage identification.

Considering that the beam members of the historic wooden structure are mainly resistant to bending in the direction of the grain, only the mechanical properties of the wood beam in the direction of the grain are considered due to the research object being the damaged wood beam. Combining the research results of wood damage detection at home and abroad, this paper presents a study to analyze the feasibility and effectiveness of curvature mode technology for wood damage detection. The damage location is determined by the sudden change of the curvature mode diagram, then the recognition effect under different damage locations and damage degrees is discussed, and finally the damage degree determination theory of wood beam based on curvature mode is derived. The modal test of an artificially damaged wooden beam is simulated. The validity of the finite element simulation method and the applicability of the derived theory are verified. Thereby, the purpose of detecting the damage position and determining the degree of damage can be achieved.

8.2 FEASIBILITY ANALYSIS OF CURVATURE MODAL BASED WOOD BEAM DAMAGE IDENTIFICATION

8.2.1 FEASIBILITY ANALYSIS PROCESS

To determine the curvature mode the feasibility analysis can be used to identify the damage of wood beams. In this paper, the damage model of wood beam is established by finite element simulation method, in which the damage is simulated by changing the magnitude of the elastic modulus E of the local element [6], and the degree of damage is determined according to the degree of decrease in the elastic modulus. The comparison of the curvature mode diagrams of the wooden beam model with the same damage location, different damage degrees, same damage degree, and different damage locations is carried out to illustrate that the curvature mode is a sensitive parameter to wood beam damage, which is feasible to apply for damage identification of wood beams.

8.2.2 FINITE ELEMENT ANALYSIS MODEL

The wooden beam model is established by the finite element method. The size of the wooden beam model is 1,100 mm (length) × 80 mm (width) × 120 mm (depth), and the wooden beam model has a modulus of elasticity (parallel to the grin) of 7,500 MPa (*Pinus sylvestris*), whose Poisson's ratio is 0.35, with the density of about 450 kg/m^3. In this paper, the finite element software ABAQUS is used to simulate the wooden beam model, which is simulated by SOLID45 unit type. There are 45 nodes and 44 units in the longitudinal Z direction of the wood beam specimen. There are two units in the vertical Y direction and in the lateral X direction of the specimen, respectively, which is shown in Figure 8.2. The modal analysis process is solved by the Lanczos eigenvalue solver.

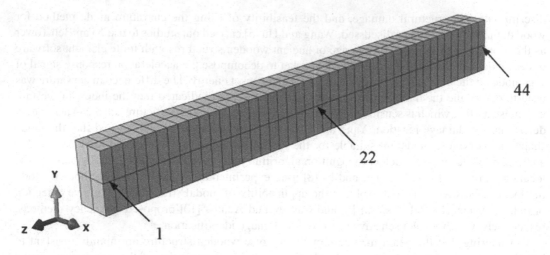

FIGURE 8.2 The finite element model of a wood beam.

In this paper, the 22-unit is first set as the damaged unit, and the elastic modulus E of the 22-unit is reduced by 5%, 10%, 20%, 30%, and 40%, respectively. By contrast, the degree of sudden change of the curvature mode shape pattern with different damage degrees is identified. Then, the modulus of elasticity of 11 unit, 15 unit, and 22 unit is reduced by 30%, the sensitivity of the curvature mode to the identification of different damage positions is determined by the comparison of curvature mode shapes.

8.2.3 ANALYSIS OF THE RELATIONSHIP BETWEEN CURVATURE MODE AND DAMAGE DEGREE OF WOODEN BEAM MODEL WITH DIFFERENT DAMAGE DEGREES AT A SINGLE SITE

In order to study the relationship between the curvature mode of the wooden beam model and the damage degree when the different damage degrees of the single damage, the 22 elements are set as the damaged unit, whose modulus of elasticity is reduced by 5%, 10%, 20%, 30%, and 40%, respectively, to simulate the corresponding degree of damage. Analysis of the damage degree judgment used in the low-order mode due to the abrupt chance in the first-order curvature mode diagram is very obvious, and the high-order modes are not easy to obtain under realistic conditions [7]. The comparison charts of the first two-order curvature modes obtained by simulation are shown in Figures 8.3 and 8.4. The displacement in the figure, without unit, is the normalized ratio of the mode. Based on the displacement modal measurement, the curvature mode is obtained by the center difference [3].

As can be seen from the curvature mode curve of the finite element model local element elastic modulus decreases at the position of the 22 defect unit, shown in Figures 8.3 and 8.4. The degree of sudden chance of the curvature mode at the damaged unit increases with the decrease of the local element's modulus of elasticity. Through research and analysis, it can be concluded that the curvature mode can be used to estimate the change of the local elastic modulus of the structure, which is very sensitive to damage identification. The damage situation can be identified by the curvature mode map, namely, the position of the abrupt chance which is the location of the damage and the degree of sudden chance which is the degree of damage.

8.2.4 ANALYSIS OF THE RELATIONSHIP BETWEEN THE CURVATURE MODE AND THE DAMAGE LOCATION OF THE WOODEN BEAM MODEL AT DIFFERENT DAMAGE LOCATIONS

In order to study the relationship between the curvature mode and the damage location of the wooden beam model at different locations of the single damage, 11 unit, 15 unit, and 22 unit are,

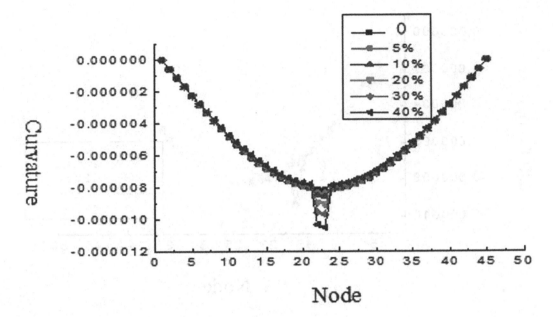

FIGURE 8.3 Contrast graph of first-order curvature modal.

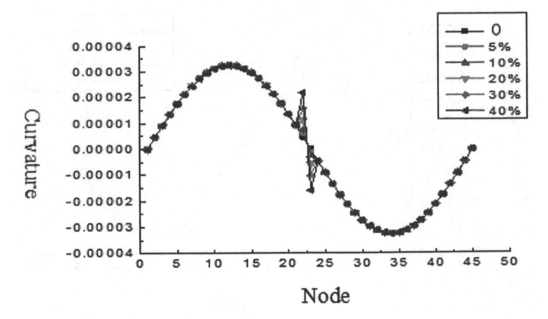

FIGURE 8.4 Contrast graph of second-order curvature modal.

respectively, set as damage units, the damage degree of which is set to 30%. The simulation results are shown in Figures 8.5 and 8.6.

It can be seen from the figure that the curvature of the model will be abrupt at the damage unit of different positions, namely, it obviously deviated from the originally smooth curvature mode curve. Through research and analysis, it can be concluded that the curvature mode can show the position of the element's elastic modulus changed, and the position of the sudden change of curvature is the damaged position. It also indicates that the curvature mode technique is feasible for the damage detection of wood beams.

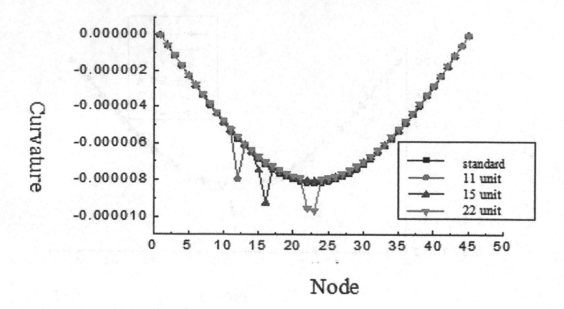

FIGURE 8.5   Contrast graph of first-order curvature modal.

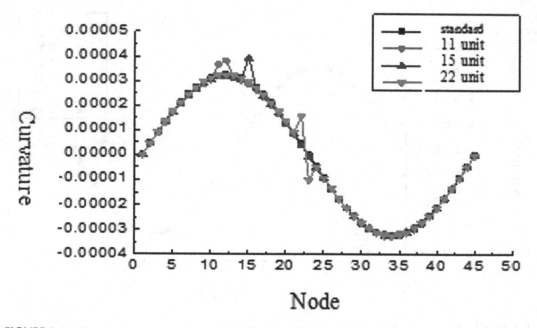

FIGURE 8.6   Contrast graph of second-order curvature modal.

8.3   THEORETICAL DERIVATION OF DAMAGE DEGREE OF WOODEN BEAMS BASED ON CURVATURE MODE

8.3.1   BASIC ASSUMPTIONS

Taking the simply supported beam as an example, the following basic assumptions are proposed: assuming that the simply supported beam is free vibration without damping, ignoring the influence of shear deformation, only the lateral displacement of each point of the beam is considered.

The beam is an equal-section straight rod, and the reduction of beam bending stiffness, *EI,* is used to simulate the damaged beam. Therefore, the bending stiffness of the beam is arbitrarily changed along the span *L* with the position *x*, which is *EI(x)* [12]. The beam is a homogeneous material with a mass per unit length, $\bar{m}$. The beam conforms to the plane assumption during the bending process and the material of which maintains the line elasticity.

8.3.2 Expression of Lateral Displacement of Beam Free Vibration under Different Supporting Conditions

8.3.2.1 Displacement Equation for Free Vibration of a Simple Supported Beam [11]

$$y(x,t) = C \sin \frac{n\pi x}{l} \cdot \sin \left(\frac{n^2 \pi^2}{l^2} \sqrt{\frac{EI}{\bar{m}}} t + \phi \right)$$

(8.1)

8.3.2.2 Displacement Equation for Free Vibration of the Fixed Support Beam [11]

$$y(x,t) = B_1 \left[\left(\cos \lambda x - \cosh \lambda x \right) - \frac{\cos \lambda l - \cosh \lambda l}{\sin \lambda l - \sinh \lambda l} \right.$$

$$\left. \left(\sin \lambda x - \sinh \lambda x \right) \right] \cdot \sin \left((\lambda l)_n^2 \sqrt{\frac{EI}{ml}} t + \phi \right)$$

(8.2)

where

$$\left(\lambda l \right)_1 = 4.73, \ \left(\lambda l \right)_2 = 7.853$$

$$\left(\lambda l \right)_n = \frac{\pi}{2}(2n+1) \quad (n = 3,4,5,\ldots)$$

8.3.3 The Expression Derivation of Damage Degree of a Wooden Beam under Different Supporting Conditions

The degree of damage, *D*, is defined as

$$D = 1 - \frac{\overline{EI}}{EI}$$

(8.3)

where $\overline{EI}$ = the stiffness after the damage, *EI* = the stiffness before the damage.

From the above formula, the relationship between the stiffness after damage and the degree of damage can be derived as follows:

$$\overline{EI} = (1 - D)EI$$

(8.4)

Substituting Eq. (8.4) into Eq. (8.1), the displacement expression of the damage site is

$$y(x,t) = C \sin \frac{n\pi x}{l} \cdot \sin \left(\frac{n^2 \pi^2}{l^2} \sqrt{\frac{(1-D)EI}{\bar{m}}} t + \phi \right)$$

(8.5)

The curvature mode is obtained by the central difference method [3].

The conversion formula for displacement and curvature is

$$\Phi_i'' = \frac{y_{i-1} - 2y_i + y_{i+1}}{l^2} \tag{8.6}$$

where y_i = the displacement of point i, which is the displacement obtained by Eq. (8.1).

The coefficient of sudden chance of curvature is defined as

$$V_q = \frac{\Delta_i}{\delta_i} = \frac{|\Phi_i'' - \Phi_{i-1}''|}{|\Phi_i'' + \Phi_{i-1}''|/2} \tag{8.7}$$

where i = Damage location, Φ_i'' = Curvature mode, Δ_i = Maximum curvature difference at the damage location, and δ_i = Maximum curvature difference at the damage location.

Substituting Eq. (8.6) into Eq. (8.7), the coefficient of sudden chance of curvature is expressed as

$$V_q = \frac{2(3y_{i-1} + y_{i+1} - 3y_i - y_{i-2})}{y_{i+1} + y_{i-2} - y_{i-1} - y_i} \tag{8.8}$$

The theoretical formula for determining the degree of damage is

$$\left\{ \begin{array}{l} V_q = \dfrac{2(3y_{i-1} + y_{i+1} - 3y_i - y_{i-2})}{y_{i+1} + y_{i-2} - y_{i-1} - y_i} \\[3mm] y_i(x,t) = C \sin \dfrac{n\pi x}{l} \cdot \sin\left(\dfrac{n^2 \pi^2}{l^2} \sqrt{\dfrac{EI(x)}{\overline{m}}} t + \phi \right) \\[4mm] \overline{EI}(x) = \begin{cases} (1-D)EI & \text{Damaged} \\[2mm] EI & \text{Non-Damaged} \end{cases} \end{array} \right. \tag{8.9}$$

Similarly, the curvature mutation coefficient is combined with the displacement equations of the fixed beam at both ends, and the theoretical formula for determining the damage degree of the fixed beam at both ends is

$$\left\{ \begin{array}{l} V_q = \dfrac{2(3y_{i-1} + y_{i+1} - 3y_i - y_{i-2})}{y_{i+1} + y_{i-2} - y_{i-1} - y_i} \\[3mm] y_i(x,t) = B_1 \left[(\cos \lambda x - \cosh \lambda x) - \dfrac{\cos \lambda l - \cosh \lambda l}{\sin \lambda l - \sinh \lambda l} \right. \\[4mm] \left. (\sin \lambda x - \sinh \lambda x) \right] \cdot \sin((\lambda l)_n^2 \sqrt{\dfrac{EI}{ml}} t + \phi) \\[4mm] \overline{EI}(x) = \begin{cases} (1-D)EI & \text{Damaged} \\[2mm] EI & \text{Non-Damaged} \end{cases} \end{array} \right. \tag{8.10}$$

where $(\lambda l)_1 = 4.73$ $(\lambda l)_2 = 7.853$, $(\lambda l)_n = \dfrac{\pi}{2}(2n+1)$ $(n = 3,4,5,\ldots)$

Equation (8.10) is the formula for determining the degree of damage of the fixed support beam. The method of use is basically the same as the damage determination formula of the simply supported beam. The experimental data and the required sudden change coefficient of curvature of the test are obtained by modal test. The displacement equation is used to calculate the theoretical curvature of sudden chance coefficient (including the unknown D). Subsequently, the corresponding equation is established by using that experimental value, which is equal to the theoretical value, and by solving that equation we can obtain the damage degree D.

8.3.4 VERIFICATION OF FINITE ELEMENT EXAMPLE

In this section, taking the simply supported beam as an example, the relevant data is directly extracted by using the model in which the damaged unit is 22 unit and the damaged degree is 30% in Section 1.3, to verify the applicability of the damage judgment formula of the simply supported beam.

The modal analysis pays more attention to the lower-order mode, because it is easy to obtain, relatively stable, and can reflect the dynamic characteristics of the component [7]. Therefore, the first-order curvature mode curve selected to calculate is shown in Figure 8.7. The formula for judging the damage degree of a simply supported beam is Eq. (9). The unknowns in the displacement equation have C, n, x, t, and the values of the unknowns are determined one by one below. As can be easily seen from Figure 8.7, the damaged site is 22 unit, $i = 22$. It is thus possible to determine the value of x in the displacement equation, since the unit division length is 25 mm, so $x = 525$ mm. Since the first-order mode is used in this paper, $n = 1$. Now assume that $t = 0$, at this time in the unvibrated state, the displacement y should be 0, so the constant $\phi = 0$.

When first-order frequency is 107 Hz, then, $t = 0.058$ s, $l = 1,100$ mm, $E = 7,500$ MPa, and $I = \dfrac{bh^3}{12}$. Uncertain unknowns only have a constant C, which can be calculated as

$$y_{20} = 0.2 \cdot C,\ y_{21} = 0.22 \cdot C,\ y_{22} = 0.6 \cdot \sin\left(8.4\sqrt{1-D}\right) \cdot C,\ y_{23} = 0.28 \cdot C$$

The unknown constant C can be eliminated by substituting y_{ll} to equation, and the damage degree D calculated is 0.36.

There is a certain error from the calculation results because the difference between the assumptions made at the time of derivation and the actual situation, which can be amplified when the difference is converted to curvature, and the approximate calculation in the calculation process also brings errors.

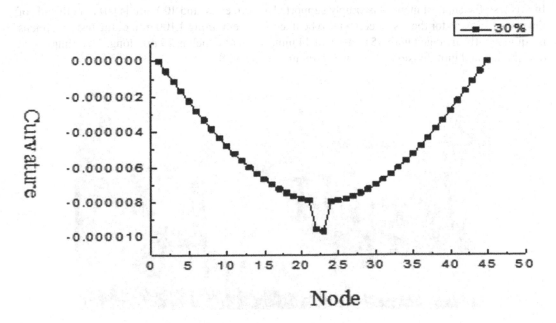

FIGURE 8.7 First-order curvature modal diagram of 22 element with 30% damage degree.

8.4 MODAL TEST AND DATA ANALYSIS

8.4.1 Test Overview

8.4.1.1 Modal Test Equipment

The modal analysis technology is used to detect wood beam damage, which generally requires three equipments: excitation equipment, sensing system, and analysis system [3]. In this test, the collected signal is analyzed and processed by the INV3060 network distributed acquisition analyzer of China Orient Institute of Noise & Vibration. The exciter adopts the applied force hammer YD-5T121109 of the Institute of China Orient Institute of Noise & Vibration, and the receiver is the United States PCB company's acceleration sensor 3801D3FB20G. The main equipments of the test are shown in Figure 8.8.

8.4.1.2 Preparation of Test Specimen

Considering that the damage of the test specimens is close to the damage of the existing wooden structures, *Pinus sylvestris* is used as the research object in this experiment. The specimens were 1.3 m long with a cross section of 80 mm × 120 mm. It is assumed that the specimen has no initial damage and all the damages are made by hand. The slotting method is used because the damage will lead to the reduction of the effective cross-sectional area of the member [12], and the damaged condition is slotted at the lower side of the test specimen, which can reduce the effective cross-sectional area and simulate the damage of timber beam. In addition, slotting reduces the moment of inertia of the section, which further reduces the stiffness. If slotting is performed on the lower side of the beam, b is constant, and h is decreasing, according to the formula of moment of inertia, $I = \dfrac{bh^3}{12}$. Assuming 20% damage, then $I' = \dfrac{bh'^3}{12} = 0.8I = 0.8\dfrac{bh^3}{12}$. Simplified, $h' = \sqrt[3]{0.8}h$, $h = 120$, namely, $h' = 111$ mm, The groove depth is 9 mm. The specific dimensions of the test specimens are shown in Figure 8.9. The basic conditions of the test specimens are shown in Table 8.1.

8.4.2 Test Preparation and Test Design

8.4.2.1 Test Specimen Support Method and Measuring Point Arrangement

In this test, the support method is simply supported at both ends, and 100 mm is left at both ends of the test specimens for the test specimens to be fixed. The remaining 1,100 mm of the test specimens is equidistantly arranged with 45 nodes and 44 units, each of which is 25 mm long. The sample size distribution and unit division are shown in Figures 8.10 and 8.11.

FIGURE 8.8 Major equipment diagram for test.

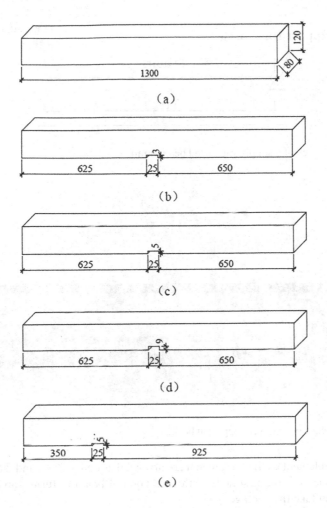

FIGURE 8.9   Specific dimension diagram of specimens. (a) Specimen 1, (b) specimen 2 (22 node damage), (c) specimen 3 (22 node damage), (d) specimen 4 (22 node damage), and (e) specimen 5 (11 node damage).

TABLE 8.1
Basic Situation of the Specimens

Specimen	Degree of Damage (%)	Damage Location
1	0	None
2	5	Midspan
3	10	Midspan
4	20	Midspan
5	10	¼

The selection of the number of measuring points, the position of the measuring points, and the direction of measurement should consider two requirements: (1) It is possible to clearly show the deformation characteristics of all modes in the test frequency band and the deformation differences between the modes after deformation; (2) Ensure that the structural points of interest are among the selected measurement points [13]. In order to achieve better test results, this paper adopts multi-point input and multi-point output method to carry out modal test. A total of three accelerometers

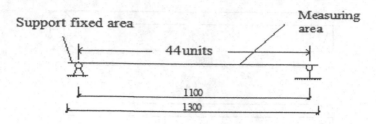

FIGURE 8.10   Test sketch of a simply supported beam (mm).

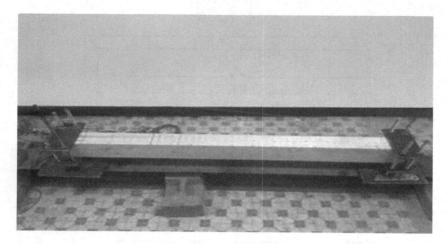

FIGURE 8.11   Test chart of a simply supported beam.

are arranged to complement each other, which are arranged at nodes 5, 19, and 29, respectively. Use the hammer to knock from the first node to the last node of beam in turns, tap each node at least three times, and then take the average.

8.4.2.2   Selection of Excitation Pattern

This test uses force hammer excitation, which is suitable for small and medium-sized structures. When hammering, the top cap is in impact contact with the test structure, and the impact transmitted to the structure is almost semi-sinusoidal. In the use of the hammer, it is necessary to avoid multiple impacts caused by the rebound, because this multiple impact signal will cause difficulties in the signal processing stage. When striking, do not use excessive force in order to avoid local deformation of the structure beyond the elastic range.

8.4.3   TEST RESULTS AND DATA PROCESSING

8.4.3.1   Analysis of Curvature Modal of Specimen

8.4.3.1.1   Natural Frequency of Test Specimen

The natural frequency is the most easily obtained modal parameter by modal test, and the test result is relatively accurate. At the same time, the natural frequency is the basis for obtaining other modal parameters, which is representative. Due to the limited experimental conditions, the high-order modes of this modal test are not easy to obtain, and only the first two modes are analyzed [7]. The first two natural frequencies of the test specimen 1–5 are shown in Table 8.2.

Comparing the test specimens 1–4, it can be seen that the natural frequencies of the same order decrease as the degree of damage increases, but the difference is not obvious. Comparing 1, 3, and 5,

TABLE 8.2
Natural Frequency

Specimen	First-Order Frequency (Hz)	Second-Order Frequency Frequency (Hz)
1	115.686	255.878
2	114.287	255.146
3	112.698	253.259
4	108.853	250.546
5	113.569	252.482

it is found that the different damage locations are not obvious on the natural frequency of the test specimen. It can be said that the natural frequency can easily discriminate whether the structure is damaged or not, but the damage position cannot be discriminated. And the change of the natural frequency needs to be compared with the standard test specimen, which is of difficulty to realize under realistic conditions, so it is unreliable to use the natural frequency index to discriminate the damage of the beam.

8.4.3.1.2 Analysis of Relationship between Curvature Mode and Damage Degree of Wooden Beam Specimens with Different Damage Degrees at a Single Site

After the experimental modal analysis, the first-order and second-order curvature modes of the intact wood beam test piece No. 1 and the damaged wood beam test piece Nos. 2, 3, and 4 were obtained, as shown in Figures 8.12 and 8.13.

Figures 8.12 and 8.13 show the change of curvature mode diagram with the increase of damage degree when the wooden beam specimen is damaged in a single place. As can be seen from the figure, similar to the finite element analysis, the curvature mode diagram has an obvious sudden change at the damage position of the wooden beam, so that the damage location can be accurately determined, and the degree of abrupt change of the curve increases with the degree of damage of the

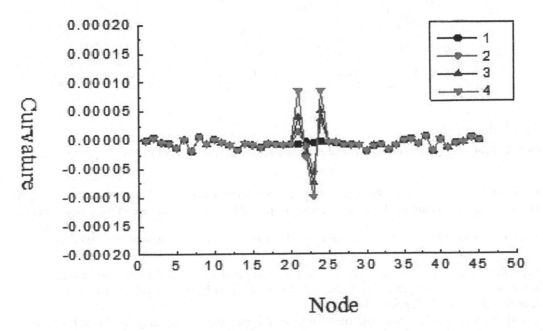

FIGURE 8.12 First-order curvature modal contrast diagrams of specimens with different damage degrees.

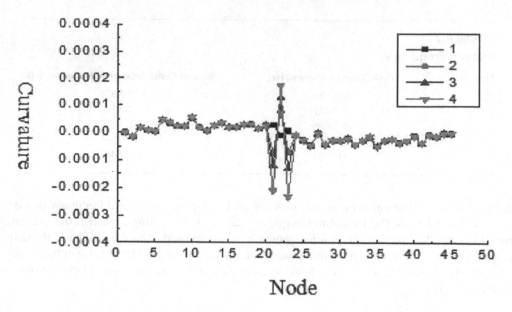

FIGURE 8.13   Second-order curvature modal contrast diagrams of specimens with different damage degrees.

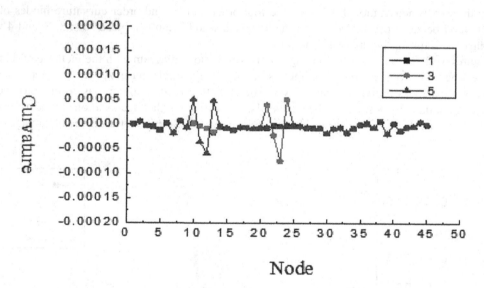

FIGURE 8.14   First-order curvature modal contrast diagrams of specimens with different damage locations.

wooden beam. The degree of the sudden change can reflect the degree of damage to a certain extent. Therefore, the curvature mode can be used to estimate the degree of damage of the wooden beam.

8.4.3.1.3   *Analysis of Relationship between Curvature Mode and Damage Location of Wooden Beam Specimens with Different Damage Locations at a Single Site*

After the experimental modal analysis, the first-order and second-order curvature modes of the intact wood beam test piece No. 1 and the damaged wood beam test piece Nos. 3 and 5 were obtained, as shown in Figures 8.14 and 8.15.

Figures 8.14 and 8.15 illustrate the change of the curvature mode diagram when the wooden beam specimen with the same damage degree is at different damage locations. As can be seen

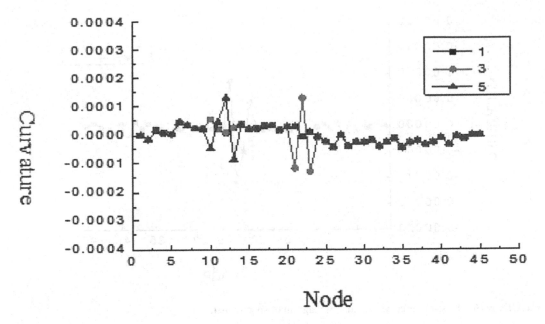

FIGURE 8.15 Second-order curvature modal contrast diagrams of specimens with different damage locations.

from the figure, similar to the finite element analysis result, when the damage location changes, the sudden change position of the curve changes accordingly. Therefore, the damaged position of the wooden beam test specimen can be easily found according to the curvature mode of the wooden beam test specimen.

8.4.3.2 Verification of Damage Degree Determination Formula

8.4.3.2.1 Verification of the Theoretical Formula for the Determination of Single Damage

From the derived the theoretical formula for determining the damage degree of the simply supported beam Eq. (8.9), the data obtained by the modal test of the test specimen 4 is extracted, and the first-order curvature mode curve is shown in Figure 8.16.

By plotting the curvature modal diagram, it can be determined that the damage position is 22 unit, $i = 22$, so that the value of x in the displacement equation can be determined because the unit division length is 25 mm, so $x = 525$ mm. Since the first-order mode is selected, $n = 1$. Now assume that $t = 0$, at this time in the unvibrated state, the displacement y should be 0, so $\phi = 0$. When the first-order frequency is 108.9 Hz, then $t = 0.058$ s. $l = 1,100$ mm, $E = 7,500$ MPa, and $I = \dfrac{bh^3}{12}$, and the V_q value obtained from the modal test is substituted into the Eq. (8.9) then $D = 0.29$ is obtained, and the damage degree of the test specimen is set to 20%, and the error is relatively large. The same method was used to determine the damage degree of each test specimen: specimen 2: $D = 0.08$; specimen 3: $D = 0.16$; specimen 5: $D = 0.15$.

8.4.3.2.2 Error Analysis

There is a certain error in the solution of the degree of damage. Besides the errors caused by the support, manual operation, and the defects of the test specimen itself, the assumption of formula deduction also causes errors. When formula is deduced, it is assumed that the beam is undamped-free vibration, but it is impossible to achieve under realistic conditions. The damping will attenuate the amplitude continuously, which will reduce the sudden change of the adjacent two nodes to some

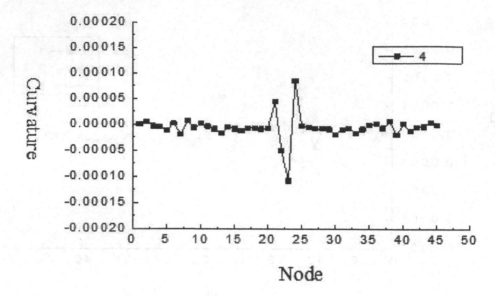

FIGURE 8.16   First-order curvature modal diagram of specimen 4.

extent. The V_q value obtained by the test is relatively small; eventually, this will cause a certain error between the judgment result and the actual set damage degree.

8.5   CONCLUSIONS

The following conclusions may be drawn through the numerical simulation of the damaged wooden beam and the small-scale modal test research and result analysis of the artificially simulated damaged wooden beam.

1. Compared with the curvature mode of the intact wooden beam, if the wooden beam is damaged, the curvature mode will have an abrupt change at the damage position, and the degree of the sudden change increases with the degree of damage. Therefore, the curvature mode is very sensitive to the damage identification of the wooden beam, which can be used to quantitatively estimate the location and degree of damage of the wooden beam.
2. According to the knowledge of free vibration of beams in structural dynamics, the degree of damage is introduced into it, and combined with the defined sudden change coefficient of curvature, the theoretical formula for determining the degree of damage of wooden beams under different supporting conditions is derived, which is verified through finite element examples and the modal test and has certain applicability under certain basic assumptions.
3. When the finite element method is used for non-destructive testing of damaged wood beam, the damage and defects of wood beam can be well simulated by reducing the local element's modulus of elasticity of the finite element model.
4. Both the finite element modal analysis and the experimental modal analysis are consistent with the curvature modes. It is proved that the curvature modal analysis is an effective method for detecting damage of wood beams.

REFERENCES

1. D Li and Q Lu. 2002. Curvature modal analysis of bending structures. *Journal of Tsinghua University (Natural Science Edition)*, 42(2), pp. 224–227.
2. L Liu. 2007. Research on damage identification of beam bridge based on curvature modal analysis. Southwest Jiaotong University.

3. D Li and Q Lu. 2001. *Experimental Modal Analysis and Its Application*. Beijing: Science Press.
4. H Xu and L Wang. 2011. Quantitative damage identification of wooden beams using curvature modal technology. *Journal of Vibration, Measurement and Diagnosis*, 31(1), pp. 110–114.
5. X Wang and W Hu. 2014. Damage identification of ancient wood structures based on difference of energy curvature of wavelet packet. *Journal of Vibration and Shock*, 33(7), pp. 153–159+186.
6. X Yang and Y Ishimaru. 2002. Application of modal analysis by transfer function to non-destructive testing of wood II: Modulus of elasticity evaluation of sections of differing quality in a wooden beam by the curvature of the flexural vibration wave. *Journal of Wood Science*, 49, pp. 140–144.
7. C Hu and MT Afzal. 2006. A statistical algorithm for comparing mode shapes of vibration testing before and after damage in timbers. *Journal of Wood Science*, 52, pp. 348–352.
8. FC Choi and J Li. 2007. Application of modal-based damage-detection method to locate and evaluate damage in timber beams. *Journal of Wood Science*, 53, pp. 394–400.
9. D Capecchi and F Vestroni. 1999. Monitoring of structural systems by using frequency data. *Journal of Earthquake Engineering and Structural Dynamics*, 28, pp. 447–461.
10. P Cawley and RD Adams. 1979. The location of defects in structure from measurements of nutural frequencies. *Journal of Strain Analysis*, 14(2), pp. 49–57.
11. J Liu and X Du. 2005. *Structural Dynamics*. Beijing: Machinery Industry Press, pp. 152–154.
12. Y Wang. 2010. Curvature modal analysis for damage identification of simply supported beams. Southwest Jiaotong University.
13. X Yang and J Jin. 2000. Frequency equation and characteristic valuc of bending vibration beam under different supporting conditions. *Journal of Shenyang Institute of Aeronautical Technology*, 4, pp. 1–6.

9 Validation of Proposed SHM Model Based on Inverse Dynamic Approach with Limited Noisy Dynamic Responses by Experimental Study

D. Bandyopadhyay
Jadavpur University

J. S. Ali
Aliah University

CONTENTS

9.1   INTRODUCTION

Structures in civil engineering arena often experience various types of distress during its design life. The damage may occur due to normal wear and tear, age, improper design, inferior construction, environmental aggression, accidental loading, etc. In addition, natural disaster of different types such as earthquake, flood, cyclone, etc. also leaves behind their sign on the civil structures in terms of damages. Structural health monitoring (SHM) for assessing the condition of the age-old structures and disaster driven damaged structures has become significantly important for the safety and durability of structures. Monitoring of structural health in terms of integrity and stability of existing structures is quite useful in improving the safety, timely intervention for repair and rehabilitation, and better utilization of the structure including avoidance of catastrophic failure. The benefit of SHM includes optimal maintenance cost with constant reliability instead of increasing maintenance cost and decreasing reliability of structures.

System identification techniques, such as adopting inverse dynamic approaches, are important tools to monitor the structural health. The objective of dynamic system identification is to update the mathematical models of the dynamics of structure by using measured responses. SHM technique based on dynamic responses is a global manner and seems to be a better proposition with limited measured data. In the proposed model, it is assumed that the damage changes the structural characteristics considerably, so that the dynamic responses are influenced by these changed parameters. Further, the uncertainties in the measured data might lead to unreliable identification. Statistical system identification techniques with probabilistic concepts may address the SHM problem with uncertainties associated in a much better way with the limited observations, model uncertainties, and other environmental and operational variations.

Different approaches of SHM problem based on dynamic responses have been attempted by several researchers. These attempts may be classified based on different ways. System identification approaches were started long ago (Adam et al. 1978, Sanayei and Onipede 1991, Casas and Aparicio 1994, Liu 1995), but the problems of practical limitations of limited data and different uncertainties associated were not adequately addressed. Later a few have considered the availability of limited dynamic response, measured at a few selected degrees of freedom (Wang and Haldar 1994, Sanayei and Scampoli 1999), and the feasibility of practical application. In recent times, statistical system identification approaches are employed for the SHM (Katafygiotis et al. 1998, Beck and Au 2001, Xia and Hao 2003, Yuen et al. 2006) with due consideration of the uncertainties associated with the measured dynamic data. Stubbs et al. (1992) presented a modal strain energy (MSE) based SHM approach for structural damage localization. Stubbs and Kim (1996) verified the practical feasibility of an MSE based method for localization and severity estimation of damage with the post-damage information as available. Carrasco et al. (1997) discussed the application of MSE change to identify the location and extent of damage in a space truss. They claimed that the change of the magnitude of the indicator can be related to the overall magnitude of the damage. Park et al. (2001) discussed another damage index based on MSE. The index was extracted from the periodical measurement data of the visually inspected cracks on a concrete box-girder bridge. They found that the environmental factors are likely to affect the accuracy of the results of the proposed model. Wang et al. (2010) studied MSE as a correlation indicator, called MSE correlation (MSEC) to identify damage for truss bridges. The MSE to damage sensitivity relationship was derived and discussed. The efficiency of MSEC based model was verified by several examples having various damage scenarios. It was found that the method has high computational efficiency if the measurements are correct. However, when measurement is contaminated with noise, the method may not be reliable with false alarms. Cornwell (1999) proposed and examined the application of an MSE based method on a plate-like structural model. It was reported that the method requires only mode shape data before and after the damage. It was also reported from the experimental validation that the method is effective even for the damage extent as low as 10% of its undamaged parameters.

The objective of the proposed paper is to develop an SHM technique based on MSE using limited experimental noisy modal data, measured at feasible selected degrees of freedom. The modal data considered both natural frequency and mode shape to identify the existence of damage, which is mainly the local reduction of stiffness and mass. There are several uncertainties associated with the assumed finite element (FE) model. On the other hand, noise in measurement is inevitable. This may lead to an unreliable identification of damage in structures. The proposed method is demonstrated with various numerical examples. Then the model is validated with the experimentally obtained noisy dynamic responses and subsequently extracted modal data. It is observed that the efficiency of the proposed damage identification algorithm depends on the finite element model and the measured dynamic responses.

9.1.1 Theoretical Formulation

The generalized eigenvalue problem of motion of finite element model can be expressed as

$$[K]\{\varphi\} = \lambda[M]\{\varphi\}, \text{ where } \lambda = \omega^2 \tag{9.1}$$

where $[M]$, $[K]$, λ, and ω indicate the global mass matrix, global stiffness matrix, eigenvalue, and natural frequency, respectively. Equation (9.1) can be solved for λ and $\{\varphi\}$. The eigenvector $\{\varphi\}$ is the mode shape and eigenvalue λ is the square of the associated natural frequency in radians per second.

Elemental MSE (MSE) is defined as the product of the elemental stiffness matrix and the second power of its mode shape component (Shi et al. 1998). For the jth element and ith mode, the elemental MSEs before and after the occurrence of damage are given as

$$\text{MSE}_{ij} = \Phi_i^T K_j \Phi_i,$$
(9.2)

$$\text{MSE}_{ij}^d = \Phi_i^{d^T} K_j \Phi_i^d$$
(9.3)

where MSE_{ij} is the undamaged jth elemental MSE corresponding to the ith mode shape, and similarly, MSE_{ij}^d is the damaged jth elemental MSE corresponding to the ith mode shape; K_j is the jth elemental stiffness matrix, and Φ_i is the ith mode shape.

The superscript d denotes the damaged state. As the damage elements are required to be identified, the undamaged elemental stiffness matrix K_j is used as an approximation of the damaged state for the expression of MSE_{ij}^d as shown in Eq. (9.3).

The elemental MSE change ratio (MSECR) proposed by Shi et al. (1998) is

$$\text{MSECR}_{ij} = \frac{\left| \text{MSE}_{ij}^d - \text{MSE}_{ij} \right|}{\text{MSE}_{ij}}$$
(9.4)

The structural damage is assumed as a loss of stiffness in one or more elements of the structure only but not of their masses. The damage in terms of the reduction of stiffness is represented by a small perturbation of the original system in the proposed model. The stiffness matrix K^d can be expressed as follows:

$$K^d = K + \sum_{j=1}^{L} \Delta K_j = K + \sum_{j=1}^{L} \alpha_j K_j, \left(-1 < \alpha_j \ll 0 \right)$$
(9.5)

Similarly, the eigenvalue λ_i^d and the mode shapes Φ_i^d at the ith mode of the damaged system are expressed as follows.

$$\lambda_i^d = \lambda_i + \Delta \lambda_i$$
(9.6)

$$\Phi_i^d = \Phi_i + \Delta \Phi_i$$
(9.7)

where α_j = coefficient defining a fractional reduction in the jth elemental stiffness matrix; and L = Total number of elements in the system. The change in the elemental MSE (MSEC) of the jth element in the ith mode is expressed as

$$\text{MSEC}_{ij} = \Phi_i^{d^T} K_j \Phi_i^d - \Phi_i^T K_j \Phi_i$$
(9.8)

Putting the Eq. (9.7) into Eq. (9.8), it becomes

$$\text{MSEC}_{ij} = 2\Phi_i^T K_j \Delta \Phi_i$$
(9.9)

The equation of motion of an un-damped dynamic system for a small perturbation becomes

$$\left[(K + \Delta K) - (\lambda_i + \Delta\lambda_i) M \right](\Phi_i + \Delta\Phi_i) = 0 \qquad (9.10)$$

Neglecting the second-order terms, Eq. (9.10) becomes

$$(K - \lambda_i M) \Delta\Phi_i = \Delta\lambda_i M \Phi_i - \Delta K \Phi_i \qquad (9.11)$$

Now, $\Delta\Phi_i$ in Eq. (9.12) can be expressed as a linear combination of the mode shapes of the original system as proposed by Fox and Kapoor (1968).

$$\Delta\Phi_i = \sum_{k=1}^{n} d_{ik}\Phi_k \qquad (9.12)$$

where d_{ik} = Scalar factors and n = Total No of modes of the original system.

Substituting Eq. (9.12) into Eq. (9.11), and multiplying by Φ_r^T to both sides of Eq. (9.11), we have

$$\sum_{k=1}^{n} d_{ik}\Phi_r^T (K - \lambda_i M)\Phi_k = \Delta\lambda_i \Phi_r^T M\Phi_i - \Phi_r^T \Delta K\Phi_i \qquad (9.13)$$

When $r \neq i$, Eq. (9.13) is simplified into the following with the orthogonal relationship,

$$d_{ir} = -\frac{\Phi_r^T \Delta K\Phi_i}{\lambda_r - \lambda_i}, \text{where} \quad r \neq i \qquad (9.14)$$

When $r = i$, d_{rr} equals 0, from the orthogonal relationship $\Phi_i^T M\Phi_i = I$. Therefore, Eq. (9.12) can be written as

$$\Delta\Phi_i = \sum_{r=1}^{n} -\frac{\Phi_r^T \Delta K\Phi_i}{\lambda_r - \lambda_i}\Phi_r, \text{ where } \quad r \neq i \qquad (9.15)$$

Substituting Eq. (9.15) into Eq. (9.9), the MSEC_{ij} becomes

$$\text{MSEC}_{ij} = 2\Phi_r^T K_j \left(\sum_{r=1}^{n} -\frac{\Phi_r^T \Delta K\Phi_i}{\lambda_r - \lambda_i}\Phi_r \right), \text{ where } \quad r \neq i \qquad (9.16)$$

Or,

$$\text{MSEC}_{ij} = \sum_{p=1}^{L} -2\alpha_p \Phi_i^T K_j \sum_{r=1}^{n} \frac{\Phi_r^T K_p\Phi_i}{\lambda_r - \lambda_i}\Phi_r, \text{ where } \quad r \neq i \qquad (9.17)$$

In the proposed model a simple model with uniformly distributed noise as follows:

$$d'_{ij} = (1 + \sigma\%)d_{ij} \qquad (9.18)$$

where d and d' are the noise free and noise contaminated values and σ is a random number that varies between zero and maximum expected error as assumed in practical cases.

9.1.2 EXPERIMENTAL EXAMPLE

A single bay two storied steel portal frame as shown in Figure 9.1 is considered for the experiment. The frame is properly fixed on the base plate of a uni-directional shaking table. Eight number of accelerometers are placed to measure the response time signals. The statistical variations in length, width, and thicknesses are measured. Details of the frame with node numbers and member numbers are shown below. The dimensions showing the nodal degrees of freedom, sensor location scheme, and the actual mounted accelerometers on the frame are shown in Figure 9.1a–c respectively. A Uniaxial Shake table with different rpm has been used for exciting the structure at different amplitudes, and data have been acquired in time domain and analyzed subsequently with the operational modal analysis.

The properties with respect to its length, width, thickness, and cross-sectional area of different members of the considered frame at its undamaged state are shown in Tables 9.1 and 9.2.

The various damaged states considered, including single element damage and multiple element damage of different extents, are shown in Figure 9.2, where UD stands for Undamaged Frame; 1D3 stands for Single Element Damage at element 1; 2D13 stands for Double Element Damage at elements 1 and 3; Damage state 3D134 stands for Double Element Damage at elements 1, 3, and 4 (Figure 9.3).

The signal analyzer boxes and accelerators used for the experiment are shown in Figure 9.4.

Eight numbers of sensors are connected to the portal frame and responses are recorded by the two A4404 –SAB systems which are pocket sized four channel vibration analyzers. Two numbers of such data acquisition systems are used having a total of eight numbers of channels. Responses are

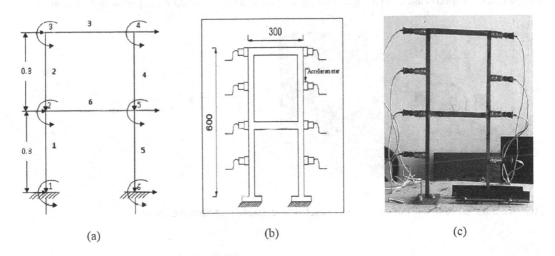

| (a) | (b) | (c) |

FIGURE 9.1   (a) Degree of freedom, (b) scheme of sensor location, and (c) actual frame with sensor.

TABLE 9.1
Properties of the Undamaged Frame

Member	Length (mm)	Width (mm)	Thickness (mm)
1	300	19.16	6.05
2	300	19.16	6.05
3	6.05	300	19.16
4	300	19.16	6.05
5	300	19.16	6.05
6	6.05	300	19.16

TABLE 9.2
Member Properties of Frame (UD)

Member	Column	Beam
Member number	1, 2, 4, 5	3, 6
Cross section area (m²)	0.0001159	0.0001159

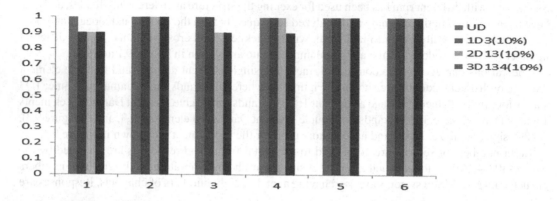

FIGURE 9.2   Various case of single and multiple damage states considered for experimental validation.

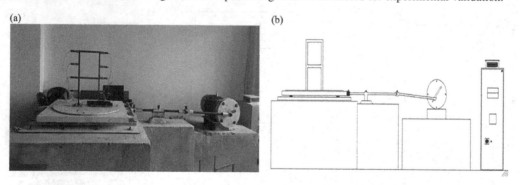

FIGURE 9.3   (a) Actual arrangement and (b) schematic diagram of Shaking Table.

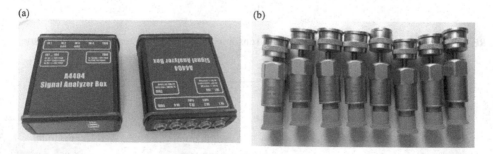

FIGURE 9.4   (a) Signal analyzer box and (b) acceleration transducer are used for experimental validation.

recorded through the data accusation systems (ADASH) as shown in Figure 9.4a and then further processed to get the mode shapes and natural frequencies using ARTeMIS Modal Pro. Typical time domain signals are shown in Figures 9.5 and 9.6.

The measured first three mode shapes for single element damage are shown in Figure 9.7.

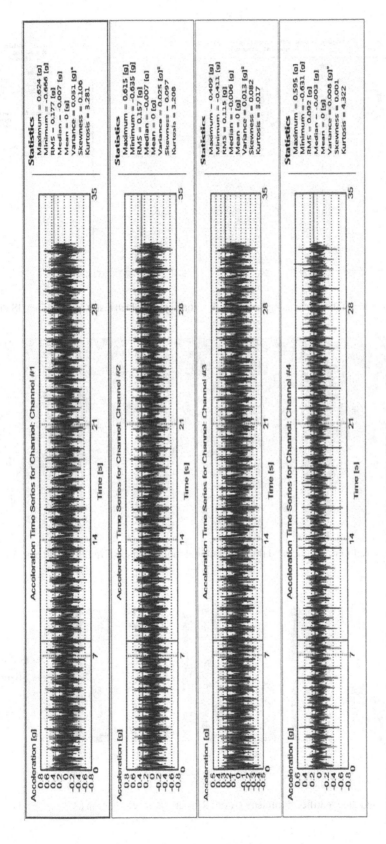

FIGURE 9.5 Typical signals recorded at four transducers through A4404-SAB in case of damage case.

1D3

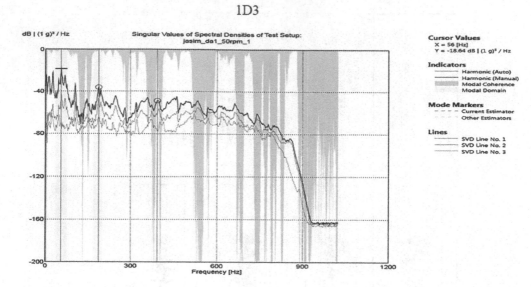

FIGURE 9.6 Estimation of the signal using frequency division duplex (FDD) result from ARTeMIS Pro (for 1D1).

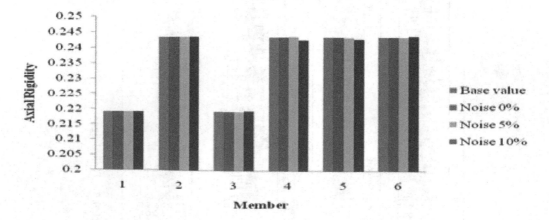

FIGURE 9.7 Measured first, second, and third operational mode shapes of the Portal Frame for 1D3 (10%).

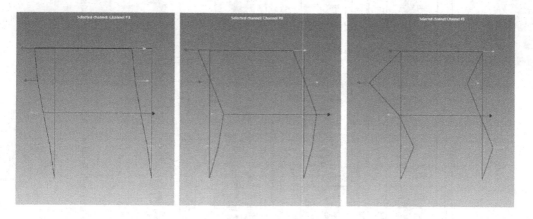

FIGURE 9.8 True and the identified parameter (Axial Rigidity, axial elongation (AE)) of the damaged Portal Frame 1D3.

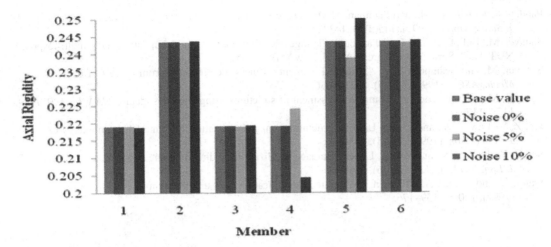

FIGURE 9.9 True and the identified parameter (Axial Rigidity, AE) for the damaged Portal Frame 2D34.

From the above measured mode shapes and experimental natural frequencies the damage is predicted to validate the MSE based damage identification model (Figures 9.8 and 9.9).

9.2 CONCLUSIONS

The present investigation is focused on the validation of the numerical model on condition monitoring of a frame, particularly the diagnosis of damage in terms of reduction of structural parameters using limited dynamic response data measured at a few selected locations. Equation error approach is employed in the finite element framework adopting system identification technique. The uncertainty associated with the measurement is duly considered in the model for its practical application. Computational algorithm based on MSE is validated with the experimentally obtained data of the portal frame example to establish the efficacy of the proposed method. The experimental study also provides an insight on a few basic aspects of dynamic testing and subsequent extraction of the modal data by operational modal analysis. The proposed method of condition monitoring is validated with the experimentally obtained data. The proposed method is able to identify the damage from limited number of noisy data with moderate accuracy. The condition of each element may be updated and monitored continuously as the measured data are available. It is also studied that the accuracy of the proposed models increases with the increase of the number of measured modes & decreases with the increase of noise level. The experimental study with Portal Frames validates the proposed method and demonstrates the potential of the numerical model for its practical implementation.

REFERENCES

Adam, R. D., Clawley, P., Pye, C. J., and Stone, B. J., "Avibrating technique for non-destructively assessing the integrity of structures", *J. Mech. Sci.*, (1978), 20(2), 93–101.

Bayissa, W. L. and Haritos, N., "Structural damage identification in plates using spectral strain energy analysis", *J. Sound Vibr.*, (2007), 307, 226–249.

Casas, J. R. and Aparicio, A. P., "Structural damage identification from dynamic test-data", *J. Struct. Eng., ASCE*, (1994), 120(8), 2437–2450.

Doebling, S. W., Hemez, F. M., Peterson, L. D., and Farhat, C., "Improved damage location accuracy using strain energy-based mode selection criteria", *AIAA J.*, (1998), 35, 693–699.

Maia, N. M. M., Silva, J. M. M., Almas, E. A. M., and Sampario, R. P. C., "Damage detection in structures: From mode shape to frequency response function methods", *Mech. Syst. Signal Process.*, (2003), 7(3), 489–498.

Pandey, A. K, Biswas, M., and Samman, M.M., "Damage detection from change in curvature mode shapes", *J. Sound Vibra.*, (1991), 145(2), 321–332.

Sanayei, M., Bell, E. S., and Javdekar, C. N., "Bridge deck finite element model updating using multi-response NDT data". Structures Congress, ASCE, New York, (2005).

Sanayei, M. and Scampoli, S. F., "Structural element stiffness identification from static test data", *J. Eng. Mech., ASCE,* (1991), 117(5), 1021–1036.

Sanayei, M. and Onipede, O., "Damage assessment of structures using static test data", *AIAA J.*, (1991), 29, 1174–1179.

Shi, Z. Y., Law, S. S., and Zhang, L. M., "Structural damage localization from modal strain energy change", *J. Sound Vibr.*, (1998), 218(5), 825–844.

Shi, Z. Y., Law, S. S., and Zhang, L. M., "Damage localization by directly using incomplete mode shapes", *J. Eng. Mech.*, (2000), 126(6), 656–660.

Wang, D. and Haldar, A., "Element level system identification with unknown input", *J. Eng. Mech.*, ASCE, (1994), 120(5), 159–176.

10 Grouting Compactness Assessment in Post-Tensioning Tendon Ducts Using Piezoceramic Transducers and Wavelet Packet Analysis

X. T. Sun, D. Li, and W. X. Ren
Hefei University of Technology

CONTENTS

10.1 INTRODUCTION

The post-tensioning method, which is widely used in bridge construction, can protect the prestressing reinforcement from environmental corrosion, enhance the bearing capacity and durability of prestressed concrete bridge. Grout compactness is the key factor to realize the function of post-tensioning technology. However, in the actual construction process, there are many factors may result in grouting defects.

Many non-destructive testing (NDT) methods, including ultrasonic echo, impact echo, ultrasonic tomography and ground penetrating radar, had been proposed to detect the defects of grouting in post-tensioning tendon duct (PTTD). A commonly used approach in NDT has been the grouting quality with scanning ultrasonic echo methods. One notable disadvantage of this method, however, is that the indication to grouting defects is not reliable enough. Lu et al. (2018) detected the defect of grouting by ultrasonic tomography. The results showed that ultrasonic tomography was a suitable method to evaluate the quality of grouting in post-tensioning structure, but the accuracy of this method need to be studied further. Olson et al. (2011) discussed the feasibility of scanning impact echo method in grouting quality evaluation. The results showed that the scanning impact echo method can improve the detection efficiency and reflect the grouting status in the bridge deck. In the recent research, Li and Long (2018) applied Hilbert Huang transform (HHT) method to analyze the signal when the impact echo method was used to detect the grouting compactness in PTTD. The results showed that HHT method could distinguish the grout compactness with the index of the thickness frequencies. Terzioglu et al. (2018) applied NDT methods including ground penetrating radar, impact echo, ultrasonic tomography, and ultrasonic echo to detect the location and severity of grouting defects. In summary, those NDT methods mentioned above require complex devices and operation, which is hard to implement in practice.

In recent years, piezoelectric materials have been widely used in structural health monitoring (SHM) due to their unique properties. Kawiecki (1998) simulated the damage of the beam by adding mass blocks to the concrete beam and placed two PZT patches at both ends of the beam as actuators and sensors. The experimental results showed that the piezoelectric ceramics had good sensitivity, repeatability, and were suitable for health monitoring of structures. By realizing the excellent performance of piezoelectric ceramics, many researchers had attempted to apply piezoelectric ceramics to the health monitoring in various areas of civil engineering. Jennifer (2008) used piezoelectric ceramic patches as distributed ultrasonic sensors to detect, locate, and characterize structural damage in plates. This research realized the application of piezoelectric ceramics on the plate-like structure. Wu and Qing (2009) combined piezoelectric transducers (PZT) and fiber Bragg grating (FBG) to form a hybrid distributed sensor network to detect the damage of the composite plate. The results showed that the distinguished advantage of this hybrid detection system was that sensors will not interfere with each other because of different working principles. Zou et al. (2010), Zou and Cui (2011), and Cui and Zou (2012) applied PZTs to the monitoring of the grout quality of the bolt.

Because of their fragility, the application of PZTs in civil engineering is limited. Song et al. (2007) from the intelligent materials and structures laboratory at Houston University carried out a health monitoring experiment on two T-beams of a 6.1-m-long reinforced concrete highway bridge. In this study, he proposed the concept of "smart aggregate" (SA). One piezoceramic patch with marble protection, called an SA, was embedded into the structure as a basic health monitoring element. The experimental results showed that this method can effectively identify the existence and severity of cracks inside the concrete structure. In later years, many researches (Song et al. 2008, Gu et al. 2010, Zhang et al. 2018, Xu et al. 2018) applied SAs to other fields of civil engineering.

In the field of monitoring grouting compactness in the duct, Jiang et al. (2016) applied piezoceramic transducers to monitor grouting compactness in a PTTD on a cross section. One transducer was bonded on the prestressing steel bar, which was used as an actuator; PZT patch sensors, mounted on the outside surface, were used as sensors. Experimental results indicated that the proposed method has the potential to monitor the grouting compactness in the tendon duct in real time. Then, Jiang et al. (2017) conducted the finite element simulation of monitoring the grouting compactness of tendon duct based on piezoelectric ceramics, which provided guidance for the experiment and application. Tian et al. (2017) applied the PZT ring actuator to adapt the prestressing steel bar better. However, those researches are still on a cross section, which required more transducers. Moreover, they are more concerned on monitoring the grouting level during the process of grouting rather than detecting defects after grouting.

In this paper, a method based on PZTs, along with active sensing approach and wavelet packet analysis, was used to detect the grouting defect. The proposed approach arranged sensors alongside the length direction of the PTTD, which could detect a larger area with fewer sensors. In addition, grouting defects include not only partially grouting but also cavities inside the tendon duct. Defects with different dimension were fabricated to simulate the grouting defect in actual engineering. In this research, we designed and manufactured four specimens to reappear four different grouting conditions. The wavelet packet energy was applied to compare the stress waves received by PZT sensors. Each specimen was maintained for 10 days. Experimental results demonstrated that this method can detect and distinguish grouting defects.

10.2 METHODOLOGY

10.2.1 Piezoelectric Effect and SA Transducers

Piezoelectric ceramics can convert electrical and mechanical energy into each other, which is called piezoelectricity. The positive piezoelectric effect is producing charge when the surface of piezoelectric ceramics is subjected to pressure; in the converse piezoelectric effect, piezoelectric material will deform when it was polarized by electric field. Piezoelectric materials can be used as both actuators and sensors because of the positive and converse piezoelectric effects.

In this paper, the diameter and height of the transducer is 20 and 20 mm, respectively. The piezoelectric patch inside the SA is $15 \times 15 \times 1$ mm, as shown in Figure 10.1.

10.2.2 Wavelet Packet-Based Analysis

Wavelet packet analysis had been used in the field of structure health monitoring for many years (Han et al. 2005, Song et al. 2007, Ren et al. 2008, Xu et al. 2017). To evaluate the signal energy values, wavelet packet analysis was applied to calculate the signal energy.

In the proposed method, the voltage signal V_k is decomposed by n-level wavelet packet decomposition into 2^n signal sets, namely

$$V_k = V_{k,1} + V_{k,2} + \cdots + V_{k,2^n-1} + V_{k,2^n} \left(k = 1, 2, \ldots, N_s \right). \tag{10.1}$$

where V_k = the signal of the k-th sensor, $V_{k,i}$ = the time series after wavelet packet decomposition, i = the frequency band ($i = 1, 2, \ldots, 2^n$), and N_s = the number of sensors.

$V_{k,i}$ can be expressed as

$$V_{k,i} = \left[V_{k,i,1}, V_{k,i,2}, V_{k,i,3}, \ldots, V_{k,i,m} \right] \tag{10.2}$$

where m is the number of sampling data.

$$e_{k,i} = \sum_{j=1}^{m} V_{k,i,j}^2. \tag{10.3}$$

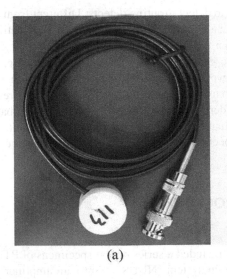

(a)

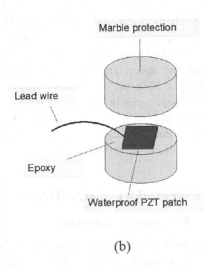

(b)

FIGURE 10.1 The structure of a PZT. (a) PZT with a Bayonet-Neil-Concelman (BNC) connector; (b) schematic.

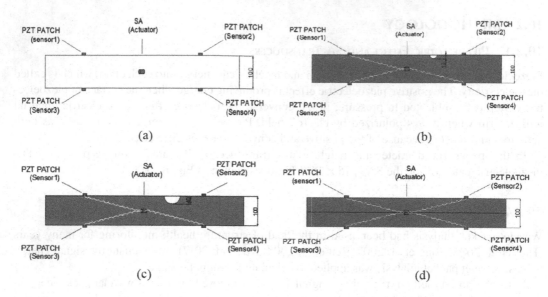

FIGURE 10.2 Test specimen. (a) 0% grouting, (b) a defect with a diameter of 10 mm between actuator and sensor 2, (c) a defect with diameter of 20 mm between actuator and sensor 2, and (d) 100% grouting.

where $e_{k,i}$ is the energy of the i-th frequency band.

$$E_k = \sum_{i=1}^{2^n} e_{k,i}^2 \qquad (10.4)$$

where E_k is the energy of the signal received by the k-th sensor.

10.2.3 DETECTION PRINCIPLE

In this research, active sensing approach was applied to detect grouting defects. Different from the electro-mechanical impedance method (EIM) (Zhou et al. 2018), the active sensing approach utilize the laws of wave propagation to realize defect detection.

As shown in Figure 10.2, in order to simulate the grouting defects in actual engineering, we fabricated four specimens. Specimen 1 reflects the working condition of no grouting. In actual engineering, PTTDs need to bear the impact of concrete in the grouting process, which may result in a certain depression in the surface of the duct. The depression and bubbles in the slurry probably cause cavity inside the tendon duct. Based on this situation, we fabricated defects with diameters of 20 and 40 mm between actuator and sensor 2 on specimens 2 and 3. Specimen 4 reappeared the working condition of full grouting.

10.3 EXPERIMENTAL EQUIPMENT AND PROCEDURES

10.3.1 EXPERIMENTAL EQUIPMENT

As shown in Figure 10.3, the experimental equipment included a series of test specimens of PTTD with different grouting defects, PZTs, a data acquisition system (NI-USB-6366), an amplifier for SA, and a laptop.

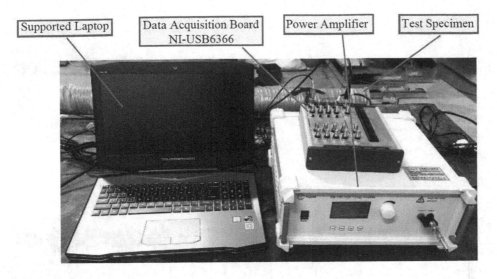

FIGURE 10.3   Test setup.

TABLE 10.1
Parameters of the Excited Swept Sine Wave Signal

Amplitude (V)	Start Frequency (kHz)	Stop Frequency (kHz)	Period (s)
10	100	200	0.5

10.3.2   EXPERIMENTAL PROCEDURE

The specimen was 1,000 mm long, 100 mm in diameter, and 2 mm in thickness. Circular glass was used to seal both ends of the duct, which prevent the grouting materials from losing. A grouting hole and a loophole were fabricated to guarantee the grouting procedure. Four specimens were fabricated for the grouting defects to reappear in actual engineering. The proposed detection methods were applied to detect the specimen, which was maintained for 10 days. The data acquisition board (NI-USB6366) was used to generate a swept sine signal and receive the signal from the sensors. The swept sine signal was amplified 50 times by the power amplifier, and was then applied to motivate the actuator. As shown in Table 10.1, the period and the amplitude of the signal are 0.5 s and 10 V, respectively, the frequency range of the swept sine signal is from 100 to 200 kHz. The sampling frequency of each channel is 2 MHz.

Condition I (0% grouting). The signal emitted by the actuator cannot be received by the sensors.

Condition II (A defect with diameter of 20 mm between actuator and sensor 2). Signal propagation paths are blocked by defects, and signals detected by sensors 1, 3, and 4 have higher energy than those from sensor 2.

Condition III (A defect with a diameter of 40 mm between actuator and sensor 2). Signals are more difficult to be detected by sensor 2, and signal from sensor 2 is weaker than the one of condition II.

Condition IV (Full grouting). Signals detected by each sensor have the strongest signal intensity.

10.4   EXPERIMENTAL RESULTS AND ANALYSIS

10.4.1   TIME-DOMAIN ANALYSIS

Time-domain waveforms in these four working conditions were shown in Figure 10.4. At the 0% grouting condition, due to the existence of no propagation medium, sensor 2 could not receive the stress wave so that all received signals were noise signals. In condition II, signal propagation paths

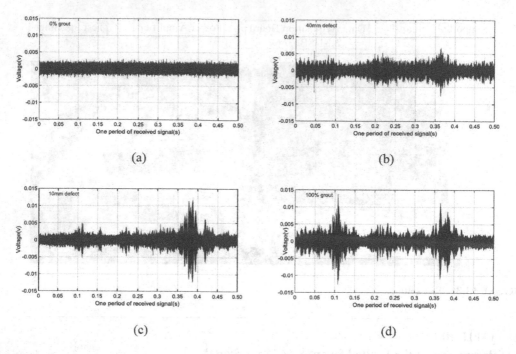

(a)   (b)

(c)   (d)

FIGURE 10.4   Time-domain signal of the PZT 2 sensor in four conditions. Condition I for (a), II for (b), III for (c), and IV for (d).

were blocked by defects, so the signals detected by sensors 1, 3, and 4 had a higher energy than signal from sensor 2. In condition III, the signal detected by sensor 2 is weaker than condition II. In condition IV, signals detected by each sensor have a strong signal intensity.

10.4.2   WAVELET PACKET-BASED ENERGY ANALYSIS

In Figure 10.5, the energy of signals received by sensor 2 under different working conditions was shown. It is clear that the energy value associated with the case of full grout is highest, and the energy associated with the case of 20 mm defect is greater than the case of 40 mm defect, and the

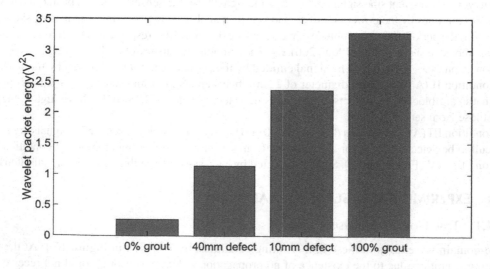

FIGURE 10.5   Energy indices for the four conditions.

condition of 0% grout reports the lowest energy. Compared with the time-domain wave, the wavelet packet-based energy could reflect the change of signal more accurately and sensitively.

10.5 CONCLUSION AND PERSPECTIVE

In practice, grouting compactness problems of PTTD include not only the empty or partially filled tendon duct but also many cavities inside the tendon duct. All those kinds of defects may cause prestressing reinforcement corroding, which will reduce structural service life. In this paper, a stress wave-based active sensing approach using PZTs and wavelet packet energy to detect grouting defects was proposed. A PZT installed on the prestressing reinforcement was used as an actuator to generate stress waves while piezoelectric patches were used as sensors to receive stress waves. The experimental results demonstrated that the grouting defects could be detected and distinguished by comparing the wavelet packet energy of received signals. The defect forms of grouting compactness in the experiment of this study were more comprehensive, and the arrangement of actuator and sensors allowed detecting a wider range with fewer PZTs. Moreover, it's worth noting that the distance between defect and the sensors may have some effect on the energy of the received signal, and defects often filled with water in practice. In the future work, cavities filled with water and the position of defects could be studied.

REFERENCES

Cui, Y., Zou, D.H. 2012. Assessing the effects of insufficient rebar and missing grout in grouted rock bolts using guided ultrasonic waves. *Journal of Applied Geophysics*, **79**, pp. 64–70.

Gu, H.C., Moslehy, Y., Sanders, D., Song, G.B. 2010. Multi-functional smart aggregate-based structural health monitoring of circular reinforced concrete columns subjected to seismic excitations. *Journal of Smart Materials and Structures*, **19**, pp. 2888–2898.

Han, J.G., Ren, W.X., Sun, Z.S. 2005. Wavelet packet based damage identification of beam structures. *Journal of International Journal of Solids and Structures*, **42**, pp. 6610–6627.

Jennifer, E.M. 2008. Detection, localization and characterization of damage in plates with an in situ array of spatially distributed ultrasonic sensors. *Journal of Smart Materials and Structures*, **13**(3), pp. 1–15.

Jiang, T.Y., Kong, Q.Z., Wang, W.X., Huo, L.S., Song, G.B. 2016. Monitoring of grouting compactness in a post-tensioning tendon duct using piezoceramic transducers. *Journal of Sensors*, **16**(8), p. 1343.

Jiang, T.Y., Zheng, J.B., Huo, L.S., Song, G.B. 2017. Finite element analysis of grouting compaceness monitoring in a post-tensioning tendon duct using piezoceramic transducers. *Journal of Sensors*, **17**, p. 2239.

Kawiecki, G. 1998. Feasibility of applying distributed piezotransducers to structural damage detection. *Journal of Intelligent Material System and Structures*, **3**, pp. 189–197.

Li, T., Long, S.G. 2018. Grout assessment of plastic ducts in prestressed structures with an HHT-based method. *Journal of Construction and Building Materials*, **180**(2018), pp. 35–43.

Lu, J.B., Tang, S.H., Dai, X.D., Fang, Z. 2018. Investigation into the effectiveness of ultrasonic tomography for grouting quality evaluation. *KSCE Journal of Civil Engineering*, 22, pp. 1–8.

Olson, L.D., Tinkey, Y., Miller, P. 2011. Concrete bridge condition assessment with impact echo scanning. *Proceeding of ASCE 2011 Geohunan International Conference*, Hunan, China, pp. 59–66.

Ren, W.X., Sun, Z.S., Xia, Y., Hao, H., Deeks, A.J. 2008. Damage identification of shear connectors with wavelet packet energy: Laboratory test study. *Journal of Structural engineering ASCE*, **134**(5), p. 832–841.

Song, G.B., Gu, H.C., Mo, Y.L. 2008. Smart aggregates: Multi-functional sensors for concrete structures-a tutorial and a review. *Journal of Smart Materials and Structures*, **17**(3), pp. 1–17.

Song, G.B. Gu, H.C., Mo, Y.L., Hsu, T.T.C., Dhonde, H. 2007. Concrete structural health monitoring using embedded piezoceramic transducers. *Journal of Smart Materials and Structure*, **16**, pp. 959–968.

Song, G.B., Gu, H.C., Mo, Y. L., Hsu, T.T.C., Dhonde, H. 2006. Concrete structural health monitoring using embedded piezoceramic transducers. *Journal of Smart Materials and Structures*, **15**(2), pp. 1837–1845.

Terzioglu, T., Karthik, M.M., Hurlebaus, S., Hueste, M.B.D., Maack, S., Woestmann, J., Wiggenhauser, H., Krause, M., Miller, P.K., Olson, L.D. 2018. Nondestructive evaluation of grout defects in internal tendons of posttensioned girders. *Journal of NDT and E International*, **99**, pp. 23–35.

Tian, Z., Huo, L.S., Gao, W.H., Song, G.B., Li, H.N. 2017. Grouting monitoring of post-tensioning tendon duct using PZT enables time-reversal method. *Journal of Measurement*, **122**, pp. 513–512.

Wu, Z.J., Qing, X.P. 2009. Damage detection for composite laminate plates with a distributed hybrid PZT/ FBG sensors network. *Journal of Intelligent Material Systems and Structures*, **20**, pp. 1069–1077.

Xu, B., Chen, H.B., Xia, S. 2017. Numerical study on the mechanism of active interfacial debonding detection for rectangular CFSTs based on wavelet packet analysis with piezoceramics. *Journal of Mechanical System and Signal Processing*, **86**, pp. 108–121.

Xu, K., Deng, Q.S., Cai, L.J., Ho, S.C., Song, G.B. 2018. Damage detection of a concrete column subjected to blast loads using embedded piezoceramic transducers. *Journal of Sensors*, **18**, p. 1377.

Zou, D.H., Cheng, J.L., Yue, R.J., Sun, X.Y. 2010. Grout quality and its impact on guided ultrasonic waves in grouted rock bolts. *Journal of Applied Geophysics*, **72**, pp. 102–106.

Zou, D.H., Cui, Y. 2011. A new approach for field instrumentation in grouted rock bolt monitoring using guided ultrasonic waves. *Journal of Applied Geophysics*, **75**, pp. 506–512.

Zhang, J., Li, Y., Du, G.F., Song, G.B. 2018. Damage detection of L-shaped concrete filled steel tube (L-CFST) columns under cyclic loading using embedded piezoceramic transducers. *Journal of Sensors*, **18**, p. 2171.

Zhou, P., Wang, D.S., Zhu, H.P. 2018. A novel damage indicator based on the electro-mechanical impedance principle for structural damage identification. *Journal of Sensors*, 18, p. 2199.

11 The Analysis of the Temperature Effect on Frequencies of a Footbridge

De-hui Tang, Jun Teng, and Wei-hua Hu
Harbin Institute of Technology (Shenzhen)

CONTENTS

11.1 INTRODUCTION

The evaluation of structural performance based on the variation of dynamic characteristics is widely applied in structural health monitoring. Among the characteristics, the natural frequencies of structure are easy to obtain with high accuracy from the ambient vibrations. However, the changes of frequencies result from the early-stage damages are reported to be covered by environmental and operational effect such as temperature, wind, and the traffic load.

Kim [1] found that the effect of traffic load on modal frequencies of a small-span bridge reached 5.4%, but they believed that the effect of traffic load on large-span bridge could be ignored. Siringoringo [2] discovered that during the typhoon, the frequencies of low-order mode decrease with the increase of wind and the maximum variation of frequencies reached 4.6%. Xia [3] found that the frequency of a two-span concrete slab drops by 0.23% for every 1°C rise in temperature. Under general operating conditions, the influence of temperature on frequency is far greater than that of wind and traffic, and in theory, the effect of temperature on modal frequency should be linear.

However, in most bridge monitoring cases, there is a non-linear negative correlation between temperature and bridge frequency [4,5]. And meanwhile, a small number of bridges are found and that there is a positive correlation between temperature and frequencies [6]. Therefore, it is meaningful to investigate the mechanism of temperature effect on modal frequencies.

In general, the changes of elastic modulus of concrete or steel caused by temperature are believed as the dominated factor that influence frequencies [7]. However, it could not explain the positive correlation between temperature and frequency in one day during the summer observed by Mosavi [6].

In this paper, elastic modulus, boundary condition, axial force, and the temperature gradient of bridge are considered as four main variables that lead to frequency changes caused by temperature. Numerical simulation and long-term monitoring of the Rainbow Bridge are used to reveal the mechanism of Temperature Effect on Frequency.

11.2 QUANTITATIVE ANALYSIS OF THE EFFECT OF TEMPERATURE ON FREQUENCY

11.2.1 ELASTIC MODULUS

The frequencies of bending mode of the simply supported beam are deduced by Clough [8]:

$$\omega_n = \frac{n\pi}{l^2}\sqrt{\frac{EI}{\bar{m}}} \quad (n = 1,2,...) \tag{11.1}$$

where ω_n is the frequency of the nth bending mode. l, EI, $\bar{m}$ are the length of the beam, bending stiffness, and unit length mass, respectively.

Xia et al. [7] give the simplified quantitative formula of frequency change caused by temperature:

$$\frac{\delta\omega_n}{\omega_n} = \frac{1}{2}(\theta_\alpha + \theta_E)\delta T \tag{11.2}$$

where θ_α and θ_E are the linear expansion coefficient and rate of change of elastic modulus with temperature of material, and T represents the temperature. For the reinforced concrete, the θ_α is far less than the θ_E. However, the rate of change of various modal frequencies with temperature is different. Commonly, the frequency of higher order modes has a higher rate of change with temperature which could not be explained by Eq 11.2, for θ_E is a constant.

It is because most of the bridges could be simplified as a multi-span elastic-support beams. When the stiffness of the elastic supports is not large enough, the θ_E turns to a function about n. The details would be stated in the following subsection.

11.2.2 BOUNDARY CONDITION

As shown in Figure 11.1, the vertical stiffness of the elastic supports X_1, X_2, ..., X_k is S_1, S_2, ..., S_k. Ignoring the shear deformation, the un-damped free vibration could be expressed as

$$EI\frac{\partial^4 Y_i(x,t)}{\partial x^4} + \bar{m}_i\frac{\partial^2 Y_i(x,t)}{\partial t^2} = 0 \tag{11.3}$$

where $Y_i(x,t)$ represents the vertical displacement of the ith beam segment and the $X_{i-1} < x < X_{i-1}$.

Partial differential equations are solved by the separation variable method. Let the

$$Y_i(x,t) = \varphi_i(x)q_i(t) \tag{11.4}$$

where the $\varphi_i(x)$ is the modal function of the ith beam segment, and $q_i(t)$ represents the generalized coordinates of the beam.,

Substitute Eq. (11.2) into Eq. (11.1) gives

$$\varphi_i(x) = A_i\cos\left(\lambda_i\,x\right) + B_i\sin\left(\lambda_i\,x\right) + C_i\cosh\left(\lambda_i\,x\right) + D_i\sinh\left(\lambda_i x\right)$$

$$\lambda_i = \sqrt[4]{\omega^2\frac{\bar{m}_i}{EI_i}} \tag{11.5}$$

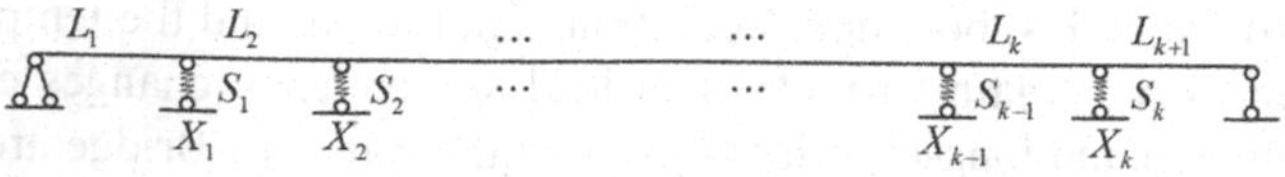

FIGURE 11.1   Multi-span elastic support beam model with k elastic supports.

where A_i, B_i, C_i, and D_i are the real constants of the ith beam segment. The $Y_i(x,t)$ can then be expressed as

$$Y_i(x,t) = \left[A_i\cos(\eta_i\lambda_i\,x) + B_i\sin(\eta_i\lambda_l\,x) + C_i\cos(\eta_i\lambda_l\,x) + D_i\sin(\eta_i\lambda_l\,x) \right] q_i(t)$$

$$\lambda_l = \sqrt[4]{\omega^2\frac{\overline{m}_l}{EI_l}}, \eta_i = \frac{\lambda_i}{\lambda_l} = \sqrt[4]{\frac{\overline{m}_iEI_l}{\overline{m}_lEI_i}}.$$

(11.6)

Considering the boundary conditions of beams, the analytical solutions of partial differential equations can be obtained.

$$S_i \to 0, \omega_n = \left(\frac{n\pi}{l}\right)^2\sqrt{\frac{EI}{\overline{m}}}$$

$$S_i \to +\infty, \omega_n = \left(\frac{\gamma_n\pi}{l}\right)^2\sqrt{\frac{EI}{\overline{m}}}$$

(11.7)

$$\text{Otherwise, } \omega_n = \left(\frac{\varphi_n\pi}{l}\right)^2\sqrt{\frac{EI}{\overline{m}}}$$

where γ_n is a dimensionless constant of the nth mode, and φ_n is an implicit function about $\frac{S_i}{EI}$.

Therefore, Eq. (11.2) could change to

$$\frac{\delta\omega_n}{\omega_n} = \frac{1}{2}(\theta_\alpha(1-2\phi_\alpha) + (1-2\phi_E)\theta_E)\delta T$$

(11.8)

where the ϕ_α and ϕ_E are dimensionless constant. And when $S_i \to 0$ or $S_i \to +\infty$ these two constants equal to 0.

11.2.3 AXIAL FORCE

Generally, the elongation of bridges under temperature rise can be adequately released by setting the expansion joints. However, in fact, the axial force caused by temperature may exist in bridges for the friction between the abutment and the box girders.

For multi-span elastically supported beams with axial forces, as shown in Figure 11.2, the differential equation of vibration can be expressed as

$$EI\frac{\partial^4 Y_i(x,t)}{\partial x^4} + N\frac{\partial^2 Y_i(x,t)}{\partial x^2} + \overline{m}_i\frac{\partial^2 Y_i(x,t)}{\partial t^2} = \sum_{i=1}^{k} S_i\,\delta(x - X_i, t)$$

(11.9)

where $\delta(X_i, t)$ represents the displacement of the ith support.

The solution of the frequencies could be

$$\omega_m = \left(\frac{mK\pi}{l}\right)^2\sqrt{\frac{EI}{\overline{m}}}\left(1 - \frac{NI^2}{(mK\pi)^2\,EI}\right)(K = 1, 2, \ldots)$$

(11.10)

for a simply supported beams, $K = 1$.

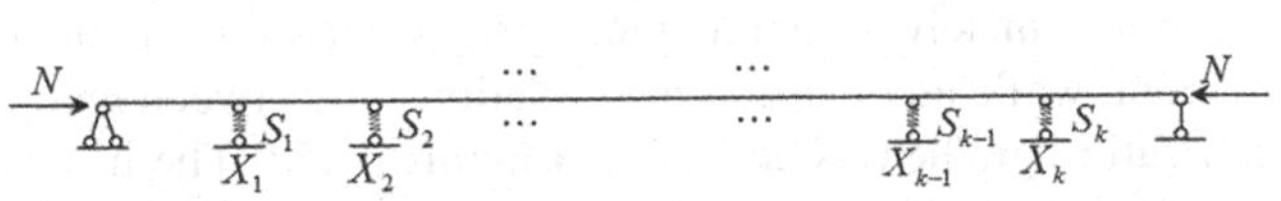

FIGURE 11.2 Multi-span elastic support beam model with axial force.

(a)   (b)

FIGURE 11.3   Front and lateral views of the Rainbow Bridge: (a) Front and (b) lateral view.

11.3   THE CONTINUOUS DYNAMIC MONITORING SYSTEM IN A FOOTBRIDGE

The Rainbow Bridge (Figure 11.3) over the Dasha River in Shenzhen is located between the Peking University Shenzhen Graduate School and the gymnasium of University Town. The arch bridge is composed of three main parts: two arches with an inclination angle of 14.4°, 34 pre-stressed cables, and a deck supported by two box girders.

Figure 11.4a shows the continuous dynamic system providing high-quality data. Acceleration signals in both the deck and arch are continuously acquired with a sampling ratio of 2,000 Hz, and then are connected to a low-pass anti-aliasing filter of 10 Hz, as well as being down-sampled to 20 Hz by the spline interpolation. The cable acceleration signal is applied with a low-pass filter of 25 Hz, and then down-sampled to 50 Hz. The eight K-type thermocouples are mounted at the corresponding position in Figure 11.4b to monitor the variation of temperature field of the bridge.

11.4   THE ANALYSIS OF THE EFFECT OF TEMPERATURE ON FREQUENCY IN A FOOTBRIDGE

Parts of modes of the Rainbow Bridge are found to be dominated by arch while the others are coupled by the deck and arch. In order to investigate the effect of temperature on frequencies of physical mode, the automatic operational modal analysis based on the covariance driven stochastic subspace identification is applied on the acceleration signals from the deck and arch [9]. Meanwhile, it is found that the coupled modes and arch dominated modes have an opposite trend with the variation of temperature.

As shown in Figure 11.5a, the points are divided into two parts: the blue one (points to a low temperature gradient) and the red one (points to a high temperature gradient).

It could be seen that the surface temperature has an obvious non-linear positive correlation with the frequency of coupled modes (third-order vertical bending mode of deck and third-order transversal bending mode of arch) in high temperature gradient (greater than 2°C). However, the surface temperature has a negative relationship with the frequency of coupled modes in low temperature gradient.

From Figure 11.5b, there is no obvious difference between the relationship of surface temperature and frequency of arch dominated mode in high and low temperature gradients. As shown in Figure 11.5c, there is an approximately linear relationship between surface temperature and frequency in low temperature gradient, while an obvious linear relationship between temperature gradient and frequency in high temperature gradient is observed in Figure 11.5d. The positive correlation effect of temperature gradient on frequency (the slope of fitting curve is 0.88E-2) is much larger than the negative correlation effect of overall temperature rise (the slope of fitting curve is 0.916E-3).

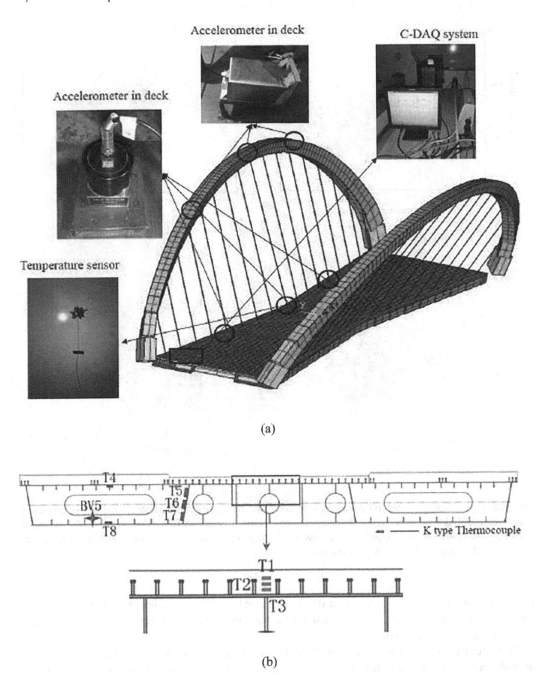

FIGURE 11.4 The continuous dynamic system: (a) Structural health monitoring system; (b) the layout of the thermocouples.

The possible reason for the above phenomenon is that when the temperature gradient is relatively small (lower than 2°C), the rise or fall of temperature can be regarded as the overall temperature rise or fall. In this case, the change of elastic modulus caused by temperature dominates the variation of frequency which could be seen in Figure 11.5c. However, when the temperature gradient is large, the smaller axial elongation may be limited by the bearing friction that would result in axial force. Under the combined action of axial force and the rise of deck caused by temperature gradient,

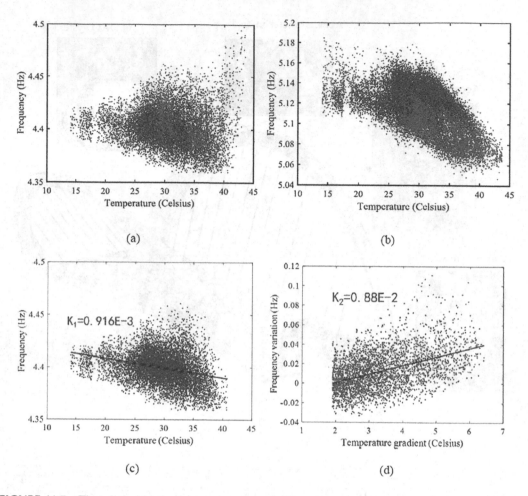

FIGURE 11.5 The relationship between temperature and frequency: (a) Temperature and the frequency of deck-arch coupled dominated mode; (b) temperature and the frequency of arch dominated mode; (c) the relationship of temperature and frequency with low temperature gradient; and (d) the relationship of temperature gradient and frequency with high temperature gradient.

the second-order effect would lead to a stiffer deck and higher frequencies. As for the arch, the ends are completely fixed by the piles foundation; therefore, the rising axial force caused by temperature lead to lower frequencies (Eq. 11.10).

To better illustrate this phenomenon, the daily and annual frequencies of the coupled modes and arch dominated modes are shown in Figure 11.6. From Figure 11.6c, d, as the temperature goes down, all the frequencies decrease. The negative relationship between frequency and temperature in 10 days is observed in a coupled mode (Figure 11.6a) while the positive one is found in the arch dominated mode (Figure 11.6b).

It should be noticed that the daily frequency has a better correlation with the temperature gradient than with the surface temperature.

11.5 CONCLUSION

In this paper, the frequency expression of a multi-span continuous beam is derived, and the long-term frequency and temperature fields of the Rainbow Bridge are analyzed.

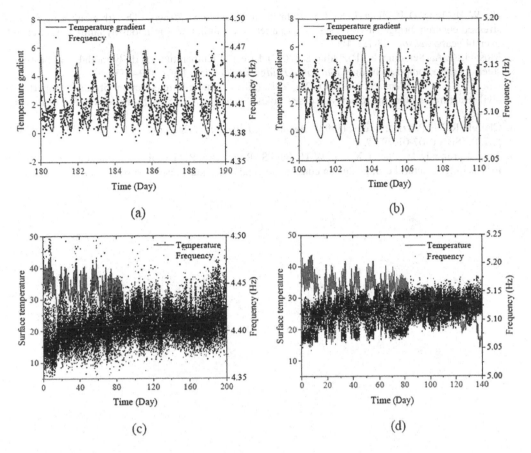

FIGURE 11.6 Daily and annual frequencies of the coupled and arch dominated modes: (a) Temperature gradient and the deck-arch coupled dominated mode; (b) temperature gradient and the arch dominated mode; (c) surface temperature and the deck-arch coupled dominated mode; and (d) surface gradient and the arch dominated mode.

The long-term monitoring results show that

1. The annual frequency change of bridge deck and arch is mainly dominated by elastic modulus;
2. The daily frequency change of the bridge deck is mainly dominated by the temperature gradient, which may result from the second-order effect caused by the axial force and the rise of the bridge deck.

REFERENCES

1. Kim C., Jung D., Kim N., et al. (2003). Effect of vehicle weight on natural frequencies of bridges measured from traffic-induced vibration. *Earthquake Engineering and Engineering Vibration*, 2(1), 109–115.
2. Siringoringo, D. M., Fujino, Y. (2008). System identification of suspension bridge from ambient vibration response. *Engineering Structures*, 30(2), 462–477.
3. Xia, Y., Hao, H., Zanardo, G., Deeks, A. (2006). Long term vibration monitoring of an RC slab: temperature and humidity effect. *Engineering Structures*, 28(3), 441–452.
4. Magalhaes, F., Cunha, A., Caetano, E. (2012). Vibration based structural health monitoring of an arch bridge: from automated OMA to damage detection. *Mechanical Systems and Signal Processing*, 28, 212–228.

5. Hu, W. H., Tang, D. H., Teng, J., Said, S., Rohrmann, R. G. (2018). Structural health monitoring of a pre-stressed concrete bridge based on statistical pattern recognition of continuous dynamic measurements over 14 years. *Sensors*, 18(12), 4117.
6. Mosavi, A. A., Seracino, R., Rizkalla, S. (2012). Effect of temperature on daily modal variability of a steel-concrete composite bridge. *Journal of Bridge Engineering*, 17(6), 979–983.
7. Xia, Y., Chen, B., Weng, S., Ni, Y. Q., Xu, Y. L. (2012). Temperature effect on vibration properties of civil structures: a literature review and case studies. *Journal of Civil Structural Health Monitoring*, 2(1), 29–46.
8. Clough, R. W., Penzien, J., eds. (1993). *Dynamics of structures*, 2nd edition, McGraw-Hill, New York, p. 738. ISBN 0-07-011394-7.
9. Teng, J., Tang, D. H., Zhang, X., Hu, W. H., Said S., Rohrmann, R. G. (2019). Automated modal analysis for tracking structural change during construction and operation phases. *Sensors*, 19(4), 9.

12 Numerical Simulation of Precast Concrete Structure with Cast-In-Situ Monolithic Joint

Du Yongfeng, Chi Peihong, and Fang Dengjia
Lanzhou University of Technology

CONTENTS

12.1 INTRODUCTION

Precast concrete structure promotes the development of building industrialization, but its seismic performance has attracted wide attention (Wu and Feng, 2018; Yang et al., 2018). At present, the research on seismic performance of precast concrete structures at home and abroad mainly focuses on experimental research (Ghosh and Cleland, 2012; Korkmaz and Tankut, 2005). By designing different specimens, the effects of connection modes and geometric parameters on the bearing capacity, ductility, and energy dissipation capacity of joints are studied, and the corresponding calculation theory and design method are established. However, due to the limitations of experimental conditions, funds, and other factors, this method is not suitable for large-scale deployment. With the development of computer technology and modern numerical methods, combined with finite element method, seismic disaster simulation of precast concrete structures is carried out, parametric analysis is carried out, and damage evolution law is studied. Structural system with good seismic performance is proposed and practical engineering design is guided, which is the future development trend. At present, most of the numerical simulation of seismic performance of precast concrete structures still use the analysis method of cast-in-place concrete structures. According to the different modeling ideas, it can be divided into the analysis method based on beam-column element and the analysis method based on three-dimensional solid element (Li and Wang, 2014). Based on the Beam-Column Joint Element two-dimensional macro-node model in the finite element software OpenSees, Tan et al. (2015) simulated the precast joints. Hawileh, Rahman, and Tabatabai (2010) established the nonlinear finite element model of prefabricated beam-column joints using ANSYS, and simulated the performance of joints under cyclic loads. The beam column uses solid elements, and the contact elements are used to simulate the interface between the joints. Fan and Lu (2011) used the finite element software Strand7 to establish a two-span assembled frame model and carried out dynamic response analysis. In this paper, the mechanical mechanism of the interface of the precast structure with integral post-pouring connection is analyzed. The finite element software ABAQUS is used to simulate the mechanical properties of the post-pouring column, and the simulation results are verified by model tests.

12.2 RESEARCH OBJECT

The research object of this paper is the precast column with post-pouring connection. The section size of the column is 200×200 mm and the bottom beam is 200×300 mm, the stirrup is HPB300, the longitudinal reinforcement is HRB400, and the section reinforcement of the member is shown in Figure 12.1.

12.3 NUMERICAL SIMULATION OF AN INTERFACE MODEL OF POST-POURING ZONE OF PRECAST MEMBERS

12.3.1 MECHANISMS OF INTERFACE STRESS IN POST-POURING ZONE OF PREFABRICATED COMPONENTS

Concrete is a kind of artificial mixing material which is made of cement as cementitious material, water, sand, and stone; chemical admixtures; and mineral admixtures when necessary, and mixes evenly according to appropriate mix proportion. Because in the process of pouring

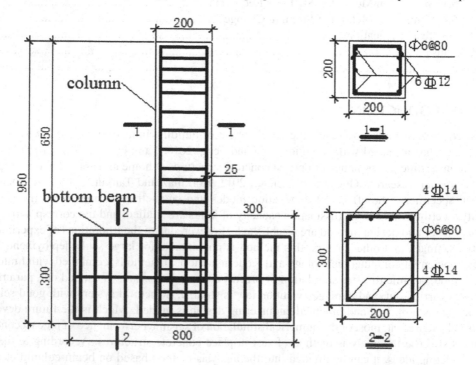

FIGURE 12.1 Section size and reinforcement diagram of components.

and compacting concrete, the coarse aggregate with larger mass in concrete will sink due to self-weight and bubbles will rise, so the surface of concrete will concentrate a large number of bubbles, the coarse aggregate with skeleton function will be far away from the upper surface, and the surface of component will form a weak surface. Because the post-poured concrete is poured on this weak surface and the hydration of cement between the new and old concrete is not obvious, the bond force between the precast components and the interface of the post-poured area is weak.

The aggregate biting effect of concrete is that the coarse aggregate in concrete is encapsulated by a cement paste and hydrated to form a hard cement stone. However, the interface between precast components and post-poured concrete is easy to produce voids, the interface is dry, and impurities will weaken the biting effect of aggregate, which will inevitably lead to its shear strength lower than that of the whole pouring structure.

This paper considers that the interface connection between concrete and prefabricated parts in post-pouring area is actually the problem of the bonding force between new and old concrete. The bonding force between new and old concrete at the interface is weakened, and the biting effect of aggregate is reduced, thus forming a weak layer with a smaller thickness.

12.3.2 INTERFACE MODEL OF POST-POURING ZONE OF PREFABRICATED COMPONENTS

In order to simulate the weakening effect, the general performance of the assembled frame is lower than that of the cast-in-place frame by using the joint connection method of post-pouring. At present, the commonly used simulation method of the precast structure is to reduce the bearing capacity or stiffness of the whole cast-in-place structure. This simulation method can make the simulation data agree well with the test results, but it cannot reflect the difference of the failure mode, stress, and strain between the two structures (Yang, 2016).Considering the sleeve connection of reinforcing bars in post-pouring area, which is mature in technology and continuous in reinforcing bars, the weak interface connection between post-pouring area and precast members is the main difference between precast and cast-in-situ frame. In order to better simulate the characteristics of precast assembly structure, considering the mechanism of interface stress, a layer of strength is set between precast members and concrete post-pouring area. The simulation of the lower concrete layer is shown in Figure 12.2.

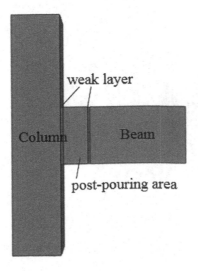

FIGURE 12.2   Schematic simulation of post-pouring area.

12.4   FINITE ELEMENT SIMULATION

12.4.1   MATERIAL CONSTITUTION

ABAQUS provides three concrete constitutive models: dispersion crack model, brittle crack model, and damage-plasticity model. In this paper, the damage-plasticity model is used to simulate the performance of concrete under low-cycle reciprocating loads. Fang et al. (2013) aiming at choosing the stress-strain relationship of concrete when ABAQUS is used to analyze reinforced concrete structure proposed some suggestions, they are used to calculate the constitutive properties of concrete in this paper.

The simulation of hysteretic behavior of reinforcing bar provided by ABAQUS finite element platform belongs to a pure reinforcing bar hysteretic constitutive model without considering the slip of the reinforcing bar. In order to better simulate the mechanical behavior of joint under reciprocating load, Clough reinforcing bar model (Clough, 1996is used to develop the secondary application program of this material model in ABAQUS.

12.4.2   ESTABLISHMENT OF FINITE ELEMENT MODEL

Combining with the mechanical characteristics of the research object, concrete adopts C3D8R solid element, and longitudinal reinforcement and stirrup adopt three-dimensional truss linear element T3D2. The interaction between components of post-poured column is simulated by setting a contact relationship. Embedded region is used between reinforcement cage and concrete components, Tie is used between concrete components, and the weak layer of post-poured connection of prefabricated members is simulated by reducing the strength or stiffness of concrete. The meshing of the finite element model is shown in Figure 12.3.

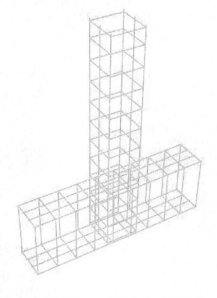

(a) Concrete mesh generation   (b) Reinforcement mesh generation

FIGURE 12.3   Precast column mesh generation.

12.5 ANALYSIS OF MODEL RESULTS

12.5.1 SIMULATION METHOD OF STIFFNESS REDUCTION

By reducing the stiffness of weak layer concrete, the precast column is numerically simulated, mainly through the method of "stiffness weakening." The main factors affecting the stiffness include the elastic modulus of materials and the section form of components. Considering that the section form does not change, the stiffness change of the post-pouring area is simulated by changing the elastic modulus of the weak layer of the precast column. Taking the elastic modulus of weak layer concrete as $0.5E$, $0.6E$, $0.7E$, $0.8E$, $0.9E$, and $1.0E$, six models are established for comparative analysis. Figure 12.4 shows the hysteretic curves of precast columns considering different stiffness weakening.

Through the comparison of Figure 12.4, it can be found that the hysteretic curve of precast column with the stiffness weakening at the joints is basically the same as that of the cast-in-place structure. The elastic modulus increases gradually from $0.5E$ to $1.0E$, but the difference is not significant. When the modulus of elasticity is $0.5E$, the hysteretic curve is irregular and the stiffness degradation is obvious in the later stage of displacement loading. Therefore, when stiffness weakening simulation is used, the stiffness should be above $0.6EI$.

12.5.2 SIMULATION METHOD OF STRENGTH CHANGE

By reducing the strength of weak layer concrete, the numerical simulation of precast columns is carried out. The bond strength of new and old concrete has a great influence on the mechanical performance of precast concrete joints. Because of the existence of bond surface, the strength of precast concrete joints is reduced (Zhang, 2015). The ultimate compressive strength of concrete is assumed to be the compressive strength of concrete, and the bearing capacity remains unchanged after reaching the ultimate strength. The ultimate tensile strength of concrete is reduced to reach the ultimate tensile strength after reduction, and the bearing capacity rapidly decreases to zero. The ultimate tensile strength of a weak layer of concrete is 0.5 f_t, 0.6 f_t, 0.7 f_t, 0.8 f_t, 0.9 f_t, and 1.0

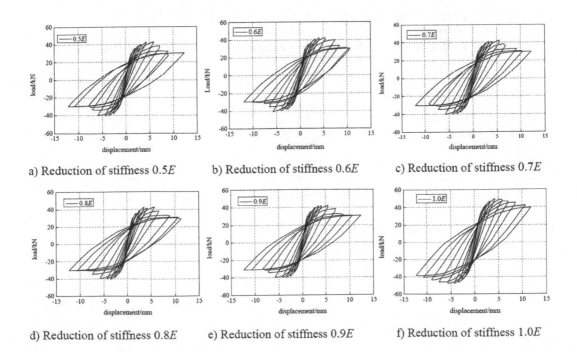

a) Reduction of stiffness $0.5E$ b) Reduction of stiffness $0.6E$ c) Reduction of stiffness $0.7E$

d) Reduction of stiffness $0.8E$ e) Reduction of stiffness $0.9E$ f) Reduction of stiffness $1.0E$

FIGURE 12.4 Hysteretic curve comparison.

f_t. Six models are established for comparative analysis. Figure 12.5 shows the hysteretic curve of precast column considering different strength reduction.

The precast column is composed of prefabricated members and post-pouring zone. The bond strength of old and new concrete has a great influence on the mechanical properties of the assembled concrete joints. Through the comparison of Figure 12.5, it can be seen that the tensile strength has a great influence on the bearing capacity of the component. When the tensile strength is less than 0.8, the bearing capacity of the component is obviously worse than that of the cast-in-place component.

12.5.3 Test Verification

In order to verify the rationality of finite element simulation, the simulation results of concrete tensile strength of 0.8 are compared with the test results. As shown in Figure 12.6, cracks occur at the interface between the bottom of the column and the bottom beam, horizontal cracks occur at the bottom of the column to the top of the column at varying degrees, and serious cracks occur at the post-pouring interface. The location and extent of plastic damage in the model simulated by finite element method are in good agreement with the experimental results. The comparison of load-displacement curves between test and simulation is shown in Figure 12.7. From the comparison of bearing capacity, the solid model reduces by 0.51% compared with the test value of the maximum positive bearing capacity of the precast column, and the solid model reduces by 2.91% compared with the test value of the maximum negative bearing capacity of the precast column. The numerical simulation results are close to the test value, and the hysteretic curve is in good agreement.

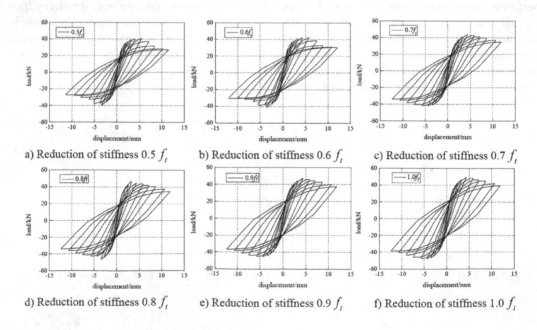

a) Reduction of stiffness 0.5 f_t b) Reduction of stiffness 0.6 f_t c) Reduction of stiffness 0.7 f_t

d) Reduction of stiffness 0.8 f_t e) Reduction of stiffness 0.9 f_t f) Reduction of stiffness 1.0 f_t

FIGURE 12.5   Hysteretic curve comparison.

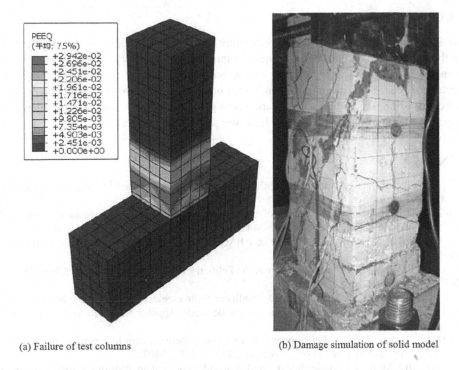

(a) Failure of test columns   (b) Damage simulation of solid model

FIGURE 12.6   Failure location and range of post-poured connection column.

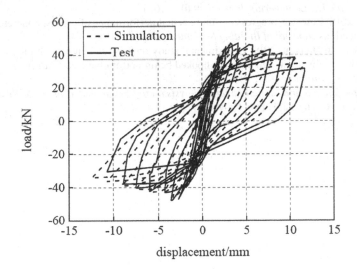

FIGURE 12.7   Hysteretic curve.

12.6   CONCLUSION

It is feasible to simulate the precast concrete structure by setting a weak layer with a smaller thickness between the post-pouring area and the connecting interface of prefabricated members. The weakening of concrete stiffness has less influence on the bearing capacity of members, and the reduction of concrete tensile strength has a greater influence on the bearing capacity of members. When the tensile strength is 0.8, the numerical simulation is in good agreement with the experimental results. The hysteresis curves of the specimens are basically similar, and the failure modes are the same.

REFERENCES

Clough R. W. 1996. *Effect of Stiffness Degradation on Earthquake Ductility Requirements*. Berkeley: University of California.

Fan L., Lu X. 2011. Seismic response analysis of a jointed precast concrete frame structure, *Journal of China University Mining & Technology*, 40(04): 544–548+555.

Fang Z., Zhou H., Lai S., et al. 2013. Choose of ABAQUS stress-strain curve, *Building Structure*, 43(S2): 559–561.

Ghosh S. K., Cleland N. 2012. Observations from the February 27, 2010, earthquake in Chile, *PCI Journal*, 57(01): 52–75.

Hawileh R. A., Rahman A., Tabatabai H. 2010. Nonlinear finite element analysis and modeling of a precast hybrid beam-column connection subjected to cyclic loads, *Applied Mathematical Modelling*, 34(09): 2562–2583.

Korkmaz H. H., Tankut T. 2005. Performance of a precast concrete beam-to-beam connection subject to reversed cyclic loading, *Engineering Structures*, 27(09): 1392–1407.

Li H., Wang D. 2014. Multi-scale finite element modeling and numerical analysis of reinforced concrete structure, *Journal of Architecture and Civil Engineering*, 31(2): 20–25.

Tan P., Li Y. Kuang Z., et al. 2015. Seismic behavior of isolation connection in assembled seismic isolation structure, *China Civil Engineering Journal*, 48(02): 10–17.

Wu G., Feng D.-C. 2018. Research progress on fundamental performance of precast concrete frame beam-to-column connections, *Journal of Building Structures*, 39(02): 1–16.

Yang H., Guo Z.-X., Yin H., et al. 2018. Experimental study on seismic behavior of precast concrete beam-to-column connections with high-strength hooked beam bottom bars, *Journal of Southeast University (Natural Science Edition)*, 48(06): 979–986.

Zhang R. 2015. *Experimental and Finite Element Analysis on the Assembled Monolithic Reinforced Concrete Frame Structure*. Changsha: Hunan University.

13 Simulation Analysis of a Bridge with a Nonlinear Tuned Mass Damper Using Incremental Harmonic Balance Method

Feng Gu and Chiu Jen Ku
Shantou University

CONTENTS

13.1 GENERAL INSTRUCTIONS

Recently, many scholars have studied the dynamic response of bridges under moving loads. Zhang et al. (2010) proposed an improved method for considering the energy dissipation of material damping when studying the dynamic response of simply supported beams under moving loads, and the influence of speed is analyzed. The results of Newmark-β numerical calculations show the variation of the deflection of simply supported beams with the moving forces under variable speed or uniform motion loads. Zhang et al. (2013) established a dynamic response analysis model of simply supported bridges under a series of harmonic loads, and derived the dynamic response solution based on modal superposition method. The calculation results show that resonance occurs when the harmonic frequency is close to the fundamental frequency of the simply supported beam. Museros et al. (2013) studied the free vibration of a simply supported beam bridge under moving loads. The maximum resonance, elimination, and resonance vertical acceleration problems were discussed by numerical calculation methods. Installing a nonlinear Tuned Mass Damper (TMD) on a bridge is a way to weaken the vibration of the bridge, and relevant scholars have also studied this. Most of the researches use the method of numerical simulation. When dealing with nonlinear problems, the calculation speed of numerical simulation is slow and the calculation accuracy is difficult to control. What's more, only discrete points can be got and can't provide complete analytical solutions. However, the incremental harmonic balance method (IHBM) can effectively solve these shortcomings.

The IHBM originally proposed by Lau and Cheung (1981) can be used with various approximate procedures such as the Rayleigh-Ritz method or the finite element method to obtain the incremental solution for nonlinear vibrations of elastic systems. Lau and Cheung (1984, 1986), and Lau et al. (1984) applied the IHBM to nonlinear periodic vibrations of thin elastic plates and shallow shells. Chen et al. (2001) combined the IHBM with the finite element method to analyze a wide class of plane structures including beams, shallow arches, and frames. In the field of bridge vibration control, Sze et al. (2005) used the IHBM to analyze the vertical vibration of a beam subjected to axial moving forces, in which the force responses of the internal resonance between the first two modes were studied. To further consider the stability of the periodic solution, Huang et al. (2011) applied the IHBM to analyze nonlinear vertical vibration of a beam subject to periodic axial moving forces. It was shown that the periodic solution obtained from the IHBM are in good agreement with the results obtained from numerical integration. Pirmoradian et al. (2015) used the IHBM to investigate the dynamic stability and resonance conditions of a Timoshenko beam excited by a sequence of identical moving masses. Zhou et al. (2016) analyzed a dynamic model of the Euler-Bernoulli beam with an intermediate nonlinear support under concentrated harmonic excitation by using the multi-dimensional IHBM.

In this paper, the model of a bridge subjected to periodic moving forces installed with a nonlinear TMD is established. The IHBM is implemented to analyze the nonlinear behaviors with the influences of the magnitude and velocity of the periodic moving forces, mass rate (the ratio of TMD mass to bridge mass), nonlinear stiffness, damping, and location of nonlinear TMD. Some interesting nonlinear phenomena have also been observed.

13.2 MODELING OF THE NONLINEAR TMD BRIDGE UNDER PERIODIC MOVING FORCES

The bridge is model as a Euler-Bernoulli elastic beam, which is installed with a nonlinear TMD, as shown in Figure 13.1. The bridge is modeled as a simple supported beam. The TMD installed in the bridge has a nonlinear spring component. In Figure 13.1, X and Y represent the longitudinal and lateral coordinates of the beam, respectively. The length of the bridge is L, and A represents the cross-sectional area of the beam. The mass per unit length of the whole beam is $m = \rho \times A$, E is the elastic modulus of the beam, I is the moment of inertia of the beam section, and ρ is the mass density of the beam. F is the periodic moving forces with a velocity of V. When one force reaches the end point, another force of the same magnitude starts from the starting point with the same velocity forming periodic forces. X_0 represents the locations of the nonlinear TMD. K_i and C represent the stiffness and damping of the nonlinear TMD spring and damper, respectively. K_b and C_b represent the stiffness and damping of the bridge, respectively. Y_b is the y-direction displacement of the bridge, and Y_l is the y-direction displacement of the nonlinear TMD.

According to the definition in the literature (Alexander and Schilder 2009), the nonlinear spring forces and damping forces are expressed respectively as follows:

$$f_k = K_1 Y + K_2 Y^3, \quad f_c = C_1 Y \tag{13.1}$$

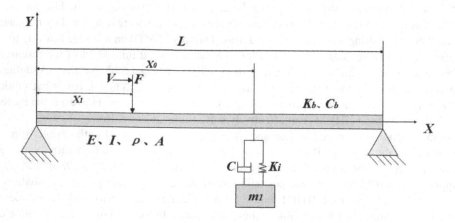

FIGURE 13.1 A nonlinear TMD bridge subjected to periodic moving forces.

The kinetic energy, potential energy, and dissipated energy of the nonlinear TMD bridge are expressed respectively as

$$T = \frac{1}{2}\rho A \int_0^L \left(\frac{\partial Y_b}{\partial t}\right)^2 dX + \frac{1}{2}m_1\left(\frac{\partial Y_1}{\partial t}\right)^2$$

$$U = \frac{1}{2}EI \int_0^L \left(\frac{\partial Y_b}{\partial t}\right)^2 dX + \frac{1}{2}K_bY_b^2 + \frac{1}{2}K_1(Y_b - Y_1)^2 + \frac{1}{2}K_2(Y_b - Y_1)^4 \qquad (13.2)$$

$$R = \frac{1}{2}\int_0^L C_b\left(\frac{\partial Y_b}{\partial t}\right)^2 dX + \frac{1}{2}C_1\left(\frac{\partial Y_b}{\partial t} - \frac{\partial Y_1}{\partial t}\right)^2$$

Substituting Eqs. (13.1–13.2) into the Lagrange equation:

$$\frac{\partial}{\partial t}\left(\frac{\partial T}{\partial \dot{X}}\right) - \frac{\partial T}{\partial X} + \frac{\partial U}{\partial X} + \frac{\partial R}{\partial \dot{X}} = F \qquad (13.3)$$

The equations of motion for the nonlinear TMD bridge can be derived as follows:

$$m_1\ddot{Y}_1 + c_1(\dot{Y}_1 - \dot{Y}_b) + k_1(Y_1 - Y_b) + k_2(Y_1 - Y_b)^3 = 0$$

$$EI\frac{\partial Y_b^4(X,t)}{\partial X^4} + \rho A\frac{\partial Y_b^2(X,t)}{\partial t^2} + C_b\frac{\partial Y_b(X,t)}{\partial t} + K_bY_b(X,t) = \qquad (13.4)$$

$$(-m_1g - m_1\ddot{Y}_1) \times \delta(X - X_0) - F \times \delta(X - Vt)$$

To improve the accuracy of the numerical calculation, the following dimensionless variables are defined as follows:

$$x = \frac{X}{L}, y_1 = \frac{Y_1}{L}, y_b = \frac{Y_b}{L}, v = \frac{V}{L},$$

$$L\delta(x - x_0) = \delta(X - X_0), L\delta(x - vt) = \delta(X - Vt) \qquad (13.5)$$

Taking the first three terms of the vibration in the simply supported beam into account:

$$y_b = \sum_{i=1}^{n} \Phi_i y_{bi}, n = 3 \qquad (13.6)$$

After substituting Eqs. (13.5–13.6) into Eq. (13.4), multiplying both sides of the resulting equation by the corresponding beam eigenfunction and integrating the equation with respect to x from 0 to 1 give rise to a set of second-order nonlinear ordinary differential equations with coupled terms, which can be written as

$$\ddot{y}_1 + \frac{c_1}{m_1}(\dot{y}_1 - \dot{y}_b) + \frac{k_1}{m_1}(y_1 - y_b) + \frac{k_2L^2}{m_1}(y_1 - y_b)^3 = 0$$

$$\sum_{i=1}^{3}\left(\int_0^1 \Phi_i\Phi_j\,dx\right)\ddot{y}_{bi} + \frac{c_b}{\rho A}\sum_{i=1}^{3}\left(\int_0^1 \Phi_i\Phi_j\,dx\right)\dot{y}_{bi} + \frac{EI}{\rho AL^4}\sum_{i=1}^{3}\left(\int_0^1 \Phi_i''\Phi_j\,dx\right)y_{bi} + \frac{k_b}{\rho A}\sum_{i=1}^{3}\left(\int_0^1 \Phi_i\Phi_j\,dx\right)y_{bi}$$

$$= \frac{1}{\rho A}(-m_1g - m_1\ddot{y}_1L) \times \Phi_j(x_0) - \frac{1}{\rho A}F \times \Phi_j(vt) \qquad (13.7)$$

In which j is taken as 1, 2, and 3. Equation (13.7) can be written in the following matrix form:

$$M\ddot{Y} + C\dot{Y} + KY + f(\ddot{y}, \dot{y}, y) = F(t) \tag{13.8}$$

13.3   APPLICATION OF IHBM TO ANALYZE THE EQUATIONS OF MOTION

Assuming $\omega = \pi v, \tau = \omega t$, Eq. (13.8) becomes

$$\omega^2 M\ddot{Y} + \omega C\dot{Y} + KY + f(\ddot{y}, \dot{y}, y) = F(\tau) \tag{13.9}$$

where M, C, and K are mass, damping, and stiffness matrices of vibration system, respectively.

$$M = I_{4\times 4}, Y = \begin{bmatrix} y_1 & y_{b1} & y_{b2} & y_{b3} \end{bmatrix}^T \tag{13.10}$$

$$C = \begin{bmatrix} \dfrac{c_1}{m_1} & -\dfrac{c_1 \times \sin(\pi x_0)}{m_1} & -\dfrac{c_1 \times \sin(2\pi x_0)}{m_1} & -\dfrac{c_1 \times \sin(3\pi x_0)}{m_1} \\[2ex] 0 & \dfrac{c_b}{\rho A} & 0 & 0 \\[2ex] 0 & 0 & \dfrac{c_b}{\rho A} & 0 \\[2ex] 0 & 0 & 0 & \dfrac{c_b}{\rho A} \end{bmatrix} \tag{13.11}$$

$$K = \begin{bmatrix} \dfrac{k_1}{m_1} & -\dfrac{k_1 \times \sin(\pi x_0)}{m_1} & -\dfrac{k_1 \times \sin(2\pi x_0)}{m_1} & -\dfrac{k_1 \times \sin(3\pi x_0)}{m_1} \\[2ex] 0 & \dfrac{\pi^4 EI + k_b L^4}{\rho A L^4} & 0 & 0 \\[2ex] 0 & 0 & \dfrac{16\pi^4 EI + k_b L^4}{\rho A L^4} & 0 \\[2ex] 0 & 0 & 0 & \dfrac{81\pi^4 EI + k_b L^4}{\rho A L^4} \end{bmatrix} \tag{13.12}$$

$$F(\tau) = \begin{bmatrix} 0 \\[1ex] \dfrac{-2m_1 g \sin(\pi x_0) - 2F\sin(\tau)}{\rho A} \\[2ex] \dfrac{-2m_1 g \sin(2\pi x_0) - 2F\sin(2\tau)}{\rho A} \\[2ex] \dfrac{-2m_1 g \sin(3\pi x_0) - 2F\sin(3\tau)}{\rho A} \end{bmatrix} \tag{13.13}$$

$$f(\ddot{y},\dot{y},y)=\begin{bmatrix} \dfrac{k_2 L^2}{m_1}[y_1 - y_{b1}\sin(\pi x_0) - y_{b2}\sin(2\pi x_0) - y_{b3}\sin(3\pi x_0)]^3 \\[4mm] \dfrac{2\omega^2 m_1 \ddot{y}_1 L \sin(\pi x_0)}{\rho A} \\[4mm] \dfrac{2\omega^2 m_1 \ddot{y}_1 L \sin(2\pi x_0)}{\rho A} \\[4mm] \dfrac{2\omega^2 m_1 \ddot{y}_1 L \sin(3\pi x_0)}{\rho A} \end{bmatrix} \qquad (13.14)$$

Y can be written as Fourier formula, $Y = SA$. In which

$$S = \begin{bmatrix} 1 & \cos\tau & \sin\tau & \dots & \cos(N\tau) & \sin(N\tau) \end{bmatrix},$$

$$A = \begin{bmatrix} a_{i0} & a_{i1} & b_{i1} & \dots & a_{iN} & b_{iN} \end{bmatrix}^T \qquad (13.15)$$

Substituting $Y = Y_0 + \Delta Y$, and $\omega = \omega_0 + \Delta\omega$ into Eq. (13.9), it becomes

$$\omega_0^2 M\Delta\ddot{Y} + \omega_0 C\Delta\dot{Y} + K\Delta Y + M_n\Delta\ddot{Y} + C_n\Delta\dot{Y} + K_n\Delta Y =$$

$$F(\tau) - f(Y_0,\dot{Y}_0,\ddot{Y}_0) - (\omega_0^2 M\ddot{Y}_0 + \omega_0 C\dot{Y}_0 + KY_0) - (2\omega_0 M\ddot{Y}_0 + C\dot{Y}_0)\Delta\omega \qquad (13.16)$$

In which

$$M_n = \frac{\partial f_n(Y_0,\dot{Y}_0,\ddot{Y}_0)}{\partial\ddot{Y}}, \quad C_n = \frac{\partial f_n(Y_0,\dot{Y}_0,\ddot{Y}_0)}{\partial\dot{Y}}, \quad K_n = \frac{\partial f_n(Y_0,\dot{Y}_0,\ddot{Y}_0)}{\partial Y} \qquad (13.17)$$

Multiplying $\delta(Y)^T$ on both sides of Eq. (13.16), and then integrating Eq. (1.16) from 0 to 2π:

$$K_m \times \Delta A = R_{m1} \times A_0 + R_{m2} + R_{m3} \times A_0 \times \Delta\omega \qquad (13.18)$$

In which

$$K_m = \int_0^{2\pi} S^T[\omega_0^2 M\ddot{S} + \omega_0 C\dot{S} + KS + M_n\ddot{S} + C_n\dot{S} + K_n S]d\tau,$$

$$R_{m1} = -\int_0^{2\pi} S^T(\omega_0^2 M\ddot{S} + \omega_0 C\dot{S} + KS)d\tau,$$

$$R_{m2} = \int_0^{2\pi} S^T[F(\tau) - f_n(\ddot{Y}_0,\dot{Y}_0,Y_0)]d\tau, \qquad (13.19)$$

$$R_{m3} = -\int_0^{2\pi} S^T(2\omega_0 M\ddot{S} + C\dot{S})d\tau,$$

The flowchart of the IHBM algorithm is shown in Figure 13.2, and the parameters of the nonlinear TMD bridge system are shown in Table 13.1:

As shown in Figure 13.3, the amplitude and frequency of vibration of bridge with a nonlinear TMD analyzed by the IHBM are the same as the results analyzed by incremental Newmark method, which can be proved to the correctness of the IHBM.

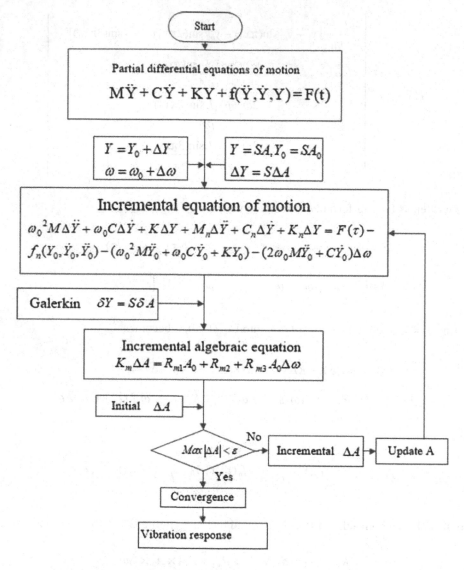

FIGURE 13.2   The flowchart of the IHBM algorithm.

TABLE 13.1
Nonlinear TMD and Bridge Parameter Table

	Value	Symbol
Length of beam	1	L (m)
Width of beam section	02	b (m)
Height of beam section	0.02	h (m)
	4×10^{-5}	A (m^2)
Height of beam section	7,850	ρ (kg/m^3)
	1.33×10^{-11}	I (m^4)
Density of beam	200	E (GPa)
		(*Continued*)

TABLE 13.1 (*Continued*)
Nonlinear TMD and Bridge Parameter Table

	Value	Symbol
Moment of inertia	0	
Young's modulus	0.5	K_b (N/m)
Stiffness of the beam	0.157	C_b (kg/s/m)
	0	m_1 (kg)
Damping of the beam	1×10^4	K_1 (N/m)
Mass of TMD		K_2 (N/m)
First-order stiffness of TMD	1	C_1 (kg/s/m)
	30	
Second-order stiffness of TMD	0.1	F (N)
	0.7	V (m/s)
Damping of TMD amplitude of forces	9.8	X_1 (m)
		g (N/kg)
Velocity of moving forces		
Location of TMD		
Gravitational acceleration		

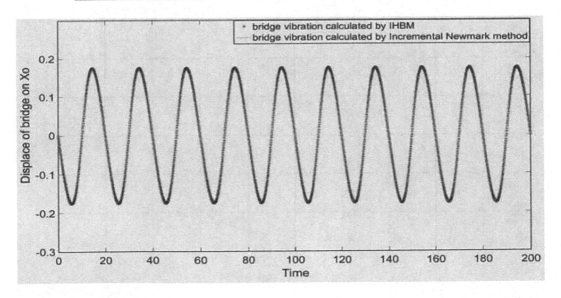

FIGURE 13.3 Displacement response of bridge vibration with nonlinear TMD at X_0 analyzed by IHBM and incremental Newmark method.

13.4 INVESTIGATION OF PARAMETER INFLUENCE ON DYNAMIC CHARACTERISTICS

a. Nonlinear responses of the beam with magnitude of the periodic moving forces.

According to Figure 13.4, each moving force corresponds to three vibration peaks, and the vibration peak increases linearly with the increase of the periodic moving forces.

b. Nonlinear responses of the beam with velocity of the periodic moving forces.

According to Figure 13.5, the vibration peak increases gradually with the increase of velocity. When the velocity of moving forces reaches 9.16 m/s, the vibration peak increases quickly until it reaches its maximum of 3.373 m. After that, the vibration peak decreases quickly.

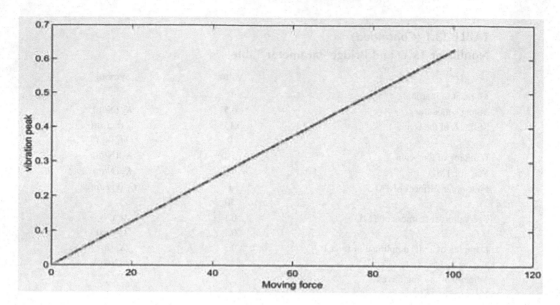

FIGURE 13.4   The bifurcation diagram of system with magnitude of the periodic moving forces.

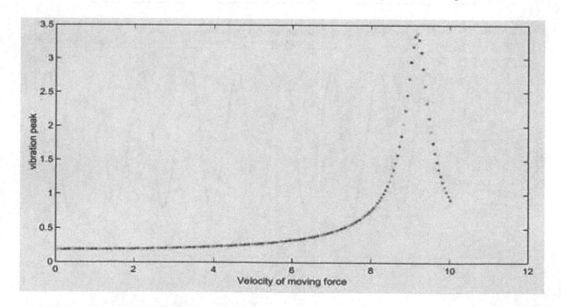

FIGURE 13.5   The bifurcation diagram of system with velocity of forces.

c. Nonlinear responses of the beam with mass rate of nonlinear TMD.
   According to Figure 13.6, the vibration peak doesn't change as the mass rate u of nonlinear TMD increases.

d. Nonlinear responses of the beam with nonlinear stiffness of nonlinear TMD.
   According to Figure 13.7, as the stiffness K_1 changes, the vibration is piecewise constant, piecewise points are about $K_1 = 4,500,000$, $11,000,000$ N/m. And there is a jump phenomenon in about $K_1 = 961,000$ N/m, which causes by the nonlinear of TMD. The vibration peak doesn't change as the stiffness K_2 increases.

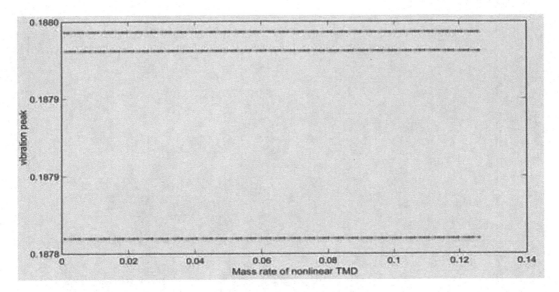

FIGURE 13.6   The bifurcation diagram of system with mass rate *u* of nonlinear TMD.

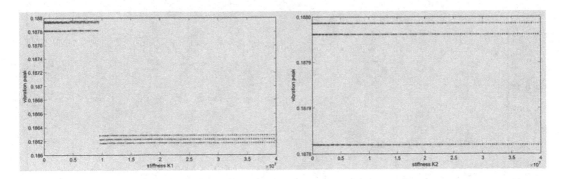

FIGURE 13.7   The bifurcation diagram of system with stiffness K_1 and K_2 of nonlinear TMD.

  e. Nonlinear responses of the beam with damping of nonlinear TMD.
      According to Figure 13.8, the vibration peak doesn't change as the damping of nonlinear TMD increases.
  f. Nonlinear responses of the beam with location of nonlinear TMD.
      According to Figure 13.9, the vibration peak increases from zero gradually as the nonlinear TMD moves from one side of the bridge to the middle of the bridge. When the nonlinear TMD reaches the middle of the bridge, the vibration is the largest. As the nonlinear TMD continues to move, the vibrations weaken gradually until zero.

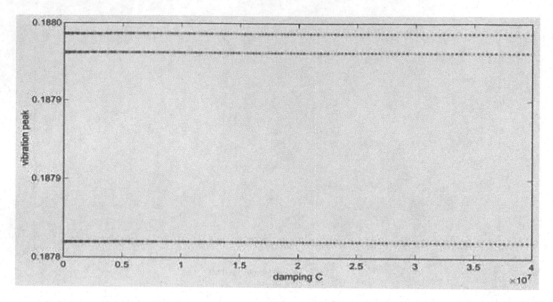

FIGURE 13.8   The bifurcation diagram of system with damping of nonlinear TMD.

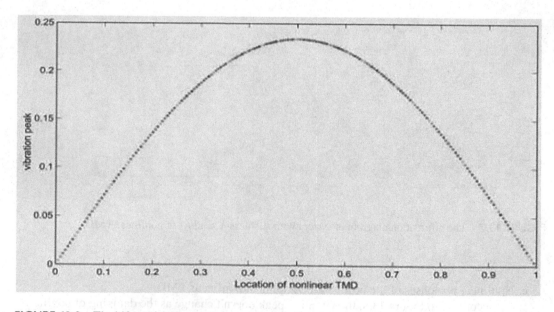

FIGURE 13.9   The bifurcation diagram of system with the location of nonlinear TMD.

13.5   CONCLUSION

The IHBM can effectively solve the vibration response problem of nonlinear TMD bridge systems subjected to periodic moving forces on beam. The vibration peak increases linearly with the increase of the periodic moving forces; The vibration peak increases gradually with the increase of velocity until it reaches a peak, after that the vibration peak decreases quickly; The change of mass rate u, nonlinear stiffness K_2, and damping C of the nonlinear TMD doesn't affect the vibration peak; As the stiffness K_1 increases, the vibration is piecewise constant, and there is a jump phenomenon in about $K_1 = 961,000\,\text{N/m}$.

REFERENCES

Alexander, N.A. and Schilder, F. 2009. Exploring the performance of a nonlinear tuned mass damper. *Journal of Sound and Vibration*, **319**, p. 445–462.

Chen, S.H. Cheung, Y.K. and Xing, H.X. 2001. Nonlinear vibration of plane structures by finite element and incremental harmonic balance method. *Nonlinear Dynamics*, **26**, p. 87–104.

Huang, J.L. Chen, S.H. Su, R.K.L. and Lee, Y.Y. 2011. Nonlinear analysis of forced responses of an axially moving beam by incremental harmonic balance method. *AIP Conference Proceedings*, **1233**, p. 941–946.

Lau, S.L. and Cheung, Y.K. 1981. Amplitude incremental variational principle for nonlinear vibration of elastic systems. *Journal of Applied Mechanics ASME*, **48**, p. 959–964.

Lau, S.L. and Cheung, Y.K. 1984. Nonlinear vibration of thin elastic plates part 2: Generalized incremental Hamilton's principle and element formulation. *Journal of Applied Mechanics ASME*, **51**, p. 845–851.

Lau, S.L. and Cheung, Y.K. 1986. Incremental Hamilton's principle with multiple time scales for nonlinear aperiodic vibrations of shells. *Journal of Applied Mechanics ASME*, **53**, p. 465–466.

Lau, S.L. Cheung, Y.K. and Wu, S.Y. 1984. Nonlinear vibration of thin elastic plates part 1: Internal resonance by amplitude-incremental finite element. *Journal of Applied Mechanics ASME*, **51**, p. 837–844.

Museros, P. Moline, E. and Martínez-Rodrigo, M.D. 2013. Free vibrations of simply-supported beam bridges under moving loads: Maximum resonance, cancellation and resonant vertical acceleration. *Journal of Sound and Vibration*, **332**(2), p. 326–345.

Pirmoradian, M. Keshmiri, M. and Karimpour, H. 2015. On the parametric excitation of a Timoshenko beam due to intermittent passage of moving masses: Instability and resonance analysis. *Acta Mechanica*, **226**, p. 1241–1253.

Sze, K.Y. Chen, S.H. and Huang, J.L. 2005. The incremental harmonic balance method for nonlinear vibration of axially moving beams. *Journal of Sound and Vibration*, **281**, p. 611–626.

Zhang, J.F. Li, X.Z. Song, L.Z. et al. 2013. Analysis on vertical dynamic response of simply-supported bridge subjected to a series of moving harmonic loads. *Applied Mechanics and Materials*, p. 361–363.

Zhang, H.D. Zhao, R.B, Liu, Y.Q. et al. 2010. Dynamic behavior of a simply supported beam with the complex damping subjected to loads moving with variable speeds. *Tenth International Conference of Chinese Transportation Professionals (ICCTP)*, p. 3215–3226.

Zhou, S.H. Song, G.Q. and Ren, Z.H. 2016. Nonlinear analysis of a parametrically excited beam with intermediate support by using multi-dimensional incremental harmonic balance method. *Chaos, Solitons & Fractals*, **93**, p. 207–222.

14 Analysis on the Internal Explosion Effects of Single-Layer Spherical Reticulated Shell

Shiqi Fu and Xuanneng Gao
Huaqiao University

CONTENTS

14.1 INTRODUCTION

Single-layer reticulated spherical shell is a typical large-span space structure, which is mostly used in landmark buildings. Once terrorist bombing attack to those buildings occurred, it would cause irreversible casualties, property damage, and serious social panic. Therefore, the research of the internal explosion effect of large-span space structure has a great significance (Du et al., 2008).

Explosive impact load belongs to a kind of typical impact load, and it has the characteristics of high strength and short pulse impact. The dynamic responses of structures subjected to explosive impact load is very complex. Gao and Wang (2009, 2010), Gao and Li (2015) established the numerical calculation model of steel reticulated shell structure, used proper orthogonal decomposition (POD) method to analyze the shock waves of steel reticulated shell under internal explosion, and then solved the space-time difference of overpressure distribution. Li and Liu (2006, 2008), Li and Lu (2014) summarized the dynamic responses of steel structure under explosive loading by studying the Johnson-Cook constitutive model. On this basis, the influences of damage accumulation effect and strain rate effect on steel were analyzed, and the dynamic responses and failure mode of columns and the whole structure were studied. Wu and Hao (2005, 2007) analyzed the influence of scaled distance on the dynamic responses of single-layer frame structure under blast loading. The results shown that if the scaled-distance was small, the impact of shock waves on the structural responses was greater; conversely, the impact of earthquake on the structural responses was greater. At present, a lot of studies on the performance of steel structures under explosion have been carried out and some useful results have been achieved, but there was still a certain gap. For example, the internal explosion effects of single-layer spherical reticulated shell.

In this paper, an explicit dynamic analysis software LS-DYNA was used to build a K8 single-layer reticulated spherical shell with a span of 40 m. The numerical simulation of the internal explosion effects in the spherical reticulated shell was carried out and analyzed. From the engineering view of explosion prevention and resistance, the concept of space height was put forward and the space height coefficient was defined. Further, the influence of space height on the internal explosion effects was discussed to provide the optimum structure type for the explosion-resistant design on a large-span space structure.

14.2   NUMERICAL MODEL

14.2.1   FINITE ELEMENT MODEL

As shown in Figure 14.1, the K8 single-layer spherical reticulated shell is selected as the structural calculation model. The structural span B, the vector height f, and the height of lower supporting structure H are 40, 8, and 10 m, respectively. The ground was considered rigid, and the bottom of the supporting column was fixed to the ground. Besides, the major structure and building envelope were connected by connecting components. A $42 \times 42 \times 42$ m air region was established to surround the whole structure, and the boundary was defined as the transmission boundary to simulate the explosion in an infinite area. The 300 kg trinitrotoluene, commonly known as TNT, explosive was defined by the volume fraction method in the structure, which was located at the center above the ground 1.4 m. In the whole process of explosion, the Arbitrarily Lagrange-Euler (ALE) algorithm was adopted to consider the interaction between the air shock waves and both the building envelope and the ground.

14.2.2   MATERIAL MODEL

The steel is simulated by Johnson-Cook constitutive model (Li, 2016). Its equation was as follows.

$$\sigma = (A_1 + A_2 \varepsilon^n)(1 + A_3 \ln \dot{\varepsilon}^*) \tag{14.1}$$

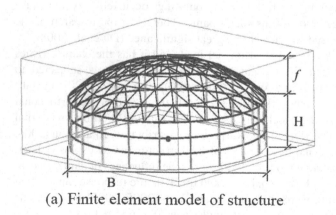

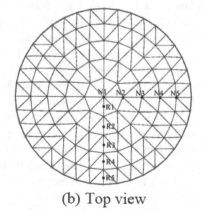

(a) Finite element model of structure   (b) Top view

FIGURE 14.1   Numerical calculation model.

TABLE 14.1

Johnson-Cook Constitutive Model Parameters of the Q235 [12]

A_1	A_2	N	A_3
320.7556×10^6	582.102×10^6	0.3823	0.0255

TABLE 14.2

Material Parameters of Explosive

P	D	P_{CJ}	A	B	R_1	R_2	ω	E_0	V
kg/m³	m/s¹	GPa	GPa	GPa	-	-	-	J/m³	-
1,630	6713	18.5	540.9	9.4	4.5	1.1	0.35	8×10^9	1.0

TABLE 14.3

Material Parameters of Air

P	C	C	C	C	C	C	C	E
kg/m³	-	-	-	-	-	-	-	J/m³
1.29	0	0	0	0	0.4	0.4	0	2.5×10^5

where σ, ε, and $\dot{\varepsilon}^*$ represent the equivalent flow stress, the equivalent plastic strain, and the relative strain rate, respectively. A_1, A_2, A_3, and n are undetermined parameters. For the Q235 steel, they are shown in Table 14.1.

The explosive employed the high-energy explosive (MAT_HIGH_EXPLOSIVE_ BURN). The Jones-Wilkins-Lee (JWL) state control equation was as follows.

$$p = A\left(1 - \frac{\omega}{R_1 V}\right)e^{-R_1 V} + B\left(1 - \frac{\omega}{R_2 V}\right)e^{-R_2 V} + \frac{\omega F_0}{V} \tag{14.2}$$

where A, B, R_1, R_2, and ω are input parameters and V is the relative volume of explosive. The parameters of the explosive are as shown in Table 14.2.

The air was simulated by MAT_NULL. The state equation was EOS_LINEAR_POLYNOMIAL as below.

$$p = C_0 + C_1 m + C_2 m^2 + C_3 m^3 + (C_4 + C_5 m + C_6 m^2)E \tag{14.3}$$

where $m = \rho/\rho_0 - 1$, E is the internal energy per unit volume, and ρ and ρ_0 are the density of air and the relative density of air, respectively. The material parameters of air are shown in Table 14.3.

14.3 INTERNAL EXPLOSION EFFECTS

14.3.1 LOADING EFFECTS OF INTERNAL EXPLOSION

As shown in Figure 14.2, the propagation laws of shock waves could be divided into three steps when the explosion occurs inside the structure. First, reflection waves begin to appear at the junction of the shock waves and the rigid ground, and they are superimposed with the incident waves to

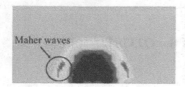 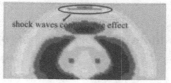 

FIGURE 14.2 Propagation law of shock waves.

form the Maher waves reflection waves. Subsequently, Maher waves and incident waves contact the building envelope, and shock waves convergence effect occurs at the top of structure and the corner between the wall and shell because of the barrier of the building envelope. Finally, the connecting components between the building envelope and the major structure quit work because they reach the failure strain, and then the structure occurs during explosion venting. In other words, some shock waves propagate to the outside of the structure, while others are reflected back to the inside of the structure. This moment, the distribution of shock waves in the structure becomes disordered, and the superposition of shock waves strengthens or weakens each other (Fu et al., 2018; Li and Ma 1992; Orienko, 2011).

According to the above analysis, A, B, and C represent the top of the spherical reticulated shell and the corners between the wall and shell and the column bottom, respectively. Their time history curve of shock waves was extracted and the pressure distribution of shock waves was analyzed. As shown in Figure 14.3, the first shock wave at A is not affected by the building envelope theoretically, and the overpressure peak is 0.098 MPa. The empirical formula shows that the overpressure of shock wave at A is 0.126 MPa. The simulation results are 22% smaller than the empirical formula. For explosive shock waves, this calculation accuracy is very good, which shows that the parameters of numerical model in this paper are reasonable and credible. Besides, A is the closest to the explosion

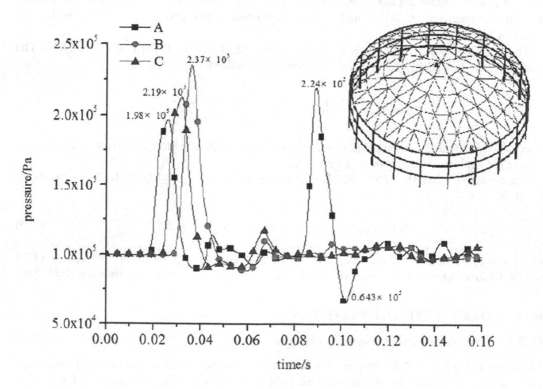

FIGURE 14.3 Pressure distribution of shock waves.

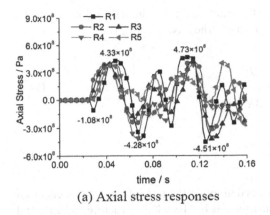

(a) Axial stress responses

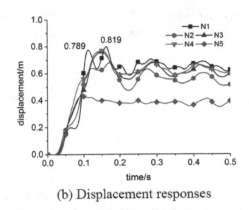

(b) Displacement responses

FIGURE 14.4   Internal explosion responses.

point and is first subjected to the shock waves, then subjected to the reflected wave that the pressure peak is greater. The pressure peak of B and C is obviously larger than that of A, which indicates that the superposition effect and corner convergence effect of shock waves are significant. These laws are similar to the results in (Wang and Gao, 2016).

14.3.2 Internal Explosion Responses

In order to study the internal explosion responses of the K8 single-layer spherical reticulated shell, the stress of the typical element and the displacement of the typical node were selected as the analysis objects. The specific position was shown in Figure 14.1b.

Figure 14.4a is the time history curve of the element stress. As shown in Figure 14.4a, the internal explosion stress responses of spherical reticulated shell had a similar change trend. The stress peaks of each element were not much different. Among them, the stress responses of R1, R2 near the top of structure and R5 near the support were larger, and the maximum appeared at R1, while the stress responses of the other members were smaller. Comparing Figure 14.4a with Figure 14.3, it could be implied that the response time of stress peak obviously lagged behind that of the pressure peak of shock waves.

Figure 14.4b is the time history curve of the node displacement. As shown in Figure 14.4b, the node displacement of N1 began to increase at 0.024s, and then the node displacement of N2–N5 gradually began to change. Due to the secondary impact effect of reflection waves, the displacement responses of both N1 and N2 near the dome of spherical reticulated shell increased rapidly twice, and then tended to be stable. The maximum of the peak value of displacement response appeared at N1, which could reach 700–800 mm. It far exceeded the design limit of large-span space structure. About 0.045 seconds, the shock waves converged at the corner between the wall and shell, and then began to reflect into the structure. The pressure on the inner surface of the structure in N5 was lower than that on the outer surface, and was greatly influenced by the lower supporting structure. So, the peak value of displacement response of N5 was smaller than other characteristic points.

14.4   DISCUSSIONS

14.4.1 Space Height Coefficient

According to Sections 3.1 and 3.2, it is shown that the internal explosion effects of structure are quite complicated. Previous studies have shown that the pressure peak of shock waves under internal explosion and its effect on the structure are closely related to the spatial size of the structure.

In order to explore the influence of the spatial size of the structure on the internal explosion effects, a space height coefficient of spherical reticulated shell was defined as

$$\alpha = \frac{\dfrac{B^2 \pi}{4} \cdot H}{\dfrac{B^2 \pi}{4} \cdot H + \dfrac{\pi}{3} \cdot (3R - f) \cdot f^2} \tag{14.4}$$

$$R = \frac{f^2 + B^2/4}{2f} \tag{14.5}$$

where α represents the space height coefficient, which is defined as the ratio of the volume enclosed by lower supporting structure to the total volume surrounded by the whole structure. B, f, H, and R are the span, rise height, height of supporting structure, and radius of spherical reticulated shell, respectively.

Using the above numerical model, $\alpha = 0.487$, 0.587, 0.655, 0.704, 0.740, and 0.769 were adopted, and the corresponding heights of supporting structure were 4, 6, 8, 10, 12 and 14 m, respectively. The calculation results and analysis were as follows.

14.4.2 WALL NO OPENINGS

According to Figure 14.3, it is revealed that the maximum of the stress response and displacement response of spherical reticulated shell under internal explosion mainly appear near the dome. So, R1 and N1 are taken as characteristic element and node, respectively, and the influence of space height coefficient on the internal explosion responses of spherical reticulated shells was analyzed.

Figure 14.4a is the relationship between α and both the peak value of stress response and the corresponding time. As shown in Figure 14.4a, with the increase of α, the peak value of stress response of R1 decreases gradually at first and then increases rapidly. At about 0.7, the peak value of stress response increases rapidly with the increase of α. Combining with the time curve to the peak value of stress response, if α is smaller, it could be seen that the peak value of stress response is larger but the corresponding time is early. It is indicated that if the space height is smaller, the stress response of the structure is mainly controlled by the first shock waves. If α is 0.55–0.70, the peak value of stress response is minimum and the corresponding time is earlier. It is indicated that the effect on the structure is the attenuated first shock waves due to the increase of space height. If α is greater than 0.7, the peak value of stress response increases rapidly and the time for R1 to reach the peak value is reduced later. It is indicated that secondary shock waves play a major role. Because the secondary shock waves are often much larger than the first shock waves, the peak value of stress response increases on the contrary. Therefore, the stress response of spherical reticulated shell under internal explosion could be effectively reduced by choosing appropriate space heights.

Figure 14.4b is the relationship between α and both the peak value of displacement response and the corresponding time. As shown in Figure 14.4b, if α is less than 0.6, the peak value of displacement response is smaller and the corresponding time is earlier. It is implied that the displacement response of N1 node is mainly controlled by the first shock waves. If α is greater than 0.6, the peak value of displacement response of N1 increases rapidly with the increase of α. It is indicated that the peak value of displacement response of the structure is mainly affected by the secondary shock waves. However, with the increase of the space height, the pressure peak of shock waves decreases continuously, and the peak value of displacement response at N1 increases slowly or even decreases (Figure 14.5).

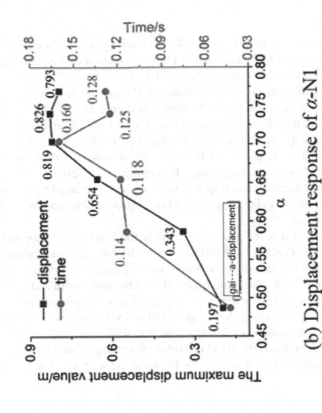

(b) Displacement response of α-N1

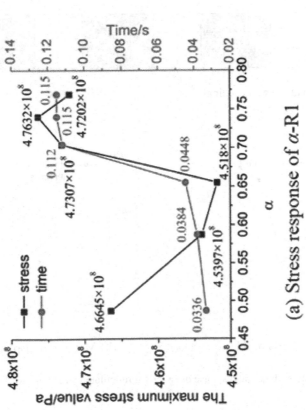

(a) Stress response of α-R1

FIGURE 14.5   Dynamic responses of spherical reticulated shell with wall no openings.

14.4.3 Wall Openings

In order to simulate the actual structure, the peak values of stresses and displacements of R1 and N1 were extracted under opening the wall (as shown in Figure 14.6), respectively, and compared with those without holes. The results were shown in Figure 14.7.

Figure 14.7a, b are the peak value of stress response of R1 and the peak value of displacement response of N1 under two cases, respectively. As shown in Figure 14.7, if the α is less than 0.65, with or without holes have little effect on the response of structure, and the peak value of stress response and displacement response are mainly controlled by the first shock waves. If α is greater than 0.65, the peak value of stress response and displacement response under the wall with openings is significantly lower than those without openings. Especially, if α is greater than 0.70, the peak value of stress response and displacement response will be greatly reduced by the wall openings. It is indicated that the wall openings have the effect of explosion venting, and the secondary shock waves will be greatly reduced.

From the above analysis, if α is less than 0.65, namely the height of the supporting structure H is less than 8 m, the wall openings have little effect on the structural response under internal explosion. However, with the increase of space height, the influence of secondary shock waves on the internal explosion responses of spherical reticulated shell becomes more and more serious. The wall with holes could effectively reduce the again impact of secondary shock waves on spherical reticulated shell. Therefore, without considering the effect of explosion venting, it is reasonable to design single-layer spherical reticulated shell with α less than 0.65, namely the height of supporting structure H should be less than 8 m. If the effect of explosion venting is considered, the space height coefficient α should be greater than 0.70, namely the height of the supporting structure H should be greater than 10 m.

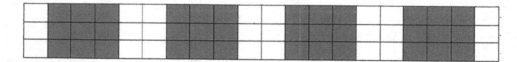

FIGURE 14.6   Sketch map of the wall openings.

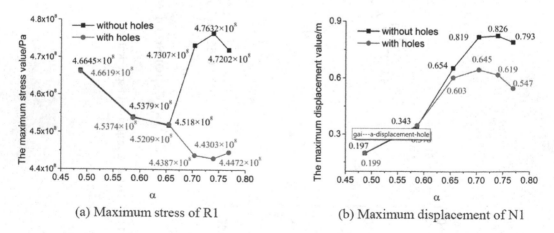

(a) Maximum stress of R1   (b) Maximum displacement of N1

FIGURE 14.7   Comparisons of dynamic response of spherical reticulated shell with wall openings and wall no openings.

14.5 CONCLUSIONS

The finite software ANSYS/LS-DYNA was used to establish the numerical model of K8 single-layer spherical reticulated shell under internal explosion, both the loading effects of internal explosion and the internal explosion responses of spherical reticulated shell were analyzed, and the influence of different space heights on the internal explosion effects of spherical reticulated shell were discussed. At the same time, the following conclusions could be drawn.

1. The propagation law of shock waves and the pressure distribution of shock waves under internal explosion were complicated, and the shock waves obviously had the reflection phenomena and convergence effect. The reflected waves were usually larger than the first shock wave, which would produce an adverse again impact on the structure. The convergence effect of shock waves mainly occurred at both the top of the spherical reticulated shell and the corner between the wall and shell, which seriously affected the internal explosion effects of the structure.
2. The peak value of internal explosion responses occurred mainly at the top and near the supports of the spherical reticulated shell. The responses near the supports were greatly affected by the convergent effect of shock waves. Besides, compared with stress response, displacement response plays a controlling role.
3. Appropriate space height could effectively reduce the internal explosion responses of the structure. If the effect of explosion venting of spherical reticulated shell was considered, the space height coefficient should be less than 0.65 (the height of the supporting structure is less than 8 m). Therefore, this paper suggested that the space height coefficient of the spherical reticulated shell should be 0.55–0.65 in practice.

ACKNOWLEDGMENTS

The authors are very grateful to the National Natural Science Foundation of China (Grant no.51278208), the Science and Technology Project of Fujian Province (Grant no.2018Y0063), and the Subsidized Project for Postgraduates' Innovative Fund in Scientific Research of Huaqiao University (Grant no. 17011086003) for the financial support of this work.

REFERENCES

Du X.L., Liao W.Z, Tian Z.M, et al. 2008. State-of-the-art in the dynamic responses and blast resistant measures of the buildings under explosive loads. *Journal of Beijing University of Technology* 34(3), p. 277–287.
Fu S.Q., Gao X.N and Chen X. 2018.The similarity law and its verification of cylindrical lattice shell model under internal explosion. *International Journal of Impact Engineering*, 122, p. 38–49.
Gao X.N and Wang S.P. 2009. Dynamic response of a large space cylindrical reticulated shell under blast loading. *Journal of Vibration and Shock*, 28(10), p. 68–73.
Gao X.N and Wang S.P. 2010. Numerical simulation of dynamic response of steel-reticulated shell structure under internal explosion. *Journal of North University of China (Natural Science Edition)*, 31(6), p. 573–580.
Gao X.N. and Li C. 2015. Dynamic response of single-layer reticulated shell structure under internal explosion, *Journal of Tianjin University (Natural Science and Engineering Edition)*, 48(S1), p. 102–109.
Li C. 2016. *Failure Mechanism and Explosion Protection Method of Cylindrical Reticulated Shell Structures under Internal Explosion*, Xiamen: Huaqiao University.
Li Y.Q. and Ma S.Z. 1992. *Mechanics of Explosion*. Beijing: Science Press.
Li Z.X. and Liu Z.X. 2006. Dynamic response analysis of steel structure under blast loading, *Journal of Building Structure*, 36(S1), p. 729–732+728
Li Z.X. and Liu Z.X. 2008. Dynamic responses and failure modes of steel structure under blast loading. *Journal of Building Structure*, 29(4), p. 106–111.

Li Z.X. and Lu J.H. 2014. Reliability analysis of steel beam under uncertain blast load, *Journal of Engineering Mechanics*, 31(4), p. 112–118+133

Orienko I.P. 2011. *Explosive Physics*. Beijing: Science Press.

Wang W.B. and Gao X.N. 2016. Propagation law of shock waves in single layer spherical lattice shell under internal explosion, *Journal of Interdisciplinary Mathematics*, 19(3), p. 527–547.

Wu C. and Hao H. 2005. Modeling of simultaneous ground shock and air-blast pressure on nearby structures from surface explosions, *International Journal of Impact Engineering*, 31(6), p. 699–717.

Wu C. and Hao H. 2007. Numerical simulation of structural response and damage to simultaneous ground shock and air-blast loads, *International Journal of Impact Engineering*, 34(3), p. 556–572.

15 Damage Features from Direct Modal Strain Measurements

G. De Roeck, E. Reynders, and D. Anastasopoulos
KU Leuven

CONTENTS

15.1 INTRODUCTION

Although a lot of progress has been made in the last decades [1–3], still a number of obstacles prevent a wider use of Vibration Based Structural Health Monitoring (VBSHM) in engineering practice. First, many proposed theoretical methods for damage assessment have been demonstrated to work for simulated numerical cases, with or without noise, and for simple laboratory experiments, but have never been applied on real structures. It is also far from obvious to find a case where during the observation period damage occurs. Second, sensitivity of modal characteristics to damage can be cumbersome because of disturbing environmental influences, mainly temperature [4,5]. Moreover, measurement noise and loading configuration can jeopardize the relation between observed changes in modal characteristics and occurred damage. Monitoring over a period that covers the complete range of environmental influences is needed. Afterwards, statistical methods can filter to a large extent those influences [6,7]. Third, although primary interest is on early detection of slight damage, the sensitivity to local defects is small. Fourth, for the exploitation of continuous monitoring data, data processing and condition interpretation should be performed in a (semi-)autonomous way. Damage detection methods that don't require a (complex) twin numerical model [8], but are just based on the processed vibration data, have clear advantages.

In this keynote paper how a direct, accurate measurement of small dynamic strains by Fiber Bragg Gratings (FBGs) can meet several of the described challenges is shown. The introduction of fiber-optic sensors that can accurately measure dynamic strains and offer ease of installation, resistance in harsh environment, and long-term stability contributed to an increased interest in adopting these sensors for VBSHM applications [9,10]. FBGs share the important advantages of other fiber-optic sensors, but additionally they are easy to multiplex and they have a relatively low cost. FBGs are being used for real-time static strain monitoring of a wide range of structures, including bridges [11,12]. During ambient excitation FBGs can measure the very small operational dynamic strains with very high accuracy, from which precise natural frequencies and modal strains can be deduced [13,14]. Thanks to multiplexing, modal strains can be obtained in a very dense grid. For the applications in this paper an FBG interrogator with a tunable laser source is used, which scans each FBG sensor individually with a wavelength sweep, offering a high wavelength resolution, and therefore a high accuracy of the measured strain amplitudes. However, the sweep also introduces delays between the different FBG sensors, which are compensated with an offline synchronization technique [15].

The application of direct modal strains for damage identification and condition monitoring of a prestressed concrete beam and a railway bridge is the subject of this paper.

15.2 PRESTRESSED CONCRETE BEAM

15.2.1 TEST STRUCTURE

The test structure, the setup of the progressive damage test (PDT), the dynamic test setup, and the setup for investigating the influence of temperature on the dynamic characteristics of the beam are presented.

A prestressed concrete beam serves as the test structure (Figure 15.1). The length of the beam is 5.0 m, and it has a rectangular cross section of 24 × 20 cm with chamfered edges (Figure 15.3). High-strength concrete (HSC) of the category C70/85 was used. The longitudinal reinforcement of the beam consists of five prestressed low relaxation strands with a nominal diameter of 12.5 mm and a characteristic tensile strength $\sigma_{pk} = 1{,}860$ MPa. Three strands are positioned at the bottom of the beam and two at the top (Figure 15.2). The strands are given an initial stress of $\sigma_{p0} = 744$ MPa (50% of $\sigma_{p0;max}$), where $\sigma_{p0;max} = 0.8 \ \sigma_{pk} = 1{,}488$ MPa. Closed-loop stirrups, with a nominal diameter of 6 mm and a center-to-center distance of 150 mm, are provided to withstand the shear forces (Figure 15.2). The weight of the beam is $w \sim 580$ kg and its density $\rho = 2{,}435$ kg/m³.

FIGURE 15.1 The reinforcement of the beam before casting the concrete and the beam on its delivery day.

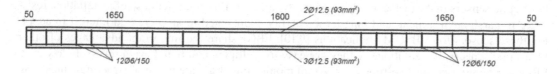

FIGURE 15.2 The reinforcement of the prestressed beam—front view (dimensions in mm).

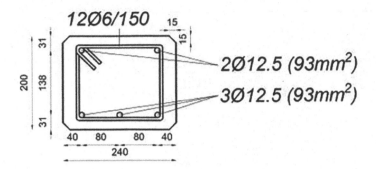

FIGURE 15.3 The reinforcement of the prestressed beam—cross section (dimensions in mm).

15.2.2 PROGRESSIVE DAMAGE TEST

The beam is subjected to a four-point PDT. The static supports are at 50 cm from the beam ends (Figure 15.4). A zone of constant bending moment is created in between two hydraulic jacks at a distance of 2 m. Five linear variable differential transformer (LVDTs) measure the vertical displacements of the beam.

A progressive damage loading scheme is adopted. There are nine loading cycles, where the load step is 30 kN for the first 2 cycles and 10 kN from the fourth cycle on. The beam failed during the ninth cycle for a load of 2P ~ 160 kN by crushing of the concrete at midspan. The maximum deflections from all LVDTs for each loading cycle are shown in Figure 15.5. The deflections of the last cycle, up to the failure of the beam, are not shown, as only the LVDT in the middle of the beam was measured for this cycle. The other LVDTs were detached to avoid damaging them during the further loading and failure of the beam. The maximum deflections at the middle of the beam, including the last loading cycle, are shown in Figure 15.6. The slope of the load-deflection curve changed after the second cycle (amplitude of 60 kN), when the beam exceeded its cracking load 2P ~ 55 kN.

15.2.3 DYNAMIC TEST SETUP

The beam is subjected to dynamic tests to identify its dynamic characteristics, i.e. natural frequencies and strain mode shapes. Dynamic tests are conducted on the undamaged and the damaged beam after the end of each loading/unloading cycle of the PDT. Changes of its dynamic characteristics will be used to identify progressive damage.

The dynamic boundary conditions approximate these for a free-free vibrating beam. This is achieved by inflated tires (Figure 15.7). The tires are placed below the beam at both ends, and when they are inflated the beam is detached from its static supports and rests solely on the flexible tires, which eliminates the probable influence of the flexibility of the supporting floor of the laboratory.

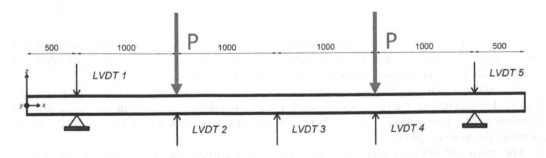

FIGURE 15.4 Configuration PDT.

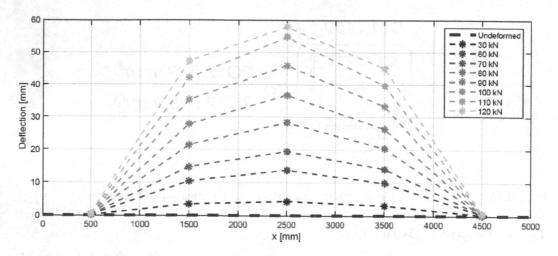

FIGURE 15.5   Deflection curves at increasing load levels.

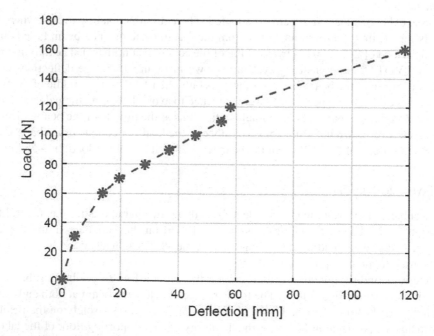

FIGURE 15.6   Midspan deflection versus load.

The beam is excited in the vertical direction by a Modal electrodynamic shaker, connected to the beam at 3.5 m from its left end. A Gaussian white noise (WN) signal is applied in the frequency range (0–400) Hz The duration of every dynamic test is 70 seconds.

The influence of temperature variations on the dynamic characteristics of the beam is also investigated. For this reason, the beam is permanently monitored during the PDT and dynamic tests are conducted for a long period of time, where it is subjected to different temperature conditions in between the loading cycles.

The amount of tests, the excitation type, and the temperature range that was achieved for each damaged state are summarized in Table 15.1.

FIGURE 15.7 Inflated tire to create free-free boundary conditions.

TABLE 15.1
Excitation Type, Number of Dynamic Tests, and Temperature Range for Each State

State	Excitation Type	No. of Tests	Temp. Range (°C)	$T_{max}-T_{min}$ (°C)
Undamaged	WN (0–400) Hz	161	22.0–32.6	10.6
After 30 kN	WN (0–400) Hz	144	21.6–34.5	12.9
After 60 kN	WN (0–400) Hz	166	18.3–34.3	16.0
After 70 kN	WN (0–400) Hz	126	19.4–34.3	14.9
After 80 kN	WN (0–400) Hz	107	22.3–34.5	12.2
After 90 kN	WN (0–400) Hz	115	23.6–34.6	11.0
After 100 kN	WN (0–400) Hz	109	23.3–34.5	11.2
After 110 kN	WN (0–400) Hz	139	22.9–34.3	11.4
After 120 kN	WN (0–400) Hz	112	22.0–34.6	12.6

In a similar research project on large prestressed concrete beams [16], dynamic strains were derived from displacements at discrete points measured with an optical camera system. In this paper, dynamic strains of the beam are measured with four chains of multiplexed FBG strain sensors from FBGs attached to the four edges of the beam, as shown in Figure 15.8. Each chain contains 20 FBG sensors (Figure 15.8). The chains are attached to the side of the top and bottom edges of the beam along its longitudinal direction through a custom clamping system (Figure 15.9). The distance between two consecutive clamping blocks is 25 cm and an FBG sensor is present in between, so measuring the average strain or macro-strain.

The strain acquisition system is the optical interrogator FAZT-I4 with relatively high wavelength accuracy (~1 pm) and precision (<0.1 pm). The FAZT-I4 has four channels that are simultaneously scanned by a tunable laser source. The sampling frequency of the FAZT-I4 is 1,000 Hz.

The position-averaged root mean square (RMS) strain value of the dynamic strain measurements is ~0.04 $\mu\varepsilon$, similar to the strain levels that many civil structures, like bridges, exhibit during ambient or operational excitation.

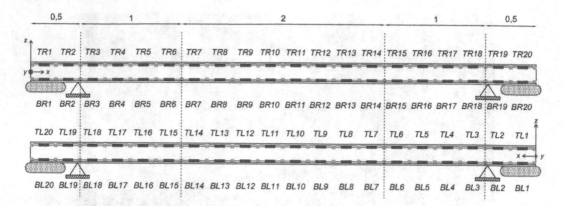

FIGURE 15.8 Layout of four chains of FBGs.

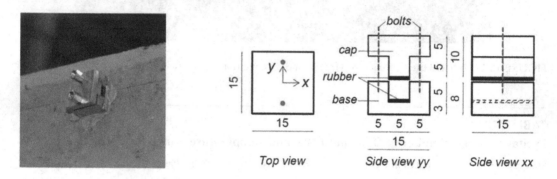

FIGURE 15.9 Clamping block.

15.3 STRAIN-BASED MODAL ANALYSIS

Dynamic strain data obtained from the FBG sensors are used for identifying the dynamic characteristics of the beam. The system identification is performed with MACEC, a MATLAB toolbox for experimental and operational modal analysis (OMA) [18]. The data-driven stochastic subspace identification stochastic subspace identification (SSI)-data technique [17,18] is applied, which is an output-only identification technique. The model order is chosen in steps of 2 and the maximum order is 100. The half number of Hankel block rows i is 40. The static or DC offset is removed first from all measured signals. Furthermore, a fourth-order Butterworth filter with a high-pass frequency of 4 Hz is applied to all channels to remove the influence of the small temperature fluctuations on the strain measurements.

15.3.1 Influence of Damage versus Temperature on Natural Frequencies

The evolution of the natural frequencies, throughout the PDT, for the first two modes (B1 and B2) is shown in Figure 15.10. The presented values are the ones measured for a uniform beam temperature of 24°C. A gradual reduction of the natural frequency values is observed, corresponding to the increased level of damage. The trend is similar for higher modes. The percentile reduction of the natural frequency values is about 3.5% for modes B1 and B2, when the undamaged natural frequencies are compared with the natural frequencies after the eighth loading cycle (amplitude of 120 kN).

The identified natural frequencies versus the corresponding temperature are presented in Figure 15.10, for all loading cycles and modes B1 and B2. The trend is similar for the other modes.

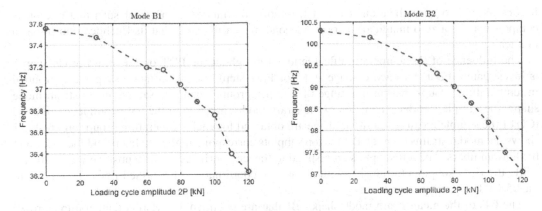

FIGURE 15.10   Reduction of natural frequencies at an increasing load level.

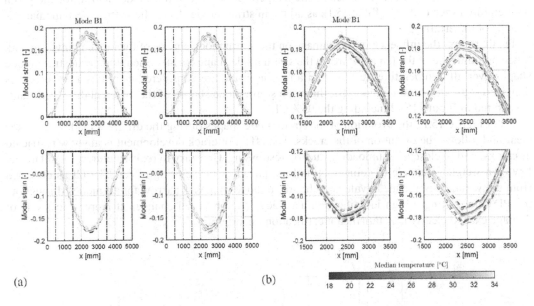

FIGURE 15.11   (a) Mean strain mode shape B1 with 95% CI, after the second loading cycle (60 kN). (b) Detail of zone 1,500–3,500 mm.

For each loading cycle, a straight line is fitted between the natural frequency values and the temperature at which they have been identified, to describe the relationship between the natural frequencies and the temperature for the specific damaged state of the beam. For all loading steps the temperature range is between 10°C and 15°C, allowing for a clear trend between natural frequency and temperature.

Figure 15.10 reveals a clear linear increase of the natural frequencies with an increase in temperature. The relationship appears to be relatively consistent among the different modes and damaged states. The influence of temperature on the natural frequencies is quite large when compared with that of damage.

15.3.2   Influence of Damage versus Temperature on Strain Mode Shapes

A normalization scheme is applied here, which allows for a consistent re-scaling of the strain mode shapes so that they can be directly compared within the different loading cycles and temperatures.

It consists of two normalization steps: first, rotating the mode strain vector such that one of its components has a zero imaginary part and second, re-scaling such that its Euclidean norm has a unit length.

The influence of the temperature fluctuations throughout the PDT on the strain mode shapes is investigated. First, for each damage state (or PDT step), the strain mode shapes are grouped in temperature groups that have a range of 2°C per group. Then, the sample mean strain mode shape, the sample standard deviation of the strain mode shapes, and the 95% confidence interval (CI) of the sample mean strain mode shape are obtained for each temperature group. Figure 15.11 shows the modal strains of mode B1: the top subplots correspond to the top fibers and the bottom to the bottom fibers. The left subplots correspond to the side of the beam with positive y-coordinates while the right subplots to the side with negative y-coordinates. The dashed lines correspond to the 95% CI.

The CIs of the mean strain mode shapes B1 that are obtained at different temperatures are all completely overlapping, as shown in detail in Figure 15.11b. The same observation applies for all modes. So there is no clear influence of the temperature on the strain mode shapes, for the given temperature range of 16°C (Table 15.1), as all mean strain mode shapes lie within the same narrow bounds of uncertainty.

Furthermore, the influence of the damage on the strain mode shapes of the beam is investigated. Since no influence of the temperature on the strain mode shapes is identified, the mean strain mode shapes over all (temperature) tests within a loading cycle are considered as representative of this damaged state. For bending mode B1, the mean strain mode shapes of the different damaged states are plotted in Figure 15.12 including the 95% CI.

From Figure 15.12 the evolution of the strain mode shapes during the different loading cycles is clear and follows the evolution of the cracks. The effect of crack development is anyhow restricted as cracks tend to close after unloading due to prestressing. Contrary to real structures, the relatively low own weight is unable to counteract, at least partially, this closing effect. The CI as obtained from the variance of the mean values of the strain mode shapes is narrow for all modes throughout the PDT, indicating a high identification accuracy. The observed changes at increasing levels of damage exceed clearly the statistical identification uncertainty.

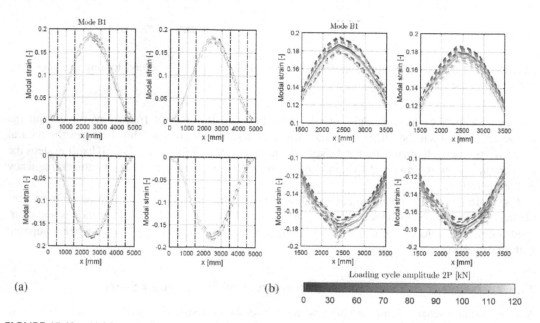

(a)   (b)

FIGURE 15.12   (a) Mean strain mode shapes with 95% CI. (b) Details of zone 1,500–3,500 mm.

FIGURE 15.13   Side view and typical span of the Arbre viaduct.

Strain amplitude changes are observed in the vicinity of the damaged area (between the points of the load application) for all modes. Changes in the amplitude of the strain mode shapes are already visible after the end of the second loading cycle with the force amplitude of 60 kN, when the beam reached its crack load, indicating early crack identification. The largest changes of the modal strain amplitude correspond to the damaged state after the end of the eighth loading cycle at 120 kN (70% of the failure load at 160 kN), when the FBG chains were detached to avoid damaging them during the next PDT loading cycle. More specifically, the amplitude of the bottom strain mode shapes is increasing with an increase in the level of damage. On the contrary, the amplitude of the top strain mode shapes is decreasing in a uniform pattern at the vicinity of the damaged zone. These changes of amplitude are related to a shift of the neutral axis of the beam. From the modal strains of each mode shifts of the neutral axis can be directly derived. This is an interesting possibility for non-model based damage identification.

15.4   LONG-TERM MONITORING OF THE 30th SPAN OF THE ARBRE VIADUCT

The Arbre Viaduct is a 2,002 m long railway bridge in the Walloon region of Belgium (Figure 15.1). The viaduct is part of the 88 km High Speed Line 1 (HSL1) that runs from Brussels to the French border. The HSL1 is connected to the French line Ligne á Grande Vitesse Nord (LGV Nord), a 333 km high speed line that connects Paris with the Belgian border and the Channel Tunnel via Lille. The lines are operated by Eurostar, TGV, Thalys PBA, and Thalys PBKA trainsets.

The viaduct consists of 36 prestressed concrete isostatic spans (Figure 15.13) with lengths varying from 51.6 to 63.2 m (2×51.6, 9×63.2, and 25×53.2 m). Each span is supported on four neoprenes, resting on concrete piers. Figure 15.13 also shows the almost undisturbed location of the viaduct: one of the consequences is the very low level of ambient excitation.

The cross section has a U-shape (Figure 15.14a,b). Openings are provided in the box-type webs for maintenance and emergency reasons (Figure 15.14c).

The 30th span of the viaduct (Figure 15.15) is monitored with two chains of FBG strain sensors for a period of 3.5 months. The two chains are located inside the evacuation box with negative y-coordinates in Figure 15.15. One fiber is attached at the bottom of the side wall and the other on the ceiling of the path, using the previously described clamping system (Figure 15.9). The distance between two consecutive clamping blocks is 2.5 m and one FBG sensor is located in between, measuring the average strain or macro-strain between both blocks. Each chain contains 20 FBG sensors (Figure 15.16c). The first sensor of each chain (the sensor with the lowest wavelength, i.e., sensors ST1 and SB1 in Figure 15.16c) is left free to measure the temperature of the span while the other 19 are measuring the strains of the span. A thermal insulation cover is provided to the fibers (Figure 15.16a), to reduce as much as possible the temperature fluctuations that would affect the FBG measurements. The strain acquisition system is the previously described FAZT-I4 interrogator.

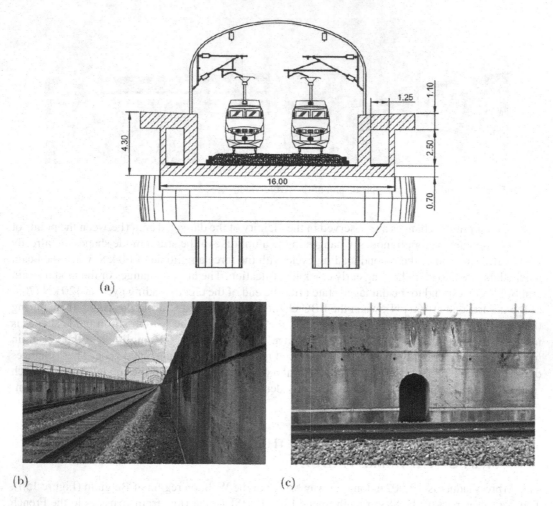

FIGURE 15.14 (a) Typical cross section of the bridge deck over a pier (dimensions in [m]). (b) The interior of the bridge deck. (c) Emergency opening.

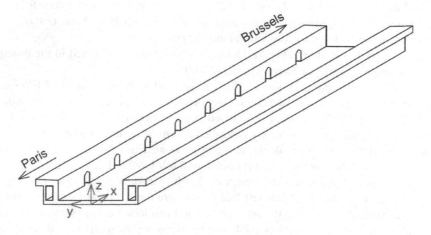

FIGURE 15.15 The monitored span.

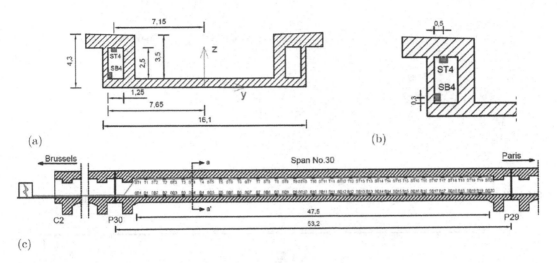

(a)

(b)

(c)

FIGURE 15.16   The setup of the FBGs. (a) Cross section a–a′, (b) detail of the cross section a–a′, and (c) front view of the FBG setup. The yellow rectangles show the clamping blocks. The blue rectangles represent the FBG sensors in between the clamping blocks. T stands for top FBG sensor and B for bottom. ST for top support and SB for bottom support (dimensions in m).

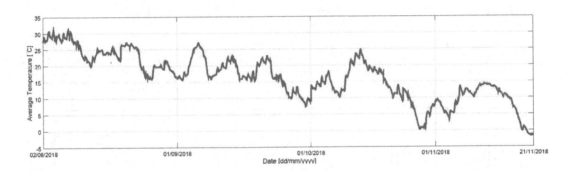

FIGURE 15.17   Variation of the mean span temperature during the monitoring period.

The temperature along the length of the span was also measured throughout the monitoring period with the use of an extra chain of 20 FBG sensors inside the thermal insulation at the ceiling of the corridor. These extra temperature sensors, not fixed to the bridge, were recording the temperature close to the strain measuring sensors of the top fiber. The mean temperature among all 20 sensors, as recorded at every time instant, is shown in Figure 15.17. The temperature variation between the temperature recordings of the sensors was less than 4°C. The maximum (mean) recorded temperature is 31°C and the minimum (mean) −2°C, so a range of 33°C.

A typical strain time history of a train traveling from Brussels to Paris is shown in Figure 15.18, as recorded from sensor T10 at the middle of the span. Compressive strains are recorded at the top of the span and tensile strains at the bottom.

For all trains traveling from Brussels to Paris the mean maximum compressive strain is about 5.5 $\mu\varepsilon$ and the mean maximum tensile strain is about 5 $\mu\varepsilon$.

It was attempted to extract from the ambient measurements the modal characteristics of the span, i.e., natural frequencies and strain mode shapes, by a strain-based OMA, using the MATLAB toolbox MACEC [18]. Unfortunately, due to the very high stiffness of the structure, the dynamic strains during ambient excitation were extremely low. The position-averaged RMS value of the processed dynamic strains (after re-sampling and filtering of the raw data) was about 0.003 $\mu\varepsilon$, which is lower

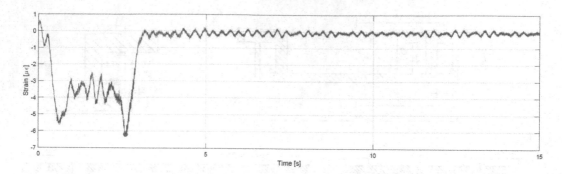

FIGURE 15.18   Strain due to a passing train measured in sensor T10 at the top of the midspan.

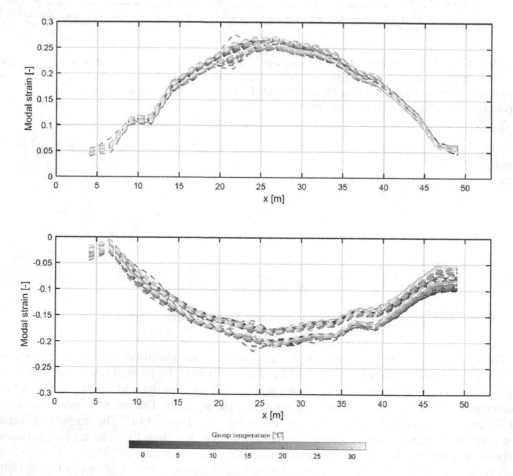

FIGURE 15.19   The 95% CI of the strain mode B1 for each temperature group. The top and bottom subplots contain the strains at the top (T) and bottom (B) of the span.

than the high-performance capabilities of the FAZT-I4 interrogator. In most cases of ambient excitation by wind, traffic in the direct neighborhood of the bridge, such extremely low ambient strains would not occur.

Therefore, strain data obtained from the recorded free response after train passage are used (see the right part of Figure 15.18). The strains are considerably higher than the ambient strains (a factor of 100), which is large enough to be measured accurately with the used interrogator. However, the free response is somewhat polluted by the forced vibrations of train vehicles running on the adjacent spans. The (weak) connection between neighboring spans is attributable to the continuity of ballast and rail.

The free response data of every train passage and for all different temperatures are used to identify the first bending mode of the span. Corresponding to the temperature range during the monitoring period (Figure 15.17), 34 temperature groups were created. For the modal strains a similar normalization as for the prestressed concrete beam has been applied. The 95% CIs of the different temperature groups of mode B1 are displayed in Figure 15.19. It can be clearly observed that the CIs for the different temperature groups overlap completely, and they all have nearly the same width. It can therefore be concluded that for the present span, there is no clear influence of the temperature on the strain mode shapes.

15.5 CONCLUSIONS

With FBG sensors very accurate static and dynamic strains can be recorded. Their accuracy outperforms that of classical strain gauges. In almost all practical cases, natural frequencies and modal strains can reliably be identified from FBG strains in operational (ambient) conditions. Multiplexing allows to measure strains at many distinct locations, which is a most wanted feature for damage localization and quantification.

In a progressive damage experiment of a prestressed concrete beam it has been demonstrated that natural frequencies are dependent on damage as well as temperature. So, natural frequencies can only be used as damage indicators after elimination of this temperature influence. On the contrary, experimental evidence shows that modal strains are largely independent on temperature, making them very appealing features for damage assessment. Moreover, clear changes of modal strains are observable at an increasing damage level. The location of these changes is directly related to the position of damage. When chains of FBGs are attached to the top and the bottom of a beam-like structure and a linear strain variation is assumed over the height, shifts of the neutral axis of the beam can be directly related to occurred damage without the need of a numerical model.

Chains of FBGs have also been applied to one span of a viaduct along the High Speed Railway Line between Brussels and Paris. Accurate dynamic strain histories were recorded during passage of the trains. However, the strains during ambient conditions were exceptionally small, due to the almost undisturbed location of the considered span. From the free response after train passage, the natural frequency and strains of the first bending mode could be identified. During the monitoring period of 3.5 months, a temperature difference of 33°C was logged. No clear influence of the temperature on the strain mode shapes could be observed, reinforcing the conclusion made for the prestressed concrete beam.

As a general conclusion, it can be stated that, for condition monitoring, arrays of optical strain sensors have a great potential, as natural frequencies and modal strains can be derived from small amplitude ambient vibrations. Moreover they are temperature independent, very sensitive to small local damage and directly interpretable in terms of damage.

ACKNOWLEDGMENTS

The research presented in this paper has been performed within the framework of the project G099014N "Robust vibration-based damage identification with a novel high-accuracy strain measurement system", funded by the Research Foundation Flanders (FWO), Belgium. The financial support of FWO is gratefully acknowledged. The authors also wish to express their gratitude to Infrabel for giving access to the Arbre viaduct.

REFERENCES

1. Doebling, S. W., Farrar, C. R., and Prime, M. B. 1998. A summary review of vibration-based damage identification methods. *The Shock and Vibration Digest*, 30(2), 91–105.
2. Brownjohn, J. M. W., De Stefano, A., Xu, Y. L., Wenzel, H., and Aktan, A. E. 2011. Vibration-based monitoring of civil infrastructure: challenges and successes. *Civil Structural Health Monitoring*, 1(3–4), 79–95.
3. Fan, W., and Qiao, P. 2010. Vibration-based damage identification methods: a review and comparative study. *Structural Health Monitoring*, 10(1), 83–111.
4. Peeters, B., and De Roeck, G. 2001. One-year monitoring of the Z24-bridge: environmental effects versus damage events. *Earthquake Engineering and Structural Dynamics*, 30(2), 149–171.
5. Deraemaeker, A., Reynders, E., De Roeck, G., and Kullaa, J. 2008. Vibration based Structural Health Monitoring using output-only measurements under changing environment. *Mechanical Systems and Signal Processing*, 22(1), 34–56.
6. Reynders, E., and De Roeck, G. 2013. Robust structural health monitoring in changing environmental conditions with uncertain data. *In Proceedings of the 11th International Conference on Structural Safety and Reliability*, ICOSSAR 2013, New York, NY, 16–20 June 2013.
7. Reynders, E., Wursten, G., and De Roeck, G. 2014. Output-only structural health monitoring in changing environmental conditions by means of nonlinear system identification. *Structural Health Monitoring*, 13(1), 82–93.
8. Teughels, A., and De Roeck, G. 2005. Damage detection and parameter identification by finite element model updating. *Archives of Computational Methods in Engineering*, 12(2), 123–164.
9. Lopez-Higuera, J. M., Cobo, L. R., Incera, A. Q., and Cobo, A. 2011. Fiber optic sensors in structural health monitoring. *Journal of Lightwave Technology*, 29(4), 587–608.
10. Glisic, B., and Inaudi, D. 2007. *Fibre Optic Methods for Structural Health Monitoring.* John Willey & Sons, West Sussex.
11. Chan, T. H. T., Yu, L., Tam, H. Y., Ni, Y. Q., Liu, S. Y., Chung, W. H., and Cheng, L. K. 2006. Fiber bragg grating sensors for structural health monitoring of Tsing Ma bridge: background and experimental observation. *Engineering Structures*, 28(5), 648–659.
12. Reynders, E., De Roeck, G., Bakir, P.G., and Sauvage, C. 2007. Damage identification on the Tilff bridge by vibration monitoring using optical fibre strain sensors. *ASCE Journal of Engineering Mechanics*, 133(2), 185–193.
13. Anastasopoulos, D., Moretti, P., Geernaert, T., De Pauw, B., Nawrot, U., De Roeck, G., Berghmans, F., and Reynders, E. 2017. Identification of modal strains using sub-microstrain FBG data and a novel wavelength-shift detection algorithm. *Mechanical Systems and Signal Processing*, 86A, 58–74.
14. Anastasopoulos, D., De Roeck, G. and Reynders, E. 2018. Modal strain identification using sub-microstrain fiber-optic Bragg grating data for damage detection of prestressed concrete structures. In KU Leuven Department of mechanical engineering, editor, *Proceedings of ISMA2018 and USD2018*, 3605–3620, Leuven, Belgium.
15. Maes, K., Reynders, E., Rezayat, A., De Roeck, G., and Lombaert, G. 2016. Offine synchronization of data acquisition systems using system identification. *Journal of Sound and Vibration*, 381, 264–272.
16. Unger, J.F., Teughels, A., and De Roeck, G. 2005. Damage detection of a prestressed concrete beam using modal strains. *Structural Engineering*, 131(9), 1456–1463.
17. Peeters, B., and De Roeck, G. 1999. Reference-based stochastic subspace identification for output-only modal analysis. *Mechanical Systems and Signal Processing*, 13(5), 855–878.
18. Reynders, E., Schevenels, M., and De Roeck, G. 2014. *MACEC 3.3: a Matlab toolbox for experimental and operational modal analysis*, BWM-2014-06, KU Leuven, internal report.

16 Displacement Estimation by Multi-Rate Data Fusion of Strain and Acceleration Data

K. Gao, S. Weng, and H.P. Zhu
Huazhong University of Science and Technology

Y. Xia
The Hong Kong Polytechnic University

X.S. Yu, G.Z. Qu, P.C. Yin, and J.F. Yan
China Railway Siyuan Survey and Design Group Co., LTD

CONTENTS

16.1 INTRODUCTION

Dynamic displacement of a supertall structure usually results from external loads, such as winds and earthquakes, and it provides useful information for structural health monitoring (SHM) since the direct relation between the displacement and the structural deformation.[1] Therefore, accurate displacement measurement is important to the deformation control and damage detection of supertall structures.[2]

Structural dynamic displacement is usually obtained by direct measurement or indirect estimation methods. Direct methods measure structural displacement by linear variable differential transducers (LVDTs), global position systems (GPSs) and vision-based systems. Those direct methods usually work on bridges, whereas suffer limitations on applications of supertall structures. Specifically, LVDTs are hard to install on supertall structures due to the requirement of firm supports next to the structure. The GPS-measured displacement is not sufficiently accurate when high accuracy of displacement is required.[2] The measurement accuracy of vision-based methods depends on a good visual condition, which is often affected by the occlusion of the scaffold or nearby structures.

Indirect estimation methods use other types of data like acceleration or strain to derive displacement. Previous researches transferred acceleration to dynamic displacement via double integration of the acceleration data, while accumulative error of the double integration causes a baseline drift, especially for long-term SHM.[3] Another limit of the acceleration-based approach is that the acceleration can barely estimate the pseudo-static displacement that is also contained in dynamic displacement. Displacement is also derived from strain data via the strain-based mode shape,[4] whereas extracting the mode shapes of a supertall structure is subject to modal analysis errors.

Recently, data fusion of direct and indirect methods is developed to improve the measurement accuracy of dynamic displacement.[1] For example, data fusion of GPS-measured displacement and acceleration is based on Kalman filtering.[5] On the other hand, data fusion of different indirect methods has also been conducted. Park[6] fused mode shape-based displacement with acceleration to obtain dynamic displacement of bridges. Those data fusion methods that use direct methods or mode shapes are still limited to be used in supertall structures.

This study develops a displacement estimation method for a supertall structure by fusing strain-derived displacement and acceleration via Kalman filtering. Different from previous researches, the strain-displacement formula in this study is derived directly from the structural geometric deformation without requiring mode shapes and thus avoids the modal analysis errors of supertall structures. The proposed method accurately estimates dynamic displacement including high-frequency and pseudo-static components by fusing high-sampling rate acceleration and low-sampling rate strain via a multi-rate Kalman filtering with smoothing. The accuracy of the proposed data fusion method is verified through a simulation of a supertall structure. The performance and applicability of the proposed method is then validated by a field test on a real 335 m tall supertall structure.

16.2   STRAIN-DERIVED DISPLACEMENT FORMULA

A supertall structure has a large slenderness ratio with a fixed bottom and can be simulated with a cantilever beam. The beam model is divided into n segments, and the i-th segment connects Point $i - 1$ and Point i. Each segment is divided into a number of micro-units. Strain sensors are located at $n + 1$ points (from Points 0 to n). The strain-derived displacement is derived from the longitudinal strain data by the following three steps.[7]

The first step is to calculate the angular rotation of the micro-unit (Figure 16.1) via the strain difference on the two sides of the micro-unit as

$$d\theta = \frac{(\varepsilon_l - \varepsilon_r)dh}{s} = \frac{\Delta\varepsilon dh}{s} \tag{16.1}$$

where $d\theta$ is the angular rotation, ε_l and ε_r are longitudinal strains at two sides, $\Delta\varepsilon = \varepsilon_l - \varepsilon_r$ is the strain difference and s and dh are the width and length of the micro-unit, respectively.

The second step is to calculate the horizontal displacement of the segment by integrating the angular rotation of the micro-units within the segment via diagram multiplication as

$$\Delta x_i = \int_0^{h_i} \bar{M} \, d\theta = \frac{1}{s} \int_0^{h_i} \bar{M}\Delta\varepsilon \, dh \tag{16.2}$$

where Δx_i is the horizontal displacement of the i-th segment, h_j is the length of segment. $\bar{M}_i$ is the bending moment at Point i.

$$\bar{M}_i = f_{\text{unit}} \sum_{j=i+1}^{n} h_j = \sum_{j=i+1}^{n} h_j \tag{16.3}$$

where $f_{\text{unit}} = 1$. The integration is calculated by multiplication of areas of the strain difference and the bending moment as

$$\Delta x_i = \frac{h_i}{6s_i} \left(2\bar{M}_{i-1}\Delta\varepsilon_{i-1} + \bar{M}_{i-1}\Delta\varepsilon_i + \bar{M}_i\Delta\varepsilon_{i-1} + 2\bar{M}_i\Delta\varepsilon_i \right) \tag{16.4}$$

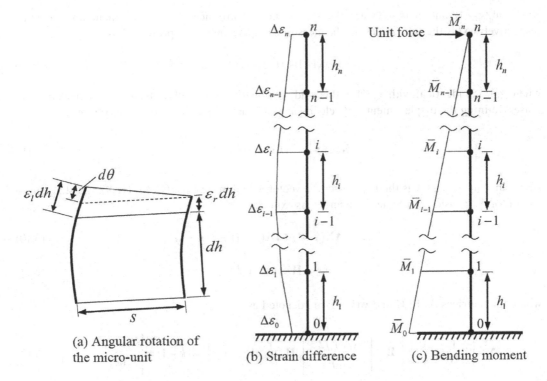

(a) Angular rotation of the micro-unit   (b) Strain difference   (c) Bending moment

FIGURE 16.1   Cantilever beam model.

Finally, the third step is to calculate the horizontal displacement of Point n by summing up the horizontal displacements of all segments below Point n as

$$x_{sd} = \frac{1}{6} \sum_{i=1}^{n} \left[\frac{h_i}{s_i} (2\bar{M}_{i-1}\Delta\varepsilon_{i-1} + \bar{M}_{i-1}\Delta\varepsilon_i + \bar{M}_i\Delta\varepsilon_{i-1} + 2\bar{M}_i\Delta\varepsilon_i) \right] \qquad (16.5)$$

where x_{sd} is the strain-derived displacement of Point n. The horizontal displacement of the supertall structure at different heights can also be achieved by altering the location of the unit force to the required point.[7]

16.3   DISPLACEMENT ESTIMATION USING KALMAN FILTERING

The Kalman filtering algorithm includes two parts. One is the state-space model that constructs the relationship between the input and the output. The other is a recursive filtering algorithm is an optimal estimator based on the minimum root-mean square principle.[1]

16.3.1   State-Space Model

Denote the displacement, velocity and acceleration at time step $k-1$ as $x(k-1)$, $(k-1)$ and $(k-1)$, respectively. The relationships of displacement, velocity and acceleration are as follows:

$$x(k) = x(k-1) + \dot{x}(k-1)\Delta t + 0.5\ddot{x}(k-1)\Delta t^2 + w_d(k-1) \qquad (16.6)$$

$$\dot{x}(k) = \dot{x}(k-1) + \ddot{x}(k-1)\Delta t + w_v(k-1) \qquad (16.7)$$

where $w_d(k-1)$ and $w_v(k-1)$ are the system noise introduced in displacement and velocity, respectively. Δt is the time interval. The measured displacement is represented as

$$x_m(k) = x(k) + v(k) \tag{16.8}$$

where $x_m(k)$ is identical with $x_{sd}(k)$ in this study. $x(k)$ is the real displacement. $v(k)$ is measurement noise. Defining the displacement and velocity as state variables to build the state vector

$$\mathbf{X}(k) = \begin{bmatrix} x(k) & \dot{x}(k) \end{bmatrix}^T \tag{16.9}$$

where the superscript T is the transpose of a vector. Combining Eqs. (16.6)–(16.9), the state-space model for data fusion of acceleration and x_{sd} is expressed as

$$\mathbf{X}(k) = \mathbf{A}\mathbf{X}(k-1) + \mathbf{B}\ddot{x}(k-1) + \mathbf{w}(k-1) \tag{16.10}$$

$$x_{sd} = \mathbf{H}\mathbf{X}(k) + v(k) \tag{16.11}$$

where the matrices $\mathbf{A}$, $\mathbf{B}$, $\mathbf{H}$ and $\mathbf{w}(k-1)$ are denoted as

$$\mathbf{A} = \begin{bmatrix} 1 & \Delta t \\ 0 & 1 \end{bmatrix}, \mathbf{B} = \begin{bmatrix} (\Delta t)^2/2 \\ \Delta t \end{bmatrix}, \mathbf{H} = \begin{bmatrix} 1 & 0 \end{bmatrix}, \mathbf{w}(k-1) = \begin{bmatrix} w_d(k-1) \\ w_v(k-1) \end{bmatrix} \tag{16.12}$$

The system noise vector $\mathbf{w}(k-1)$ and the measurement noise $v(k)$ are Gaussian white noise with covariance matrix of $\mathbf{Q}$ and $\mathbf{R}$ respectively as

$$\mathbf{Q} = q \begin{bmatrix} (\Delta t)^3/3 & (\Delta t)^2/2 \\ (\Delta t)^2/2 & \Delta t \end{bmatrix}, \mathbf{R} = \frac{r}{\Delta t} \tag{16.13}$$

where q and r are the variances of acceleration and displacement noises.

16.3.2 RECURSIVE FILTERING ALGORITHM

The recursive filtering algorithm has two processes, i.e., estimating and updating processes. In the estimating process, the state vector at time step k is predicted as

$$\bar{\mathbf{X}}(k) = \mathbf{A}\hat{\mathbf{X}}(k-1) + \mathbf{B}\ddot{x}_m(k-1) \tag{16.14}$$

where $\bar{\mathbf{X}}(k)$, $\hat{\mathbf{X}}(k-1)$ and $x_m(k-1)$ are the predicted state vector, the filtered state vector and the measured acceleration, respectively. The covariance matrix is an error evaluation of the prediction, which is calculated by

$$\bar{\mathbf{P}}(k) = A\hat{\mathbf{P}}(k-1)\mathbf{A}^T + \mathbf{Q} \tag{16.15}$$

where $\bar{\mathbf{P}}(k)$ and $\hat{\mathbf{P}}(k-1)$ are the covariance matrices of the prediction error and the filtering error. In the updating process, the strain-derived displacement in Eq. (16.5) corrects the predicted displacement

$$\hat{\mathbf{X}}(k) = \bar{\mathbf{X}}(k) + \mathbf{K}(k)\left[x_{sd}(k) - \mathbf{H}\bar{\mathbf{X}}(k) \right] \tag{16.16}$$

Therefore, the filtered displacement is a data fusion of acceleration and strains. The Kalman gain $\mathbf{K}(k)$ serves as an optimal weight which is calculated by

$$\mathbf{K}(k) = \bar{\mathbf{P}}(k)\mathbf{H}^T \left[\mathbf{H}\bar{\mathbf{P}}(k)\mathbf{H}^T + \mathbf{R} \right]^{-1} \tag{16.17}$$

The covariance matrix of the filtered estimation is calculated by

$$\hat{\mathbf{P}}(k) = \left[\mathbf{I} - \mathbf{K}(k)\mathbf{H}^T \right] \bar{\mathbf{P}}(k) \tag{16.18}$$

The estimating process and updating process repeat alternatively at every time step to obtain the optimal displacement.

In practice, acceleration and strain are measured at different sampling rates. Therefore, a multi-rate Kalman filtering approach is required. In the multi-rate Kalman filtering, there is only estimating process performed when acceleration data are available while strain data are absent. The updating process works together with the estimating process when both acceleration and strain data are available. In the case, smoothing technique is needed to diminish the estimation error results due to the absence of updating processes. The smoothing used in this study is a fixed-interval smoothing.[1] The smoothed displacement is obtained by

$$\tilde{\mathbf{X}}(k^*) = \hat{\mathbf{X}}(k^*) + \mathbf{G}(k^*) \cdot \left[\tilde{\mathbf{X}}(k^* + 1) - \bar{\mathbf{X}}(k^* + 1) \right] \tag{16.19}$$

where $\hat{\mathbf{X}}(k^*)$ is the smoothed state vector, and $k^* = k + p - 1, k + p - 2, \ldots k$ (p is a round number of f_d/f_s). $\mathbf{G}(k^*)$ is the smoothing gain calculated by

$$\mathbf{G}(k^*) = \hat{\mathbf{P}}(k^*)\mathbf{A}^T \bar{\mathbf{P}}(k^* + 1)^{-1} \tag{16.20}$$

The accuracy of the estimated displacement is evaluated using the normalized root-mean square error (NRMSE) as

$$\text{NRMSE} = \sqrt{\sum_{k=1}^{N} \left(x_{\text{est}}(k) - x_{\text{ref}}(k) \right)^2} \Bigg/ \sqrt{\sum_{k=1}^{N} \left(x_{\text{ref}}(k) \right)^2} \times 100\% \tag{16.21}$$

where $x_{\text{est}}(k)$ is the estimated displacement and $x_{\text{ref}}(k)$ denotes the reference value.

16.4 SIMULATION ON A SUPERTALL STRUCTURE

Wuhan Yangtze Navigation Centre (WYNC) is a 66-floor frame-core wall supertall structure in Wuhan, China, with a height of 335 m.[8] Figure 16.2 shows the structure and its finite element model. The cross section of the structure is a square with an outer frame of 50×50 m and an inner core wall of 30×30 m. The excitation imposed on the top floor generates displacement including both dynamic and pseudo-static components. The sampling rate is 100 Hz in the simulation. The displacement response calculated from the numerical model via dynamic analysis is regarded as reference.

Firstly, the displacement at Floor 66 is calculated by the strain data using the method described in Section 2. According to sensor locations, the model of WYNC is divided into segments, and each measured storey is regarded as a micro-unit. The measured strains on two sides of the same floor derive the angular rotation of the micro-units, then the displacement of the segment is calculated by diagram multiplication. Afterwards, horizontal displacement at Floor 66 is obtained by integrating the displacements of all segments. The NRMSE of the strain-derived displacement is 5.02%.

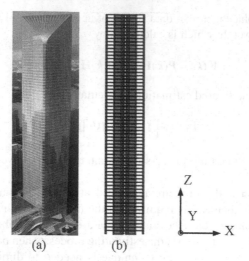

FIGURE 16.2   WYNC: (a) site view and (b) numerical model.

Next, the strain-derived displacement and the acceleration are fused via the proposed Kalman filtering approach. Displacement estimated by only acceleration using approach proposed by Lee et al.[9] is also used for comparison. Figure 16.3 compares the displacements estimated by different methods. The acceleration-derived displacement has large errors in the pseudo-static component. The strain-derived displacement is close to the reference, and the data fusion displacement matches the reference the best. The NRMSE of the data fusion displacement is 3.64%, which is more accurate than strain-derived displacements.

Afterwards, Gaussian white noises of 5%–20% are added to the strain and acceleration data to investigate the influence of measurement noise to the data fusion method. Table 16.1 compares the NRMSEs of displacement estimations at different noise levels. The NRMSEs of the acceleration-derived displacements are around 65.5%. The NRMSEs of the strain-derived displacement increases from 5.03% to 8.72% as the noise increases, while the NRMSEs of the data fusion displacement increases slightly from 4.06% to 5.05%.

In practice, the sampling rate of acceleration is usually higher than strain. Therefore, the multirate Kalman filtering with smoothing is used to fuse the acceleration data and strain with different

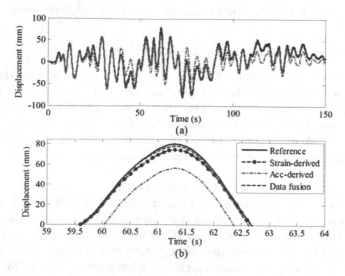

FIGURE 16.3   Comparison of displacements.

TABLE 16.1

Comparison of the NRMSE of Displacements Derived by the Acceleration-Derived, Strain-Derived and Data Fusion Methods in Different Noise Cases

	Noise of Both Strain and Acceleration			
Method (%)	5	10	15	20
Acceleration-derived (%)	65.45	65.48	65.50	65.53
Strain-derived (%)	5.03	5.80	7.35	8.72
Data fusion (%)	4.06	4.13	4.58	5.05

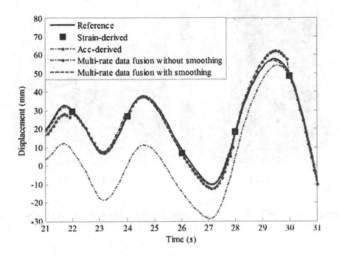

FIGURE 16.4 Comparison of displacements.

sampling rates. The sampling rate of acceleration and strain is set to 100 and 0.5 Hz. A 10% Gaussian white noise is used to simulate a practical working noise. Figure 16.4 compares different displacement estimations. The strain-derived displacement contains only a few displacement points of the dynamic displacement. The acceleration-derived displacement fails to estimate the pseudo-static component, which leads to large errors. The multi-rate data fusion displacement approximates the exact displacement with a slight error, and the smoothed displacement is in good agreement with the reference.

16.5 FIELD TEST ON A SUPERTALL STRUCTURE

A field test on the 335 m WYNC (Figure 16.5) is implemented to verify the applicability of the proposed method on real supertall structures. Strain sensors are installed at floors 5, 11, 16, 25 and 34. Accelerometer and GPS system are installed at floor 34. The sampling rates of the acceleration, strain and GPS are 10, 1 and 10 Hz, respectively. The displacement at floor 34 is derived from strains and then estimated by data fusion of the strain-derived displacement and acceleration. The data fusion displacement is compared with the GPS measurement.

The structure is under ambient excitation, so the vibration amplitude is very small. As shown in Figure 16.6, the GPS-measured displacement is irregular with an amplitude range of about ±10 mm. On the other hand, the displacement estimated by the proposed data fusion approach indicates the cyclic motion of the supertall with a cycle period of about 4.3 seconds and a tiny 0.5 mm amplitude. The data fusion displacement is more reasonable and closes to the real displacement than the GPS-measured displacement. This is because neither cyclic motion nor small vibration amplitude is indicated from the GPS-measured displacement. Therefore, the data fusion method is able to estimate tiny dynamic displacement, whereas the GPS is unable to measure.

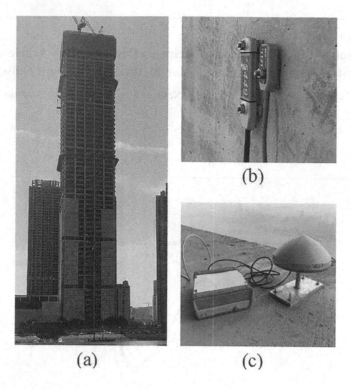

FIGURE 16.5   (a) field view of WYNC, (b) strain and acceleration sensors, (c) GPS system.

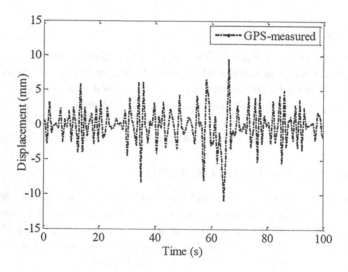

FIGURE 16.6   GPS-measured displacement.

In terms of frequency domain, the comparison of power spectral density (PSD) obtained using Welch's method[10] for the GPS-measured displacement and data fusion displacement is shown in Figure 16.7. It clearly shows that the first- and second-order frequencies can be directly detected from data fusion displacement, whereas the GPS-measured displacement is unable to detect even the first frequency. Consequently, the proposed method is more advantageous than GPS system both in the time and frequency domains in field application (Figure 16.8).

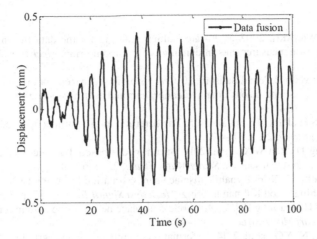

FIGURE 16.7 Data fusion displacement.

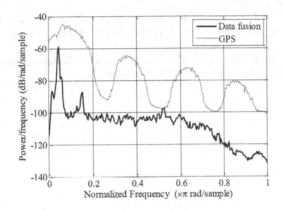

FIGURE 16.8 Comparison of welch PSD for two kinds of displacements.

16.6 CONCLUSIONS

The proposed multi-rate Kalman filtering approach combines high-frequency acceleration and low-frequency strain to accurately obtain dynamic displacement including both high-frequency and pseudo-static components.

Results of the numerical simulation show that the proposed data fusion method accurately estimates displacement in high noise cases. In addition, the proposed method is effective for acquiring dynamic displacement even when the sampling rate of the strain data is low. Results of the field test on a 335 m-high supertall structure indicates the data fusion displacement is more accurate than the GPS-measured displacement both in the time and frequency domains.

16.7 FUNDING

This work is supported by the grants from Basic Research Program of China (contract number: 2016YFC0802002), the National Natural Science Foundation of China (NSFC, contract number: 51629801 and 51778258), Research Grants Council of the Hong Kong Special Administrative Region, China (Project No. PolyU 152621/16E) and the Research Funds of China Railway Siyuan Survey and Design Group Co., LTD. (2018D001).

REFERENCES

1. Smyth, A. and Wu, M.L. 2007. Multi-rate Kalman filtering for the data fusion of displacement and acceleration response measurements in dynamic system monitoring. *Mech Syst Signal Process*, 21(2), p. 706–723.
2. Almeida S.C., Oliveira C.C. and Batista J. 2016. A vision-based system for measuring the displacements of large structures: Simultaneous adaptive calibration and full motion estimation. *Mech Syst Signal Process*, 72, p. 678–694.
3. Park K.T., Kim S.H., Park H.S., et al. 2005. The determination of bridge displacement using measured acceleration. *Eng Struct*, 27, p. 371–378.
4. Wang Z.C., Geng D., Ren W.X., et al. 2014. Strain modes based dynamic displacement estimation of beam structures with strain sensors. *Smart Mater Struct*, 23, p. 645–663.
5. Kim K.Y. and Sohn H. 2016. Dynamic displacement estimation by fusing LDV and LiDAR measurements via smoothing based Kalman filtering. *Mech Syst Signal Process*, 82, p. 339–355.
6. Park J.W., Sim S.H. and Jung H.J. 2014. Wireless displacement sensing system for bridges using multi-sensor fusion. *Smart Mater Struct*, 23, p. 1–12.
7. Xia Y., Zhang P., Ni Y.Q., et al. 2014. Deformation monitoring of a supertall structure using real-time strain data. *Eng Struct*, 67, p. 29–38.
8. Gao F, Chen P, Xia Y, et al. 2019. Efficient calculation and monitoring of temperature actions on supertall structures. *Eng Struct*, 193, p. 1–11.
9. Lee H.S., Hong Y.H. and Park H.W. 2009. Design of an FIR filter for the displacement reconstruction using measured acceleration in low-frequency dominant structures. *Int J Numer Meth Eng*, 82, p. 403–434.
10. Xu Y., Brownjohn J., Hester D., et al. 2017. Long-span bridges: Enhanced data fusion of GPS displacement and deck accelerations. *Eng Struct*, 147, p. 639–651.

17 Concrete Crack Image Recognition Based on DBSCAN and KPCA

Mingxin Gao and Yang Liu
Harbin Institute of Technology

CONTENTS

17.1 INTRODUCTION

Bridge structure is an important part of the transportation system, and its structural safety plays a crucial role in the safety and smoothness of transportation. However, during the actual operation process, the health condition of bridge structure faces lots of threats from different aspects, such as material aging, overload, and erosion environment (James et al., 2013; Sluzek, 2010). These potential threats may cause different degrees of damages to the key position of a bridge structure. So, it is necessary to check the appearance of the bridge structure regularly.

Cracks detection is one of the important parameters of bridge health detection (Zhu et al., 2011; Valenca et al., 2012; Nishikawa et al., 2012; James et al., 2013). On the one hand, crack is a sign of problem in the health of a bridge, indicating that some part of the structure is about to or has been damaged. On the other hand, the occurrence of cracks will cause the internal stress of the bridge structure to reorganize, creating an unknown safety hazard. Traditional crack detection mainly relies on experienced technicians, it is subjective, inefficient, and time consuming (Benmansour and Cohen, 2009; Li et al., 2017). And the result of this type of detection technique is not only affected by the location of bridge structure but also by the natural climate. With the continuous development of scientific research technology and the mutual application of interdisciplinary, utilizing image recognition technology to process the photo obtained from the civil structure (Pynn et al., 1999; Mathurin and Velinsky, 2000; Tung et al., 2002; Sunil and Paul, 2006; Zou et al., 2012; Valenca et al., 2012), it has gradually become a hot research direction for crack detection neighborhood.

In this study, we use image processing technique (IPT) to build an architecture for detecting concrete cracks from images. The first objective of this article is to build a robust architecture that can detect the edge of cracks in a bridge structure. The second objective is to construct a monitoring factor that can describe the variation of concrete crack width by utilizing kernel principal component analysis (KPCA) algorithm. The main advantage of the proposed IPT-based detection of concrete cracks is

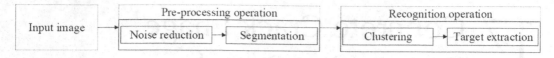

FIGURE 17.1 Example of image recognition technology architecture.

that it automatically extracts the edge of concrete cracks and monitors the variation of concrete crack width compared with the traditional detection approaches. This research's content is described as follows. Section 2 presents the synopsis of the proposed approach for concrete cracks detection. Section 3 introduces all the methodologies of the proposed architecture. Section 4 presents the experimental results of the proposed IPT-based concrete cracks detection. Section 5 concludes this article.

17.2   IPTs USED FOR CONCRETE CRACKS DETECTION

IPTs can be defined as the processing of perceiving target and obtaining the understanding of the content, such as whether the targets appear, where they are located, and how their geometrical dimensions change. The integration of optical equipment and IPTs is a promising prototype for bridge health detection. Normally, image recognition technology is composed of preprocessing operation (noise reduction, morphological operation, etc.) and recognition operation (clustering and target extraction). Figure 17.1 shows an example of image recognition technology architecture. Starting from an input image with three channels (RGB level) or one channel (Grayscale), the first step of the architecture is to perform denoising operation. Then, finding the targets of interest in the image and segmenting them to eliminate the irrelevant information. Based on the above information extracted from raw image, the next step of the architecture is to divide the pixels in the image into corresponding categories according to a specific rule. Finally, on the basis of the feature information (color information, texture information, shape information, spatial region information, etc.), extract the object in the image space.

In theory, the noise of image is unpredictable. However, through image denoising the image quality can be improved while also increasing the signal-to-noise ratio of the image. In practical applications, the selected noise feature not only should be able to describe the image itself but also could be clearly distinguished from other categories. Naturally, the visualized features of objects in the image convert into non-visualized feature by some algorithm or mathematical operation, so that the computer has the vision to achieve the purpose of object recognition in the image.

17.3   METHODOLOGY

This section explains the overall architecture, the algorithm used in this study, and the backgrounds of specific algorithm. Figure 17.2 demonstrates the IPT-oriented framework for concrete cracks detection. Grayscale converts the raw image from RGB level to grayscale. Edge detection uses the Canny operator to detect crack branch in the image that with grayscale. Image segmentation aggregates the feature information from the above operation. It consists of two parts: pixel points clustering and target extraction. Crack monitor describes the edge contour and the variation of concrete crack width.

17.3.1   GRAYSCALE PROCESSING

The initial images in the actual project are all natural-color image of the RGB type. In order to improve the processing result of the image later, reduce the size of the image itself, and speed up the operation, it is necessary to convert the initial photo into grayscale image during the preprocessing of digital images. The color of any pixel in the natural-color image is determined by three components R, G, and B. Each component has a lot of 256 values that can be selected. That is to say there are more than 16.7 million colors in the natural-color image. Obviously, there are 256

Raw image

↓

Grayscale processing

↓

Edge detection

Canny opretor

↓

Image segmentation

↓

Crack monitor

↓

Result

FIGURE 17.2 The IPT-oriented framework for concrete cracks detection.

grayscale variations per pixel in the grayscale image, and the amount of grayscale image calculation is significantly lower than that of a true color image (Figures 17.3 and 17.4).

There are four grayscale processing methods: weighted average, mean, component, and max method. In this study, we directly use the rgb2gray function in MATLAB® software corresponds to the weighted average method.

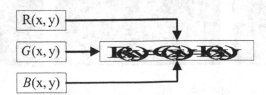

$R(x, y)$

$G(x, y)$

$B(x, y)$

FIGURE 17.3 Grayscale processing.

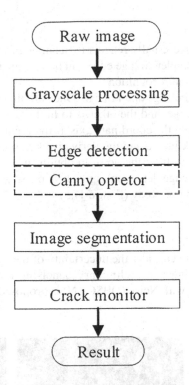

198	192	189	190	196	190	185
192	187	189	193	194	191	188
191	188	191	195	193	192	192
192	189	189	191	192	192	193
197	194	193	193	192	191	191
197	195	196	195	193	193	194
198	198	198	198	196	197	198
200	200	201	199	200	201	201
200	201	202	201	202	202	201
199	201	203	201	201	199	198
198	200	202	201	200	198	195
198	199	201	200	201	197	194

FIGURE 17.4 Grayscale image.

17.3.2 EDGE DETECTION

The edge of an object in the image usually reflects the change of feature information or a certain development trend. Whether a complete image edge can be obtained has an important influence on subsequent image segmentation and recognition.

There are two ways to detect the edge of an image. One is to find the maximum value of the first derivative as the point of crack edge, and the other is to find the zero point of the function image of the second derivative as crossing the coordinate axis as the point of crack edge. However, due to the presence of noise in the crack image and the derivative being more sensitive to noise, the corresponding filter should be used to filter the noise to obtain the complete crack edge as possible. In order to improve the effect of edge detection as much as possible, the Canny operator is used to calculate the edge of the target in this study (Figure 17.5).

17.3.3 IMAGE SEGMENTATION

The main work of this section is to consider the uncertainty of the crack shape of the bridge. Based on the connected domain marking and preliminary denoising crack image, the Density-Based Spatial Clustering of Application with Noise (DBSCAN) is proposed to realize the segmentation of the bridge crack image (Figure 17.6).

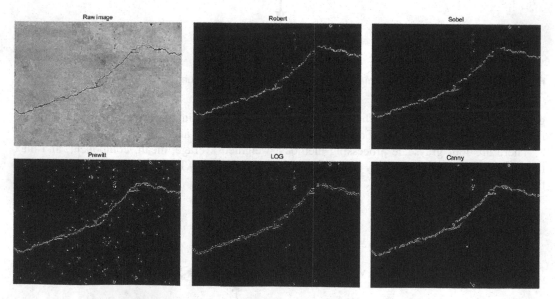

FIGURE 17.5   Edge detection: comparison of five edge detection operators.

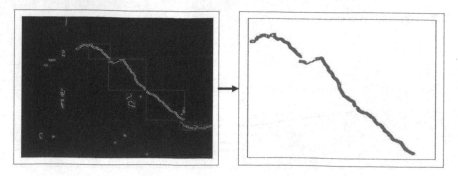

FIGURE 17.6   Image segmentation based on density clustering.

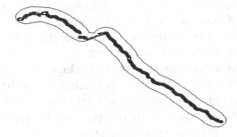

FIGURE 17.7 Crack monitor: crack profile description.

Compared with the traditional distance-based clustering algorithm (such as K-means), the DBSCAN does not need to preset the clustered categories. So, it can find different shape categories and remove noise by input filtering parameters if needed.

Image segmentation based on density clustering belongs to the image segmentation method combining specific theoretical methods and segmentation ideas. This method can divide each pixel in the image into the corresponding feature space and divide it into different categories according to the unique properties of different pixel points. The category then maps the structure to the image and produces the result of the image segmentation.

17.3.4 CRACK MONITOR

As mentioned earlier, a complete crack skeleton can be extracted separately after processing the crack image. The work that needs to be done now is to explore a way to describe the trend of crack width in bridges. In view of the key parts of the existing cracks and the width of the cracks being still in the development stage, a "non-image" expression is sought to describe the development trend of bridge crack width (Figure 17.7).

As the width of the crack increases, the number of data points in the sample will gradually increase, and the range of the distribution will expand outward. Similarly, the reconstruction error of each data point relative to the total space of the sample will also change.

Therefore, the range of KPCA description of the boundary of the crack image will be correspondingly enlarged, and the value corresponding to the contour line will also have a changing trend, so that the visual image change of the crack width can be changed by "non-image" - numerical value. The way to represent is to play a monitoring and early warning role for the development of crack width in key parts of the bridge.

17.4 TESTING IMAGES

To validate the proposed IPT-based concrete cracks detection from the previous section, abundant tests are conducted, and the results are presented. According to the unique structural density system of the object in the image, DBSCAN automatically classifies the feature points while excluding the noise points, and finally extracts the category system of interest in the image to achieve the purpose of image segmentation. Note that KPCA describes the cracks profile by constructing the maximum value in the sample's overall space.

17.5 CONCLUSIONS

An IPT-based approach for detecting cracks on concrete bridge image was proposed. The architecture of IPTs consist of Grayscale processing, Edge detection, Image Segmentation, and Crack monitor.

The main advantage of the proposed IPT-based detection of concrete cracks is that it can automatically extract the edge of concrete cracks and monitor the variation of concrete crack width compared with traditional detection approaches. Therefore, this approach can greatly save labor input and improve the efficiency of bridge management.

At present, there are many achievements in the technology of concrete crack recognition. But, one common limitation of almost all IPT-based approaches is that the incapability of perceiving internal deep feature information due to the natural environment. In future, the IPT for concrete crack detection will be developed to detect different types of damage information, such as cracks, delamination, holes, and corrosion of concrete structures.

REFERENCES

Benmansour, F., Cohen, L.D., 2009. Fast object segmentation by growing minimal paths from a single point on D or 3D images. *Journal of Mathematical Imaging and Vision* 33, p. 209–221.

James, T., Vivek, K. and Anthony, Y., 2013. Automating the crack map detection process for machine operated crack sealer. *Automation in Construction*. 31, p. 10–18.

Li, G., Zhao, X., Du, K., Ru, F., Zhang, Y., 2017. Recognition and evaluation of bridge cracks with modified active contour model and greedy search-based support vector machine. *Automation in Construction*. 78, p. 51–61.

Mathurin, R., Velinsky, S.A., 2000. Simulated annealing for the optimal trajectory planning of an automated crack sealing machine. *Proceedings ASME Design Technical Conferences*.

Nishikawa, T., Yoshida, J., Sugiyama, T., Fujino, Y., 2012. Concrete crack detection by multiple sequential image filtering. Comput.-Aided Civ. *Infrastructure*. 27, p. 29–47.

Pynn, J., Wright, A., Lodge, R., 1999. Automatic identification of cracks in road surfaces. *Conference Publication on Image Processing and Its Applications*. vol. 465, IEEE, Piscataway.

Sluzek, A., 2010. Novel machine vision methods for outdoor and built environments. *Automation Construction*. 19, p. 291–301.

Sunil, K.S., Paul W.F., 2006. Automated detection of cracks in buried concrete pip images. *Automation in Construction*. 15, p. 58–72.

Tung, P., Hwang, Y., Wu, M., 2002. The development of a mobile manipulator imaging system for bridge crack inspection. *Automation in Construction*. 11, p. 717–729.

Valenca, J., Dias-da-Costa, D., Julio, E.N.B.S., 2012. Characterisation of concrete cracking during laboratorial tests using image processing. *Construction and Building Materials*. 28, p. 607–615.

Zhu, Z., Stephanie, G. and Ioannis, B., 2011. Visual retrieval of concrete crack proper-ties for automated post-earthquake structural safety evaluation. *Automation in Construction*. 20, p. 874–883.

Zou, Q., Cao, Y., Li Q., Mao, Q., Wang, S., 2012. CrackTree: Automatic crack detection from pavement images. *Pattern Recognitions Letters*. 33, p. 227–238.

18 Pedestrian Induced Vibration of Slab

Yan-an Gao
Huaiyin Institute of Technology

CONTENTS

18.1 INTRODUCTION

The vibration problems of those slender structures induced by human excitation are still a big challenge. Some uncertainties with human locomotive behaviors and body dynamics properties result in the complexity in describing a pedestrian. In order to explore the effect of pedestrian excitation on structure, Rainer et al.[1] earlier measured the dynamic load factors (DLF) during walking, running and jumping based on an instrumented platform. The force-time history of walking foot[2] showed that ground reaction force from walking excitations occupies an M shape, which can be fitted by Fourier function.[3] Although the ground reaction force model can describe the excitations from locomotive pedestrians, the contributions from pedestrian dynamic properties can't be included. For finding the influencing mechanism of human body characteristics, pedestrian is modeled as a mass-spring (MS) system[4] to obtain the variation of structural dynamic characteristics. An MS model is assumed to maintain an upright pedestrian, and the damping from human is ignored. Later, an inverted pendulum (IP) model[5,6] is further introduced to simulate the stepping behaviors of locomotive pedestrian. But the instability of this model is a big challenge to simulate some practical walking behaviors of human. Further research confirms that a bipedal model[7] is more applicable for reproducing the gait behaviors of pedestrian. A human-structure dynamic interaction model based on the bipedal system is proposed to describe the effect of walking pedestrian on footbridge[8] and lively slab[9]. However, the dynamic stability with walking locomotion in this model is seriously dependent on a constant energy feedback mechanism. To improve the bipedal stability and broaden its application scopes, Gao et al.[10–13] introduced a self-driven mechanism, and it makes the bipedal model to describe the more complicated walking behaviors with crowd. It further describes the pedestrian-induced lateral vibration with a slender footbridge.[14] At present, the studies on pedestrian induced variations of vibratory properties of slab are rarely explored. In this paper, the pedestrian is simplified as a lump mass-damp-spring system with one lump mass and two damping-spring legs. The slab is simply supported and the modal superposition method[15] is applied to describe the dynamic behaviors of slab. In addition, the pedestrian model is introduced to analyze the multi-layout crowd effects. It quantitatively tries to describe the relationship between crowd size and the variations of structural frequency, damping and vibration enduring crowd excitations.

18.2 GOVERNING EQUATION

It is assumed that pedestrian remains in an upright state as in Figure 18.1. The acting force from the pedestrian is replaced by the pseudo-ground reaction force[3]. In addition, following assumptions are set up. First, all pedestrians remain upright standing on a slab. Second, all pedestrians are assumed to have the same parameters such as lump mass m_p, leg stiffness k_{leg} and leg damp c_{leg}. In addition, only the first mode of the structure is considered and the contributions from higher modes are ignored.

The governing equation of the slab with crowd action can be obtained as

$$\ddot{W} + 2\xi_{cp}\omega_{cp}\dot{W} + \omega_{cp}^2 W = \frac{m_p g \mu}{M_1}\left(1 + \sum_{j=1}^{4}\alpha_j \sin\left(2\pi i f_p t\right)\right) \tag{18.1}$$

where

$$2\xi_{cp}\omega_{cp} = 2\xi_s\omega_s + \frac{c_{leg}}{M_s}\sum_{q=1}^{N}\Phi_{1,1}^{q} \tag{18.2}$$

$$\omega_{cp} = \sqrt{\omega_1^2 - \frac{2\delta k_{leg}}{M_s}\sum_{q=1}^{N}\Phi_{1,1}^{q}} \tag{18.3}$$

where N means the number of pedestrians or crowd size; ξ_{cp} and $\omega_{cp} = 2\pi f_{cp}$ are damping ratio and circular frequency of the crowd-slab system, respectively. $\mu = \sum_{q=1}^{N}\phi_1\left(x_q, y_q\right)$ is the distribution coefficient with the coordinate $\left(x_q, y_q\right)$ of pedestrian. The structural frequency ratio between f_{cp} with crowd and f_1 without crowd is given as Eq. (18.4). The structural damping ratio between ξ_{cp} with crowd and ξ_1 without crowd is given as Eq. (18.5).

$$\frac{f_{cp}}{f_1} = \sqrt{1 - \frac{2\delta k_{leg}}{\omega_1^2 M_1}\chi_N} \tag{18.4}$$

$$\frac{\xi_{cp}}{\xi_1} = \left(\frac{f_{cp}}{f_1} + \frac{c_{leg}}{2\xi_1\omega_{cp}M_1}\chi_N\right) \tag{18.5}$$

where $\chi_N = \sum_{q=1}^{N}\phi_1^2\left(x^q, y^q\right)$ is the distribution coefficients of crowd on slab. It is noted that the frequency f_{cp} and damping ratio ξ_{cp} of structure with crowd are related with the crowd size and their distribution situations. The more intensive crowd distribution or larger crowd size would result in lower frequency and higher damping.

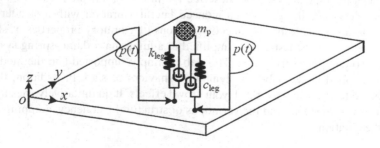

FIGURE 18.1 Mechanical diagram of mass-damp-spring slab system.

The maximum response of the system under steady excitations can be obtained as

$$\ddot{W}_{max} = \frac{\beta m_p g}{M_1} \sum_{j=1}^{4} \alpha_j \qquad (18.6)$$

where M_1 is the modal mass of slab, α_j is the DLF and g means the gravity acceleration. The β is defined as a dynamic amplification factor (DAF), which denotes the response ratio induced by dynamic excitation and static load.

$$\beta = \frac{\mu \left(f_p / f_{cp} \right)^2}{\sqrt{\left[1 - \left(f_p / f_{cp} \right)^2 \right]^2 + \left[2 \xi_{cp} \left(f_p / f_{cp} \right) \right]^2}} \qquad (18.7)$$

For verifying the introduced model, a flexible 12-m square slab[16] is applied to endure walking pedestrian excitation. The structural parameters are as follows: Poisson ratio $v = 0.2$, damping ratio $\xi = 1\%$, Young's modulus $E = 3.0 \times 10^{10} \text{N/m}^2$ and density $\rho = 2,400 \text{ kg/m}^3$. The slab thickness h is just 0.15 m and its four sides are simply supported. A typical pedestrian model is adopted with a mass of $m_p = 80$kg, leg length $l_0 = 1.0$ m, stiffness $k_{leg} = 20$ kN/m and damping ratio $\xi_{leg} = 8\%$. Four layouts are analyzed and they have uniform distribution in Figure 18.2a, interval distribution in Figure 18.2b, triangle distribution in Figure 18.2c and diagonal line distribution in Figure 18.2d. Pedestrians are arranged standing on the square lattices without overlap. The slab is divided to $n_x \times n_y$ grids with the numbers of n_x and n_x in x and y directions, respectively. Pedestrians are always represented with the largest scale in two directions according to these distribution formats.

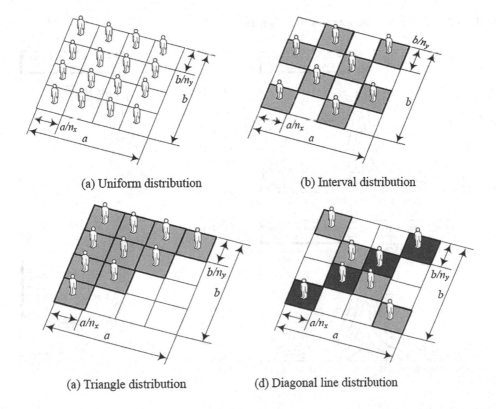

(a) Uniform distribution   (b) Interval distribution

(a) Triangle distribution   (d) Diagonal line distribution

FIGURE 18.2 Four layout formats of crowd on slab.

Figure 18.3a gives the frequency ratio of slab with crowd to without crowd effect under different layouts of crowd on structure. The values of M_p/M_s denote the mass ratios of crowd to structure in the tail end points rather than the whole curves. The frequency ratios in subfigure (a) are decreased along with the increase of grid density, which indicates that the denser crowd would cause the greater deterioration on structural vibration. The layout of uniform distribution has a largest change on the slab frequency, and diagonal line distribution has a slightest change on the frequency. The parts of reasons are that the uniform distribution has the larger mass ratio between crowd and structure. However, the damping ratios of slab including crowd effect to without crowd have some opposite effects in subfigure (b). The larger grid densities or mass ratios would result in the stronger damping capacity of slab. Especially, the uniform distribution has the most remarkable effect.

The excitation frequency of crowd on structural responses is plotted in Figure 18.4. The red and blue lines mean with and without crowd effect, respectively. Compared with the β peaks without crowd effect, the peaks with crowd effects are lowered and they weaken the vibrations of slab.

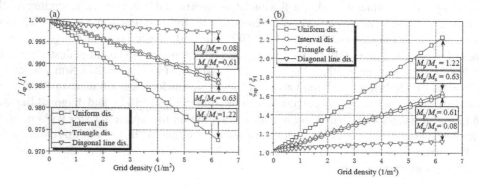

FIGURE 18.3 Effects of crowd size on slab properties.

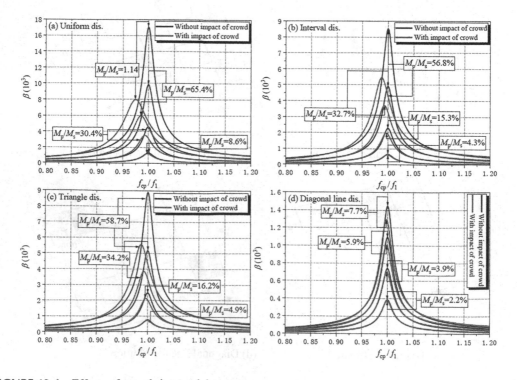

FIGURE 18.4 Effects of crowd size on slab responses.

This is due to the presence of crowd the fundamental frequency of slab is reduced, which effectively results in bypassing the resonance when the excitation ratio is $f_{cp}/f_1 = 1$. The frequencies with the peaks with crowd effect are left shifted due to the decrease of fundament frequency under crowd contributions. The DAF β peaks are increased along with the increase of M_p/M_s to all distributions. Comparing other distribution formats, the uniform distribution triggers the higher peak values of DAF.

The above layouts reveal that the distribution situations of crowd on slab have some significant effects on structural properties and responses. The uniform distribution is the worst to changing frequency of structure, and exciting its vibration. However, it also has the most significant improvement of structural damping capacity. It is noted that the crowd can distinctly reduce the vibration than the situation without crowd effect under resonant excitations. In order to explore the crowd effect on the properties of slab, some crowds with constant mass ratios to structure have uniformly radiant layout with only single round from the center point to the boundary of slab (Figure 18.5). The perpendicular distance of pedestrian to the center point is $\eta a/2$ or $\eta b/2$. A position coefficient η is introduced to represent pedestrian location on the slab and it ranges from 0 to 1. The effects of crowd layout on structural properties are showed in Figure 18.6. It is noted that the most drastic effects do not occur in the center position ($\eta = 0$). The ratio of f_{cp}/f_1 is first decreased and then increased along with an increase of position coefficient η in Figure 18.6a. The larger ratio of M_p to M_s would result in the lower frequency of slab. The worst effects on the slab frequency approximately occur in the situation that crowd locate the position $\eta = 0.25$. The effects of crowd layout on structural damping are

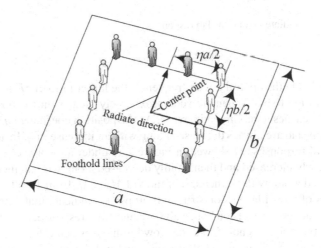

FIGURE 18.5 Radiate layout diagram of crowd on slab.

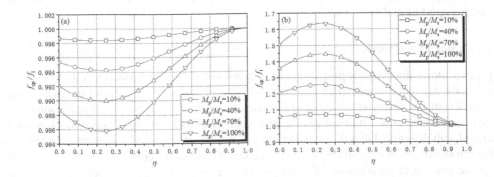

FIGURE 18.6 Effects of radiate layout on slab properties.

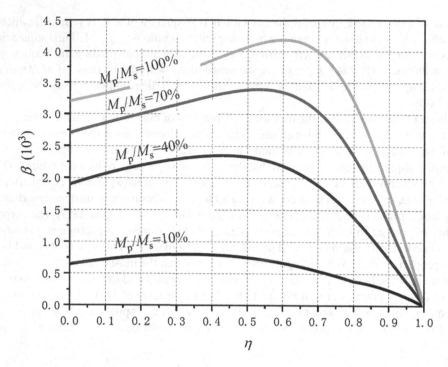

FIGURE 18.7   Effects of radiate layout on slab responses.

plotted in Figure 18.6b. On the contrary, with frequency, the larger ratio of M_p to M_s would result in the higher damping ratio of slab. When crowd approximately locates on the position $\eta = 0.25$, the damping ratio of slab reaches a maximum increase. As the position coefficient η is larger than 0.25, the effects of crowd on slab are always decreased along with an increase of η. In addition, the effects of crowd on structural responses are showed in Figure 18.7 under a resonance excitation of $f_p = f_1$. The DAF β is first slowly increased and then rapidly decreased along with an increase of η. In addition, β is also increased along with an increase of ratio of M_p to M_s from 10% to 100%.

The investigations of crowd layouts on structural properties indicate that a uniform distribution of crowd has the most adverse effects on both slab frequency and response, but greatly improves the damping capacity of the slab. It is noted that the crowd influences can effectively reduce structural response than the response without crowd effects on structural dynamic properties. It is noted that the most adverse position of crowd on structural properties will not be in the center point of slab. The layout approaching the boundary position can effectively reduce the influences of crowd on structural properties and responses.

18.3   CONCLUSIONS

The paper explores the dynamic properties and responses of slab under crowd excitations. Numerical example indicates that the pedestrian can deteriorate the fundamental frequency but can improve the damping of slab. Crowd layouts on structural properties indicate that a uniform distribution of crowd has the most adverse effects on both slab frequency and response, but greatly improves the damping capacity of slab. It is noted that the crowd influences can effectively reduce structural response than without crowd effects on structural dynamic properties. The most adverse position of crowd on structural properties would not occur in the center point of the slab. The layout approaching the boundary position can effectively reduce the influence of crowd on structural properties and responses.

ACKNOWLEDGEMENTS

The research is supported by the National Natural Science Foundation of China on No. 51808247, Natural Science Research in Jiangsu Colleges and Universities on No.18KJB560003.

REFERENCES

1. Rainer J H and Pernica G, 1986, Vertical dynamic forces from footbridges, *Canadian Acoustics*, 14 (2), 12–21.
2. Ebrahimpour A, Hamam A, Sack R L and Patten W N, 1996, Measuring and modeling dynamic loads imposed by moving crowds, *Journal of Structural Engineering*, 122 (12), 1468–1474.
3. Li Q, Fan J, Nie J, Li Q and Chen Y, 2010, Crowd-induced random vibration of footbridge and vibration control using multiple tuned mass dampers, *Journal of Sound and Vibration* 329, 4068–4092.
4. Zhou D and Ji T, 2006, Dynamic characteristics of a beam and distributed spring-mass system, *International Journal of Solids and Structures* 43, 5555–5569.
5. Hof A L, Gazendam M G J and Sinke W E, 2005, The condition for dynamic stability, *Journal of Biomechanics* 38, 1–8.
6. Macdonald, 2009, Lateral excitation of bridges by balancing pedestrians, *Proceedings of the Royal Society A* 465, 1055–1073.
7. Yang Q S, Qin J W and Law S S, 2015, A three-dimensional human walking model, *Journal of Sound and Vibration* 357, 437–456.
8. Qin J W, Law S S, Yang Q S and Yang N, 2013, Pedestrian-bridge dynamic interaction, including human participation, *Journal of Sound and Vibration* 332, 1107–1124.
9. Perez M C and Lorenzana A, 2017, Walking model to simulate interaction effects between pedestrians and lively structures, *Journal of Engineering Mechanics* 143 (9), 1–9.
10. Gao Y A and Yang Q S, 2018, A theoretical treatment of crowd-structure interaction, *International Journal of Structural Stability and Dynamics* 18 (1), 1–17.
11. Gao Y A, Wang J and Liu M, 2017, The vertical dynamic properties of flexible footbridges under bipedal crowd induced excitation, *Applied Sciences* 7 (7), 677.
12. Gao Y A, Yang Q S and Dong Y, 2018, A three-dimensional pedestrian-structure interaction model for general applications, *International Journal of Structural Stability and Dynamics* 18 (9), 1–27.
13. Gao Y A, Yang Q S and Qin J W, 2017, Bipedal crowd-structure interaction including social force effects, *International Journal of Structural Stability and Dynamics* 17 (7), 1–31.
14. Yang Q S and Gao Y A, 2018, A theory treatment of pedestrian-induced lateral vibration of structure, *Journal of Dynamic Systems, Measurement, and Control* 140, 1–13.
15. Catbas F, Pakzad S, Racic V, Pavic A and Reynolds P, Topics in dynamics of civil structures, *Proceedings of 31st IMAC, A Conference on Structural Dynamics*, Springer, Berlin.
16. Han H X and Zhou D, 2017, Mechanical parameters of standing body and applications in human-structure interaction, *International Journal of Applied Mechanics*, 9 (2), 1–30.

19 Piezoelectric Admittance-Based Damage Detection via Data Compression and Reconstruction

H. Li, D. Ai, and H. Zhu
Huazhong University of Science and Technology

H. Ge
Meijo University

H. Luo
Huazhong University of Science and Technology

CONTENTS

19.1 INTRODUCTION

Since conventional electromechanical admittance (EMA) technique is unable to satisfy the demand for meticulous structural damage detection in the structural health monitoring (SHM) field, inter-disciplines such as wireless sensing, signal processing and data driven theories are investigated and extended to study/engineering applications. As a matter of fact, data compression and reconstruction have raised the attentions of researchers recently, while there are several challenges for the practical application of piezoelectric ceramic transducer (PZT) and monitoring networks, including a substantial amount of sensor data required to implement damage detection analyses in SHM systems.

Sohn and Farrar (2000) proposed a pattern recognition strategy for structural vibration signal and applied the approach to a bridge column and a surface-effect fast boat. Park et al. (2008) combined principal components analysis (PCA) based data compression with K-means clustering algorithm to realize a wireless SHM system. Duan and Kang (2014) compared the data recovery performance among the continuous basis pursuit (CBP), semidefinite programming (SDP) and spectral iterative hard threshold (SIHT), and investigated the precision and time consuming via a numerical three-story frame model. Bao et al. (2011, 2013) extended the compressive sampling (CS) based data compression and data loss recovery to SHM applications. In addition, an adaptive sparse time-frequency analysis method was suggested to identify the time-varying cable tension forces of bridges (Bao et al. 2017). Furthermore, compressed sensing theory was adopted to compress structural response data from a reinforced concrete (RC) structure, and the damage detection process within the reconstructed data showed high accuracy and effective compression (Jayawardhana et al. 2017).

In this study, an innovative approach of structural damage detection combined piezoelectric admittance technique with Orthogonal Matching Pursuit (OMP) based data reconstruction method was proposed and investigated. A damage detection experiment of simply-supported steel beam was conducted to validate the feasibility of the proposed approach, where six PZT sensors were arranged to obtain measurements. Parameter analysis with respect to sparsity level K was conducted to explore the correlation with total error (TE) and compression ratio (CR), thus obtaining an optimal sparsity level. During the damage evaluation stage, root-mean-square deviation (RMSD) was utilized to diagnose structural health status.

19.2 MAIN THEORIES

19.2.1 PIEZOELECTRIC CERAMIC TECHNIQUE

As an intelligence material, applications of PZTs have been widely accepted in sensors and actuators utilized for structural damage detection and health monitoring based on the direct and converse piezoelectric effect of piezoelectric ceramic.

In order to take the inter-reaction of the bonding layer between the PZT patch and the structure into consideration, by which PZT sensors can be stuck to the measured structure, an improved one-dimensional admittance model is proposed. As shown in Figure 19.1, the bonding layer and the structure are considered as a single-degree-of-freedom system (Xu and Liu 2002),

$$Y = j\omega \frac{b_a l_a}{h_a} \left(\bar{\varepsilon}_{33}^{\sigma} + \frac{d_{31}^2 \bar{E}_{11}^E Z_a(\omega)}{\xi Z_s(\omega) + Z_a(\omega)} \frac{\tan(\kappa l_a)}{\kappa l_a} - d_{31}^2 \bar{E}_{11}^E \right) \qquad (19.1)$$

where j is $(-1)^{1/2}$; ω denotes the angular frequency of excitation; b_a, l_a and h_a denote the width, length and thickness of the PZT patch, respectively; $\bar{\varepsilon}_{33}^{\sigma} = \varepsilon_{33}^{\sigma}(1 - \delta j)$ is the complex electric permittivity of the PZT patch at a constant stress; δ is the dielectric loss factor of the PZT patch; η is the structural mechanical loss factor; $Z_s(\omega)$ and $Z_a(\omega)$ denote the mechanical admittance of the structure and the PZT patch, respectively; d_{31} is the coupling PZT constant; $\bar{E}_{11}^E = E_{11}^E(1 + j\eta)$ is the complex Young's modulus of the PZT patch at a constant electric field; $\kappa = \omega\sqrt{\rho/\bar{E}_{11}^E}$ is the wave number, ρ is the density of the patch; ξ modifies the function of the structure in the dynamic interaction.

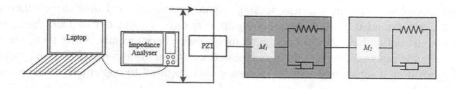

FIGURE 19.1 One-dimensional interaction model of a PZT structure considering the influence of bonding layer.

19.2.2 COMPRESSED SENSING THEORY

CS mainly contains a sparse representation of the collected signal, design of measurement matrix and signal reconstruction. Prerequisite of CS is that the original signal presents a good sparsity itself or shows a sparse characteristic in some transform (Davis et al. 1997, Candes 2006, Donoho 2006). Generally speaking, original experimental signal doesn't have a sparse characteristic itself, no matter in the time or frequency domain. To find the sparse land of the original signal, transforms such as wavelet basis, Fourier basis and cosine basis are suggested to implement on the original signal.

Introducing an $n \times n$ matrix $\Psi = [\psi_1, \psi_2,..., \psi_n]$, where ψ_i is a column vector, discrete time signal $X(i)$, $i = 1, 2,..., n$ in $\mathcal{R}^n$ can be transformed into α, which is the corresponding basis coefficient matrix in the orthogonal basis Ψ (Candes et al. 2006, Candes and Wakin 2008).

$$X = \Psi\alpha \tag{19.2}$$

Within a linear projection of X, a data vector Y of length m can be obtained:

$$Y = \Phi X \tag{19.3}$$

where Φ is called the measurement matrix.

Substituting Eq. (19.2) into Eq. (19.3), Y is obtained as follows:

$$Y = \Phi X = \Phi\Psi\alpha = \Theta\alpha \tag{19.4}$$

where Θ is called the sensing matrix.

If the basis coefficients α in Eq. (19.4) have K ($K \ll N$) non-zero values, decay exponentially and approximate zero after sorting, the signal is called K-sparse in domain (Candes et al. 2006, Candes and Wakin 2008).

Equation (19.4) is known as a non-deterministic polynomial problem, and the signal could be reconstructed directly. If the sensing matrix Θ satisfies a so-called restricted isometry property (RIP) (Candes 2006, Donoho, 2006):

$$1 - \delta_k \leq \frac{\|\Theta v_k\|_2^2}{\|v_k\|_2^2} \leq 1 + \delta_k \tag{19.5}$$

for all K-sparse vectors v_k, where δ_k is an isometry constant with $\delta_k \in (0,1)$, can the K-sparse signals be accurately reconstructed via the optimization problem.

The basis coefficients α can be reconstructed by solving the l_1 optimization problem:

$$\hat{\alpha} = \arg\min \|\tilde{\alpha}\|_1 \ \text{ such that } \Theta\alpha = y \tag{19.6}$$

19.2.3 ORTHOGONAL MATCHING PURSUIT

The prerequisite of compressed sensing results in sparse characteristic in some basis of the original signal. If the signal turns to own K-sparse characteristic when projected onto an appropriate orthogonal basis, the accuracy of the signal reconstruction can be ensured. Generally, Gaussian random matrix and Bernoulli random matrix are selected as common measurement matrices.

Mallat and Zhang (1993) presented an iterative algorithm based on Matching Pursuit (MP), where orthogonal projection is expressed as

$$P_V f = \sum_n a_n x_n \tag{19.7}$$

where P_V is the orthogonal projection operator onto V.

$$f = \sum_{i=1}^{k} a_i x_{n_i} + R_k f = f_k + R_k f \tag{19.8}$$

where f_k is the current approximation and $R_k f$ is the current residual.

Introducing the kth-order model for $f \in H$,

$$f = \sum_{n=1}^{k} a_n^k x_n + \mathrm{R}_k f, \text{ with } \langle \mathrm{R}_k f, x_n \rangle = 0, \ n = 1,\ldots,k \tag{19.9}$$

where the superscript k in a_n^k denotes the different weights on different model orders. The above kth-order model is updated to a model of $(k+1)$th-order,

$$f = \sum_{n=1}^{k+1} a_n^{k+1} x_n + \mathrm{R}_{k+1} f, \text{ with } \langle \mathrm{R}_{k+1} f, x_n \rangle = 0, \ n = 1,\ldots,k+1 \tag{19.10}$$

An auxiliary model for the relationship between x_{k+1} and all the previous x_k ($k = 1, 2, \ldots, n$) is required to perform the transform (Pati et al. 1993, Needell and Vershynin 2010).

$$x_{k+1} = \sum_{n=1}^{k} b_n^k x_n + \gamma_k, \text{ with } \langle \gamma_k, x_n \rangle = 0, \ n = 1,\ldots,k \tag{19.11}$$

According to Eqs. (19.7) and (19.8), it can be obtained that

$$\sum_{n=1}^{k} b_n^k x_n = P_{V_k} x_{k+1} \tag{19.12}$$

$$\gamma_k = P_{V_k^\perp} x_{k+1} \tag{19.13}$$

Thus the correct form can be updated as

$$a_n^{k+1} = a_n^k - a_k b_n^k, \ n = 1,\ldots,k \tag{19.14}$$

$$a_{k+1}^{k+1} = \alpha_k = \frac{\langle \mathrm{R}_k f, x_{k+1} \rangle}{\langle \gamma_k, x_{k+1} \rangle} = \frac{\langle \mathrm{R}_k f, x_{k+1} \rangle}{\|\gamma_k\|^2} = \frac{\langle \mathrm{R}_k f, x_{k+1} \rangle}{\|x_{k+1}\|^2 - \sum_{n=1}^{k} b_n^k \langle x_n, x_{k+1} \rangle} \tag{19.15}$$

The residual $R_{k+1} f$ satisfies $R_k f = R_{k+1} f + \alpha_k \gamma_k$, and

$$\|R_k f\|^2 = \|R_{k+1} f\|^2 + \frac{|\langle R_k, x_{k+1} \rangle|^2}{\|\gamma_k\|^2} \tag{19.16}$$

19.3 VALIDATION OF THE APPROACH USING MONITORING DATA FROM AN EXPERIMENTAL SIMPLY-SUPPORTED BEAM

19.3.1 DESCRIPTION OF THE SIMPLY-SUPPORTED BEAM EXPERIMENT

In this article, an experiment of a steel beam with $500 \times 35 \times 5$ mm (length × width × thickness) rectangular section was implemented to validate the proposed approach. The steel beam was simply supported on both the ends within block rubbers on a desk. Six PZT patches were bonded to the beam as shown in Figure 19.2. PZT#1 and PZT#6 were 200 mm apart, PZT#2 and PZT#5 were 100 mm apart, and PZT#3 and PZT#4 were 50 mm apart from the midspan, respectively.

Four different cases were designated to investigate the feasibility of the proposed approach for identifying structural defects when encountering sensor faults. Case 1: to serve as a reference, all the sensors were perfectly bonded to the steel beam except PZT#4, which was partially bonded as approximately three-quarter area of the patch was painted with high strength epoxy, while the steel beam was undamaged; Case 2: PZT#5 was scratched with a knife as slight damage while PZT#6 was damaged to several fissures as severe damages relatively; Case 3: eight coins each with a diameter of 19 mm, a thickness of 1.67 mm and a mass of 3.20 g were placed in the middle of the beam span, as a kind of slight structural damage caused by added mass; Case 4: a notch with the dimension of $35 \times 4 \times 2$ mm (length × width × thickness) was cut on the midspan, as a kind of slightly severe structural damage. The whole experimental process was under approximately 20°C, and all the admittance measurements were collected via a minitype SHM system consisting of a laptop and Agilent 4294A analyzer within a 1 V voltage excitation. The measurement frequency ranged from 80 to 480 kHz.

19.3.2 ADMITTANCE DATA RECONSTRUCTION USING OMP ALGORITHM

Above all, linear normalization processing was performed onto all the original admittance data to increase the stability of dataset,

$$x' = \frac{x - \min(x)}{\max(x) - \min(x)} \tag{19.17}$$

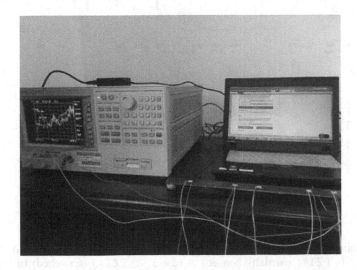

FIGURE 19.2 Measuring system for a steel beam with damaged sensors by Agilent 4294A admittance analyzer.

The comparison between original discrete admittance signal $X(t)$ and reconstructed signal $Y(t)$ in different cases are shown in Figure 19.3. All the measurements are on the condition that sparsity level K ranges from 400 to 2,100, while the number of observation alters correspondingly. Figure 19.3 shows partial admittance variation tendencies along with frequency in different cases. Three kinds of sensor damage severities, including healthy, scratched and damaged states are set up to consider sensor faults.

Concretely, signals from PZT#1–#3, PZT#5, and PZT#6 which are perfectly bonded in case 1 are utilized as a baseline, while a partially bonded sensor PZT#4 is designated as a reference to provide the situation suffering from bonding issues. In cases 2, 3 and 4, PZT#1–#3 are regarded as healthy sensors in a so-called regular monitoring situation, while PZT#5 and PZT#6 represent damaged sensors, which can be utilized to investigate the identification effectiveness when encountering sensor faults.

19.3.3 DATA COMPRESSION EFFECTS

CR is introduced here to measure the data compression performance:

$$CR = \frac{S_{orig}}{S_{comp}}$$ (19.18)

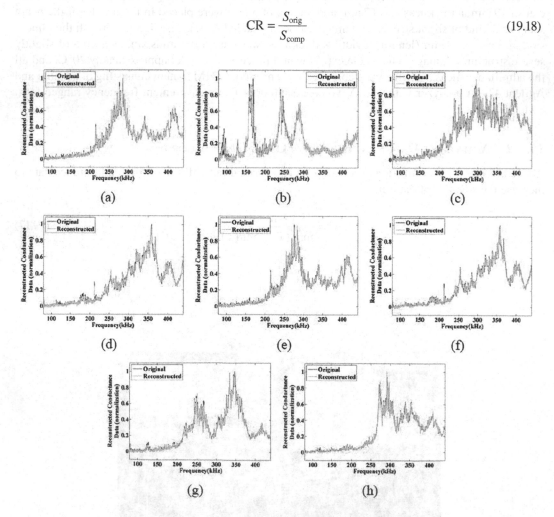

FIGURE 19.3 Comparison of original admittance data with reconstructed data: (a) PZT#2 (perfectly bonded) in case 1, (b) PZT#4 (partially bonded) in case 1, (c) PZT#5 (scratched) in case 2, (d) PZT#6 (damage) in case 2, (e) PZT#2 (undamaged) in case 3, (f) PZT#6 (damaged) in case 3, (g) PZT#1 (undamaged) in case 4 an (h) PZT#6 (damaged) in case 4.

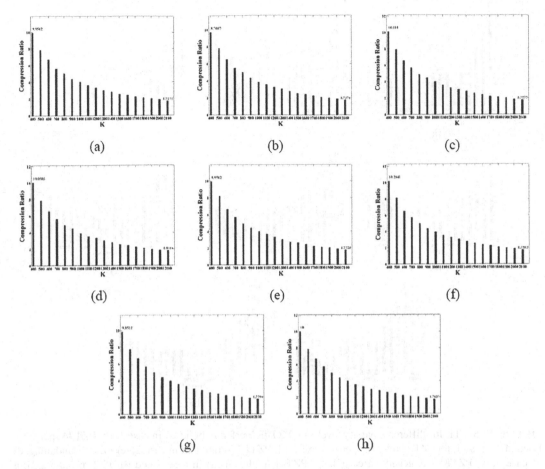

FIGURE 19.4 CR for different sparsity levels: (a) PZT#2 (perfectly bonded) in case 1, (b) PZT#4 (partially bonded) in case 1, (c) PZT#6 (undamaged) in case 1, (d) PZT#6 (damage) in case 2, (e) PZT#1 (undamaged) in case 3, (f) PZT#6 (damaged) in case 3, (g) PZT#3 (undamaged) in case 4 and (h) PZT#5 (scratched) in case 4.

where S_{orig} and S_{comp} denote the original data size and compressed data size, respectively.

Figure 19.4 shows the corresponding CRs for different sparsity levels: For $K = 400$, the CR can be up to around 10, while low to approximately 1.7 for $K = 2,100$. CR turns out to be a nonlinear descending along with sparsity level.

19.3.4 Influence of Sparsity Level on Reconstruction Accuracy

Figure 19.5 demonstrates the corresponding TEs for different sparsity levels to find out the optimal sparsity for each PZT sensor data processing. The majority of the TEs range from 1.67% to 2.79%, except the partially bonded PZT#4 data, whose TE is 6.84% in case 1, 7.24% in case 3 and 6.69% in case 4, respectively, namely that reconstructed admittance data with bonding defect is considered to cause distortions. As a consequence, it is difficult to determine whether the reconstructed data from PZT#4 can be utilized to identify the structural damage.

It can be deduced that the TE value can reach to the lowest level when the sparsity K is in the range of 1,400 to 1,600 for all the admittance data.

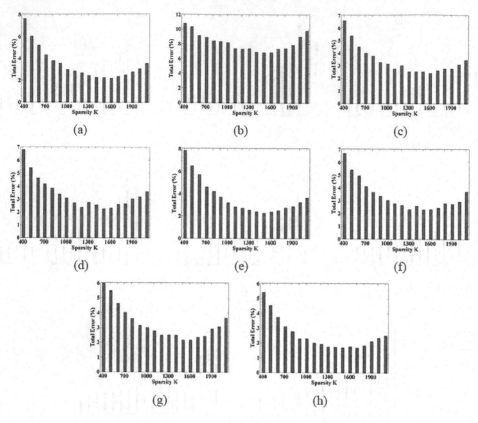

FIGURE 19.5 TE for different sparsity levels: (a) PZT#3 (perfectly bonded) in case 1, (b) PZT#4 (partially bonded) in case 1, (c) PZT#5 (undamaged)in case 1, (d) PZT#5 (scratched) in case 2, (e) PZT#3 (undamaged) in case 3, (f) PZT#5 (scratched) in case 3, (g) PZT#1 (undamaged) in case 4 and (h) PZT#6 (damaged)in case 4.

19.3.5 DAMAGE EVALUATION USING RMSD

The RMSD for the real part of the admittance data is utilized as a damage indicator, expressed as

$$\text{RMSD}(\%) = \sqrt{\frac{\sum_{i=1}^{N}\left(G_i^1 - G_i^0\right)^2}{\sum_{i=1}^{N}\left(G_i^0\right)^2}} \times 100 \qquad (19.19)$$

where G_i^0 denotes the undamaged admittance signature at the i-th measurement point, and G_i^1 denotes the damaged admittance signature at the corresponding point.

The baseline of the RMSD values is calculated before a structural damage is introduced. Figure 19.6a shows that the RMSD index for the original PZT#2 data between case 1 and 3 is 0.071, while the reconstructed data denotes 0.0761, namely that the error is 7.183%. The RMSD index for PZT#3 turns out to be more accurate since the error between the original and reconstructed data denotes only 1.499%.

Furthermore, the difference of the RMSD index between the mass-added and notch-damaged cases can be clearly distinguished, and a distinct increase of RMSD index can be found from case 3 to 4 in both Figure 19.6a, b. The reference of RMSD index for PZT#5 and PZT#6 is set out, and

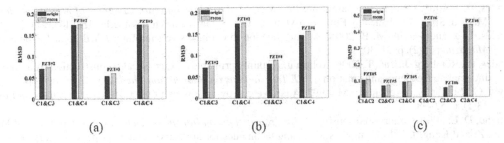

(a)   (b)   (c)

FIGURE 19.6 Comparison of RMSD indices depending on original data and reconstructed data: (a) PZT#2 and PZT#3, (b) PZT#2 and PZT#4 and (c) PZT#5 and PZT#6.

the comparisons of RMSD between original and reconstructed data are shown in Figure 19.6c. The RMSD of PZT#6 in the notch-damaged case is 0.4424, almost eight times more than that in the mass-added case, recognizing which situation the structure is in. However, there is little difference of RMSD for PZT#6 among the reference case, the mass-added case and the notch-damaged case. In addition, Figure 19.6c shows that the error of the reconstructed PZT#6 data in case 3 denotes 17.082%, which indicates the sensor may be damaged and cannot offer effective feedback, whereas those of scratched PZT#5 are all under 6%, which demonstrates that the sensor data are still available, as a result of merely a slight deterioration.

19.4   CONCLUSIONS

This article investigates an innovative approach of damage detection based on PZT technique integrated with OMP based data reconstruction algorithm for SHM. Above all, the original data are collected utilizing PZT technique in a minitype SHM system. Then a linear projection is implemented for the transmitted data x in the above system onto y by a random matrix. A sparsity-adaptive OMP based algorithm improved via parameter analyses is utilized to reconstruct the normalized admittance data. Both the reconstruction accuracy and compression effect of the proposed approach are investigated via an analysis of the admittance data from an experimental simply-supported beam in an SHM system. Noting that the reconstructed data displayed uniform accordance with the original ones, regardless of the sensor states. Moreover, the reconstruction error can be decreased to an acceptable degree through parametric analysis aimed at sparsity level. As to the damage identification effects of the proposed approach, the results demonstrate that the RMSD index can distinguish slight structural damage (mass addition) from slightly severe defects (a notch) to some extent.

Considering that the optimal reconstruction error should be explored through more effective approach, which may result in false alarms in structural damage detection, a future study will be investigated. Furthermore, structural damage quantification based on the proposed approach of PZT technique and data reconstruction need more attention and research. Besides, wireless sensing technology of PZT will be introduced to make the data acquisition system more convenient, while an embedded module design of the algorithm procedures will be developed, so that it can be put into engineering application.

REFERENCES

Bao, Y., Beck, J. L., Li, H. 2011. Compressive sampling for accelerometer signals in structural health monitoring. *Structural Health Monitoring*, 10(3), p. 235–246.

Bao, Y., Li, H., Sun, X., et al. 2013. Compressive sampling-based data loss recovery for wireless sensor networks used in civil structural health monitoring. *Structural Health Monitoring*, 12(1), p. 78–95.

Bao, Y., Shi, Z., Beck, J. L., et al. 2017. Identification of time-varying cable tension forces based on adaptive sparse time-frequency analysis of cable vibrations. *Structural Control and Health Monitoring*, 24(3), p. 1–17.

Candes, E. J. 2006. Compressive sampling. *Proceedings of the International Congress of Mathematicians*, Madrid, Spain, Published by European Mathematical Society, Zurich, Switzerland, p. 1433–1452.

Candes, E. J. and Wakin, M. B. 2008. An introduction to compressive sampling. *IEEE Signal Processing Magazine*, 25(2), p. 21–30.

Candes, E., Romberg, J., Tao, T. 2006. Robust uncertainty principles: exact signal reconstruction from highly incomplete frequency information. *IEEE Transactions on Information Theory*, 52(2), p. 489–509.

Davis, G., Mallat, S., Avellaneda, M. 1997. Adaptive greedy approximations. *Constructive Approximation*, 13(1), p. 57–98.

Donoho, D. L. 2006. Compressed sensing. *IEEE Transactions on Information Theory*, 52(4), p. 1289–1306.

Duan, Z. and Kang, J. 2014. Compressed sensing techniques for arbitrary frequency-sparse signals in structural health monitoring. *Sensors and Smart Structures*, 9061, p. 90612W.

Jayawardhana, M., Zhu, X., Liyanapathirana, R. 2017. Compressive sensing for efficient health monitoring and effective damage detection of structures. *Mechanical Systems and Signal Processing*, 84(1), p. 414–430.

Mallat, S. G. and Zhang, Z. 1993. Matching pursuits with time-frequency dictionaries. *IEEE Transactions on Signal Processing*, 41(12), p. 187–200.

Needell, D. and Vershynin, R. 2010. Signal recovery from incomplete and inaccurate measurements via regularized orthogonal matching pursuit. *IEEE Journal of Selected Topics in Signal Processing*, 4(2), p. 310–316.

Park, S., Lee, J., Yun, C., et al. 2008. Electro-mechanical impedance-based wireless structural health monitoring using PCA-data compression and k-means clustering algorithms. *Journal of Intelligent Material Systems and Structures*, 19, p. 509–20.

Pati, Y. C., Rezaiifar, R., Krishnaprasad, P. S. 1993. Orthogonal matching pursuit: Recursive function approximation with applications to wavelet decomposition. *Proceedings of the 27th Annual Asilomar Conference on Signals, Systems, and Computers*, p. 40–44.

Sohn, H. and Farrar, C. R. 2000. A statistical pattern recognition paradigm for vibration-based structural health monitoring. *Proceedings of Smart Engineering System Design, Neural Networks, Fuzzy Logic, Evolutionary Programming, Complex Systems, and Data Mining*, 10, p. 1001–6.

Xu, Y.G. and Liu, G. R. 2002. A modified electro-mechanical impedance model of piezoelectric actuator-sensors for debonding detection of composite patches. *Journal of Intelligent Material Systems and Structures*, 13(6), p. 389–396.

20 Several Damage Indices Based on Transmissibility for Application in Structural Damage Detection

Jianping Han and Hongyu Zhang
Lanzhou University of Technology

Zhiqiang Hou
The Fifth Construction Co., Ltd. of China TIESIJU Civil Engineering Group

CONTENTS

20.1 INTRODUCTION

Civil engineering structures undergo gradual degradation or suffer from the natural and/or man-made hazards such as earthquakes, hurricanes and explosions during their service life. It is necessary to detect and assess the service condition of structures by appropriate Structural Health Monitoring (SHM) approaches (Brownjohn 2006, Farrar and Worden 2010, Ou and Li 2010). Many vibration-based SHM approaches have been developed, which means modal parameters such as modal frequencies, modal damping ratios and mode shape vectors are identified based on the response of infrastructures, and a structural condition is evaluated based on these identified modal parameters (Farrar et al. 2001, Yan et al. 2007, Limongelli 2019). However, the difficulties such as uncontrolled input, complex excitation device, higher cost and assumption on uncertain input have an adverse effect on the further applications of this kind of approach.

Recently, the transmissibility functions (TFs), as a mathematical representation of the output-to-output relationship, which have been proven to be an effective and potential approach to identify the damage or assess the severity of damage (Maia et al. 2011, Devriendt et al. 2010, Zhou et al. 2016). A transmissibility-based indicator, which is similar to the response vector assurance criterion, was defined to detect damage as well as to quantify the severity in a relative way (Maia et al. 2011). Devriendt et al. (2010) explored a damage detection approach for combining transmissibility measurements under different test cases. Zhou et al. (2016) developed an approach to detect structural damage using TF together with the Mahalanobis distance, hierarchical clustering and similarity analysis. The suitability of the proposed approach was verified by simulated results on a ten-floor structure and experimental tests on a free-free beam. Several comprehensive reviews have been published to review the concept of TFs, challenges and possible trends of TF-based system identification (SI) and condition assessment in the field of SHM (Deraemaeker and Worden 2011, Chesné and Deraemaeker 2013, Yan et al. 2019).

In this paper, the damage detection procedures based on the transmissibility with different distance measures such as Cosh distance, Euclidean distance, Itakura distance and Mahalanobis distance were proposed. And, simply supported steel H-beam structures with and without different damage states were excited by the specified hammer and the acceleration response data of the beams at different locations were recorded. Additionally, the acceleration responses of the undamaged beam were taken as benchmark for comparison, and the acceleration responses of the beam with different damage states were processed. Two corresponding damage indices such as Damage Detection Index (DDI) and Advanced Resulting Index (ARI) were discussed, and the thresholds were set to detect and assess the structural condition.

20.2   FUNDAMENTALS OF TRANSMISSIBILITY

For a linear multi-degree of freedom system subjected to external dynamic forces $f(t)$, the dynamic equilibrium equation of motion is

$$M\ddot{x}(t) + C\dot{x}(t) + Kx(t) = f(t) \tag{20.1}$$

where M, C and K are the mass, damping and stiffness matrices of the system, respectively. The vector $x(t)$ contains displacement responses of the system at all degrees of freedom (DOF)s. $f(t)$ is the external excitation vector of the system.

And then, the direct transmissibility between the test point i and the reference point j can be defined as

$$T_{i,j}(\omega) = \frac{X_i(\omega)}{X_j(\omega)} \tag{20.2}$$

where $X_i(w)$ and $X_j(w)$ denote the fast Fourier transform (FFT) amplitude of system response x_i and x_j under the action of harmonic given coordinates, respectively.

However, for large-scale structures, it is difficult to achieve effective excitation in many cases, and only response signals under natural or dynamic excitation can be obtained. The cross-power spectrum and power spectrum can be used to replace X_i and X_j in Eq. (20.2) when using the data of output response, and it can be represented as

$$T_{i,j}(\omega) = \frac{X_i(\omega)}{X_j(\omega)} = \frac{X_i(\omega) \times X_j(\omega)}{X_j(\omega) \times X_j(\omega)} = \frac{G_{ij}}{G_{jj}} \tag{20.3}$$

where G_{ij} and G_{jj} denote cross-power spectrum and auto-power spectrum of system output responses, respectively.

20.3   DISTANCE MEASURE–BASED DAMAGE DETECTION INDICATORS

The TF of adjacent points of the structure changes with damage. The healthy state and damage information of the structure can be assessed by analyzing the variability of the transmissibility function before and after the damage. Recently, the approach of combining TF between two adjacent points before and after damage with the distance measures has been explored to detect the structural condition.

20.3.1   Cosh Distance

Cosh distance as a symmetric distance measure and the best error criterion in discrete spectral all-pole modeling gives an approach to measure the global difference between two signal spectra (Qiao et al. 2012), which can be defined as

$$d_C(X,Y) = \frac{1}{2N} \sum_{i=1}^{N} \left[\frac{x_i}{y_i} - \log\frac{x_i}{y_i} + \frac{y_i}{x_i} - \log\frac{y_i}{x_i} - 2 \right] \tag{20.4}$$

where N is the dimensional number of Cosh space, $X = [x_1, x_2, \ldots, x_N]$ and $Y = [y_1, y_2, \ldots, y_N]$ denote the FFT coefficient vectors of arbitrary points in N-dimension Cosh space.

When the detected sequence is the same or close to the benchmark sequence, i.e., $G_{ij} \approx G_{jj}$ or $G_{ij} = G_{jj}$ and $C(T_i, T_j) \approx 0$, it means there is no damage in the detection case. On the contrary, when the benchmark sequence is not similar to the detected sequence, or there is a deviation between them, i.e., $G_{ij} \neq G_{jj}$ and $C(T_i, T_j) \neq 0$, it means the damage may occur in the structure.

20.3.2   Euclidean Distance

The Euclidean distance $d_E(X,Y)$ is defined as the square root of the sum of the squares of the difference of coordinates between two points as shown in Eq. (20.5) (Gower 1982).

$$d_E(X,Y) = \left(\sum_{i=1}^{N} (x_i - y_i)^2 \right)^{\frac{1}{2}} \tag{20.5}$$

It is clear from Eq. (20.5) that the Euclidean distance will be in the range of $[0,\infty]$. If the value of $d_E(X,Y)$ equals to 0, it means there is no damage in the detection case. On the contrary, if the value increases, it means the damage may occur in the structure.

20.3.3   Itakura Distance

Vectors of arbitrary points in N-dimensional space of Itakura are defined as $X = [x_1, x_2, \ldots, x_N]$ and $Y = [y_1, y_2, \ldots, y_N]$, respectively. The Itakura distance can be defined as (Itakura and Umezaki 1987, Zheng and Mita 2009)

$$d_I^2(X,Y) = \sum_{i=1}^{N} \frac{X_i^2}{Y_i^2} - 1 \tag{20.6}$$

If the detected sequence is the same or close to the benchmark sequence, $d_I^2(X,Y) \approx 0$. If $d_I^2(X,Y) \gg 0$, it means that the damage may occur in the structure.

20.3.4 MAHALANOBIS DISTANCE

Mahalanobis distance was combined with transmissibility to detect structural damage (Figueiredo et al. 2011, Zhou et al. 2015). The Mahalanobis distance can be defined as

$$d_M(X,Y) = \sqrt{\{x_i - y_i\}^T [\Sigma]^{-1} \{x_i - y_i\}}$$

(20.7)

where Σ is the covariance matrix of eigenvectors, the superscript T denotes transposed. Vector $X = [x_1, x_2, \ldots, x_n]$ and $Y = [y_1, y_2, \ldots, y_n]$ are the sample observations of two TFs, respectively.

According to Eq. (20.7), the larger the distance between two adjacent measuring points, the greater the variation of the structure before and after damage, which indicates that the damage extent of components between adjacent measuring points is more serious.

20.4 DAMAGE INDICATORS

20.4.1 DAMAGE DETECTION INDEX

Taking the Cosh distance as an example, the DDI was defined as

$$\text{DDI}(\omega) = \frac{d_{\text{Cosh}}^t}{\max\left(d_{\text{Cosh}}^u\right)}$$

(20.8)

where $\max\left(d_{\text{Cosh}}^u\right)$ denotes the maximum d_{Cosh}^u of the intact or undamaged structure, and d_{Cosh}^t is the Cosh distance for each test condition. If DDI ≈ 1 (or in log scale DDI ≈ 0), it means the tested structure can be considered as undamaged. And if there is a deviation between the benchmark and detected sequences, that is DDI $\neq 1$ (or in log scale DDI > 0), it means that the tested structure may be damaged.

20.4.2 ADVANCED RESULTING INDEX

The maxima of Cosh distance of the intact or undamaged structure are close, so the peak value of DDI is used for comparing. Therefore, ARI was introduced by extracting the peak values of DDI,

$$\text{ARI} = \max_{\omega}(\text{DDI})$$

(20.9)

If ARI ≈ 1 (or in log scale ARI ≈ 0), it means that the tested structure has the same or close condition as the benchmark structure. If ARI > 1 or ARI < 1 (or in log scale ARI > 0), it means the tested structure may be damaged.

20.5 THE NUMERICAL ANALYSIS OF A SIMPLY SUPPORTED STEEL H-BEAM

20.5.1 NUMERICAL MODEL DESCRIPTION

In order to verify the effectiveness and feasibility of the proposed damage detection indicators, a finite element (FE) model of a simply supported H-beam (Figures 20.1 and 20.2) with 14 beam elements, 15 nodes and 3 degrees of freedom at each node will be used. The acceleration response of each node under simulated hammering load was recorded.

The elastic modulus reduction of some elements was taken to simulate the damage of the beam. The case 1 is the benchmark beam without damage. For the other cases, the damage was only introduced in elements 11 and 12 with different reduction of elastic modulus as shown in Table 20.1.

FIGURE 20.1 The numerical model of the simply supported steel H-beam.

| 1 | 2 | 3 | 4 | 5 | 6 | 7 | 8 | 9 | 10 | 11 | 12 | 13 | 14 |

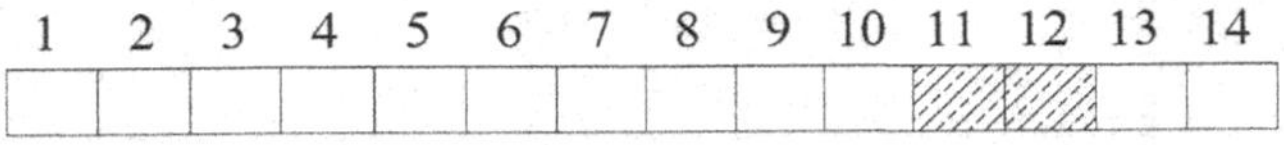

FIGURE 20.2 The elements of the simply supported steel H-beam.

TABLE 20.1
Damage Cases of the Simply Supported Steel H-Beam

Case	Description	State
1	Reference condition (100% of the original modulus of elasticity (E))	Undamaged
2	Elastic modulus reduction 5% (95%E)	Damaged
3	Elastic modulus reduction 10% (90%E)	
4	Elastic modulus reduction 20% (80%E)	
5	Elastic modulus reduction 30% (70%E)	
6	Elastic modulus reduction 40% (60%E)	
7	Elastic modulus reduction 50% (50%E)	

Note that the acceleration responses of the simulated beam were contaminated by 5% Gaussian white noise to consider the sensitivity of damage to a random noisy signal.

20.5.2 APPLICABILITY OF THE DISTANCE MEASURE–BASED DAMAGE DETECTION INDICATORS

The transmissibility for all measurements under intact and damaged cases were taken as input parameters of Eqs. (20.4) to (20.8) and then the damage indices were calculated, respectively. The different damage detection indicators (in logarithmic) corresponding to damage states with 5% and 50% elastic modulus reduction with 5% Gaussian white noise are shown in Figure 20.3.

From Figure 20.3, it can be concluded that the distance measure-based damage detection indicators which combine the transmissibility with different measurements can effectively detect the damage that has occurred. Comparing the calculated results of DDI with and without 5% noise, it can be seen that noise has a little influence on the calculated results.

20.5.3 THE RESULTS OF ARI

The ARI values under different damage cases and different elements with and without 5% noise are shown in Figures 20.4 and 20.5, respectively.

From Figures 20.4 and 20.5, it can be found that Itakura distance is most sensitive to damage, followed by Euclidean distance. The Itakura and Mahalanobis distances can detect the damage location more easily than the others. By comparing the results with and without 5% noise, similarity can be found, which indicates the approaches have a good anti-noise performance, especially the indicator based on Cosh distance.

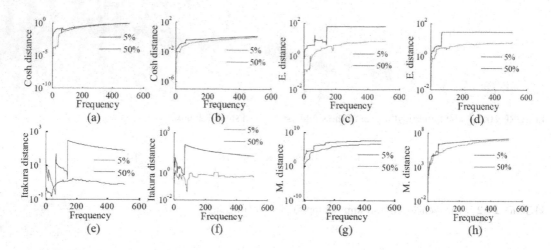

FIGURE 20.3   The (a) Cosh distance, (c) Euclidean, (e) Itakura and (g) Mahalanobis distance under damage cases 2 and 7 without noise, the (b) Cosh distance, (d) Euclidean, (f) Itakura and (h) Mahalanobis distance under damage cases 2 and 7 with 5% noise.

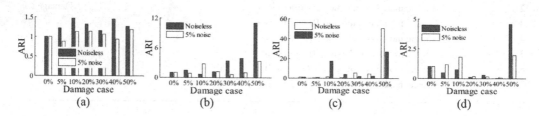

FIGURE 20.4   The ARI values of (a) Cosh distance, (b) Euclidean, (c) Itakura and (d) Mahalanobis distance under different damage cases.

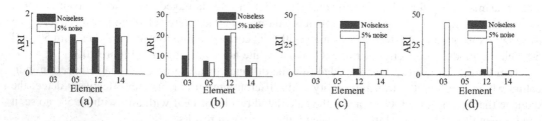

FIGURE 20.5   The ARI values of (a) Cosh distance, (b) Euclidean, (c) Itakura and (d) Mahalanobis distance at different elements.

20.6   EXPERIMENTAL STUDY OF A SIMPLY SUPPORTED STEEL H-BEAM

As shown in Figures 20.6 and 20.7, a dynamic test of a simply supported steel H-beam which is the same as the aforementioned finite element model has been conducted. The beam was divided into seven elements along the longitudinal direction and one accelerometer was arranged on each element.

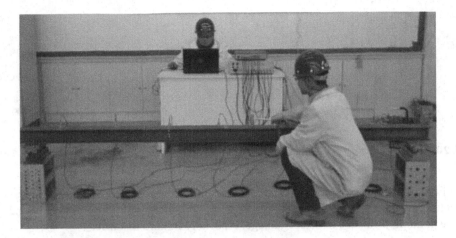

FIGURE 20.6   The dynamic test of a simply supported steel H-beam.

FIGURE 20.7   Accelerometer layout for the dynamic test of a simply supported steel H-beam.

TABLE 20.2

Test Cases of the Simply Supported Steel H-Beam

Case	Description	State Condition
1	Reference (without drill)	Undamaged
2	One 8 mm drill	Damaged
3	Two 8 mm drills	
4	Three 8 mm drills	
5	Four 8 mm drills	

20.6.1   TEST SCHEME

The INV306U hardware system and COINV DASP V10 data acquisition software were used to collect and process both horizontal and vertical acceleration response data from seven high-frequency wired sensors. As shown in Table 20.2, the actual damage of the beam was simulated by the web drilling at the element 2. The borehole diameter is 8 mm, and the number of holes was set to 1, 2, 3 and 4, corresponding to the test cases 2, 3, 4 and 5.

The node 2 of the simply supported steel H-beam structures with and without damage was excited by the specified hammer with 30 N and a duration of 0.1 seconds. In order to filter out the noise or false components in the test signal, the frequency domain method was used to digitally filter the acceleration response data of each channel. Then, Band-Pass Filter (BPF) was used to filter the raw data with the bandwidth ranging from 15 to 800 Hz.

20.6.2   APPLICABILITY OF THE DISTANCE MEASURE–BASED DAMAGE DETECTION INDICATORS

The procedure to obtain the detection indicators corresponding to different damage states is the same as Section 5.2. The indicators (in logarithmic) corresponding to the test cases 2 and 5 are shown in Figure 20.8.

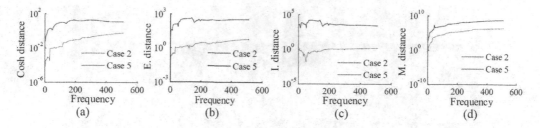

FIGURE 20.8   The (a) Cosh, (b) Euclidean, (c) Itakura and (d) Mahalanobis distance under cases 2 and 5.

From Figure 20.8, it can be concluded that the results of the four-distance measure-based damage detection indicators under different test cases are consistent with those in Section 5.2. All of them have good performance to detect damage, especially Itakura distance.

20.6.3   The Results of ARI

The ARI values of each damage case are shown in Figure 20.9. Take a typical case 5 as an example, the ARIs at nodes 1–7 of the beam are shown in Figure 20.10.

From Figures 20.9 and 20.10, the ARI values of the test case 4 are significantly larger than the remaining ARI values. And the ARI values at node 2 are much larger than the other locations, which indicates that the damage indicators based on four distance measures can effectively identify the damage location.

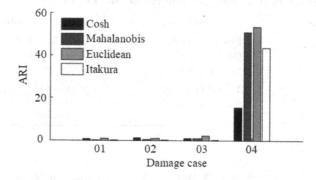

FIGURE 20.9   ARI values under different damage cases.

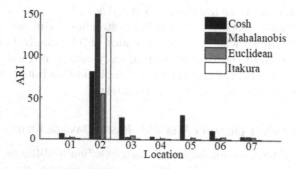

FIGURE 20.10   ARI values at different locations.

20.7   CONCLUSION

Comparison of the results of numerical studies and experimental investigation of the simply supported steel H-beam with different damages leads to the following conclusions:

1. The four approaches show very good and effective performance in detecting damage. Obvious changes of the damage detection indicators combining TFs with the different distance measures were noted due to structural damage, which indicates that the transmissibility-based damage detection indicators can be applied for damage assessment in the SHM field.
2. Two damage detection indices DDI and ARI can effectively detect the location and severity of structural damage. For SHM, they were not widely efficient to step forward in the damage detection, namely to identify the type of damage for all test cases.
3. Comparing the numerical results of damage detection with and without 5% noise, it can be seen that noise has little influence on the calculated results, which indicates that the approaches are also reliable to minimize the influence of measurement noise.

REFERENCES

Brownjohn, J. M. W. 2006. Structural health monitoring of civil infrastructure. *Philosophical Transactions of the Royal Society A: Mathematical, Physical and Engineering Sciences*, 365(1851), p. 589–622.

Chesné, S., and Deraemaeker, A. 2013. Damage localization using transmissibility functions: A critical review. *Mechanical Systems and Signal Processing*, 38(2), p. 569–584.

Deraemaeker, A., and Worden, K. 2011. *New Trends in Vibration Based Structural Health Monitoring*. CISM International Centre for Mechanical Sciences. Springer, Wien New York.

Devriendt, C., De Sitter, G., and Guillaume, P. 2010. An operational modal analysis approach based on parametrically identified multivariable transmissibilities. *Mechanical Systems and Signal Processing*, 24(5), p. 1250–1259.

Farrar, C. R., and Worden, K. 2010. An introduction to structural health monitoring. In *New Trends in Vibration Based Structural Health Monitoring* (p. 1–17). Springer, Vienna.

Farrar, C. R., Doebling, S. W., and Nix, D. A. 2001. Vibration–based structural damage identification. *Philosophical Transactions of the Royal Society of London. Series A: Mathematical, Physical and Engineering Sciences*, 359(1778), p. 131–149.

Figueiredo E., Park G, Farrar C. R., et al. 2011. Machine learning algorithms for damage detection under operational and environmental variability. *Structural Health Monitoring: An International Journal*, 10(6), p. 559–572.

Gower, J. C. 1982. Euclidean distance geometry. *The Mathematical Scientist*, 7(1), p. 1–14.

Itakura, F., and Umezaki, T. 1987. Distance measure for speech recognition based on the smoothed group delay spectrum. *IEEE International Conference on Acoustics, Speech, and Signal Processing. (In ICASSP'87)*. 12, p. 1257–1260.

Limongelli, M. P. 2019. *Seismic Structural Health Monitoring: From Theory to Successful Applications*. Springer, Berlin.

Maia, N. M. M., Almeida, R. A. B., Urgueira, A. P. V., and Sampaio, R. P. C. 2011. Damage detection and quantification using transmissibility. *Mechanical Systems and Signal Processing*, 25(7), p. 2475–2483.

Ou, J. P., and Li, H. 2010. Structural health monitoring in mainland China: Review and future trends. *Structural Health Monitoring: An International Journal*, 9(3), p. 219–231.

Qiao, L., Esmaeily, A., and Melhem, H. G. 2012. Signal pattern recognition for damage diagnosis in structures. *Computer-Aided Civil and Infrastructure Engineering*, 27(9), p. 699–710.

Yan, Y. J., Cheng, L., Wu, Z. Y., and Yam, L. H. 2007. Development in vibration-based structural damage detection technique. *Mechanical Systems and Signal Processing*, 21(5), p. 2198–2211.

Yan, W.-J., Zhao, M.-Y., Sun, Q., and Ren, W.-X. 2019. Transmissibility-based system identification for structural health monitoring: Fundamentals, approaches, and applications. *Mechanical Systems and Signal Processing*, 117, p. 453–482.

Zheng, H. T., and Mita, A. 2009. Localized damage detection of structures subject to multiple ambient excitations using two distance measures for autoregressive models. *Structural Health Monitoring*, 8(3), p. 207–222.

Zhou, Y.-L., Figueiredo, E., Maia, N. M. M., Sampaio, R. P. C., and Perera, R. 2015. Damage detection in structures using a transmissibility-based Mahalanobis distance. *Structural Control and Health Monitoring*, 22(10), p. 1209–1222.

Zhou, Y.-L., Maia, N. M. M., Sampaio, R. P. C., and Wahab, M. A. 2016. Structural damage detection using transmissibility together with hierarchical clustering analysis and similarity measure. *Structural Health Monitoring: An International Journal,* 16(6), p. 711–731.

21 Substructuring-Based Damage Assessment of a Steel Railway Bridge Using Operational Modal Data

L. He
Fuzhou University

E. Reynders
University of Leuven (KU Leuven)

V. Zabel
Bauhaus-Universität Weimar

G. C. Marano and B. Briseghella
Fuzhou University

G. De Roeck
University of Leuven (KU Leuven)

CONTENTS

21.1 INTRODUCTION

For existing steel bridges, assessment of their working conditions is a critical issue to be addressed due to the increasing number of aged structures and the catastrophic results of the possible in-service failures (Kuehn et al. 2008). As a global method for assessment, vibration-based structural health monitoring has gradually grown from its infancy to maturity in the past decades. Experimental modal data of the structure that are extracted from operational conditions under ambient excitations had been shown to provide an accurate and reliable reference for the relevant damage identification process (Doebling et al. 1998; Carden and Fanning 2004). In this regard, FE model updating technique is often implemented in order to provide an accurate estimation of the structural condition (Brownjohn et al. 2003; Zordan et al. 2014; Liu et al. 2016; Fa et al. 2016; He et al. 2018).

In this paper, damage assessment is conducted for an aged steel bridge in Europe based on operational modal data collected from the previous experimental campaign. By formulating the model updating process into an optimization procedure, the sensitivity-based method is applied to solve the relevant numerical problems. Robustness of the solution is the main concern of the current study. In this respect, a substructuring-based updating approach is proposed and implemented. The identification results show satisfactory agreement to the on-site inspection. In the following sections, the structure and tests are first introduced; the methodology that lies behind the numerical work is then explained for the specific case study, and finally discussion about the updating results are provided.

21.2   THE BRIDGE AND TESTS

The structure is a steel-plate-girder bridge located in Großheringen, Germany, built in 1931. It is a part of the railway line between Berlin and Munich, also known as the Saaletalbrücke bridge. The whole structure consists of five simply supported spans. The first span of the length of 35 m was chosen as the subject of the current case study (on the right-hand side in Figure 21.1). Ambient vibration tests were performed on the studied span in May 2010 by a joint research team from KU Leuven University (KUL) and the Bauhaus-University Weimar. A hybrid measurement system consisted of a novel wireless sensor network, and a classical wired acquisition system was deployed to measure the acceleration responses under ambient excitations of the upper and bottom flanges of the main girders, respectively (He et al. 2011).

Operational modal analysis was performed on the collected data using the Stochastic Subspace Identification algorithm (SSI) as implemented in the MACEC software (Peeters and De Roeck 1999; Reynders et al. 2011). There are totally three global modes identified between 0 and 10 Hz, with their modal data summarized in Figure 21.2. The experimental results are very accurate as characterized by the smoothness of the mode shape plots and the relatively small standard deviations of the natural frequencies (being less than 0.05 Hz), which provide a reliable reference for the following damage assessment process.

21.3   FE MODEL UPDATING

21.3.1   THE UPDATING PROCESS

Damage assessment of the bridge is conducted by using the Finite Element updating technique. Since the undamaged/initial state of the bridge does not exist, a refined FE model including totally 79,240 four-node shell elements was built to replace the undamaged bridge. An extensive survey was conducted on-site in order to check the dimensions of all the members, especially regarding the variation of thickness of the riveted sections of the main girders and cross beams, and to compare with the original technical drawings. Numerical modal analysis performed on the refined

FIGURE 21.1   A steel-plate-girder-railway bridge in Großheringen, Germany.

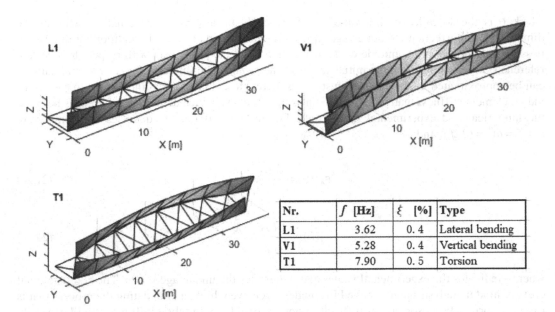

Nr.	f [Hz]	ξ [%]	Type
L1	3.62	0. 4	Lateral bending
V1	5.28	0. 4	Vertical bending
T1	7.90	0. 5	Torsion

FIGURE 21.2   The experimental modal data.

shell model and the results obtained are thought to be reliable, as shown by the very high Modal Assurance Criterion (MAC) values calculated between the numerical mode shapes and the identified experimental counterparts (being higher than 0.95 for all three modes).

The updating process of damage assessment is illustrated by the diagram in Figure 21.3. In particular, the following nonlinear least-square problem is formulated and solved (Teughels 2005):

$$\theta^* = \min_{\theta} \sum_i g_i^2(\theta) = \min_{\theta} \sum_i w_{\varepsilon,i}\left(\varepsilon_{z,i}(\theta)\right)^2, \text{ being } w \text{ the positive weighting factors} \qquad (21.1)$$

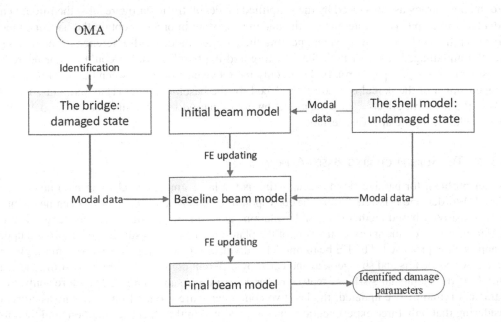

FIGURE 21.3   The updating process diagram.

where θ is the so-called design variables. They could be any model parameters, whereas the dimensionless correction factors are generally preferable to avoid the ill-conditioning due to the possible differences in magnitude of the parameters. The factors are set with respect to a certain reference values, being often the initial values of the physical parameters. Since structural damage can be approximated by the changes in stiffness, herein a scale factor k is introduced to each individual element stiffness matrix. $\varepsilon_{z,i}$ is known as the residuals, which define the differences between the numerical and experimental modal data. For instance, if only frequencies are considered, $z = \lambda = \omega^2 = (2\pi f)^2$ and

$$\varepsilon_{\lambda,i}(\theta) = \frac{\lambda_i(\theta) - \tilde{\lambda}_i}{\tilde{\lambda}_i} \tag{21.2a}$$

$$\varepsilon_{\lambda,i}(\theta) = \frac{\lambda_i(\theta)}{\lambda_{0,i}} - \frac{\tilde{\lambda}_i}{\tilde{\lambda}_{0,i}} \tag{21.2b}$$

where ~ indicates the experimental values and 0 indicates the undamaged state. The computational costs of modal analysis upon a refined FE model is relatively high, and updating the process that is run on a normal PC is sometimes difficult, if not impossible. A simplified FE model will generally be plausible, only if the introduced numerical modeling errors will not overlay the changes of the modal data influenced by the parametrized damage (Friswell 2006). In this regard, a simplified FE model including only 396 two-node beam elements and 16 link elements is built, which is termed the initial beam model (see Figure 21.3).

The optimization problem is solved by using the well-established sensitivity-based method (Mottershead et al. 2011). The sensitivity matrix or the Jacobian matrix is defined by

$$J(\theta)_{ij} = \frac{\partial g_i(\theta)}{\partial \theta_j} \tag{21.3}$$

The sensitivity has to be evaluated at each iteration and its values varied at different points. To minimize the inaccuracy as introduced by the simplified model on the updating results, the initial beam model has to be updated in reference to the undamaged state in order to obtain the baseline model (Titurus et al. 2003). Moreover, when updating the baseline model with reference to the damaged state, the undamaged states of both the structure and the baseline model could be considered in the residuals (see e.g. Eq. (21.2b)). In detail, only the relative changes between the undamaged and damaged states of the modal data are reproduced for the baseline model (Hao and Xia 2002). As a result, the negative effects due to the modeling errors caused by the simplified FE model will be further reduced for the updating results.

21.3.2 THE SUBSTRUCTURING-BASED APPROACH

Another problem for parametric model updating is the large amount of element parameters to be updated (Worden and Friswell 2009). The corresponding numerical problem often became daunting for the sensitivity-based methods. Possible solutions are damage parameterization, regularization, etc. Herein, based on the specific structure of the plate-girder bridge, a substructuring-based updating approach is proposed. The FE beam model is divided into several substructures: main girders, bracings, cross beams and stringers, as indicated by different colors in Figure 21.4. A unique scale factor k of the elemental stiffness matrix is applied to each substructure. Therefore, only three parameters remain to be updated: the last two components are grouped into one substructure, by considering that only three experimental values are adopted in the objective function (and therefore only three equations), in order to avoid solving an underdetermined system.

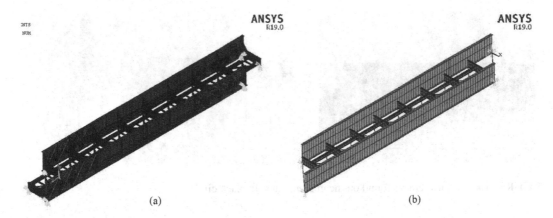

FIGURE 21.4 The refined FE model built with shell elements (a) and the simplified FE model built with beam elements (b).

21.4 THE SOLUTIONS

The solution procedure is implemented in MATLAB® by applying the embedded trust-region-reflective algorithm. The weighting factors are put to unity. The natural frequencies of the undamaged state are 3.715 Hz, 5.391 Hz and 8.468 Hz, respectively, based on the shell model. The optimization problems are solved twice to obtain the baseline and final models, respectively, with the initial values $k_i = 1$ ($i = 1,2,3$) and the specified upper and lower bounds.

Figure 21.5 plots the residuals as defined by Eq. (21.3) traced through the solution procedures: (21.2a) for baseline and (21.2b) for the final models. To obtain the baseline model, only six iterations are needed and the e converges to almost zeros. To obtain the final model, seven iterations are needed. And the natural frequencies of the final model are 3.561 Hz, 5.277 Hz and 8.035 Hz, respectively, in turn for L1, V1 and T1. As compared to the experimental data, the relative differences are 2% for L1 and T1 and almost 0% for V1. The final solutions of k_i equal to 0.96, 0.80 and 1.05 in turn for the main girders, bracings and the combined substructure of stringers and cross beams. The results suggest a stiffness reduction of almost 20% on the bracings, meanwhile the stiffness of main girders and cross beams remain almost unchanged together with the stringers. On-site inspection found structural deterioration mainly on the bracings; especially cracks were discovered at several gusset plates connecting the bracings and the main girders (see Figure 21.6). By contrast, the remaining structure is still in good condition, including the supports' conditions. It suggests the results of the current study are reasonable and acceptable.

To investigate the robustness of the solutions, the condition number of the sensitivity matrix at the final solution point is calculated using the definition Abs(s_1/s_m), with s being the smallest and

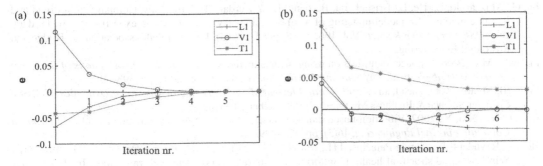

FIGURE 21.5 Residuals during the solution procedure: baseline (a) and final (b) models.

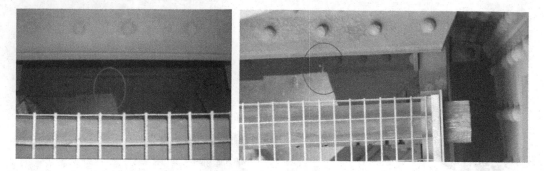

FIGURE 21.6 Typical cracks found on the gusset plates (in black circles).

maximum singular values (Friswell and Mottershead 2013). The condition number is found to be 13.05, suggesting a well-conditioned problem.

21.5 CONCLUSIONS

Vibration-based damage assessment is conducted for an aged steel plate-girder railway bridge. The highly accurate operational modal results are found vital to the success of the application. By means of an FE model updating based on a simplified beam model, reduction of the stiffness is identified for the bracings for about 20%. The implemented substructuring-based approach helps to guarantee the robustness of the numerical optimization procedure. On-site inspection also confirms the findings of the current case study.

ACKNOWLEDGEMENT

Financial support of the RFCS research project CT-2009-00027 "FAtigue Damage controL and assESSment for railway bridges" is acknowledged for the experimental campaign. The first author needs to thank the financial supports from the Fuzhou municipal under the research project "2017-G-67" and those from the Fuzhou University under the research project "XRC-1675".

REFERENCES

Brownjohn, J.M.W., Moyo, P., Omenzetter, P. and Lu, Y., 2003. Assessment of highway bridge upgrading by dynamic testing and finite-element model updating. *Journal of Bridge Engineering*, 8(3), pp. 162–172.
Carden, E.P. and Fanning, P., 2004. Vibration based condition monitoring: a review. *Structural Health Monitoring*, 3(4), pp. 355–377.
Doebling, S.W., Farrar, C.R. and Prime, M.B., 1998. A summary review of vibration-based damage identification methods. *Shock and Vibration Digest*, 30(2), pp. 91–105.
Fa, G., Mazzarolo, E., He, L., Briseghella, B., Fenu, L., & Zordan, T., 2016, May. Comparison of direct and iterative methods for model updating of a curved cable-stayed bridge using experimental modal data. In *IABSE Symposium Report* (Vol. 106, No. 8, pp. 538–545). International Association for Bridge and Structural Engineering.
Friswell, M.I., 2006. Damage identification using inverse methods. *Philosophical Transactions of the Royal Society A: Mathematical, Physical and Engineering Sciences*, 365(1851), pp. 393–410.
Friswell, M. and Mottershead, J.E., 2013. *Finite Element Model Updating in Structural Dynamics* (Vol. 38). Springer Science & Business Media, Berlin/Heidelberg, Germany.
Hao, H. and Xia, Y., 2002. Vibration-based damage detection of structures by genetic algorithm. *Journal of Computing in Civil Engineering*, 16(3), pp. 222–229.
He, L., Reynders, E., García-Palacios, J.H., Marano, G.C., Briseghella, B. and De Roeck, G. 2018, Aug. Wireless-based structural health monitoring of a short-span concrete highway bridge. In *Proceedings 10th International Conference on Short and Medium Span Bridges*, Quebec City, Quebec, Canada.

He, L., Reynders, E., Hsu, T.Y. and De Roeck, G., 2011. Analysis of dynamic coupling between spans of two multi-span bridges using ambient vibration measurements. *Proceedings of Eurodyn*, 2011, pp. 1552–1558.

Kuehn, B., Lukic, M., Nussbaumer, A., Günther, H.P., Helmerich, R., Herion, S.H.K.M., Kolstein, M.H., Walbridge, S., Androic, B., Dijkstra, O. and Bucak, Ö., 2008. Assessment of Existing Steel Structures: Recommendations for Estimation of Remaining Fatigue Life (No. BOOK). European Commission's Science and Knowledge Service, Joint Research Center (JRC).

Liu, T., Zhang, Q., Zordan, T. and Briseghella, B., 2016. Finite element model updating of Canonica bridge using experimental modal data and genetic algorithm. *Structural Engineering International*, 26(1), pp. 27–36.

Mottershead, J.E., Link, M. and Friswell, M.I., 2011. The sensitivity method in finite element model updating: a tutorial. *Mechanical Systems and Signal Processing*, 25(7), pp. 2275–2296.

Peeters, B. and De Roeck, G., 1999. Reference-based stochastic subspace identification for output-only modal analysis. *Mechanical Systems and Signal Processing*, 13(6), pp. 855–878.

Reynders, E., Schevenels, M. and De Roeck, G., 2011. *MACEC 3.2: A MATLAB Toolbox for Experimental and Operational Modal Analysis*. Leuven University, Leuvan.

Teughels, A., 2005. Inverse modelling of civil engineering structures based on operational modal data.

Titurus, B., Friswell, M.I. and Starek, L., 2003. Damage detection using generic elements: Part I. Model Updating. *Computers & Structures*, 81(24–25), pp. 2273–2286.

Worden, K. and Friswell, M.I., 2009. Modal–vibration-based damage identification. Encyclopedia of Structural Health Monitoring.

Zordan, T., Briseghella, B. and Liu, T., 2014. Finite element model updating of a tied-arch bridge using Douglas-Reid method and Rosenbrock optimization algorithm. *Journal of Traffic and Transportation Engineering (English Edition)*, 1(4), pp. 280–292.

22 Time-Frequency Features of Continuous Metro Bridge under Various Excitations

J. Yang, X. H. He, and Y. F. Zou
Central South University
National Engineering Laboratory for High Speed Railway Construction
Joint International Research Laboratory of Key
Technology for Rail Traffic Safety

CONTENTS

22.1 INTRODUCTION

Train-bridge vibration influences the riding comfort of passengers in the train and reduces the lives of the bridge Zhai et al. (2019). Studying the frequencies of the bridge is essential for understanding the train-bridge interaction and identifying damages of the train-bridge system.

Unlike the ambient vibration or free vibration of the bridge and the train, the frequencies of the self-excited vibrations of the train-bridge system vary in time and exhibit strong nonstationary features. The Fourier Transform (FT) has been used extensively for stationary signals to perform spectrum analysis, but is not able to reveal the time-varying characteristics for nonstationary signals. The Continuous Wavelet Transform (CWT) enables the time-frequency analysis by translating and scaling the wavelet basis. To achieve good time and frequency resolutions, the selection of the wavelet basis is significant Torrence and Compo (1998). Kim et al. (2016) provided a guideline for the proper selection of the wavelet type in civil engineering applications. Chakraborty and Basu (2010) used the Modified Littlewood-Paley (MLP) basis to investigate the frequency nonstationarity in the responses of primary-secondary systems. Cantero et al. (2019) analyzed the time-varying frequencies of a laboratory vehicle-beam system by utilizing the CWT with the MLP basis. The MLP basis was also suggested to analyze a real-life composite bridge in Sweden by Cantero et al. (2016). The vehicle's speed and axle configuration as well as the nonlinear behavior of the structure could be distinguished in the time-frequency energy map. The Morlet basis enables the balance between the time and frequency resolution and has been applied widely in applications Kim et al. (2016). Tao et al. (2018) performed the CWT combined with the Morlet basis on train-induced vibrations on a set of steel-truss railway bridges. Yoon et al. (2000) analyzed the acoustic emission behavior of corroded reinforced concrete beams by conducting the CWT with the Morlet wavelet basis. The wavelet results exhibited a favorable correlation with the damage conditions of the concrete beams. Gurley et al. (2003) developed a Morlet wavelet-based coherence and bicoherence technique to detect intermittent first- and higher-order correlation between a pair of signals in the time-frequency domain. Grinsted et al. (2004) discussed the cross wavelet transform and the wavelet coherence based on the Morlet wavelet basis. The phase angle statistics could be used to gain confidence in causal relationships and test mechanistic models of physical relationships between time series. Kijewski-Correa and Kareem (2007) compared the performance of the wavelet transform by the Morlet wavelet basis with that of the empirical mode decomposition in extracting the Instantaneous Frequency (IF) from signals embedded in noise. It was concluded that the wavelet transform seeking harmonic similitude at various scales produces lower variance IF estimates than the empirical mode decomposition with Hilbert transform. Other popular techniques for identifying structural IF include short time FT Sayed et al. (2016), empirical mode decomposition coupled with the Hilbert-Huang spectrum Han and van der Baan (2013), synchrosqueezing transform (SST) Cao et al. (2016) and synchrosqueezed wavelet transform (Thakur et al. 2013, Li et al. 2017). For more information about these techniques, please refer to the comprehensive review by Tary et al. (2014).

In this application, the time-frequency representations for a long-span continuous metro bridge under various train excitation events are carried out by utilizing the CWT analysis combined with a wavelet filter, aiming to studying the nonstationary features of the bridge. The field experiment of the bridge is introduced in Section 22.2. The selection of the wavelet basis and the CWT analysis for the vibrations of the bridge under various excitations are presented in Section 22.3. In the end, conclusions are drawn in Section 22.4.

22.2 EXPERIMENTAL BRIDGE AND TRAIN

The measured bridge is a three-span continuous metro bridge with all supports fixed, as shown in Figure 22.1. The bridge has a middle span as long as 150 m. The middle support is like a 'Y' shape with the part within the shape empty. The bridge is made of concrete and has double-track box sections. 941B accelerometers are installed at the mid-span of the bridge to measure the vertical and lateral accelerations of the bridge, as shown in Figures 22.1 and 22.2a. The transmission band of the 941B accelerometer is 0.25~80 Hz. The test metro train consisted of six vehicles filled with 120 t sandbags simulating AW3 load, as shown in Figure 22.2b.

Table 22.1 provides the event information in this study and their code name used in the following context. It should be mentioned that the other train crossing with the test train for Case5 is empty.

22.3 WAVELET ANALYSIS

The measured bridge signals are recommended to be filtered into the interested frequency domain before performing the CWT analysis to remove the influence of irrelevant frequencies. In this study, the frequencies above 6 Hz are filtered out for the bridge signals by utilizing the wavelet filter with the 'db4' basis. In this way, no time shift occurs, which is usually found by using finite impulse response (FIR) or infinite impulse response (IIR) filters.

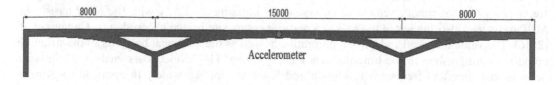

FIGURE 22.1 Sketch of the measured bridge (unit: cm).

FIGURE 22.2   (a) A 941B accelerometer inside the box bridge, (b) test metro train.

TABLE 22.1
Summary of Events

Code Name	Starting Time on Bridge	Speed (km/h)	Event
Case1	2018-11-19 16:08:01	60	Traversing bridge from TP to XH station
Case2	2018-11-19 16:05:13	60	Traversing bridge from XH to TP station
Case3	2018-11-19 16:15:18	80	Traversing bridge from TP to XH station
Case4	2018-11-19 16:19:20	60	Braking on bridge from XH to TP station
Case5	2018-11-19 16:51:36	80	Two trains crossing on bridge from TP to XH station

The CWT of a signal $x(t)$ is defined as below:

$$W_x(a,b) = \int_{-\infty}^{\infty} x(t) \frac{1}{\sqrt{a}} \, \psi * \left(\frac{t-b}{a} \right) dt \tag{22.1}$$

where $\psi(t)$ is the mother wavelet and the superscript asterisk represents its complex conjugation. The wavelet coefficient is the inner product of the signal with the mother wavelet translated by the parameter b and scaled by the parameter a.

The effect of the wavelet basis on the time-frequency results is exampled in Figure 22.3. Figure 22.3a and b have poor frequency resolutions, while Figure 22.3c has the poor time resolution. Overall, Figure 22.3d has balanced good resolutions for both the time and the frequency. Therefore, the complex Morlet basis is adopted in the following CWT analysis.

The complex Morlet wavelet basis is defined in MATLAB® by

$$\psi(t) = \frac{1}{\sqrt{\pi f_b}} e^{\left(-\frac{t^2}{f_b} + i2\pi f_c t \right)} \tag{22.2}$$

where f_b is a time-decay parameter which controls the decay of the wavelet in the time domain and the corresponding bandwidth in the frequency domain and f_c is the positive central frequency. Increasing f_b narrows the bandwidth around the central frequency f_c and results in slower decay of the wavelet in the time domain. The two parameters are chosen to be 1 and 2.3, respectively, for the signals in this study. Figure 22.4 shows the complex Morlet wavelet basis (cmor1–2.3) used in this study.

A pseudo-frequency f corresponding to the scale a can be defined by

$$f = \frac{f_c}{a\Delta t}. \tag{22.3}$$

where Δt is the sampling period of the analyzed signal.

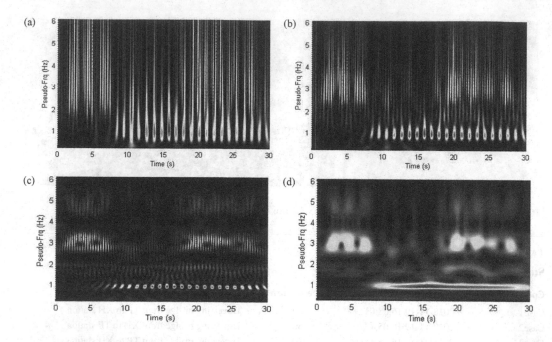

FIGURE 22.3   Time-frequency analysis of bridge lateral acceleration for Case1: (a) CWT with Mexican basis, (b) CWT with Morlet basis, (c) CWT with MLP basis and (d) CWT with complex Morlet basis (cmor1–2.3).

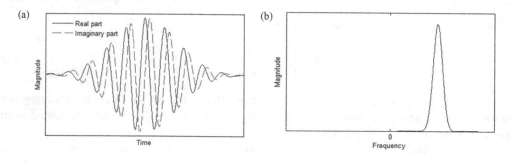

FIGURE 22.4   The complex Morlet wavelet basis (cmor1–2.3) in (a) the time domain and (b) the frequency domain.

22.3.1   EFFECT OF TRAIN SPEED

Figure 22.5 shows the forced and free accelerations of the bridge at mid-span in vertical and lateral directions and their CWT when the train traverses the bridge at the speed of 60 km/h (Case1). The time-historic accelerations in Figure 22.5 have been filtered by the low-pass wavelet filter with the stopping frequency of 6 Hz. As found in Figure 22.5a, the first bending frequency of 1.2 Hz is excited. The amplitude of the excited frequency varies and reaches the maximum when the train is near the bridge mid-span. Figure 22.5b shows that high lateral frequencies of about 3.3 and 4.6 Hz have large and periodical amplitudes when the train enters and exits the bridge, while the low frequency of 0.9 Hz appears starting at 9 seconds.

When the traversing speed of the train increases to 80 km/h (Case3), the time-frequency maps of the vertical and lateral accelerations of the bridge at mid-span are depicted in Figure 22.6. Compared with Figure 22.5a, the high vertical frequency of 2.4 Hz is also induced when the train enters and exits the bridge. In addition, the amplitude of the first bending frequency also increases at a long period.

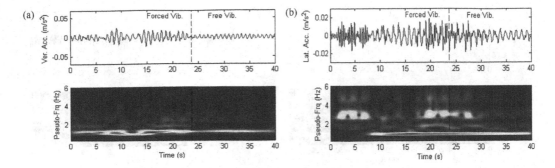

FIGURE 22.5 Filtered bridge accelerations at mid-span and their CWT for Case1: (a) vertical direction, (b) lateral direction.

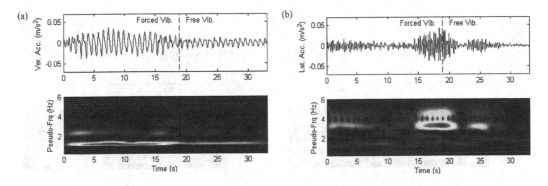

FIGURE 22.6 Filtered bridge accelerations at mid-span and their CWT for Case3: (a) vertical direction, (b) lateral direction.

Compared with Figure 22.5b, the low frequency of 0.9 Hz does not appear, and the amplitudes for high frequencies of 3.3 and 4.6 Hz increase obviously near the train exiting the bridge.

22.3.2 EFFECT OF TRAVELING DIRECTION

When the test train travels at the opposite direction as Case1 (Case2), the accelerations and time-frequency maps of the bridge are plotted in Figure 22.7. Compared with Figure 22.5a, a higher frequency of 2.5 Hz is also excited. In addition, the variation for the frequency of 1.2 Hz is more obvious. Compared with Figure 22.5b, the amplitudes for the low frequencies of 0.9 and 3.3 Hz are much lower near the train exiting the bridge. The difference between Figures 22.5 and 22.7 might be caused by the small curvature of the bridge in the lateral direction.

22.3.3 EFFECT OF TRAIN BRAKING

When the test train brakes suddenly on the bridge at the initial speed of 60 km/h (Case4), the accelerations of the bridge and their CWT are depicted in Figure 22.8. Compared with Figure 22.7a, the higher frequency of 2.5 Hz is induced into large amplitudes after the train stops on the bridge (free vibration). It can be found from Figure 22.8a that the amplitude of the bridge vertical vibration tends to decrease to zero between 17 and 23 seconds when the train is at the braking period. However, the bridge is bounced into large free vibrations when the train stops on the bridge, which is due to the release of the braking force at the vertical direction. Nevertheless, this phenomenon is not observed in the lateral vibration of the bridge, as shown in Figure 22.8b. Compared with Figure 22.7b, the frequencies higher than 0.9 Hz are basically suppressed after 17 seconds.

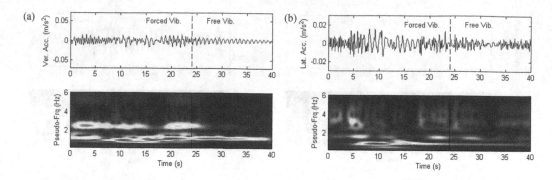

FIGURE 22.7 Filtered bridge accelerations at mid-span and their CWT for Case2: (a) vertical direction, (b) lateral direction.

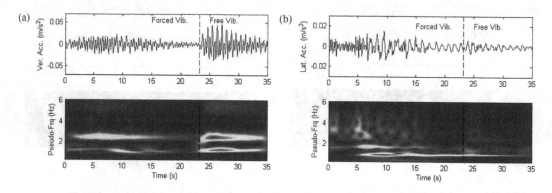

FIGURE 22.8 Filtered bridge accelerations at mid-span and their CWT for Case4: (a) vertical direction, (b) lateral direction.

22.3.4 EFFECT OF CROSSING TRAIN

When the test train crosses with another train on the bridge, the accelerations of the bridge and their CWT are plotted in Figure 22.9. It should be mentioned that two trains travel at the same speed of 80 km/h. Compared with Figure 22.6a for one train traveling at the same speed of 80 km/h, both vertical frequencies of 1.2 and 2.4 Hz are excited simultaneously between 3.5 and 12.4 seconds, and the frequency of 2.4 Hz is dominant in the free vibration of the bridge. The frequency of 2.4 Hz is the second bending mode of the bridge, whose amplitude is found to be built up during the two trains with different weights (one loaded with sandbags and the other not) exiting the bridge. Compared with Figure 22.6b, the amplitude for the frequency of 3.3 Hz reaches maximum soon after the trains enter the bridge.

22.4 CONCLUSIONS

The nonstationary features for the vibrations of a long-span continuous metro bridge under various train excitation cases in the low-frequency range (lower than 6 Hz) are revealed clearly by the CWT combined with a wavelet filter. The main specific conclusions are listed as follows:

1. A balanced resolution for both time and frequency domain can be realized by tuning the time-decay parameter and the central frequency parameter of the complex Morlet wavelet basis, which is the advantage of the complex Morlet wavelet basis.

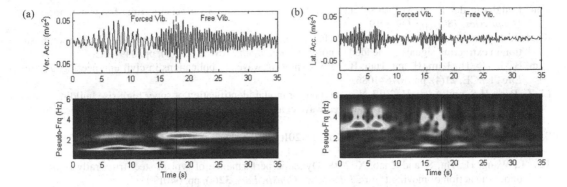

FIGURE 22.9 Filtered bridge accelerations at mid-span and their CWT for Case5: (a) vertical direction, (b) lateral direction.

2. When the train enters or exits the bridge, higher vertical and lateral frequencies of the bridge tend to be excited. The amplitudes of higher frequencies increase with the train speed no larger than 80 km/h.

3. When the test train brakes suddenly on the bridge at the initial speed of 60 km/h, the amplitude of the bridge vertical vibration tends to decrease to zero during the braking period. However, the bridge is bounced into free vibrations with large amplitudes when the train stops on the bridge, which might be due to the release of the braking force acting on the bridge at the vertical direction.

4. The amplitude for the second bending mode of the bridge (2.4 Hz) is found to be built up when the two crossing trains with different weights (one loaded with sandbags and the other not) exit the bridge at the same time.

In the future, other time-frequency techniques, e.g., the SST is planned to be utilized to improve the resolutions for the time and frequency domain further.

ACKNOWLEDGEMENTS

This study is supported by the National Natural Science Foundation of China under Grant (U1534206), the National Key Research and Development Program of China (2017YFB1201204), the Project of Science and Technology Research and Development Program of China Railway Corporation under Grant (2017T001-G) and the Postdoctoral Fund of Central South University under Grant (205443).

REFERENCES

Cantero, D., McGetrick, P., Kim, C.-W. and Obrien, E. 2019. Experimental monitoring of bridge frequency evolution during the passage of vehicles with different suspension properties. *Eng. Struct.*, **187**, pp. 209–219.

Cantero, D., Ülker-Kaustell, M. and Karoumi, R. 2016. Time–frequency analysis of railway bridge response in forced vibration. *Mech. Syst. Signal Pr.*, **76–77**, pp. 518–530.

Cao, H., Xi, S., Chen, X. and Wang, S. 2016. Zoom synchrosqueezing transform and iterative demodulation: Methods with application. *Mech. Syst. Signal Pr.*, **72–73**, pp. 695–711.

Chakraborty, A. and Basu, B. 2010. Analysis of frequency nonstationarity via continuous wavelet transform in the response of primary-secondary systems. *J. Struct. Eng.*, ASCE, **136**(12), pp. 1608–1612.

Grinsted, A., Moore, J.C. and Jevrejeva, S. 2004. Application of the cross wavelet transform and wavelet coherence to geophysical time series. *Nonlinear Proc. Geoph.*, **11**, pp. 561–566.

Gurley, K., Kijewski, T. and Kareem, A. 2003. First- and higher-order correlation detection using wavelet transforms. *J. Struct. Eng.*, ASCE, **129**(2), pp. 188–201.

Han, J. and van der Baan, M. 2013. Empirical mode decomposition for seismic time-frequency analysis. *Geophysics*, **78**(2), pp. O9–O19.

Kijewski-Correa, T. and Kareem, A. 2007. Performance of wavelet transform and empirical mode decomposition in extracting signals embedded in noise. *J. Eng. Mech.*, ASCE, **133**(7), pp. 849–852.

Kim, B., Jeong, H., Kim, H. and Han, B. 2016. Exploring wavelet applications in civil engineering. *J. Civ. Eng.*, KSCE, **21**(4), pp. 1076–1086.

Li, Z., Park, H.S. and Adeli, H. 2017. New method for modal identification of super high-rise building structures using discretized synchrosqueezed wavelet and Hilbert transforms. *Struct. Des. Tall Spec.*, **26**(3), pp. e1312.

Sayed, M.A., Kaloop, M.R., Kim, E. and Kim, D. 2016. Assessment of acceleration responses of a railway bridge using wavelet analysis. *J. Civ. Eng.*, KSCE, **21**(5), pp. 1844–1853.

Tao, T., Wang, H., Hu, S. and Zhao, X. 2018. Dynamic performance of typical steel truss-railway bridges under the action of moving trains. *J. Perform. Constr. Fac.*, **32**(4), pp. 04018053.

Tary, J.B., Herrera, R.H., Han, J. and van der Baan, M. 2014. Spectral estimation-What is new? What is next?. *Rev. Geophys.*, **52**(4), pp. 723–749.

Thakur, G., Brevdo, E., Fučkar, N.S. and Wu, H.-T. 2013. The Synchrosqueezing algorithm for time-varying spectral analysis: Robustness properties and new paleoclimate applications. *Signal Process.*, **93**(5), pp. 1079–1094.

Torrence, C. and Compo, G.P. 1998. A practical guide to wavelet analysis. *B. Am. Meteorol. Soc.*, **79**(1), pp. 61–78.

Yoon, D.-J., Weiss, W.J. and Shah, S.P. 2000. Assessing damage in corroded reinforced concrete using acoustic emission. *J. Eng. Mech.*, ASCE, **126**(3), pp. 273–283.

Zhai, W., Han, Z., Chen, Z., Ling, L. and Zhu, S. 2019. Train–track–bridge dynamic interaction: A state-of-the-art review. *Vehicle Syst. Dyn.*, **57**, pp. 1–44.

23 Experimental System and Damage Identification of Small-Scale Wind Turbine Blades

J. M. Gutierrez, R. Astroza, J. Abell, and Soto, C.
Universidad de los Andes

F. Jaramillo
University of Chile

M. Guarini
Universidad de los Andes

M. Orchard
University of Chile

CONTENTS

23.1 INTRODUCTION

Since industrial revolution, energy sources have been one of the key components of humanity, and the society has turned extremely dependent on electricity. For that reason, taking a wind turbine (WT) out of operation has a significant impact, including economical and operational issues. At the same time, if a WT collapses, the cost of the entire system, plus labor, must also be added to the economic cost of losing one generator (which includes as well long standby periods while maintenance or repairments are performed) (Pinar Perez et al. 2013).

According to Caithness Windfarm Information Forum, more than 160 accidents occurred in WT blades (WTBs) in 2016 (Caithness, n.d.). However, this statistic is not realistic, since Windaction (a group formed to provide information about WT farms) frequently reports structural failures at different locations throughout the globe, confirming that only a small percentage of these structural damages are officially reported (WindAction, n.d.). G-Cube, a specialist provider of insurance for WT projects, confirmed that annually more than 3,800 failures are presented only in WTBs (GCube, n.d.). Because of this, being able not only to detect a catastrophic failure before it occurs but also to minimize maintenance time in the windmills is crucial.

In this paper, experimental vibration data collected from WTBs at different damage states are used to estimate the evolution of the modal properties as a function of damage. To this end, the testing equipment was designed and built to carry out dynamic tests on WTBs. Extreme and fatigue-type loading was represented by resonance and low-frequency tests, respectively. Before and after each extreme and fatigue-type experiment, pull-back and impact tests were conducted and free-vibration response data were collected by an accelerometer array and used to identify the modal properties of the blades using an output-only system identification method.

23.2   EXPERIMENTAL SETUP AND TESTING PROTOCOL

23.2.1   WIND TURBINE BLADES

A set of 12 glass-fiber/epoxy-resin composite blades from a 12 m-tall 5 kW WT, from two different production batches, was used in the experiments. Each blade was 2.4 m long and had a total mass of 12 kg. Since the stacking sequence (number and orientation of fibers) and mechanical properties of the fibers were not available from the provider, this information was obtained following the procedures recommended in codes ASTM D792 (ASTM 2013), ASTM D2584 (ASTM 2018), and ASTM D3574 (ASTM 2017).

23.2.2   TESTING MACHINE

A complete system (Figure 23.1) was designed and built to conduct the required experiments to be carried out in this research. The experiments include (1) modal identification tests based on free

FIGURE 23.1 Testing machine used to conduct the experiments. White circle shows a stiffer anchoring point, where blade was connected to perform static tests using chain hoists.

vibrations, (2) dynamic tests with fatigue- and resonance-type excitations, and (3) low-force and destructive static bending tests. During all the experiments, response data from accelerometers, displacement sensors linear variable differential transformer (LVDTs), and load cells were recorded.

23.2.3 TYPES OF TESTS

Static and dynamic tests can be conducted with the built equipment. During static tests, a point load is applied to the blade and its magnitude is measured using a load cell. In addition, the displacement at two points along the blade is measured using LVDTs (Figure 23.2a). These data allow obtaining force versus displacement curves (Figure 23.2b), information that will be used in the future to calibrate finite element (FE) models.

Dynamic tests included three different types of experiments. First, free vibrations were induced by hits with a rubber mallet. Second and third, controlled frequency vibrations at fixed amplitudes (which could be selected from 4.0, 5.0, and 7.5 cm) were induced to the WTBs. Here, resonance (at a frequency about 90% of the fundamental frequency of the blade) and fatigue tests (at a frequency about 50% of the fundamental frequency of the blade) were conducted. Before and after each fatigue/resonance test, free-vibration response data were collected and used to estimate the modal properties of the blade and track their evolution as the damage progressed. More information about the test protocols for the fatigue and resonance test will be provided later in this paper. Note that nine uniaxial accelerometers (see black blocks in Figure 23.2a) measuring in the flapwise direction of the blade collected the vibration data used for system identification (Figure 23.3).

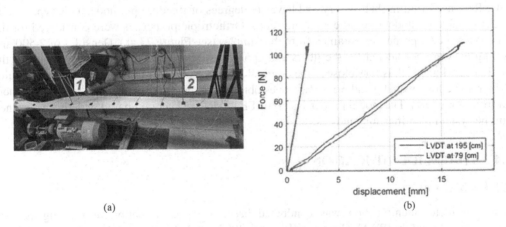

(a)   (b)

FIGURE 23.2 (a) Instrumentation during the dynamic tests (LVDTs are depicted with numbers 1 and 2, and black blocks correspond to accelerometers), (b) Force-displacement curve obtained during a static test.

FIGURE 23.3 Snapshot taken during a resonance test.

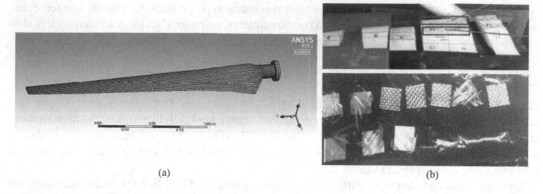

(a)   (b)

FIGURE 23.4   (a) An FE model developed in ANSYS, (b) pieces of the cut blades photo of the fiberglass after following the ASTM recommendations.

23.3   FE MODEL

One of the twelve blades was cut every 10 cm along its length to have a detailed description of the cross sections and geometry of the blades. In addition, as mentioned above, the stacking properties of the composite were obtained following the procedures recommended in the ASTM standards. This information was used to develop an FE model in the software ANSYS (ANSYS, n.d.) (Figure 23.4a). The model had a total of 7,542 four-node SHELL181 elements, which are based on the Reissner-Mindlin shell theory and have six degrees of freedom per node. This type of element is suitable to model layered composite shells. Orthotropic properties were considered for the fibers based on the properties obtained from the samples (see Figure 23.4b). Densities of 2,500 and 1,160 kg/m^3 were considered for the fibers and epoxy, respectively, obtaining a total mass for the model of 11.6 kg, which is very close to the nominal mass of the blades (12 kg). The Young modulus for fiber and epoxy were 90 and 3.8 GPa, values obtained manually calibrating force-displacement data from static tests. The modal properties of the FE model are later compared with those obtained from the system identification analysis.

23.4   SYSTEM IDENTIFICATION

23.4.1   METHOD

Modal parameter identification was conducted based on a combination of the Eigensystem Realization Algorithm (ERA) (Juang and Pappa 1985) and the Natural Excitation Technique (NExT) (James et al. 1993). This two-stage approach is referred to as NExT-ERA, in which NExT determines beforehand the cross-correlation matrices of a reference response channel with other measured responses, while later, ERA is employed to estimate the modal parameters (i.e., fundamental frequencies, mode shapes, and damping ratios). A variation of this method named multiple-reference NExT-ERA (MNExT-ERA) was employed in this research for more robust identification results, despite already having the free-vibration responses required for ERA. The general workflow of this modal identification approach is summarized in the following steps:

1. Free-vibration response was measured with the accelerometer array shown in Figure 23.2a.
2. MNExT algorithm was applied to obtain cross-correlation matrices.
3. ERA method used the cross-correlation matrices computed from MNExT to obtain the state and output matrices of the state-space model, **A** and **C**, respectively.
4. Natural frequencies and damping ratios of the blades were computed from the eigenvalues of matrix **A**, and mode shapes were computed from the eigenvectors of matrices **A** and **C**.

23.4.2 RESULTS

Free-vibration response data from eleven pristine blades (before running any fatigue/resonance test) were recorded. A total of 18 impact tests were conducted on each blade, collecting a total of 198 data sets. Stabilization diagrams were used to define the physical modes. The stability criteria for consecutive model orders consisted of a 1% error in frequency, 5% error in damping ratio, and 2% error in mode shape (evaluated using the modal assurance criterion or MAC) (see Astroza et al. 2013 for more details). The modes satisfying these stability criteria at least ten times, among the 100 model orders considered, were selected as potential physical modes of the blades. In addition, the fast Fourier transform (FFT) of all the acceleration channels was also used to select further the physical modes by keeping the identified modes with frequencies no more than 5% apart to those obtained from the peaks of the FFTs. In addition, modes with damping ratios larger than 5% were also removed. Statistics (mean values and coefficients of variation) of the identified natural frequencies and damping ratios from the 198 data sets are reported in Table 23.1. In this table, N denotes the number of times that the corresponding mode was identified from the 198 data sets.

Modes 1, 3, 4, and 6 are consistently identified from most of the test, at least from 80% of them. Contrarily, modes 2 and 5 are identified from 61% and 28% of the tests, respectively. As discussed later, these two modes are primarily associated to edgewise deformation and because of the considered instrumentation (nine uniaxial accelerometers in the flapwise direction), there is no much information about them in the recorded data used to conduct the modal identification. The identification results of the natural frequencies of the eleven blades are very consistent, and low coefficient of variations for the six modes is observed. Note that the standard IEC-61400 (IEC 2014) accepts a maximum variation of 10% in the dynamic parameters of two identically manufactured blades.

The damping ratios identified for the six modes are below 2.02%, indicating limited sources of energy dissipation in the blades tested. As expected and consistent with observations from previous studies conducted in other type of structures, larger scatter is obtained for the damping ratios. This is most likely due to the underlying mathematical model assumed in the system identification (Astroza et al. 2013). Finally, MAC values of all the identified mode shapes, taking a set of single hit results for one blade as reference mode shapes, are shown in Figure 23.5a. Very limited variability in the mode shapes of modes 1 and 3 is observed, while a large scatter is seen in modes 5 and 6. To better understand the source of the variability in the mode shapes, Figure 23.5b shows the percentage of MAC values (from the total of N identified modes shown in Table 23.1) below and above 0.95. From Figure 23.4 it is noted that for mode 4 only a few identified modes have MAC values below 0.95; however, they cause a large scatter in the standard deviation of the corresponding MAC value shown in Figure 23.5a. In the case of modes 5 and 6, a large number of modes are identified with MAC values below 0.95, indicating that the variability in the MAC values is general

TABLE 23.1

Natural Frequencies (f^{exp}) and Damping Ratios (ξ) Identified Using MNExT-ERA Method

| Mode | Frequency (f^{exp}) | | Damping Ratio (ξ) | | N |
	Mean (Hz)	C.V. (%)	Mean (%)	C.V. (%)	
1	9.0	2.1	2.02	41	185
2	20.6	3.1	1.05	31	121
3	32.7	2.6	1.01	17	185
4	72.3	1.9	1.22	12	198
5	105.6	1.9	1.38	39	56
6	126.3	1.4	1.8	22	159

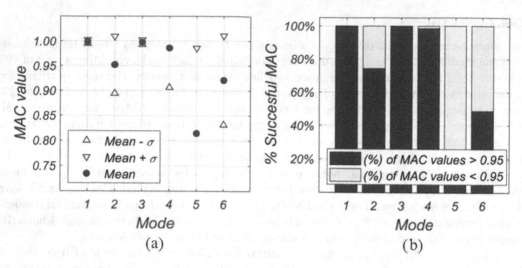

FIGURE 23.5 Mean MAC values of the blade's mode shapes (a) and ratio of the MAC values bigger than 95% of all mode identifications (b).

TABLE 23.2

Experimental and FE-Predicted Natural Frequencies (f^{exp} and f^{fem}) of the Six Identified Modes and the Corresponding MACs

Mode	f^{exp} (Hz)	f^{fem} (Hz)	$\Delta f/f^{exp}$ (%)	Direction	MAC
1	9.05	9.30	2.76	Flapwise	0.994
2	20.64	13.68	33.72	Edgewise	0.987
3	32.67	32.57	0.31	Flapwise	0.984
4	72.30	74.59	3.17	Flapwise	0.943
5	105.64	99.20	6.10	Edgewise	0.890
6	126.30	132.55	4.95	Flapwise	0.879

and not coming from some specific tests. Note that modes 5 and 6 are expected to contribute less to the total response of the blades; therefore, the signal-to-noise ratio (SNR) is expected to be less for these modes than for the lower modes.

Table 23.2 shows the comparison between the mean values of the identified natural frequencies and those obtained from the FE model described in Section 3 and the MAC values of the corresponding mode shapes. In addition, Figure 23.6 compares the identified and FE-predicted mode shapes. A good agreement between the natural frequencies is evidenced, with relative errors lower than 5% for all modes except modes 2 and 5. As mentioned above, the instrumentation of the blades did not consider accelerometers in the edgewise direction, and therefore these modes were not well characterized. In addition, the elastic properties of the fibers in the transverse direction were not properly calibrated, because static tests were only conducted in the flapwise direction.

23.5 DAMAGE IDENTIFICATION

The following two testing protocols were defined to guarantee the cumulative structural degradation (i.e., damage) in the blade, so time variant modal properties could be studied as a measure of damage progression:

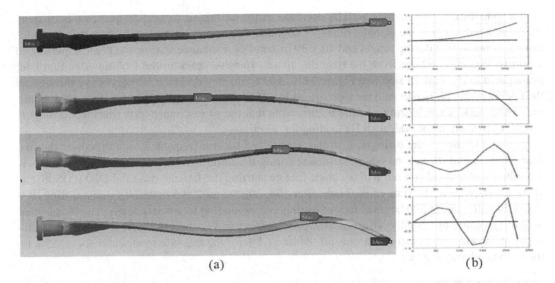

FIGURE 23.6 First four mode shapes in the flapwise direction: (a) FE model, (b) Experimentally identified.

1. First, *resonance tests* were conducted with an exciting frequency about 90% of the frequency of the fundamental mode of the blades. This test was conducted on one blade from the first fabrication batch. A sinusoidal excitation with an amplitude of 4 cm and a frequency of 8 Hz was applied for 3-minute-long periods. After each test in the protocol was completed, free vibration tests were performed, and the data recorded was employed to identify the modal properties of the blade.
2. Second, low-frequency *fatigue tests* were performed in another blade with an exciting frequency about 50% of the frequency of the fundamental mode. The main purpose in these tests was to induce a fatigue-type loading, so slowly degradation was induced in the blade instead of producing a brittle collapse after a short testing period (like resonance-type tests).

Figure 23.7 shows the histograms of the normalized natural frequencies identified for the first four flapwise modes during the resonance tests. The pristine conditions correspond to the condition of the blade before conducting any test. In this figure, state i corresponds to the condition of the blade

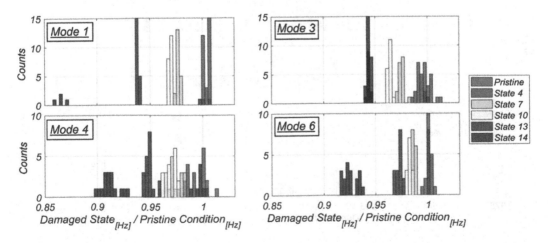

FIGURE 23.7 Histograms of the normalized natural frequencies identified for the first four flapwise modes during the resonance-type testing protocol.

after conducting *i* resonance tests. The decrease in the identified natural frequencies is evident, implying a reduction in the global stiffness of the blade due to the damage induced by the resonance tests. It is noted that during test 14 and after 39 minutes of resonance testing, the blade experienced a failure located at 67 cm from the base due to an extensive crack in the leading edge that later propagated to the bottom side (see Figure 23.8). The frequency decrease is progressive throughout the testing protocol. However, only approximately a 10% of reduction was required to reach the failure of the blade, experiencing a brittle failure. Note that the 14 resonance tests translate to roughly 19,000 cycles when considering 8 Hz loading for 40 minutes.

Figure 23.9 shows the histograms of the normalized natural frequencies identified for the first four flapwise and the second edgewise modes during the fatigue tests. Here, DS00 corresponds to the initial undamaged condition of the blade (before starting the fatigue tests) and DS04, DS08, and DS12 correspond to the state of the blade after 108, 209.5, and 298.5 hours of fatigue testing. It is noted that after about 300 hours of fatigue testing, corresponding to approximately 5 million cycles, no damage could be visually observed. However, a clear trend is observed from the identified natural frequencies, with reductions between 1.0% and 1.5% for the flapwise modes and about 0.5% for the second edgewise mode.

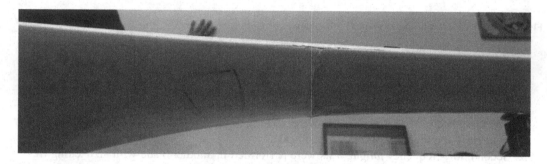

FIGURE 23.8   Failure of the blade during the 14th resonance test.

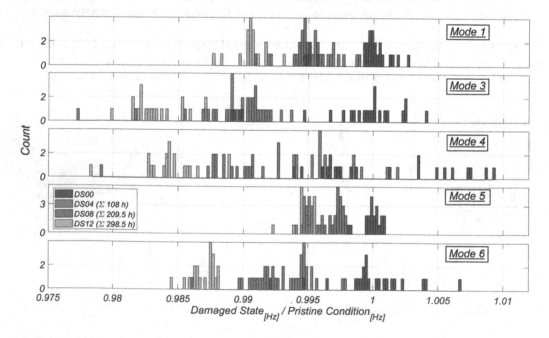

FIGURE 23.9   Histograms of the normalized natural frequencies identified for the first four flapwise and second edgewise modes during the fatigue-type testing protocol.

It is noted that the evolution of damping ratios and mode shapes as damage progressed were also investigated. The large scatter in damping ratio estimated does not allow to use it as a reliable parameter for damage detection. In addition, no clear trend in the estimated damping ratio as damage progresses was observed. On the other hand, mode shapes and quantities derived therefrom (e.g., curvature mode shapes) required a very dense instrumentation; otherwise, interpolation of the mode shapes is too rough, as observed from numerical studies not shown here because of space limitations and also pointed out by others (e.g., Dessi and Camerlengo 2015).

23.6 CONCLUSIONS

In this paper, the modal properties of blades of a 5 kW WT were identified from free-vibration response data using an output-only system identification method. Resonance- and fatigue-type tests were conducted on two different blades to analyze the evolution of the identified modal properties as damage progressed. It was concluded that decreases in natural frequencies are well correlated with the observed damage, while damping ratios, mode shapes, and quantities derived therefrom do not provide reliable features for damage identification purposes. From the resonance test, a brittle failure of the blade after a reduction in the flapwise natural frequencies of about 10% was observed.

ACKNOWLEDGEMENTS

The authors acknowledge the support provided by the Fondo de Fomento al Desarrollo Científico y Tecnológico under grant FONDEF IDeA ID17I1040.

REFERENCES

ANSYS, n.d. ANSYS engineering simulation & 3d design software. [Online] Available at: http://www.ansys. com.

ASTM D792-13. 2013. *Standard Test Methods for Density and Specific Gravity (Relative Density) of Plastics by Displacement.* ASTM International, Conshohocken, PA.

ASTM D3574-17. 2017. *Standard Test Methods for Flexible Cellular Materials - Slab, Bonded, and Molded Urethane Foams.* ASTM International, Conshohocken, PA.

ASTM D2584-18. 2018. *Standard Test Method for Ignition Loss of Cured Reinforced Resins.* ASTM International, Conshohocken, PA.

Astroza, R., Ebrahimian, H., Conte, J.P., Restrepo, J.I., and Hutchinson, T.C. 2013. "Statistical analysis of the identified modal properties of a 5-story RC seismically damaged building specimen." In *Safety, Reliability, Risk and Life-cycle Performance of Structures and Infrastructures*, eds., George Deodatis, Bruce R. Ellingwood, Dan M. Frangopol, CRC Press Taylor & Francis, Boca Raton, FL.

Caithness, n.d. www.caithnesswindfarms.co.uk [Online].

Dessi, D. and Camerlengo, G. 2015. "Damage identification techniques via modal curvature analysis: Overview and comparison." *Mechanical Systems and Signal Processing*, **52–53**, pp. 181–205.

GCube, n.d. www.gcube-insurance.com [Online].

IEC. 2014. *IEC6400-Part 23: Full-Scale Structural Testing of Rotor Blades.* International Electrotechnical Commission, Geneva, Switzerland.

James, G.H., Carrie, T.G., and Lauffer, J.P. 1993. "The Natural Excitation Technique (NExT) for modal parameter extraction from operating wind turbines." *Technical Report SAND92–1666, UC-261*, Sandia National Laboratories, Albuquerque, NM.

Juang, J.-N. and Pappa, R. S. 1985. "An Eigensystem Realization Algorithm for modal parameter identification and model reduction." *Journal of Guidance, Control, and Dynamics*, **8**(5), p. 620–627.

Pinar Perez, J., García Marquez, F., Tobias, A. and Papaelias, M. 2013. "Wind Turbine reliability analysis." *Renewable and Sustainable Energy Reviews*, **23**(1), pp. 463–472.

WindAction, n.d. www.windaction.org [Online].

24 Estimation of Nonlinear System States with Unknown Loading and Limited Measurements

J. He, X. X. Zhang, and M. C. Qi
Hunan University

CONTENTS

24.1 INSTRUCTIONS

The estimation of the states of a partially observed dynamic system is important for structural health monitoring and vibration control (Xu and He 2017). In this regard, the classical Kalman filter (KF) is a well-recognized recursive algorithm that provides unbiased and optimal state estimation for a linear dynamic system from noise-contaminated response measurements. A variety of KF-based algorithms for state estimation or damage detection have also been studied and well developed. For example, a multi-rate KF approach was proposed by Smyth and Wu (2007) for state estimation when multi-type data with different sampling frequencies were involved. With the extension use of KF as a damage detector, Bernal (2013) developed a lag shifted whiteness test (LSWT) for damage detection in the presence of changing process and measurement noise. Based on KF algorithm, an integrated method was proposed by Zhu et al. (2013) for optimal sensor placement and response reconstruction. He et al. (2015) extended this method in the active control system by employing the estimated structural responses for vibration control. Kim and Park (2017) investigated the application of KF for estimating a time-varying process disturbance in a building space.

Although the classical KF and the aforementioned KF-based methods can satisfactorily estimate structural states, the external excitation in these algorithms is assumed to be either known or modeled as a zero mean white Gaussian process. However, in many cases, it is difficult or sometimes impossible to directly measure the input force, or the Gaussian assumption is violated. Therefore, over the past years, various KF with unknown input (KF-UI) methods have been developed to circumvent the above limitations. For example, based on limited output-only measurements, Papadimitriou et al. (2011) proposed a KF-based approach for predicting the strain/stress responses. By introducing the unknown inputs into state vector, Lourens et al. (2012) developed an augmented KF for force identification in structural dynamics. A dual implementation of KF was conducted by Eftekhar Azam et al. (2015) for estimating the unknown input and states of a linear state-space model. To prevent the so-called drifts caused by measurement noises, Liu et al. (2016) proposed an improved KF-UI method based on data fusion of partial acceleration and displacement measurements. This method was further extended by Lei et al. (2016) on the basis of modal KF. More recently, through the implementation of KF-UI, Zhang and Xu (2017) presented a novelty damage identification method by utilizing the reconstructed response and excitation. A KF-based inverse approach was developed by Zhi et al. (2017) for the estimation of wind loads on tall buildings and validated via wind tunnel tests as well as field tests on Taipei 101 Tower. Based on the augmented KF and the modal expansion technique, Ren and Zhou (2017) proposed two strain estimation algorithms for unmeasured members in the truss structure. By minimizing the overall estimation errors of structural responses to a desired target level, Hu et al. (2017) proposed a KF-based approach for multi-type sensor placement and response reconstruction with unknown seismic loading. Although the KF-UI methods mentioned above can provide promising results of state estimation and loading identification, most of them are suitable to manage linear state-space models. However, nonlinearity exists widely in many civil structures, such as the initiation and growth of damage, the hysteretic characteristics of structural components and so forth.

In this paper, it is aimed to extend the classical KF approach to circumvent the aforementioned limitations for jointly estimating the structural states of nonlinear systems and the unknown inputs applied to them. Based on the scheme of the classical KF, an improved KF-UI approach is proposed and the analytical recursive solutions are derived and given. A revised form of observation equation is obtained with the aid of a projection matrix. The structural states are then estimated with limited measurements. The unknown loadings are identified at the same time by means of least squares estimation. Two sorts of nonlinear hysteretic structures are used to numerically demonstrate the effectiveness of the proposed approach.

24.2 FORMULAS OF THE PROPOSED APPROACH

The equation of motion of an n degree-of-freedom (DOF) structure subject to unknown loadings can be given as

$$\mathbf{M}\ddot{\mathbf{x}}(t) + \mathbf{F}[\dot{\mathbf{x}}(t), \mathbf{x}(t), \boldsymbol{\theta}] = \boldsymbol{\varphi}\mathbf{f}(t) + \boldsymbol{\varphi}^u\mathbf{f}^u(t) \tag{24.1}$$

where $\ddot{\mathbf{x}}(t)$, $\dot{\mathbf{x}}(t)$ and $\mathbf{x}(t)$ are the vectors of structural acceleration, velocity and displacement, respectively; $\mathbf{F}[\dot{\mathbf{x}}(t), \mathbf{x}(t), \boldsymbol{\theta}]$ represents the restoring force vector which can be expressed in the linear or nonlinear form; $\mathbf{M}$ is the mass matrix; $\boldsymbol{\theta}$ is the structural parameters; $\mathbf{f}(t)$ and $\mathbf{f}^u(t)$ are the known and unknown excitation vectors, respectively; $\boldsymbol{\varphi}$ and $\boldsymbol{\varphi}^u$ are the influence matrices associated with known and unknown excitations, respectively.

Define the state vector $\mathbf{Z}(t) = \left[\mathbf{x}(t)^T, \dot{\mathbf{x}}(t)^T \right]^T$. Then, a general expression of the state-space equation can be found as

$$\dot{\mathbf{Z}}(t) = \left\{ \begin{array}{c} \dot{\mathbf{x}}(t) \\ \mathbf{M}^{-1}\left[-\mathbf{F}[\dot{\mathbf{x}}(t), \mathbf{x}(t), \boldsymbol{\theta}] + \boldsymbol{\varphi}\mathbf{f}(t) + \boldsymbol{\varphi}^u\mathbf{f}^u(t) \right] \end{array} \right\} = \mathbf{g}\left(\mathbf{Z}(t), \mathbf{f}(t), \mathbf{f}^u(t), t \right) + \mathbf{w}(t) \tag{24.2}$$

where $\mathbf{w}(t)$ is process noise vector with zero mean and a covariance matrix $\mathbf{Q}(t)$.

Let $\hat{\mathbf{Z}}_{k|k}$ and $\hat{\mathbf{f}}_k^u$ be the estimates of $\mathbf{Z}_k$ and $\mathbf{f}_k^u$ at time $t = k \times \Delta t$ with Δt being time intervals, respectively. Equation (24.2) can be then linearized with respect to the estimates $\hat{\mathbf{Z}}_{k|k}$ and $\hat{\mathbf{f}}_k^u$ as follows:

$$\mathbf{g}\left(\mathbf{Z}_k, \mathbf{f}_k, \mathbf{f}_k^u, k\Delta t \right) \approx \mathbf{g}\left(\hat{\mathbf{Z}}_{k|k}, \mathbf{f}_k, \hat{\mathbf{f}}_k^u, k\Delta t \right) + \mathbf{U}_{k|k}\left(\mathbf{Z}_k - \hat{\mathbf{Z}}_{k|k} \right) + \mathbf{W}_{k|k}\left(\mathbf{f}_k^u - \hat{\mathbf{f}}_k^u \right) \tag{24.3}$$

where

$$\mathbf{U}_{k|k} = \left.\frac{\partial \mathbf{g}\left(\mathbf{Z}_k,\mathbf{f}_k,\mathbf{f}_k^u,k\Delta t\right)}{\partial \mathbf{Z}_k}\right|_{\mathbf{Z}_k=\hat{\mathbf{Z}}_{k|k},\mathbf{f}_k^u=\hat{\mathbf{f}}_k^u} \quad ; \mathbf{W}_{k|k} = \left.\frac{\partial \mathbf{g}\left(\mathbf{Z}_k,\mathbf{f}_k,\mathbf{f}_k^u,k\Delta t\right)}{\partial \mathbf{f}_k^u}\right|_{\mathbf{Z}_k=\hat{\mathbf{Z}}_{k|k},\mathbf{f}_k^u=\hat{\mathbf{f}}_k^u} = \begin{bmatrix} \mathbf{0} \\ \mathbf{M}^{-1}\varphi^u \end{bmatrix}; \quad (24.4)$$

Note $\dot{\mathbf{Z}}(k\Delta t) = \left(\mathbf{Z}_{k+1} - \mathbf{Z}_k\right)/\Delta t$, and then the following expression can be derived:

$$\mathbf{Z}_{k+1} = \mathbf{Z}_k + \Delta t \cdot \left[\mathbf{g}\left(\hat{\mathbf{Z}}_{k|k},\mathbf{f}_k,\hat{\mathbf{f}}_k^u,k\Delta t\right) + \mathbf{U}_{k|k}\left(\mathbf{Z}_k - \hat{\mathbf{Z}}_{k|k}\right) + \mathbf{W}_{k|k}\left(\mathbf{f}_k^u - \hat{\mathbf{f}}_k^u\right) + \mathbf{w}_k \right] \quad (24.5)$$

For practical considerations, only partial acceleration responses are considered in this study leading to the following discretized observation equation,

$$\mathbf{y}_k = \mathbf{L}\ddot{\mathbf{x}}_k + \mathbf{v}_k = \mathbf{L}\mathbf{M}^{-1}\left\{-\mathbf{F}[\dot{\mathbf{x}}_k,\mathbf{x}_k,\theta] + \varphi\mathbf{f}_k + \varphi^u\mathbf{f}_k^u\right\} + \mathbf{v}_k \quad (24.6)$$

where $\mathbf{y}_k$ is the acceleration measurements at time $t = k \times \Delta t$; $\mathbf{L}$ is the matrix associated with the locations of accelerometers; $\mathbf{v}_k$ is the measurement noise assumed to be a Gaussian white noise with zero mean and a covariance matrix $\mathbf{R}_k$.

Equation (24.6) can be also rearranged as

$$\mathbf{D}\mathbf{f}_k^u = \mathbf{h}(\mathbf{Z}_k) - \mathbf{y}_k + \mathbf{v}_k \quad (24.7)$$

in which $\mathbf{h}(\mathbf{Z}_k) = \mathbf{L}\mathbf{M}^{-1}\left\{-\mathbf{F}[\dot{\mathbf{x}}_k, \mathbf{x}_k, \theta] + \varphi\mathbf{f}_k\right\}$; $\mathbf{D} = -\mathbf{L}\mathbf{M}^{-1}\varphi^u$.

Then, the unknown input $\mathbf{f}_k^u$ can be determined by means of least squares estimation as

$$\mathbf{f}_{k,\text{LSE}}^u = \left(\mathbf{D}^{\mathsf{T}}\mathbf{D}\right)^{-1}\mathbf{D}^{\mathsf{T}}\left[\mathbf{h}(\mathbf{Z}_k) - \mathbf{y}_k + \mathbf{v}_k\right] \quad (24.8)$$

The error of the aforementioned solution can be calculated as

$$\mathbf{err} = \mathbf{D}\mathbf{f}_k^u - \mathbf{D}\mathbf{f}_{k,\text{LSE}}^u = \left(\mathbf{I} - \mathbf{D}\left(\mathbf{D}^{\mathsf{T}}\mathbf{D}\right)^{-1}\mathbf{D}^{\mathsf{T}}\right)\left[\mathbf{h}(\mathbf{Z}_k) - \mathbf{y}_k + \mathbf{v}_k\right] \quad (24.9)$$

where $\mathbf{I}$ is an identity matrix; $\mathbf{D}\left(\mathbf{D}^{\mathsf{T}}\mathbf{D}\right)^{-1}\mathbf{D}^{\mathsf{T}}$ is known as a projection matrix. As a limit, the error shown in Eq. (24.10) should tend to zero, leading to

$$\Phi\mathbf{y}_k = \Phi\mathbf{h}(\mathbf{Z}_k) + \Phi\mathbf{v}_k \quad (24.10)$$

where $\Phi = \mathbf{I} - \mathbf{D}\left(\mathbf{D}^{\mathsf{T}}\mathbf{D}\right)^{-1}\mathbf{D}^{\mathsf{T}}$.

As observed from Eq. (24.10), with the aid of a projection matrix, a revised form of observation equation is obtained. A merit of this observation equation is that the unknown input is not explicitly presented. Thus, the multiple regression problem described by Eq. (24.6) is transformed into a single regression problem, and then the principle of KF can be employed for the state estimation. Since only the linear state-space equation is considered in the classical KF, it is not suitable for nonlinear systems. To cover a more general case as mentioned in Eq. (24.2), the priori state estimate in this study is calculated as follows:

$$\hat{\mathbf{Z}}_{k+1|k} = \hat{\mathbf{Z}}_{k|k} + \int_{k\Delta t}^{(k+1)\Delta t} \mathbf{g}\left(\hat{\mathbf{Z}}_{k|k},\mathbf{f}_k,\hat{\mathbf{f}}_k^u,k\Delta t\right)dt \quad (24.11)$$

The priori estimate error is defined as the difference between $\mathbf{Z}_{k+1}$ and $\hat{\mathbf{Z}}_{k+1|k}$. Then, the following expression of priori estimate error can be obtained on the basis of Eqs. (24.5, 24.7, 24.8, 24.10, 24.11),

$$\boldsymbol{\varepsilon}_{k+1|k} = \boldsymbol{\Gamma}_1\left(\mathbf{Z}_k - \hat{\mathbf{Z}}_{k|k}\right) + \boldsymbol{\Gamma}_2\mathbf{v}_k + \Delta t\mathbf{w}_k = \boldsymbol{\Gamma}_1\boldsymbol{\varepsilon}_{k|k} + \boldsymbol{\Gamma}_2\mathbf{v}_k + \Delta t\mathbf{w}_k \qquad (24.12)$$

where

$$\boldsymbol{\Gamma}_1 = \mathbf{I} + \Delta t\mathbf{U}_{k|k} + \Delta t\left[\begin{array}{c} \mathbf{0} \\ -\left(\mathbf{L}^{\mathrm{T}}\mathbf{L}\right)^{-1}\mathbf{L}^{\mathrm{T}}\mathbf{H}_{k|k} \end{array}\right] \quad \boldsymbol{\Gamma}_2 = \Delta t\left[\begin{array}{c} \mathbf{0} \\ -\left(\mathbf{L}^{\mathrm{T}}\mathbf{L}\right)^{-1}\mathbf{L}^{\mathrm{T}} \end{array}\right] \quad \mathbf{H}_{k|k} = \left.\frac{\partial\mathbf{h}(\mathbf{Z}_k)}{\partial\mathbf{Z}_k}\right|_{\mathbf{Z}_k=\hat{\mathbf{Z}}_{k|k}}$$

$$(24.13)$$

The priori estimate error covariance matrix can be found as

$$\mathbf{P}_{k+1|k} = E\left(\boldsymbol{\varepsilon}_{k+1|k}\boldsymbol{\varepsilon}_{k+1|k}^{\mathrm{T}}\right) = \boldsymbol{\Gamma}_1\mathbf{P}_{k|k}\boldsymbol{\Gamma}_1^{\mathrm{T}} + \boldsymbol{\Gamma}_2\mathbf{R}_k\boldsymbol{\Gamma}_2^{\mathrm{T}} + \Delta t^2\mathbf{Q}_k \qquad (24.14)$$

Based on the revised observation equation as shown in Eq. (24.10), the posteriori state estimate in this study can be calculated as

$$\hat{\mathbf{Z}}_{k+1|k+1} = \hat{\mathbf{Z}}_{k+1|k} + \mathbf{G}_{k+1}\left[\boldsymbol{\Phi}\mathbf{y}_{k+1} - \boldsymbol{\Phi}\mathbf{h}(\hat{\mathbf{Z}}_{k+1|k})\right] \qquad (24.15)$$

where $\mathbf{G}_{k+1}$ is the KF gain matrix at time $t = (k + 1) \times \Delta t$. Similarly, taking the difference between Eqs. (24.5) and (24.15) leads to

$$\boldsymbol{\varepsilon}_{k+1|k+1} = \mathbf{Z}_{k+1} - \hat{\mathbf{Z}}_{k+1|k+1} = \left(\mathbf{I} - \mathbf{G}_{k+1}\boldsymbol{\Phi}\mathbf{H}_{k+1|k}\right)\boldsymbol{\varepsilon}_{k+1|k} - \mathbf{G}_{k+1}\boldsymbol{\Phi}\mathbf{v}_{k+1} \qquad (24.16)$$

where $\mathbf{H}_{k+1|k}$ can be determined using Eq. (24.13) while $\mathbf{Z}_k = \hat{\mathbf{Z}}_{k+1|k}$.

Accordingly, the posteriori estimate error covariance matrix can be computed as

$$\mathbf{P}_{k+1|k+1} = \left(\mathbf{I} - \mathbf{G}_{k+1}\boldsymbol{\Phi}\mathbf{H}_{k+1|k}\right)\mathbf{P}_{k+1|k}\left(\mathbf{I} - \mathbf{G}_{k+1}\boldsymbol{\Phi}\mathbf{H}_{k+1|k}\right)^{\mathrm{T}} + \mathbf{G}_{k+1}\boldsymbol{\Phi}\mathbf{R}_{k+1}\boldsymbol{\Phi}^{\mathrm{T}}\mathbf{G}_{k+1}^{\mathrm{T}} \qquad (24.17)$$

Based on the principle of KF technique, the optimality criterion used for determining gain matrix $\mathbf{G}_{k+1}$ is equivalent to minimizing the trace of the posteriori estimate error covariance matrix $\mathbf{P}_{k+1|k+1}$. Therefore, the gain matrix can be calculated by setting the partial derivative of $\mathrm{tr}(\mathbf{P}_{k+1|k+1})$ with respect to $\mathbf{G}_{k+1}$ to zero.

$$\frac{\partial\mathrm{tr}(\mathbf{P}_{k+1|k+1})}{\partial\mathbf{G}_{k+1}} = 2\mathbf{G}_{k+1}\boldsymbol{\Phi}\left(\mathbf{H}_{k+1|k}\mathbf{P}_{k+1|k}\mathbf{H}_{k+1|k}^{\mathrm{T}} + \mathbf{R}_{k+1}\right)\boldsymbol{\Phi}^{\mathrm{T}} - 2\mathbf{P}_{k+1|k}\mathbf{H}_{k+1|k}^{\mathrm{T}}\boldsymbol{\Phi}^{\mathrm{T}} = 0 \qquad (24.18)$$

in which $\mathrm{tr}(\cdot)$ denotes the trace operator. Then, the gain matrix can be found as

$$\mathbf{G}_{k+1} = \mathbf{P}_{k+1|k}\mathbf{H}_{k+1|k}^{\mathrm{T}}\boldsymbol{\Phi}^{\mathrm{T}}\left[\boldsymbol{\Phi}\left(\mathbf{H}_{k+1|k}\mathbf{P}_{k+1|k}\mathbf{H}_{k+1|k}^{\mathrm{T}} + \mathbf{R}_{k+1}\right)\boldsymbol{\Phi}^{\mathrm{T}}\right]^{-1} \qquad (24.19)$$

On the basis of the posteriori state estimate shown in Eq. (24.15), the unknown inputs at time $t = (k + 1) \times \Delta t$ can be identified according to Eq. (24.8) as follows:

$$\hat{\mathbf{f}}_{k+1}^{u} = \left(\mathbf{D}^{\mathrm{T}}\mathbf{D}\right)^{-1}\mathbf{D}^{\mathrm{T}}\left[\mathbf{h}(\hat{\mathbf{Z}}_{k+1|k+1}) - \mathbf{y}_{k+1}\right] \qquad (24.20)$$

It can be found that the proposed approach has a similar frame of classical KF, including time update equations as shown in Eqs. (24.11) and (24.14) as well as measurement update equations as shown in Eqs. (24.15), (24.17) and (24.19). Also, the unknown excitations can be simultaneously identified according to Eq. (24.20). The effectiveness of the proposed approach will be investigated in the following sections through two numerical examples.

24.3 NUMERICAL EXAMPLES

24.3.1 NONLINEAR HYSTERETIC STRUCTURE WITH DAHL MODEL

To investigate the performance of the proposed approach for state estimation, an eight-story shear-frame structure with a Dahl model on the first floor is considered herein. The mass and stiffness parameters of the building structure are 300 kg and 1.4×10^5 N/m for each floor, respectively. The Rayleigh damping assumption with $\alpha = 0.1789$ and $\beta = 0.0038$ is adopted.

The nonlinear restoring force (NRF) provided by Dahl model is given as

$$R_f^{\text{Dahl}} = k_d \Delta x_i + c_d \Delta \dot{x}_i + f_d z + f_0 \qquad (24.21)$$

where k_d, c_d, f_d and f_0 are the parameters describing the properties of Dahl model; Δx_i and $\Delta \dot{x}_i$ are the relative displacement and velocity between the i-th and $(i-1)$-th floor, respectively; z is a dimensionless coefficient used for the description of Coulomb friction,

$$\dot{z} = \sigma \cdot \Delta \dot{x}_i \left[1 - z \cdot \text{sgn}(\Delta \dot{x}_i) \right] \qquad (24.22)$$

in which σ is the coefficient controlling the shape of hysteresis loop, and sgn($\cdot$) denotes signum function. Here, the parameters of Dahl model are set to $k_d = 30$ N/m, $c_d = 150$ N·s/m, $f_d = 400$ N, $f_0 = 0$ N and $\sigma = 1500$ s/m.

The El-Centro earthquake with a Peak ground Acceleration (PGA) of 0.34 g is applied to the building structure, and the nonlinear structural responses are obtained by means of Runge-Kutta method with the time interval of 0.001 s. The acceleration responses at the 2nd, 3th, 6th and 8th floor are assumed to be known, and the level of 5% noise is considered. Herein, the unknown quantities to be estimated include the state vector $\mathbf{Z} = \left[x_1, \ldots, x_8, \dot{x}_1, \ldots, \dot{x}_8, z \right]^T$ and the unmeasured ground motion.

Based on the proposed approach, the unmeasured structural responses, such as the displacement and velocity, can be estimated. Since the magnitudes of displacement, velocity, and in particular, the loadings are of different orders, a normalized root mean square error (NRMSE) defined in Eq. (23) is used as a measure of deviation between the actual and estimated values. The NRMSE results in this example are shown in Table 24.1.

$$\text{NRMSE} = \frac{1}{A} \sqrt{\frac{1}{np} \sum_{i=1}^{np} \left(\xi_i^{\text{est}} - \xi_i^{\text{act}} \right)^2} \qquad (24.23)$$

in which A denotes the amplitude of the signal; np is the number of sampling points; ξ_i^{est} and ξ_i^{act} are the estimated and actual values of the i-th sample, respectively.

Obviously, the values of NRMSE are small, which means the results of joint estimation by the proposed approach are reliable. For ease of comparison, the time series of the estimated displacement and velocity responses of the top floor are shown in Figure 24.1 as an example. Figure 24.2 gives the comparison of the identified ground acceleration with the actual one. From Figures 24.1 and 24.2, it can also be confirmed that the proposed approach is capable of estimating nonlinear structural responses and unknown input with acceptable accuracy.

TABLE 24.1

NRMSE Results for Nonlinear Hysteretic Structure with Dahl Model

Estimated Quantities	NRMSE ($\times 10^{-3}$)	Estimated Quantities	NRMSE ($\times 10^{-3}$)
x_1	9.60	v_1	1.17
x_2	9.14	v_2	0.72
x_3	8.77	v_3	0.74
x_4	8.19	v_4	0.84
x_5	7.76	v_5	0.86
x_6	7.44	v_6	0.93
x_7	7.21	v_7	0.87
x_8	7.09	v_8	0.79
$f_{seismic}$	5.01	–	–

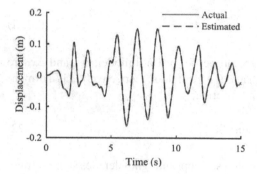 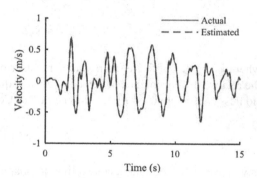

FIGURE 24.1 Comparison of structural responses of nonlinear system with Dahl model.

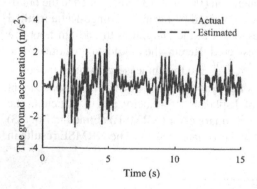 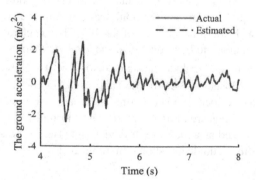

FIGURE 24.2 Comparison of random force applied to the nonlinear structure with Dahl model.

24.3.2 NONLINEAR HYSTERETIC STRUCTURE WITH BOUC-WEN MODEL

To consider the joint estimation of a building structure with multiple nonlinearities, an eight-story shear-type building with Bouc-Wen model on each floor is considered in this numerical example. The equation of motion of such nonlinear structure under earthquake excitation can be written as

$$\mathbf{M}\ddot{\mathbf{x}}(t) + \mathbf{C}\dot{\mathbf{x}}(t) + \mathbf{K}\mathbf{z}'(t) = -\mathbf{M1}\ddot{x}_g(t) \qquad (24.24)$$

where $\mathbf{z}'(t)$ is the hysteretic component given as

$$\dot{z}_i' = \Delta\dot{x}_i - \eta_i\left|\Delta\dot{x}_i\right|\left|z_i'\right|^{\mu_i-1} z_i' - \gamma_i(\Delta\dot{x}_i)\left|z_i'\right|^{\mu_i} \tag{24.25}$$

in which η_i, μ_i and γ_i are the hysteretic parameters of Bouc-Wen model on the i-th floor. Here, the following values are employed: $m_i = 125\,\mathrm{kg}$, $k_i = 20\,\mathrm{kN/m}$, $c_i = 100\,\mathrm{N\cdot s/m}$, $\eta_i = 2{,}000\,\mathrm{s}^{-2}$, $\mu_i = 2$ and $\gamma_i = 1{,}000\,\mathrm{s}^{-2}$ ($i = 1, \ldots, 8$). Instead of El-Centro earthquake, the seismic input considered in this example is Kobe earthquake with a PGA of 0.63 g. The nonlinear responses are computed by Runge-Kutta methods with the time interval of 0.001 s. The acceleration responses at the 1st, 3th, 5th, 7th and 8th floors are considered for joint estimation. In this example, the state vector to be estimated is $\mathbf{Z} = \left[x_1,\ldots,x_8, \dot{x}_1,\ldots,\dot{x}_8, z_1',\ldots,z_8'\right]^T$.

The results of NRMSE of the displacement, velocity and seismic input are shown in Table 24.2. Clearly, small values of NRMSE can be found in Table 24.2. The direct comparison of time series of the estimated displacement and velocity of the top floor is depicted in Figure 24.3. Moreover, Figure 24.4 gives the comparison of the identified ground acceleration with the actual one. Again, it can be found from Figures 24.3 and 24.4 that the differences between the estimated values and actual ones are very small.

TABLE 24.2

NRMSE Results for Nonlinear Hysteretic Structure with Bouc-Wen Model

Estimated Quantities	NRMSE ($\times 10^{-2}$)	Estimated Quantities	NRMSE ($\times 10^{-2}$)
x_1	0.50	v_1	0.09
x_2	1.67	v_2	0.34
x_3	1.31	v_3	0.12
x_4	1.19	v_4	0.39
x_5	0.55	v_5	0.04
x_6	0.87	v_6	0.33
x_7	0.99	v_7	0.13
x_8	1.04	v_8	0.07
f_{seismic}	0.17	–	–

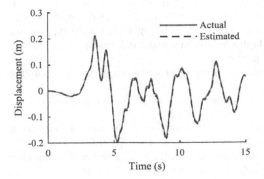 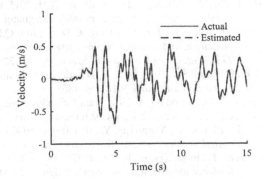

FIGURE 24.3 Comparison of structural responses of nonlinear system with Bouc-wen model.

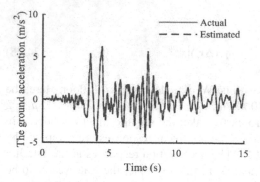 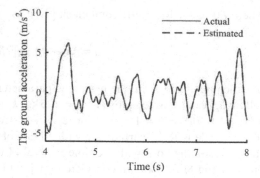

FIGURE 24.4 Comparison of random force applied to the nonlinear structure with Bouc-wen model.

24.4 CONCLUSIONS

In this paper, a KF-based approach is proposed for joint estimation of nonlinear system states and unmeasured loads. Based on the usage of a projection matrix, a modified version of observation equation is obtained. Then, analytical recursive solution of the proposed approach is derived by using KF principle. The unknown inputs are identified by Least Square Estimation (LSE) at the same time. The effectiveness of the proposed approach is validated via two sorts of nonlinear hysteretic structures. The results show that the proposed approach is capable of not only satisfactorily estimating structural states of nonlinear system but also identifying unknown loadings applied to it with acceptable accuracy.

ACKNOWLEDGEMENTS

The authors gratefully acknowledge the financial support from the National Natural Science Foundation of China under project number of Grant No. 51708198. The support from the National Natural Science Foundation of Hunan Province (No. 2018JJ3054) is also greatly appreciated.

REFERENCES

Bernal, D. 2013. Kalman filter damage detection in the presence of changing process and measurement noise. *Mech. Syst. Signal Proc.*, **39**(1–2), pp. 361–371.

Eftekhar Azam, S., Chatzi, E. and Papadimitriou, C. 2015. A dual Kalman filter approach for state estimation via output-only acceleration measurements. *Mech. Syst. Signal Proc.*, **60–61**, pp. 866–886.

He, J., Xu, Y.L., Zhang, C.D. and Zhang, X.H. 2015. Optimum control system for earthquake-excited building structures with minimal number of actuators and sensors. *Smart Struct. Syst.*, **16**(6), pp. 981–1002.

Hu, R.P., Xu, Y.L., Lu, X., Zhang, C.D., Zhang, Q.L. and Ding, J.M. 2017. Integrated multi-type sensor placement and response reconstruction method for high-rise buildings under unknown seismic loading. *The Strut. Design of Tall and Special Build.*, **27**(6), p. e1453.

Kim, D.W. and Park, C.S. 2017. Application of Kalman filter for estimating a process disturbance in a building space. *Sustainability*, **9**(10), pp. 1868.

Lei, Y., Luo, S., and Su, Y. 2016. Data fusion based improved Kalman filter with unknown inputs and without collocated acceleration measurements. *Smart Struct. Syst.*, **18**(3), pp. 375–387.

Liu, L., Zhu, J., Su, Y. and Lei, Y. 2016. Improved Kalman filter with unknown inputs based on data fusion of partial acceleration and displacement measurements. *Smart Struct. Syst.*, **17**(6), pp. 903–915.

Lourens, E., Reynders, E., DeRoeck, G., Degrande, G. and Lombaert, G. 2012. An augmented Kalman filter for force identification in structural dynamics. *Mech. Syst. Signal Proc.*, **27**, pp. 446–460.

Papadimitriou, C., Fritzen, C.P., Kraemer, P., and Ntotsios, E. 2011. Fatigue predictions in entire body of metallic structures from a limited number of vibration sensors using Kalman filtering. *Struct. Control. Health Monit.*, **18**(5), pp. 554–573.

Ren, P. and Zhou, Z. 2017. Strain estimation of truss structures based on augmented Kalman filtering and modal expansion. *Adv. Mech. Eng.*, **9**(11), pp. 1–10.

Smyth, A. and Wu, M. 2007. Multi-rate Kalman filtering for the data fusion of displacement and acceleration response measurements in dynamic system monitoring. *Mech. Syst. Signal Proc.*, **21**(2), pp. 706–723.

Xu, Y.L. and He, J. 2017. *Smart Civil Structures*. CRC Press, Boca Raton, FL.

Zhang, C.D. and Xu, Y.L. 2017. Structural damage identification via response reconstruction under unknown excitation. *Struct. Control. Health Monit.*, **24**(8), p. e1953.

Zhi, L.H., Fang, M.X. and Li, QS. 2017. Estimation of wind loads on a tall building by an inverse method. *Struct. Control. Health Monit.*, **24**(4), p. e1908.

Zhu, S., Zhang, X.H., Xu, Y.L. and Zhan, S. 2013. Multi-type sensor placement for multi-scale response reconstruction. *Adv. Struct. Eng.*, **16**(10), pp. 1779–1797.

25 Performance Analysis of Shear Connectors in Demountable Composite Bridge Deck with Steel Sheets

K. Ye and J. Zhang
Hefei University of Technology

CONTENTS

25.1 INTRODUCTION

Two kinds of traditional bridge deck are generally used in practical applications which are reinforced concrete and steel bridge decks, respectively (Porter et al. 1976, Porter and Ekberg Jr 1977, Daniels and Crisinel 1993a, b). Under the long-term action of vehicle load, especially under the condition of heavy load or even overload, the shear capacity of reinforced concrete bridge deck is insufficient, the bridge decks deteriorate seriously and the design service life is shortened obviously. Steel bridge deck has high bearing capacity and is convenient to construct; however, due to the fatigue characteristics of steel, fatigue cracks in stiffener and weld joints will greatly affect the durability and applicability of the bridge deck. At present, there are two kinds of steel-concrete composite bridge deck, one is that the bottom steel plate is flat and the other is that the bottom steel plate is corrugated. By making the plate into a wavy shape, the two webs can provide larger shear capacity, thus a better shear capacity and bending capacity can be obtained (Ong and Mansurt 1986). Instead of flat plate, the formed steel plate can not only provide better shear and bending resistance but also make the shape lighter and more beautiful (Partrick and Bridge 1994).

Generally, most shear joints are welded to steel beams by steel plates and are then cast into concrete. This means that the steel beams are completely attached to the concrete slabs, making the deconstruction and reuse of the steel components nearly impossible. To allow easy separation of the deck from the steel beam, the removable shear connector embedded in the concrete is preferred (Moynihan and Allwood 2014, Kwon et al. 2010). This paper focuses on studying the mechanical properties of those demountable shear connectors and the effects of concrete grade and collar size

239

FIGURE 25.1   The Kong Li Huai River Bridge.

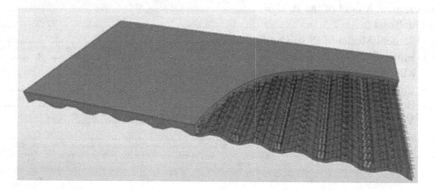

FIGURE 25.2   The structural diagram of the corrugated steel-concrete composite bridge deck.

on the shear capacity and shear performance. It will be helpful to determine the feasibility of this type of connection in composite construction and monitor the health condition of the bridge in the later use stage.

As shown in Figure 25.1, taking the Kong Li Huai River Bridge in Anhui Province as the real example, the new corrugated steel-concrete composite bridge deck in Figure 25.2 is the research object. According to the actual engineering situation, through the establishment of finite element model, the overall stress condition and unfavorable section of the bridge are studied, and the mechanical properties of the removable bridge deck are also investigated. Combined with the global and local model, the stresses in the most unfavorable position are obtained.

25.2   FINITE ELEMENT MODEL OF THE BRIDGE

The general finite element software ANSYS is used to establish the spatial calculation model of the bridge as shown in Figure 25.3. A 3-D beam element is used to simulate the longitudinal beam, main beam, and arch rib, and a 3-D two-node tie is used to simulate the hanger rod and the tie rod. The panel is not affected by gravity into a bridge after the bridge built up, so only the stress of the whole bridge model under the influence of live load, the most unfavorable conditions for symmetric full bridge load, and load standard for highway-I level are analyzed.

FIGURE 25.3 Overall model of the Kong Li Huai River Bridge.

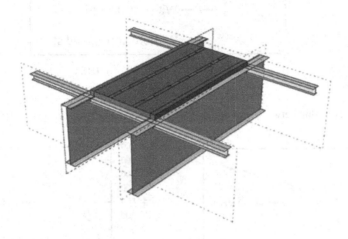

FIGURE 25.4 Local model of a bridge deck.

The results of the overall model show that the middle section of the main span is the most unfavorable section, so the 3.75 m × 7.0 m bridge panel is selected as shown in Figure 25.4, and the general finite element software ABQUS is adopted for further numerical simulation analysis. The relationship between the load and the slip of the bridge panel is depicted in Figure 25.5, and the load-slip curve of long side at the bottom of bridge deck center, valley top, and plate end and load-deflection curve with and without steel sheets are shown in Figure 25.6. To sum up, it can be seen that the relative slip between corrugated steel folding plate and concrete is very small and distributed under 0.1 mm, regardless of the transverse or longitudinal bridge direction. Therefore, it can be considered that concrete and corrugated steel folding plate are completely connected together and have good cooperative working performance. The bearing capacity of the corrugated steel-concrete composite slab is 2.59 times that of the ordinary reinforced concrete slab, and the deflection is only 36.5% of the reinforced concrete slab. The corrugated steel folding plate can significantly improve the flexural capacity of the bridge deck.

25.3 PUSH-OUT TESTS

According to the as-built drawings and standard push-out test specimen in Eurocode-4, the test specimens are fabricated in the lab. Concrete specimen has a height of 800 mm, width of 750 mm, and thickness of 150 mm. The steel plate is 8 mm thick and adopts Q235 steel for the twists and turns plate structure, as shown in Figure 25.7. The bolt used in the specimen and tensile strength test is shown in Figure 25.8. The beam is a 200 × 200 mm general H-beam. The bolts can be removed. In this paper, the full-size test in Figure 25.9 is adopted, and bolt of 180 mm was specially used instead of the 90 or 100 mm in the test.

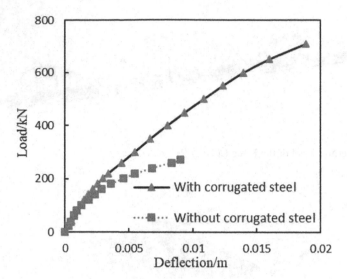

FIGURE 25.5 Relationship between load and deflection with and without corrugated steel sheets.

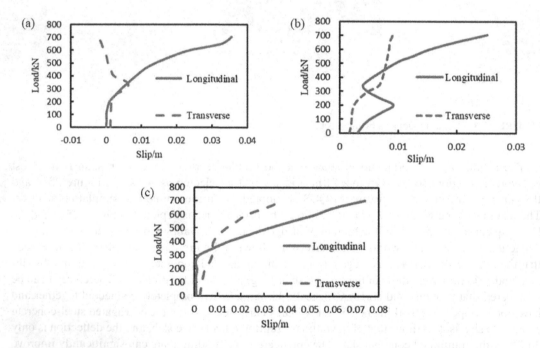

FIGURE 25.6 Load-slip curve of the bridge deck. (a) Relationship between load and slip at the bottom floor of the central of the bridge deck. (b) Relationship between load and slip diagram at the top of the central of the bridge deck. (c) Relationship between load and slip of the bridge panel end (long side).

25.4 FINITE ELEMENT MODEL OF PUSH-OUT SPECIMENS

25.4.1 FINITE ELEMENT TYPES AND NETWORKS

In order to obtain accurate results from finite element analysis, all parts of the shear connection should be properly modeled. Concrete slab, steel beam, rebar, and shear bolt are the main factors affecting shear connection performance. The interactions between the components are also very important. Geometric nonlinearity and material nonlinearity are included in the analysis.

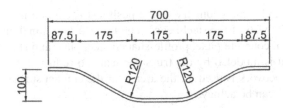

FIGURE 25.7 Profile and dimensions of corrugated sheeting.

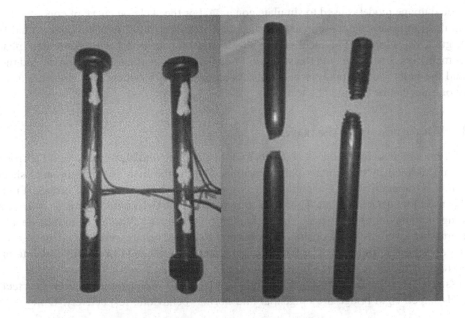

FIGURE 25.8 Bolt used in the specimen and tensile strength test.

FIGURE 25.9 Push-out tests of removable and non-removable composite bridge deck specimens.

As shown in Figure 25.10, the geometry of the push test specimen is established by assuming semi-symmetry along the web of the steel beam. The three-dimensional eight node unit (C3D8R) is used to model the stud, concrete plate, profile-shaped steel plate and steel beam, and the welded steel wire reinforcement is modeled by the truss element. To reduce computing time, apply bold lines as a whole. A fine network is used for the area near the contact surface between concrete and bolt, and accurate results can be achieved.

25.4.2 MATERIAL CONSTITUTIVE RELATION

The plastic damage model is used to simulate concrete. For the skeleton curve of concrete, the unified formula of uniaxial compression and tensile stress-strain relationship of concrete with different strength grades is adopted. The ideal elastic-plastic constitutive model is used for steel plate and reinforcement. The steel beam and the steel reinforcement use linear elastic treatment, and the steel plates and the bolts use elastoplastic materials. Then the strength of concrete and ultimate tensile capacity of bolts can be obtained by tests.

25.4.3 INTERACTION AND CONSTRAINTS

The surface contacts between the steel sheets and the concrete and between bolt and surrounding concrete are defined by the contact pair algorithm. Due to the high stiffness of steel sheet and shear bolt, it is the main plane. Concrete surface is assumed to be a subordinate surface. Finite slip method is used for surface contact. The interaction between the two contact surfaces is determined by the tangent behavior of the surface. The default normal behavior is assumed to include a "hard" contact pressure-overclose relationship. This normal behavior ensures minimal penetration from the surface to the main surface. The friction coefficient between steel plate and concrete plate is assumed to be 0.45.

It is assumed that the default frictionless and "hard" contact interactions between the steel deck and the H-shaped rigid bridge are tangential and normal behaviors, respectively. Using the same

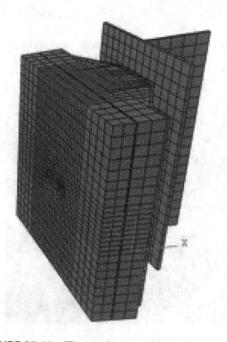

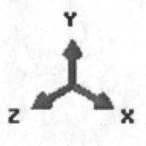

FIGURE 25.10   The semi-symmetric model of push-out test.

"contact pair" method, the top of H-type steel is defined as the main surface, and the bottom of the steel plate is defined as the secondary surface. Under internal constraints, wire reinforcement is embedded into the concrete slab.

25.4.4 BOUNDARY CONDITIONS AND LOADING

The displacement in the Y-direction of all joints on the bottom surface of the concrete slab is restricted. According to the symmetry, the displacement in the X- and Z-directions of the bolt element nodes is restricted, and only the displacement in the Y-direction is allowed. A displacement of 0.25 mm/s is applied to the loading surface of the beam.

25.4.5 VALIDATION AND ANALYSIS OF FINITE ELEMENT MODEL

After the experiment, the calculation results of the finite element model are compared with the experimental data, as shown in Figures 25.11 and 25.12. The results show that the calibrated model is in good agreement with the actual results. According to the experimental results in Figure 25.13, at the same diameter and concrete strength, the bearing capacity of removable composite specimen decreases by about 20%, and the maximum slip increases by about 35%.

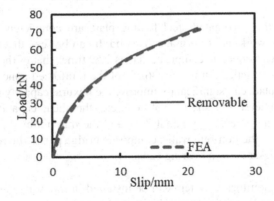

FIGURE 25.11 Comparison between the experimental results and numerical simulation of a removable specimen.

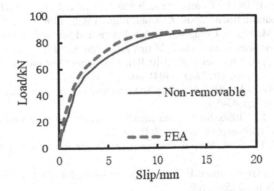

FIGURE 25.12 Comparison between the experimental results and numerical simulation of a non-removable specimen.

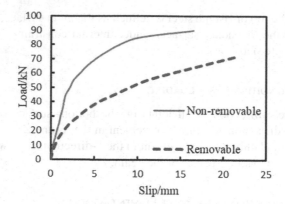

FIGURE 25.13 Comparison between the removable and non-removable push-out tests.

25.5 CONCLUSIONS

In this paper, the mechanical properties of the composite bridge deck with profiled sheets are analyzed on the basis of practical engineering. In line with the development direction of the architectural field, the bridge deck can be replaced with removable composite bridge panels. Then the mechanical properties are obtained, and the following conclusions have been drawn:

1. The concrete and the corrugated steel folding plate are completely connected together, and the cooperative working performance is good. It can be seen that the bearing capacity of the corrugated steel-concrete composite slab is 2.59 times that of the ordinary reinforced concrete slab, and the deflection is only 36.5% of the reinforced concrete slab. The corrugated steel folding plate can significantly improve the flexural capacity of the bridge deck.
2. With the same diameter and strength of concrete, the shear capacity of the removable composite specimen decreases by about 20% and the maximum slip increases by about 35%. Compared with the non-removable composite bridge panel, the removable composite bridge panel provides a certain bearing capacity.

In the future, the health monitoring system can be installed in the bridge deck, which will greatly assure safe operation of the bridge in the whole life cycle.

REFERENCES

Daniels, B.J. and Crisinel, M. 1993a. Composite slab behavior and strength analysis. Part I: Calculations with test results and parametric analysis. *J. Struct. Eng.*, ASCE, **119**(1), pp. 16–35.
Daniels, B.J. and Crisinel, M. 1993b. Composite slab behavior and strength analysis. Part II: Comparisons with test results and parametric analysis. *J. Struct. Eng.*, ASCE, **119**(1), pp. 36–49.
Kwon, G., Engelhardt, M.D., and Klingner, R.E. 2010. Behavior of post-installed shear connectors under static and fatigue loading. *J. Constr. Steel Res.*, **66**(4), pp. 532–541.
Moynihan, M.C. and Allwood, J.M. 2014. Viability and performance of demountable composite connectors. *J. Constr. Steel Res.*, **99**, pp. 47–56.
Ong, K.C.G. and Mansurt, M.A. 1986. Sheer-bond capacity of composite slabs made with profiled sheeting. *Int. J. Cem. Compos. Lightweight Concrete*, **8**(4), pp. 231–237.
Partrick, M. and Bridge, R.Q. 1994. Partial shear connection design of composite, *Eng. Struct.*, **16**(5), pp. 348–362.
Porter, M.L., Ekberg Jr, C.E., Greimann, L.F. et al. 1976. Shear-bond analysis of steel-deck-reinforced slabs. *J. Struct. Div.*, **102**(12), pp. 2255–2269.
Porter, M.L. and Ekberg Jr, C.E. 1977. Behavior of steel-deck reinforced slabs. *J. Struct. Div.*, **103**(3), pp. 663–667.

26 Bridge Damage Classification and Detection Using Fully Convolutional Neural Network Based on Images from UAVs

Jiyuan Shi, Rongzhi Zuo, and Ji Dang
University of Saitama

CONTENTS

26.1 INTRODUCTION

During the period from 1954 to 1973, economy in Japan grew high, with many infrastructures such as roads and bridges being constructed. However, many such infrastructures are over 50 years (Ministry of Land, Infrastructure, Transport and Tourism, 2016) and nowadays are facing many problems such as aging deterioration. Thus, many structures are expected to be inspected in a rapid method in the next few decades.

The traditional method is manual inspection. However, it is expensive in human cost, rental cost of special inspection vehicles, and transportation constrain time.

The methods applying UAVs can collect images quickly, and deterioration detection is available for some structure members and is difficult to access by convenient manual inspection. It will obviously fasten the process to diagnose the status of the infrastructure. To evaluate the status of inspected structures, it is necessary to mark the damages on collected images. However, this operation also costs a large amount of human processing hours.

To address this issue, a lot of methods based on image processing to detect damage on structures have been proposed. With the application of multiple layer structured Convolutional Neural networks (Deep Learning), it is possible to develop approaches with higher accuracy to detect the damages from the images taken by UAVs. Deep Learning-based image classification has been taken to detect cracks (Cha and Choi 2017). However, most of methods were proposed for laboratory photographs with fix camera to object distance and angle to recognize the damage correctly. And image segmentation method is also successful in object detection from aerospace images of several cities (Iglovikov and Shvets 2018) to identify building seismic damage. This can detect structural damage robustly from real-world images in more universal sources such as UAVs.

In this study, a damage detection method based on Fully Convolutional Neural Network (FCNN) is proposed. As training a FCNN model needs large amounts of images, images are collected and manually marked to generate the training data. Then two methods are used to prepare two different data sets for training. The two FCNN models trained from different data sets are compared to get a better one and finally use the better model to test its capability of damage detection.

26.2   STRUCTURE OF FCNN

In general, to detect the location of objects from images, an FCNN architecture consists of two parts (Figure 26.1). One part using five down-sampling blocks is based on VGG16 structure (Simonyan and Zisserman 2014) to capture context. The other part based on U-Net (Jonathan Long, Evan Shelhamer, Trevor Darrell, 2014) consists of five up-sampling blocks and using a symmetric expanding path to enable precise localization. This section explains both of the two blocks and introduces the details or backgrounds of the layers used in the study.

Table 26.1 presents the detail of each layer. The input size of the images is 224×224 pixels with three channels (Red, Green, and Blue). Down-sampling is combined from convolutional (Conv) layers connected with the rectified linear unit (ReLU) layer and MaxPooling layer, while up-sampling layer consisted of up-sampling layer and convolutional layer, which is followed by an auxiliary layer named batch normalization (BN) layer. During the process of up-sampling block, some layers are copied and then concatenated with the output of the convolutional layer. After the data was calculated through a down-sampling block, the width and height will become half. The up-sampling block will double the width and height of the input.

Convolutional is adopted in both down- and up-sampling blocks. One convolutional layer uses a set of kernels with learnable weight to perform the convolution operation. Each kernel slides on the input array with a specific step size defined as stride and the convolution is implemented by this process. The multiplications are done between the element from kernel and the element from subarray of input to get a receptive field. Then the multiplied values are summed with bias added to get a value in the output array. To maintain the output size equal to the input size, adopt zero-padding (Pad) for the input array. This process is shown in Figure 26.2. The output size of convolutional layer is calculated by the formula:

$$O = \frac{(I + 2 \times P - K)}{S} + 1 \tag{26.1}$$

where O = output size, I = input size, P = pad number, K = kernel size, and S = stride. The depth of the output is same as the depth of the convolutional layer.

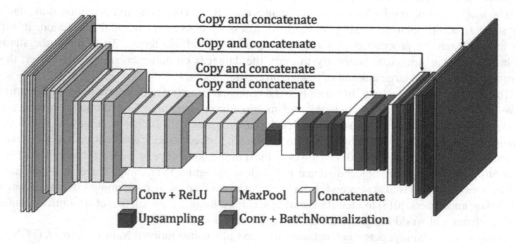

FIGURE 26.1   Overall architecture.

TABLE 26.1

The Detailed Specifications of FCNN

Layer	Type	Pad	Kernel Size	Stride	Output Size	Note
1	Input	-	-	-	$224 \times 224 \times 3$	
2	Conv+ReLU	1	$3 \times 3 \times 64$	1	$224 \times 224 \times 64$	Down-sampling block 1
3	Conv+ReLU	1	$3 \times 3 \times 64$	1	$224 \times 224 \times 64$	
4	MaxPool	0	2×2	2	$112 \times 112 \times 64$	
5	Conv+ReLU	1	$3 \times 3 \times 128$	1	$112 \times 112 \times 128$	Down-sampling block 2
6	Conv+ReLU	1	$3 \times 3 \times 128$	1	$112 \times 112 \times 128$	
7	MaxPool	0	2×2	2	$56 \times 56 \times 128$	
8	Conv+ReLU	1	$3 \times 3 \times 256$	1	$56 \times 56 \times 256$	Down-sampling block 3
9	Conv+ReLU	1	$3 \times 3 \times 256$	1	$56 \times 56 \times 256$	
10	Conv+ReLU	1	$3 \times 3 \times 256$	1	$56 \times 56 \times 256$	
11	MaxPool	0	2×2	2	$28 \times 28 \times 256$	
12	Conv+ReLU	1	$3 \times 3 \times 512$	1	$28 \times 28 \times 512$	Down-sampling block 4
13	Conv+ReLU	1	$3 \times 3 \times 512$	1	$28 \times 28 \times 512$	
14	Conv+ReLU	1	$3 \times 3 \times 512$	1	$28 \times 28 \times 512$	
15	MaxPool	0	2×2	2	$14 \times 14 \times 512$	
16	Conv+ReLU	1	$3 \times 3 \times 512$	1	$14 \times 14 \times 512$	Down-sampling block 5
17	Conv+ReLU	1	$3 \times 3 \times 512$	1	$14 \times 14 \times 512$	
18	Conv+ReLU	1	$3 \times 3 \times 512$	1	$14 \times 14 \times 512$	
19	MaxPool	0	2×2	2	$7 \times 7 \times 512$	
20	Up-sampling	-	-	-	$14 \times 14 \times 512$	Up-sampling block 1
21	Concatenate	-	-	-	$14 \times 14 \times 1024$	
22	Conv+BN	1	$3 \times 3 \times 512$	2	$14 \times 14 \times 512$	
22	Conv+BN	1	$3 \times 3 \times 256$	2	$14 \times 14 \times 256$	
23	Up-sampling	-	-	-	$28 \times 28 \times 256$	Up-sampling block 2
24	Concatenate	-	-	-	$28 \times 28 \times 512$	
25	Conv+BN	1	$3 \times 3 \times 256$	2	$28 \times 28 \times 256$	
26	Conv+BN	1	$3 \times 3 \times 128$	2	$28 \times 28 \times 128$	
27	Up-sampling	-	-	-	$56 \times 56 \times 128$	Up-sampling block 3
28	Concatenate	-	-	-	$56 \times 56 \times 256$	
29	Conv+BN	1	$3 \times 3 \times 128$	2	$56 \times 56 \times 128$	
30	Conv+BN	1	$3 \times 3 \times 64$	2	$56 \times 56 \times 64$	
31	Upsampling	-	-	-	$112 \times 112 \times 64$	Up-sampling block 4
32	Concatenate	-	-	-	$112 \times 112 \times 128$	
33	Conv+BN	1	$3 \times 3 \times 64$	2	$112 \times 112 \times 64$	
34	Up-sampling	-	-	-	$224 \times 224 \times 64$	Up-sampling block 5
35	Conv+BN	1	$3 \times 3 \times 3$	2	$224 \times 224 \times 3$	
36	Output	-	-	-	$224 \times 224 \times 3$	

Another key aspect in FCNN is the MaxPooling layer, which reduces the spatial size of the input array and is often defined as down-sampling. It uses a subarray to get the max values. Figure 26.3 shows the MaxPooling layer with a stride of two. And the output size of the array is calculated by the formula:

$$O = \frac{I - P}{S} + 1$$

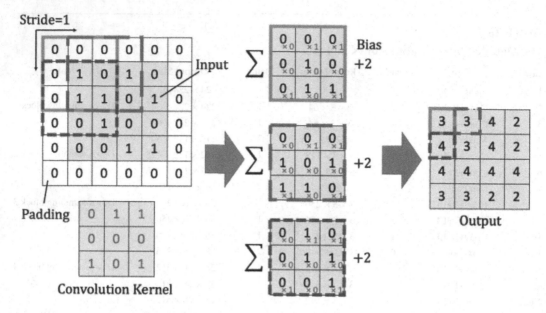

FIGURE 26.2 Convolutional layer example.

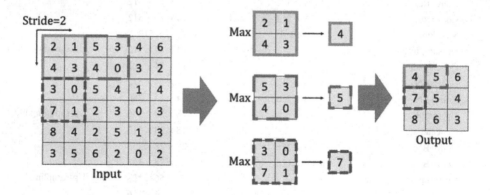

FIGURE 26.3 MaxPooling layer example.

where O = output size, I = input size, P = pooling size, and S = stride size. In this study, $P = 2$ and $S = 2$. Thus, once the input array goes past the MaxPooling layer, the width and height become half.

The Up-Sampling layer is widely used in the up-sampling blocks. It increases the width and height of the input array. Any element in the input array will be repeated in two directions (Figure 26.4). Since the number of Up-Sampling layers and MaxPooling layers are same, the input and output sizes of the images are the same.

26.3 BUILDING DATABASE

Here we used 200 high-resolution images full of corrosions collected from different steel bridges under different illuminance conditions and shooting angles. Then the corrosion parts in these images are manually marked by MATLAB® 2019a to make the annotations database. The corrosion part is thought as category B, while the undamaged part and the background are marked as category A. After that, images and annotations were divided randomly into two parts: 80% of images is for training and the other part is for testing. The data set is called data set A. To compare

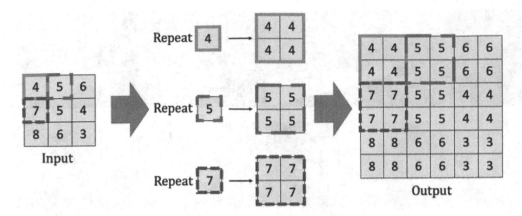

FIGURE 26.4 Up-Sampling layer example.

TABLE 26.2
The Proportion of Training and Testing Data Sets

	Training	Testing
Data set A (Full images)	160	40
Data set B (Cropped images)	9,728 (5,122 with damage)	2,719 (1,313 with damage)

the difference results of different methods from preparing the data set, we copied both training and testing data from data set A and cropped the images and annotations into 224×224 pixels to make the data set B. Table 26.2 shows the detail of training and testing data set. Since the data set is not very large, the whole process of training both data sets are similar, about 17 minutes by only using CPU i7-6700 with memory 16 GB.

26.4 EVALUATION OF TRAINING MODEL

After training the model, a testing data set is used to w the detection ability of the model to show the location of the corrosion in the images. Here we use Jaccard index (Intersection Over Union) as an evaluation metric. The result is intersection over union for similar measure between different sets A and B, and the formula is defined as follows:

$$J(A,B) = \frac{|A \cap B|}{|A \cup B|} = \frac{|A \cap B|}{|A| + |B| - |A \cap B|}$$

where $J(A, B)$ is the accuracy of detection compared from annotations and predictions from the model. A is called ground truth and means the set of pixels from annotation data set and B is the set of pixels predicted by the FCNN model.

The accuracy of the method using full images is about 82.79%, while the accuracy of the other method is 88.68%. To see and compare the results of the two methods in an easier approach, after testing the cropped images, we stitch the 224×224 images into a full image. The results are shown in Figure 26.5.

In Figure 26.5 I-(c) and I-(d), the pixels of corrosion are predicted, but the numbers of pixels are quite different. The same phenomenon can be found in Figure 26.II-(c), II-(d) and III-(c), III-(d).

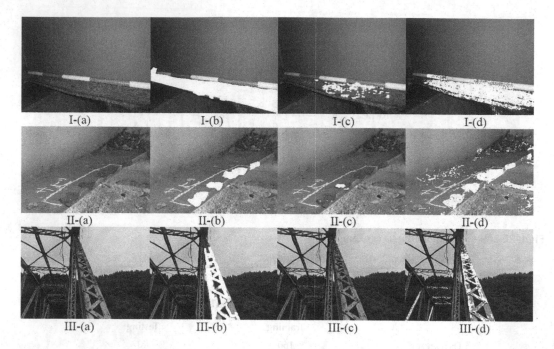

FIGURE 26.5   Examples of prediction results: (a) original image, (b) ground truth, (c) prediction result from full images, and (d) prediction result from cropped images.

The main reason is full images will make the system focus on the relationship between the corrosion part and the undamaged part instead of being only focused on the pixels of damage and the pixels adjacent to them. The model trained from cropped images is too sensitive to damage and coatings and easily predict coating degradation as the category of corrosion even though we haven't marked such class of damage in the original images. What's more, it is because of the size limitation of cropped images that some concrete pixels are not distinguished from steel and then predicted as corrosion. Conversely, the other model would not be so sensitive to damages that even ignoring some damage pixels would confuse concrete and steel.

Figure 26.6 presents some relatively failed samples of predictions. The gray boxes show the false negative errors while the black boxes show the false positive errors. The false positive errors are mainly about predicting coating degradations as corrosions. And the false negative errors happened when the corrosion is too serious that the features are different with the other level. This is because the images including such serious corrosions only account for a small percentage of the training set.

One image of UAV inspection was also used to test the availability of application of FCNN and UAV based on bridge corrosion detection. The image is with a resolution of 4,096*2,160 and separated to 18×9 small images. We can find that the corrosion pixels in the image is recognized successfully by the FCNN model. However, some other areas such as yellow plants and railings are misrecognized as corrosions. The result is shown in Figure 26.7.

26.5   CONCLUSION AND FUTURE WORK

A damage detection method based on FCNN is proposed to identify and position the corrosion from the images of bridges. And we create two different data sets in two different methods to train the model and then compare the outcomes from these two data sets. And the accuracies come up to 82.79% and 88.68%. It can generally reflect the location of the damage and to some extent show the size of the damage. Then we discussed the advantages and disadvantages of these two methods.

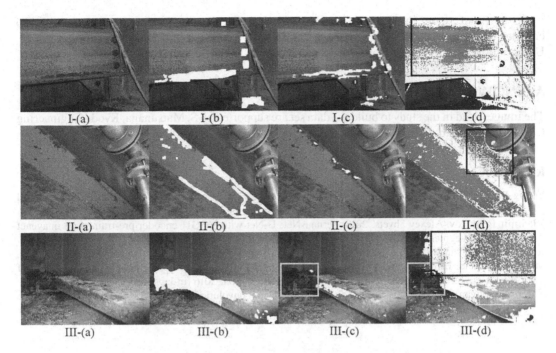

I-(a) I-(b) I-(c) I-(d)

II-(a) II-(b) II-(c) II-(d)

III-(a) III-(b) III-(c) III-(d)

FIGURE 26.6 Failed samples of prediction results: (a) original image, (b) ground truth, (c) prediction result from full images, and (d) prediction result from cropped images.

FIGURE 26.7 Damage detection using images taken from UAV. (a) original and (b) after prediction.

In the future, more images with corrosions of different damage levels would be collected to enhance accuracy. And then the variety of damage to be detected will be enriched. Future more, UAVs will be used to collect more images and videos to test the capability of the FCNN model.

ACKNOWLEDGEMENT

The images used in the study to build the data set are supported by S. Matsunaga, Kyodo Engineering Consultant Co., Ltd, Fukuoka Branch, Japan.

REFERENCES

Young-Jin Cha, Wooram Choi. 2017. Deep learning-based crack damage detection using convolutional neural networks. *Computer-Aided Civil and Infrastructure Engineering*, 32(5), pp. 361–378.
Vladimir Iglovikov, Alexey Shvets. 2018. TernausNet: U-Net with VGG11 encoder pre-trained on imagenet for image segmentation.
Jonathan Long, Evan Shelhamer, Trevor Darrell. 2014. Fully convolutional networks for semantic segmentation.
Ministry of Land, Infrastructure, Transport and Tourism. 2016. White Paper on Present state and future of social capital aging. Infrastructure maintenance information, 2016 (in Japanese).
Karen Simonyan, Andrew Zisserman. 2014. Very deep convolutional networks for large-scale image recognition.

27 Quarter Car Parameter Estimation with Application to Road Profile Evaluation Using a Smartphone

K. Xue, T. Nagayama, and D. Su
The University of Tokyo

CONTENTS

27.1 INTRODUCTION

There is a high demand to maintain and monitor road conditions such that drivers and pedestrians can travel safely and comfortably on them. Although high accurate laser-based profiler and high-speed inertia profiler have been developed for the road profile estimation, they are primarily used for important highway measurements, and it is not practical to frequently evaluate other road networks due to its high cost and installation constrains. Nowadays, the discovery of the aged and affected parts of the local roads mainly depends on the field experts, which leads to neglect of proper inspection for many municipalities due to the lack of resources or experts. The vehicle-smartphone-based road profile measurement framework that combines a global positioning system (GPS) measurements and inertial measurement unit (INS) is a promising alternative to solve these problems because of its low cost and fast communication as well as its potentially broad population coverage as common devices.

Direct road profile estimation and indirect road index assessment are commonly used in the response-based road condition evaluation (Zhao, Nagayama and Xue 2019). In direct profile estimation, observers are utilized to estimate the road profile using vehicle responses. While road roughness indices are generally used to evaluate the road condition in indirect road condition estimation, quarter car (QC) and half car (HC) models are often used to express the vehicle dynamics. In fact, vehicle model type and model parameters provide significant information in the response-based road roughness estimation not only in the direct profile estimation but also in indirect situations. For most studies, the parameters are obtained from vehicle manufacturer or from laboratory tests, which is time consuming and may not provide expected results since the parameters may not be suitable for the proposed algorithm when the vehicle is driving on the road.

This paper presents a novel vehicle parameter estimation and road profile estimation method using vehicle body accelerations measured by a smartphone. The vehicle is modeled as a QC model and the parameters are identified by an output-only algorithm. It is achieved by running a vehicle on the same road with different speeds, by assuming that the estimated profiles for different speed cases must be the same, the QC model parameter is first identified by multi-objective genetic algorithm (GA). During the GA parameter searching and the final road profile estimation, road profiles are estimated by a combination of the augmented Kalman filter (AKF), the Robbins Monro (RM) algorithm, and the Rauch-Tung-Striebel (RTS) smoothing. Drive tests with two different vehicles showed that the proposed algorithm can compensate for differences among vehicles with different drive speeds and estimate profiles.

27.2 PRELIMINARIES

27.2.1 ROAD INTERNATIONAL ROUGHNESS INDEX (IRI)

International Roughness Index (IRI), which was proposed by the World Bank as a road condition indicator, summarizes the road quality that affect drive comfort, dynamic driving loads, etc. (Sayers and Karamihas 1998). IRI is one of the most commonly used road roughness indices for evaluating road profiles. It is calculated using a QC (Golden Car) vehicle math model with known parameters, whose responses are accumulated to yield a roughness index with units of slope (mm/m). The Golden Car performs like a spatial filter, and the wave number response of the IRI QC filter is shown in Figure 27.1. If the input is a sinusoid with an amplitude that is slope, the output is the product of the input amplitude and the value obtained from the plot. Therefore, the IRI of a normal road can be regarded as a weighted slope related to the wavelength numbers.

In this paper, IRI is adopted as the main evaluation index of the road profile roughness. Because of the frequency characteristics of the Golden car model, IRI has a sensitive frequency band of [0.033 0.8] cycle/m (Figure 27.1). It has peaks at 0.065 cycle/m (wavelength 15 m) and 0.42 cycle/m (wavelength 2.4 m), and the sensitivity decreases to one-third of the sensitivity of the peaks when the frequency is lower than 0.033 cycle/m (wavelength 30 m) or larger than 0.8 cycle/m (wavelength 1.25 m). Although the frequency characteristic changes continuously with respect to the spatial frequency, the sensitivity decreases rapidly when leaving this band.

27.2.2 ROAD POWER SPECTRAL DENSITY (PSD)

The IRI is sufficient to evaluate the pavement roughness of profiles but does not allow much place for the analysis of road surface effects on vehicles, and therefore mathematical representation of road profiles is necessary for the studies of road-vehicle interactions. Generally, the analyses require

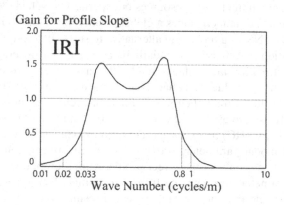

FIGURE 27.1   The IRI Golden Car filter.

a thorough knowledge of the distribution of frequencies with accompanying amplitudes, and power spectral density (PSD) is much used for this purpose. The PSD used in the road profile description is often approximated with a simple function using only a few parameters since all roads show a similar basic trend—that is the amplitude in profile elevation grows with wavelengths. By assuming that the road profiles are homogeneous and Gaussian distribution, the International Organization for Standardization (ISO) specified different classes of road condition based on a simple line fit on the logarithm of the road PSD (ISO 1995), as,

$$G_d(n) = C_{sp} n^{-w} \tag{27.1}$$

where n is the spatial frequency in cycle/m and C_{sp} and w are two constants standing for the excitation energy and road waviness, respectively. In the following, this characteristic of road will be used to exclude identified vehicle parameter with poor PSD performance.

27.2.3 SMARTPHONE-BASED SENSOR

The smartphone can record acceleration and angular velocity in three directions as well as GPS signals simultaneously; therefore, it is employed as a measurement device in this paper. Since the sampling rates of the smartphones vary from each other and fluctuate over time, offline resampling based on the time stamp must be first performed in order to achieve a constant sampling frequency (100 Hz). For the vibration data, although the sampling frequency is resampled to 100 Hz, the accuracy of accelerations and angular velocities cannot be guaranteed for the high-frequency range. Comparison of the accuracy of vibration data with high-accuracy devices showed that good consistency can be observed up to 20 Hz, which is sufficient for vehicle-based vibration applications due to the fact that the frequencies of the tire modes are designed to be approximately 15 Hz for most of the vehicles. Thus, a low-pass filter is used to cutoff the high-frequency components. On the other hand, a high-pass filter should also be applied to eliminate the effects of low-frequency integral errors and sudden drive speed changes. As a result, this is equivalent to apply different bandpass filters to the estimated profile in the spatial domain for different speeds due to the relationship that

$$f = \frac{V}{\lambda} \tag{27.2}$$

where f is frequency in cycle/second, V is speed in m/second, and λ is wavelength in m/cycle. Consequently, the upper bound estimation in the spatial domain is reduced when the drive speed is large while the lower bound estimation is increased if the drive speed is small. To compensate the speed effects, a fixed bandpass filter $[f_a/V \cdot f_b/V]$ can also be applied to the estimated profile in the spatial domain before being transformed to IRI or PSD. From Figure 27.1, f_a/V and f_b/V can be set to 0.02 and 0.8 cycles/m, respectively.

27.3 ROAD PROFILE ESTIMATION

The model of the QC and the corresponding parameters are shown in Figure 27.2. In the system, m_s is the sprung mass of the QC model and m_u is the unsprung mass. k_s and c_s are the suspension stiffness and damping ratio, respectively, and k_t is the tire stiffness. y_{ex} is the pavement excitation and V is the speed of the vehicle model. The corresponding vertical displacement of the sprung and unsprung masses are k_s and z_u, respectively. By defining the displacement vector as $x = [z_s \ z_u]^T$, the equation of motion can be written as

$$M\ddot{x}(t) + C\dot{x}(t) + Kx(t) = Py_{ex}(t) \tag{27.3}$$

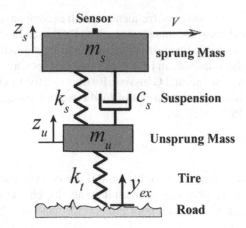

FIGURE 27.2   The QC model.

where **M**, **C**, and **K** are the mass, damping, and stiffness matrices of the QC model, respectively. **P** is the QC model's load effect vector. Because the smartphone is mounted on the sprung mass of the vehicle, defining $\mathbf{x}_c = \begin{bmatrix} \mathbf{x}^T & \dot{\mathbf{x}}^T \end{bmatrix}^T$, $y_c = y_{ex}$ and $\mathbf{u}_c = \begin{bmatrix} \ddot{z}_s & z_s \end{bmatrix}^T$, the equation of motion in the state space representation yields

$$\dot{\mathbf{x}}_c(t) = \mathbf{A}_c \mathbf{x}_c(t) + \mathbf{B}_c \mathbf{y}_c(t)$$
$$\mathbf{u}_c(t) = \mathbf{C}_c \mathbf{x}_c(t) + \mathbf{D}_c \mathbf{y}_c(t)$$
(27.4)

Defining an augmented state vector $\mathbf{x}_a = \begin{bmatrix} \mathbf{x}_c^T & y_c & \dot{y}_c \end{bmatrix}$, then an augmented state-space representation including model error and measurement noise can be described as

$$\dot{\mathbf{x}}_a(t) = \mathbf{A}_{ca} \mathbf{x}_a(t) + \varsigma$$
$$\mathbf{u}_c(t) = \mathbf{C}_{ca} \mathbf{x}_a(t) + v$$
(27.5)

To apply the discrete Kalman filter (KF) to the augmented system, Eq. (27.5) should be converted to their corresponding discrete version as

$$\mathbf{x}_{a-\text{dis}}(k+1) = \mathbf{A}_{ca-\text{dis}} \mathbf{x}_{a-\text{dis}}(k) + \varsigma_{\text{dis}}(k)$$
$$\mathbf{u}_{\text{dis}}(k) = \mathbf{C}_{ca-\text{dis}} \mathbf{x}_{a-\text{dis}}(k) + v_{\text{dis}}(k)$$
(27.6)

Assuming the posterior state estimate $\hat{\mathbf{x}}_a(k)$ and the posteriori error covariance matrix $\mathbf{P}(k)$ are initialized at time step $k=0$, the discrete KF is operated as

$$\hat{\mathbf{x}}_a^-(k+1) = \mathbf{A}_{ca-\text{dis}} \hat{\mathbf{x}}_a(k)$$

$$\mathbf{P}^-(k+1) = \mathbf{A}_{ca-\text{dis}} \mathbf{P}(k) \mathbf{A}_{ca-\text{dis}}^T + \mathbf{Q}(k)$$

$$\mathbf{G}(k+1) = \mathbf{P}^-(k+1) \mathbf{C}_{ca-\text{dis}}^T \left[\mathbf{C}_{ca-\text{dis}} \mathbf{P}^-(k+1) \mathbf{C}_{ca-\text{dis}}^T + \mathbf{R}(k+1) \right]^{-1}$$ (27.7)

$$\hat{\mathbf{x}}_a(k+1) = \hat{\mathbf{x}}_a^-(k+1) + \mathbf{G}(k+1) \left(\mathbf{u}(k+1) - \mathbf{C}_{ca-\text{dis}} \hat{\mathbf{x}}_a^-(k+1) \right)$$

$$\mathbf{P}(k+1) = \left(\mathbf{I} - \mathbf{G}(k+1) \mathbf{C}_{ca-\text{dis}} \right) \mathbf{P}^-(k+1)$$

In this KF process, the profile is both obtained as a function of time as a part of the state vector.

By assuming that the covariance matrices are diagonal and are initialized at time step $k=0$, the estimation of the $\mathbf{Q}$ and $\mathbf{R}$ using RM approximation scheme at time step k can be formulated as

$$\mathbf{Q}_{RM}(k) = \left(1-\alpha_Q\right)\mathbf{Q}_{RM}(k-1) + \alpha_Q \mathbf{G}(k)\left(\mathbf{u}_{dis}(k) - \mathbf{u}_{dis}^-(k)\right)\left(\mathbf{u}_{dis}(k) - \mathbf{u}_{dis}^-(k)\right)^T \mathbf{G}(k)^T$$

$$\mathbf{R}_{RM}(k) = \left(1-\alpha_R\right)\mathbf{R}_{RM}(k-1) + \alpha_R \mathbf{G}(k)\left(\mathbf{u}_{dis}(k) - \mathbf{u}_{dis}^-(k)\right)\left(\mathbf{u}_{dis}(k) - \mathbf{u}_{dis}^-(k)\right)^T \quad (27.8)$$

$$\mathbf{u}_{dis}^-(k) = \mathbf{C}_{ca-dis}\hat{\mathbf{x}}_a(k)$$

where the forgetting parameters α_Q and α_R are both constants smaller than 1.

The RTS smoothing is started from the last time step N of the KF, with $\hat{\mathbf{x}}_{RTS}(N) = \hat{\mathbf{x}}_a(N)$ and $\hat{\mathbf{P}}_{RTS}(N) = \hat{\mathbf{P}}(N)$, and then the smoother propagates backward from $k=N-1$ as

$$\hat{\mathbf{x}}^-(k+1) = \mathbf{A}_{ca-dis}\hat{\mathbf{x}}(k)$$

$$\mathbf{P}^-(k+1) = \mathbf{A}_{ca-dis}\mathbf{P}(k)\mathbf{A}_{ca-dis}^T + \mathbf{Q}_{RM}(k)$$

$$\mathbf{G}_{RTS}(k) = \mathbf{P}(k)\mathbf{A}_{ca-dis}^T\mathbf{P}^-(k+1)^{-1} \quad (27.9)$$

$$\hat{\mathbf{x}}_{RTS}(k) = \hat{\mathbf{x}}(k) + \mathbf{G}_{RTS}(k)\left[\hat{\mathbf{x}}_{RTS}(k+1) - \hat{\mathbf{x}}^-(k+1)\right]$$

$$\mathbf{P}_{RTS}(k) = \mathbf{P}(k) + \mathbf{G}_{RTS}(k)\left[\mathbf{P}_{RTS}(k+1) - \mathbf{P}^-(k+1)\right]\mathbf{G}_{RTS}(k)^T$$

where $\mathbf{G}_{RTS}$ is the gain of the RTS smoother. Since the velocity of the model is measured by the smartphone GPS, the profiles are then converted to the spatial domain using the relationship between the time and the distance.

27.4 MODEL IDENTIFICATION

This section provides the procedure of QC and HC model identification, which will be used to estimate the road profile described in the last section. The parameters are identified using the multi-objective GA, which minimizes the road profile PSD differences as well as the IRI differences when the vehicle travels on the same road at different speeds.

For a linear-time-invariant system, the system responses are closely related to the combination of the system parameters. For the case that bandpass filter in the spatial domain is fixed, from Eq. (27.2), if the running speed of vehicle is v_n, the frequency range of the responses used for the road profile estimation is $[f_a f_b]$Hz. Thus, different vehicle responses are used for the profile estimation when different speeds are applied to the system. If a set of parameters can be found for the system so that the same road profile can be estimated for different speeds traveling on the same road, they are considered as the best system parameters for road profile estimation.

In this paper, the QC parameters for road profile estimation are proposed to be identified by using smartphone measurements while the vehicle is driving on the same profile with different constant speeds, as shown in Figure 27.3. For different speeds, for a given set of parameters, the road profiles are first estimated using KF, RM, and RTS smoothing in the time domain and then transferred to their corresponding PSD and IRI using the smartphone GPS speed information. It is known that if the same road profile is traveled, the PSD and IRI for different speeds must also be the same.

Here, it is assumed that the vehicle travels along the same profile with different speeds, and the unknown parameters are estimated through the minimization of the differences among the individual IRI estimations as well as the smoothed PSD estimations at different speeds using the multi-objective GA. In the following, two objective functions are defined for the multi-objective GA.

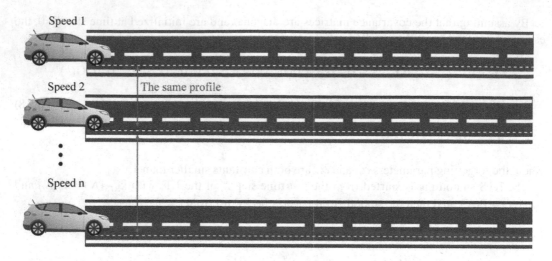

FIGURE 27.3   Vehicle travels on the same profile with different speeds.

Suppose that the total length of the road profile for the parameter identification is a. The evaluation length and the evaluation interval of IRI are L and δ_{IRI}, respectively. Then the objective function related to IRI F_1 is defined as

$$F_1 = \sum_{k=1}^{n-1} \sum_{j=k+1}^{n} \sum_{i=1}^{(a-2L)/\delta_{\text{IRI}}} \left| \text{IRI}_j(i) - \text{IRI}_k(i) \right| \tag{27.10}$$

where n is the total number of different speeds of the vehicle. Similarly, the objective function related to smoothed PSD F_2 is

$$F_2 = \sum_{k=1}^{n-1} \sum_{j=k+1}^{n} \sum_{i=1}^{(a-2L)/\delta_{\text{IRI}}} \left| \text{PSD}_j(i) - \text{PSD}_k(i) \right| \tag{27.11}$$

where N_{PSD} is the total number of spatial frequencies within the spatial bandpass filter range $[\omega_a\ \omega_b]$.

Since the normal road is assumed, the objectives defined in Eq. (27.11) will emphasize on the large wavelength components. On the other hand, because the wavelength components are weighted based on Figure 27.1, Eq. (27.10) will focus on the middle wavelength components which is in the range of [0.045 0.6] cycles/m. However, the components with small wavelength are still be underestimated. Therefore, several constrains should be defined on the estimated PSD during the multi-objective GA optimization. For the ith and jth speed road estimation, the following constrain should be satisfied:

$$\text{Constrain 1} \quad \text{slope}_i - \text{slope}_j < \varepsilon_1 \tag{27.12}$$

where slope_i and slope_j are the slopes of the smoothed PSD. For individual PSDs:

$$\text{Constrain 2} \quad R^2 > \varepsilon_2 \tag{27.13}$$

R^2 is the corresponding Coefficients of Determination to judge the fitness between the smoothed and linear fitted PSDs.

27.5 EXPERIMENT

The sixth-generation iPod Touch was selected as the measurement instrument, since there is no built-in GPS, the Bad-elf GPS for Lightening Connector was used to provide the GPS signal. An iOS application named iDRIMS, was developed, a free to download iPhone app, to measure and record the three direction accelerations, angular velocities, and the GPS signals, due to road roughness, simultaneously. In general, the sampling frequency of the acceleration and angular velocity measurements of the smartphone is not accurate, and therefore a resampling algorithm is necessary to correct the sampling frequency.

It is known that the sampling of the measurement signal is performed based on the clock of Micro Electro Mechanical System (MEMS), and the sampling frequency may deviate from the default value or slowly fluctuate due to the inaccuracy of the clock. On the other hand, while data are sampled based on the clock of a smartphone, uncertain sampling delays may occur for each sample. Therefore, by using the iDRIMS, nearly 100 Hz sampling with the MEMS side clock is first achieved, and at the same time a time stamp is given for each sample. Next, based on the time stamp, the sample is estimated every 1 second and then resampled to convert the sampling frequency to 100 Hz. Finally, when the clock on the smartphone side is synchronized with the absolute time of the GPS, it is possible to achieve the same precision time of smartphone with the absolute time of the GPS.

Field test was conducted on a 200 m ordinary road in Hokkaido city, Japan (Figure 27.4). Two different cars shown in Figure 27.4 were examined to verify the application of the proposed road estimation algorithm. For each vehicle, one smartphone was installed on the left side of the dashboard close to the front tire to measure the vehicle responses during the test. In order to eliminate the necessary coordinate transformation, the smartphone was mounted on the horizontal surface of the dashboard such that the responses of the vertical direction acceleration can be directly measured. The profile estimation accuracy of the proposed method was examined by comparing the estimates with a reference obtained by a high accurate profiler.

After the test, the vertical acceleration measurements are processed by a [0.2 15] Hz bandpass filter and then integrated to obtain vertical displacement. After the AKF, RM, and RTS smoothing, the estimated time domain profile is mapped to the spatial domain using the speed information obtained from the GPS. The profile is further interpolated at an interval of 0.05m and again processed by a bandpass filter of [0.02 0.8] cycle/m. Finally, the processed profile is converted to the corresponding PSD and IRI for accuracy comparison among each other and with the profiler. The evaluation length of IRI is 20 m for the entire target course.

For each vehicle, three different speeds of 40, 50, and 60 km/h were tested, and during driving, the left-hand side tires were adjusted to follow the same profile, which was marked by a white line shown in Figure 27.4. After identifying the QC model parameters using multi-GA, the road profiles were estimated and then transferred to the corresponding IRI. The results are shown in Figures 27.5 and 27.6, respectively. As can be seen from Figures 27.5 and 27.6, a good accuracy can be obtained by the proposed algorithm.

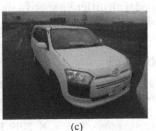

(a)   (b)   (c)

FIGURE 27.4 Experimental course and vehicles. (a) Course, (b) Land Cruiser, (c) Succeed.

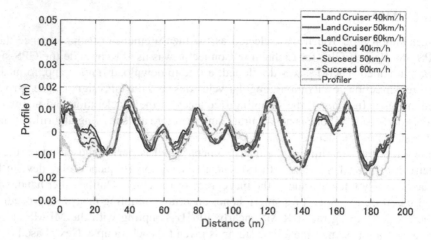

FIGURE 27.5 Comparison of the estimated profiles of two vehicles of different speeds with the profiler.

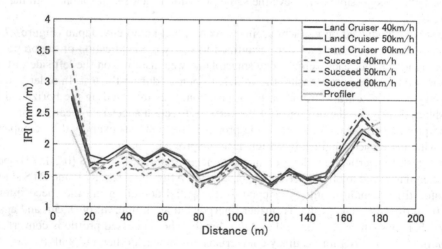

FIGURE 27.6 Comparison of the estimated IRI of two vehicles of different speeds with the profiler.

27.6 CONCLUSIONS

An output-only QC model identification for the estimation the road profile has been proposed. By using the vehicle body acceleration responses and the vehicle speed measured by a single smartphone, the road profile is estimated by combining AKF, RM, and RTS smoothing if the parameters of the QC model are known. For the model parameter identification, by running a vehicle on the same road with different speeds, the QC model parameter is identified by PSD-based and IRI-based multi-objective GA with two constrains.

REFERENCES

ISO (1995-09-01). Mechanical vibration - Road surface profiles - Reporting of measured data, ISO 8608:1995(E), International Organization for Standardization (ISO).

Sayers, M.W., Karamihas, S.M. 1998. *The Little Book of Profiling*, University of Michigan, Ann Arbor, MI.

Zhao, B., Nagayama, T., Xue, K. 2019. Road profile estimation, and its numerical and experimental validation, by smartphone measurement of the dynamic responses of an ordinary vehicle, *Journal of Sound and Vibration*, 457: 92–117.

28 Highway Bridge Weigh-in-Motion via Moving Load Identification

Xiangang Lai and A. Emin Aktan
Drexel University

Kirk Grimmelsman
Intelligent Infrastructure Systems

Matteo Mazzotti and Ivan Bartoli
Drexel University

CONTENTS

28.1 INTRODUCTION

Traffic loads on highway bridges have been steadily increasing in the past years. Actual truck axle configurations and weights may deviate significantly from standard legal design trucks depending on states, freight corridors, and seasons. Even though highway bridges are designed for live load considering standard legal trucks, the influence of heavy truck loads on the deterioration and long-term degradation of bridges remain unknown. It is well known that the overweight traffic on the road will shorten its service life as well as impact the bridge infrastructure. It is important to understand quantitatively if and how the volume and weight attributes of truck traffic may lead to accelerated deterioration and degradation of bridge components so that policy decisions on legal truck loads can be tied to scientific evidence. The effect of moving loads on bridges depends on many parameters, including truck traffic volume, gross vehicle weights, axle weight and configurations, and speeds and the dynamic characteristics of both the bridge and crossing trucks. It is expected that the effect of increasing truck loads on bridge performance may only become apparent over a long time frame after many crossings, depending also on the seasonal climate and environmental conditions. It follows that quantifying whether and how trucks impact bridge service life requires long-term observations and measurements over years.

Currently, management and enforcement of truck weights operating on the nation's bridges is primarily facilitated by truck measurements at roadside weigh stations and/or mainline pavement weigh-in-motion systems (Lydon et al. 2015, Yu et al. 2016). These systems can characterize the volume of trucks and their geometric and weight characteristics and the data can be used to adjust the live load factors employed for bridge design and load ratings. However, this data is not sufficient to quantify and understand the resulting bridge-vehicle interactions (BVI) and their subsequent effects on the service life of different bridge components.

The most promising approach for understanding how trucks impact bridges might be instrumenting bridges following the principles of bridge weigh-in-motion (BWIM). BWIM systems typically employ instrumentation installed on the bridge to capture the crossing vehicle weight and geometry. BWIM can be classified based on the different assumptions used to capture the interaction between bridge and vehicles. BWIM can use (1) static algorithms based on the Moses's algorithm (Moses 1979) which ignores the dynamic responses, (2) moving force identification (MFI) (Law and Fang 2001) which treats the truck weights as constant point loads while considering bridge dynamic response, and (3) BVI (Lu and Liu 2011) which accounts for the vehicle and bridge coupled dynamic behavior. All these approaches require a comprehensive understanding of the instrumented structure while idealization of the bridge intrinsically introduces uncertainties in the weight estimation. Therefore, digital twins which can capture the behavior of a bridge are necessary prior to estimating trucks' weights.

In this paper, MFI formulation to solve weights of moving vehicle axles was derived based on the dynamic programming and Bellman's optimality principle (Trujillo 1978). Some questions the authors attempted to respond to are: How reliable the algorithm is? What are the requirements to successfully implement it on real operating bridges? A simulated model was employed to test the applicability of the algorithm. Suggestions and recommendations were made to deploy this approach for real bridge applications based on the findings of the numerical analysis.

28.2 MFI-BWIM ALGORITHM

The implementation of BWIM including the MFI requires the knowledge of truck axle configuration and vehicle speed (Žnidaric et al. 2002, O'Brien et al. 2012), which generally involves scrutinizing the peaks in the sensor responses to estimate the time employed by the vehicles to cover a known distance and to determine the speed as well as axle configuration. Since this paper will focus on MFI based BWIM, the axle configurations and speed are assumed to be known variables.

28.2.1 STATE SPACE MODEL FORMULATION

Moving load identification treats the vehicle weights as moving loads applied on the bridge dynamic system. The dynamic system can be represented by a finite element model with n degree of freedom (DOF)

$$M\ddot{U} + C\dot{U} + KU = Ff \tag{28.1}$$

where M, C, K are the system mass, damping, and stiffness matrices, U is a vector containing all the DOFs such as displacement and/or rotations, the dots mean time derivative and second time derivative. $f\,(nf \times 1)$ represents the excitation forces as a function of time such as wheel weights, while F is the matrix operator that applies the external forces f into the DOF of the system. To solve the above equation, state space formulation is used to convert Eq. (28.1) into a first-order differential equation.

$$\dot{X} + \bar{K}X = \bar{F}f \tag{28.2}$$

where $X = \begin{bmatrix} U \\ \dot{U} \end{bmatrix}$, $\bar{K} = \begin{bmatrix} 0 & I \\ -M^{-1}K & -M^{-1}C \end{bmatrix}$, and $\bar{F} = \begin{bmatrix} 0 \\ -M^{-1}F \end{bmatrix}$. Equation (28.2) can be rewritten using the standard exponential matrix representation

$$X_{i+1} = GX_i + P_i f_i \qquad (28.3)$$

in which $G = e^{\bar{K}h}$, $P_i = \bar{K}^{-1}(G-I)\bar{F}_i$, h denotes the time step and $(i+1)$ represents the values at $(i+1)$ time step.

28.2.2 INVERSE PROBLEM SOLUTION

To solve the equation for the unknown force term f representing the moving vehicle across bridge, the goal is to find f, causing the system response to closely match the experimental measurement. In practice, the number of DOF measured experimentally is much smaller than the DOF of the whole finite element discretized model. Let's denote the set of DOF to be measured as D (m × 1), then D relates to X ($n \times 1$) with a transformation matrix Q, as

$$D = QX \qquad (28.4)$$

The actual measurements are represented by Z which has the same dimension of D. The least square error of the computation is defined as (Trujillo 1978)

$$\text{error} = \sum_{i=1}^{N}\left[\left(Z_i - QX_i, A(Z_i - QX_i)\right) + \left(f_i, \mathbf{B}f_i\right)\right] \qquad (28.5)$$

where (x, y) denotes the inner product, N means the length of the time vectors. $\mathbf{A}$ and $\mathbf{B}$ are symmetric positive definite weight matrices of the measurement term (first term) and the force term (second term). In practice, $\mathbf{A}$ is usually taken as the identity matrix and $\mathbf{B}$ is a diagonal matrix (Law and Fang 2001). The weight matrices help to mitigate difficulties associated to the smoothness of the unknown force term and to the uncertainties and noise present in field experiments.

It should be noted that the knowledge of M, C, K in Eq. (28.1) is a prerequisite to minimize Eq. (28.5). For the case of the bridge considered, as well as other structures, knowing the system characteristics can be achieved with the paradigm of structural identification (Çatbaş et al. 2013). Additionally, since the location of the vehicle relative to the bridge are changing with time and the force transformation matrix $\mathbf{F}$ is not constant but varies with time, it is necessary to track the location of the axles which can be obtained with the knowledge of axle configuration and vehicle speed.

The error expressed in Eq. (28.5) can be minimized by leveraging the structure of dynamic programing (Bellman 2013) and Bellman's Principle of Optimality (Bellman 2016), which leads to the definition of the minimum values of E for any initial state/condition c and the number of stage N. Get

$$g_N(c) = \min_{f_i}\left[\left(Z_1 - Qc, A(Z_1 - Qc)\right) + \left(f_1, \mathbf{B}f_1\right)\right] + g_{N-1}\left(Gc + Pf_1\right) \qquad (28.6)$$

leading to the recursive formula (Trujillo and Busby 1997) to calculate vectors $\mathbf{S}$ and $\mathbf{R}$. Let

$$D = \left(2\mathbf{B} + 2P^T\mathbf{R}_{i+1}P\right)^{-1}$$
$$H = \left(2\mathbf{R}_{i+1}P\right)^T \qquad (28.7)$$

then

$$\mathbf{R}_i = Q^T\mathbf{A}Q + G^T\left(\mathbf{R}_{i+1} - H^T DH/2\right)G$$
$$\mathbf{S}_i = -2Q^T\mathbf{A}Z_i + G^T\left(I - H^T DP^T\right)\mathbf{S}_{i+1} \qquad (28.8)$$

These formulae are used to compute the vectors $\mathbf{S}$ and $\mathbf{R}$ backward, where the initial condition at the last stage N is given by

$$\mathbf{R}_N = \mathbf{Q}^T \mathbf{A} \mathbf{Q}$$

$$\mathbf{S}_N = -2\mathbf{Q}^T \mathbf{A} \mathbf{Z}_N$$

(28.9)

and to calculate the response X with Eq. (28.3) and the force f (Law and Fang 2001) by

$$f_i = -\left(2B + 2P^T \mathbf{R}_{i+1} P\right)^{-1} P^T \left(\mathbf{S}_{i+1} + 2\mathbf{R}_{i+1} G X_i\right)$$

(28.10)

The L-curve method (González et al. 2008) which tries to find a balance point of the measurement and force has been shown to be effective in determining the regularization term B in MFI problem. To plot the L-curve, a set of regularization terms at a certain range are defined to calculate the solution norms and force norms. The optimal regularization parameter is based on a value that provides a good trade-off between the two norms. However, the norms do not have any physical meaning, and one cannot say what are the proper magnitudes. Other approaches such as convergence of force term as a function of B (Law and Fang 2001) and adaptive Tikhonov regularization (Li and Law 2010) are also available. However, finding the convergence of force terms requires the knowledge of true force in advance, making it impractical to bridge vehicle problem, while finding B at each iteration by adaptive Tikhonov regularization is intrinsically costly in computation.

Therefore, the approach used here is to define the difference between measurements and predicted displacements as a function of the regularization term B. Then the optimal B can be found by minimizing the difference with the optimization tool 'fminsearch' available in MATLAB®. Instead of choosing B from limited number of options, using an optimization tool could search the minimum at a much higher precision.

28.2.3 Mode Superposition

Ideally, Eq. (28.1) should be solved by optimizing Eq. (28.6) according to the MFI algorithm. However, complex structures such as highway bridges typically require digital models with tens and even hundreds of thousands of DOFs, making the iterative scheme computationally impractical. Researchers (Clough and Penzien 2003, Craig and Kurdila 2006, González et al. 2008) had shown that mode superposition method was able to transform the DOFs into principal coordinates with the assumption of linear system and uncoupled equation of motions. The transformation is given by

$$\underset{n\times 1}{\mathbf{U}} = \underset{n\times n_u}{\Phi}\ \underset{n_u \times 1}{u}$$

(28.11)

where n_u is the number of modes used to reconstruct the equation of motion that is much smaller than n, and Φ contains the normalized eigenvectors. Substituting Eq. (28.11) into Eq. (28.1) and multiplying both sides by Φ^T leads to

$$\tilde{M}\ddot{U} + \tilde{C}\dot{U} + \tilde{K}U = \Phi^T \mathbf{F} f$$

(28.12)

where

$$\tilde{M} = \Phi^T M \Phi$$

$$\tilde{C} = \Phi^T C \Phi$$

$$\tilde{K} = \Phi^T K \Phi$$

(28.13)

The dimension of the reduced mass, damping, and stiffness matrices are reduced to $n_u \times n_u$ after transformation. The transformed equations of motion can be solved by the approach described in the previous section.

28.3 TESTING WITH SIMULATED MODEL

The approach described above has been successfully applied to the inverse dynamic problem where either (1) the forces were applied at a fixed location or (2) the number of DOFs was relatively small. A skew prestressed concrete bridge was calibrated (Lai et al. 2019) leveraging static load testing measurements, and its center span superstructure was chosen as an example to test the MVI BWIM algorithm. The total number of DOFs of the model is approximately 40,000, and the number of unknowns is set to be 6, i.e., 3 axles. Twenty modal vectors (with frequencies within 6 and 25 Hz) and 12 measurements (at/near the midspan, with a sampling rate of 100 Hz) contaminated with 5% noise were employed in the analysis. The vehicle speed is assumed to be 30 m/s, a typical highway speed. The truck axle configuration is taken from one of the trucks used in the load testing, as depicted in Figure 28.1 in which one axle is lifted and idle. A small fluctuation was imposed to the applied force for the calculation to account for the dynamic effect.

The three axles are spaced by 5, 1.5 m, and the bridge is skewed westward at 16.3°. Three successive axles enter the bridge. The axle weights were identified and are presented in Figure 28.2. From the deformation responses of the bridge, the axle forces can be estimated only when the axle is on the bridge span. Consequently, the different axle forces are zero before their corresponding axle enters the bridge and once it exists. The plots in Figure 28.2 show a satisfactory agreement between

FIGURE 28.1 The truck axle configuration with assumed forces.

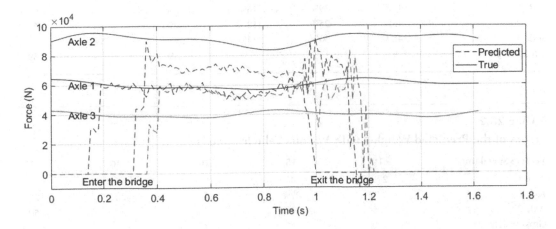

FIGURE 28.2 Predicted axle weights of a vehicle moving across the bridge.

the 'true' overall force applied by the moving truck and the predicted force while single axle forces have larger differences. The entire truck is loading the bridge from approximately 0.4–0.9 seconds. The MVI BWIM algorithm appears to underestimate the force imposed by axle 2 while overestimating the force associated to axle 3. Overall, the approach erroneously estimates the separate weights while overall capturing the total weight. This might be related to the short distance between two axles (1.5 m).

The algorithm appears promising in estimating the weights of moving vehicles. With the support of simulations, parametric studies can be performed to comprehend the influence of vehicle speed, number of measurements, sampling rate, etc. on the effectiveness of the MVI BWIM algorithm. It is essential to examine the effect of these parameters to provide some insight into real case applications and to support planning of field implementation. The effect of number of measurements, sampling rate, signal noise level, and vehicle speed on the force estimation accuracy defined as Eq. (28.14) was studied.

$$\text{Error} = \frac{|f_{true} - f_{predicted}|}{|f_{true}|} \tag{28.14}$$

Results summarizing these parametric studies are given in Table 28.1–Table 28.4. While the low sampling rate increase the errors, high sampling rate might not be necessary to improve the accuracy.

This might be caused by the fact that with high sampling rate, the high frequency modal responses are captured and mixed with the low frequency modes, while the force identification algorithm is limited to low frequency modes. Filtering high frequency of the signals might help. Vehicle speed also affects the accuracy of the MVI BWIM algorithm. As it is shown in Table 28.2, for vehicle speeds within 20 m/s (~70 km/h) and 45 m/s (160 km/h), the accuracy does not change significantly but it decreases when the vehicles travel at low speed.

When high sampling rate measurements are combined with low vehicle speed, the increase in the number of data points in the recorded signal requires a larger number of time steps in the inverse dynamic programming, which could potentially increase the numerical error at each time step and result in a larger discrepancy.

TABLE 28.1
Errors of the Predicted Weights with Various Signal Sampling Rates

Sampling Rate (Hz)	10	30	50	100	200	500
Axle 1	16%	39%	16%	6%	12%	63%
Axle 2	18%	25%	18%	20%	23%	67%
Axle 3	46%	30%	34%	39%	28%	50%
Gross weight	9%	14%	6%	4%	6%	59%

TABLE 28.2
Errors of the Predicted Weights with Various Vehicle Speeds

Vehicle Speed (m/s)	10	15	20	30	45
Axle 1	27%	14%	10%	6%	12%
Axle 2	37%	27%	24%	20%	20%
Axle 3	27%	31%	32%	39%	50%
Gross weight	22%	11%	7%	4%	6%

TABLE 28.3
Errors of the Predicted Weights with Various Number of Measurements

# of Measurements	6	8	10	12	18	24
Axle 1	54%	16%	13%	6%	6%	6%
Axle 2	45%	18%	27%	20%	21%	22%
Axle 3	51%	46%	31%	39%	36%	36%
Gross weight	35%	9%	10%	4%	4%	5%

TABLE 28.4
Errors of the Predicted Weights with Various Signal-Noise Ratios

Noise Level (%)	5	10	20
Axle 1	6%	15%	17%
Axle 2	20%	27%	28%
Axle 3	39%	31%	32%
Gross weight	4%	11%	12%

Regarding the number of measurements, it is well known it should be larger than the number of unknowns. Table 28.3 shows that 6 measurements appear insufficient for 6 wheels' weights estimation. Increasing the number of measurements does reduce the errors. The errors of gross weights reduce to 4% as the number of measurement increases to 12 (twice the number of unknowns). Afterwards, further adding measurements does not introduce appreciable improvements. For the trucks such as the trailers with up to 7 or 8 axles, a large number of measurements would be required, which would affect negatively practical BWIM implementations.

Finally, Table 28.4 shows the expected estimation errors increase of the moving load as the measurement of signal-noise ratio increases. Even with large simulated noise level, the algorithm is still able to capture the axle weights with errors within 30%. The MFI-BWIM algorithm appears to be quite robust to the unavoidable noise that can affect field measurements.

These four parameters were studied independently in this work. However, they might be mutually interdependent and could also be coupled with the bridge dynamic characteristics. Further investigations are required to strengthen the above conclusions. Furthermore, questions such as how many modes are needed to reconstruct the dynamic system are essential and should be examined, particularly using real measurements.

28.4   CONCLUSIVE REMARKS

It was shown that the MFI based BWIM has the potential to estimate truck weights over bridges. The mode superposition reduction technique can reduce the size of the inverse problem significantly, which is critical for models of infrastructure systems that have more than tens of thousands of DOFs. Dynamic programming with the Bellman's optimality principle, combined with the Tikhonov regularization, allow to solve the inverse problem. The effect of parameters such as sampling rate, vehicle speed, number of measurements, and signal-noise ratio has been discussed and shown to affect the MFI-BWIM algorithm effectiveness. For instance, it was observed that the algorithm performed better when sampling rates falling within 50–200 Hz were considered. In addition, larger number of measurements corresponded to improved predictions. As an example, MFI-BWIM leveraging 12 measurements with noise levels smaller than 10% appeared to capture fairly accurately the response of vehicles with 3 axles traveling at a normal highway speed. This study is confined to simulated results and will be expanded by incorporating actual deformation measurements of an operating bridge subject to truck crossings. Nonetheless, this work can provide some useful guidance for planning and instrumenting a BWIM on real bridges.

ACKNOWLEDGEMENTS

The authors wish to express their gratitude for the support received from FHWA, the Oregon Department of Transportation and Dr Navid Zolghadri toward this study. The authors also wish to thank Office of Global Engagement and Education Abroad in Drexel University via International Travel Award program for the supports to facilitate the presentation of this paper.

REFERENCES

Bellman, R. 2013. *Dynamic Programming*, Courier Corporation, Santa Monica, CA.
Bellman, R. 2016. *Introduction to the Mathematical Theory of Control Processes: Linear Equations and Quadratic Criteria*, Elsevier.
Çatbaş, F. N., Kijewski-Correa, T. and Aktan, A. E. 2013. *Structural Identification of Constructed Systems: Approaches, Methods, and Technologies for Effective Practice of St-Id*, Reston, VA, ASCE.
Clough, R. W. and Penzien, J. 2003. *Dynamics of Structures*, (revised), Computers and Structures, Inc., Berkeley, CA.
Craig, R. R. and Kurdila, A. J. 2006. *Fundamentals of Structural Dynamics*, John Wiley & Sons, Hoboken, NJ.
González, A., Rowley, C. and OBrien, E. J. 2008. A general solution to the identification of moving vehicle forces on a bridge. *International Journal for Numerical Methods in Engineering*, **75**(3), pp. 335–354.
Lai, X., Mazzotti, M., Grimmelsman, K., Aktan, A. E. and Bartoli, I. 2019. The effect of truck loads on highway bridges via bridge weigh in motion algorithm. *Proceedings of 12th International Workshop on Structural Health Monitoring*, Stanford, CA, 2019.
Law, S. S. and Fang, Y. 2001. Moving force identification: Optimal state estimation approach. *Journal of Sound and Vibration*, **239**(2), pp. 233–254.
Li, X. and Law, S. 2010. Adaptive Tikhonov regularization for damage detection based on nonlinear model updating. *Mechanical Systems and Signal Processing*, **24**(6), pp. 1646–1664.
Lu, Z. and Liu, J. 2011. Identification of both structural damages in bridge deck and vehicular parameters using measured dynamic responses. *Computers & Structures*, **89**(13–14), pp. 1397–1405.
Lydon, M., Taylor, S. E., Robinson, D., Mufti, A. and Brien, E. J. O. 2015. Recent developments in bridge weigh in motion (B-WIM). *Journal of Civil Structural Health Monitoring*, **6**(1), pp. 69–81.
Moses, F. 1979. Weigh-in-motion system using instrumented bridges. *Journal of Transportation Engineering*, **105**(3).
OBrien, E., Hajializadeh, D., Uddin, N., Robinson, D. and Opitz, R. 2012. Strategies for axle detection in bridge weigh-in-motion systems. *Proceedings of the International Conference on Weigh-in-Motion (ICWIM 6)*, Dallas, TX, pp. 79–88.
Trujillo, D. 1978. Application of dynamic programming to the general inverse problem. *International Journal for Numerical Methods in Engineering*, **12**(4), pp. 613–624.
Trujillo, D. M. and Busby, H. R. 1997. *Practical Inverse Analysis in Engineering*, CRC press, Boca Raton, FL.
Yu, Y., Cai, C. S. and Deng, L. 2016. State-of-the-art review on bridge weigh-in-motion technology. *Advances in Structural Engineering*, **19**(9), pp. 1514–1530.
Žnidaric, A., Lavric, I. and Kalin, J. 2002. Free-of-axle detector bridge WIM measurements on short slab bridges. *Proceedings of 3rd International WIM Conference*, Canary Islands, Spain, pp. 231–239.

29 Experimental and Numerical Studies on Dynamic Performance of a Steel Stair under Human Action

Li Junxing and Zhang Wenyuan
Harbin Institute of Technology

CONTENTS

29.1 INTRODUCTION

There is more interest in slender and lightweight staircase due to the wide use of high-strength lightweight materials and esthetic requirements, which cause an important serviceability issue by human-induced loads (Davis and Murray 2009, Santos et al. 2019). The reason for this is that the range of natural frequencies of light and slender structures often matches the dominant frequencies of the human-induced dynamic load. While high-frequent lightweight structures may also fail to satisfy stringent vibration serviceability criteria since the high-frequency components of pedestrian dynamic loads may also generate strong responses (Racic et al. 2009).

To predict the vibration of the staircase accurately, a reliable load expression is crucial. There have been scarce investigations to describe the process of ascending or descending staircases compared to walking on flat ground. These findings reach an agreement that the footfall load induced on even surfaces is quite different from that induced on stairs. Previous researches place a force plate on the stair to directly derive vertical load. Kerr and Bishop (2001) measured the first four harmonic dynamic load factors (DLFs) of a single step to compose the loads for people walking on a floor and a stair and found the differences between the loads from people on floors and stairs. Kasperski and Czwikla (2012) studied the differences between the right and the left foot loading and the effects of stair geometry on step frequency and DLFs. Recently, wireless tri-axial acceleration transducers were used to get the load time history of consecutive steps by Yongfeng et al. (2016). As reported in this thesis, the value of each order DLFs and phase angle studied were in good consistency with the research results of Kerr and Bishop. Yongfeng et al. (2016) also concluded that the footfall forces applied on the stairs were 150% and 300% of that applied on floors for ascents and descents, respectively.

The problem of human-structure interaction is another aspect that affects the prediction accuracy because the behavior of the occupied human as dynamic systems can change the dynamic characteristics of the structure. The interaction is often found in slender structure such as footbridges (Zivanovic et al. 2005), long-span floors (Mohammed and Pavic 2017), and long cantilevered structure (Setareh 2012), which proved the influence of human is not negligible. However, to the best knowledge of author, little studies have assessed whether this interaction is applicable in stair structures. Busca et al. (2014) proposed a method to analyze the effects of passive people on the modal parameters of a lowly damped staircase, and a satisfactory agreement has been found between the prediction and the experimental evidence. Subsequently, the methodology is improved by Cappellini et al. (2016) to predict the structural response of staircases due to the presence of moving people. In this work, the results show that the use of the dynamic properties of the empty structure can lead to a high overestimation of the vibration amplitudes.

Under human actions, the external steel accommodation stairway can exhibit excessive vibration as a consequence of being lightweight. Our aim with this study is to investigate the dynamic character and response under human action based on experimental testing and Finite Element (FE) modeling. A detailed experiment test was conducted for an empty structure occupied with different number of passive people as well as a single pedestrian with different walking frequencies. To verify the experimental results, an FE model of steel stair was established with ABAQUS. From the result of the study, it was evident that (1) the external steel accommodation stairway exhibits a higher first bent mode natural frequency, (2) under nearly four times the weight of the structure of the occupied passive human action, the natural frequency of the stairway shows a little change, and (3) the vibration was quite significant during the walking test.

This paper is structured as follows. Section 29.2 describes the test program. Section 29.3 presents the results of stationary tests and walking cases. In Section 29.4, an FE model of a steel stair is established to illustrate the main results of walking people tests. Finally, the main findings of this study are summarized in Section 29.5.

29.2 EXPERIMENTAL PROGRAM

29.2.1 Description of the Structure under Test

The structure studied is a stringer-type stairway that serves as an outdoor fire escape located in a public building in Harbin, Heilongjiang and is shown in Figure 29.1a. The treads are each composed of a steel plate with a thickness of 3 mm and two longitudinal stiffening ribs whose dimensions are shown in Figure 29.1b. The stringers provide support for the step plates, which is supported on the platform cantilever beam. The connection between the treads and the stringers is welded, as is the connection between the stringers and platform cantilever beam. And measurements were carried out on the stair unit connected to the first floor for safety and measurement convenience.

29.2.2 Test Procedure

Output-only methods were adopted in the experimental identification of dynamic characteristics of the staircase both for empty and occupied. To get a strong excitation of the structure, different methods were used. For the empty structure, a person was asked to suddenly jump off the stair from the fourth tread to generate the vibration on the structure. While the heel impact method was employed for the tests with, respectively, 2, 4, 6, and 8 people standing on the structure, this method has been

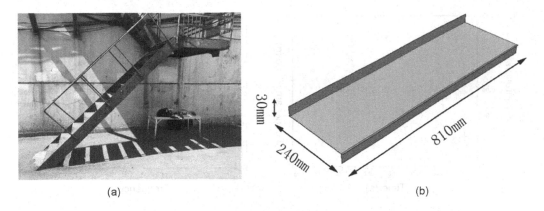

FIGURE 29.1   (a) The stairway under test. (b) The dimension of the treads.

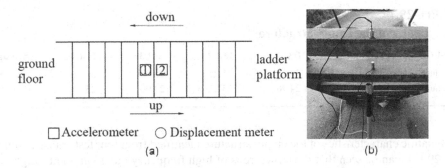

FIGURE 29.2   (a) Location of the measuring points. (b) Accelerometers and LVDT installation on the center of the treads.

used in previous researches (Blakeborough and Williams 2003, Xuhong et al. 2017). People were asked to stand at the center of the treads near the mid span of the stringers and to raise their heels approximately 50 mm to produce a sudden impact.

Walking tests were conducted to obtain both the displacement and acceleration response at the mid span. A person walked along the center of the treads at different frequencies with the aid of a metronome. According to research, the walking frequency of ascending stairs is mostly concentrated around 2 Hz, while that of descending stairs is mostly concentrated around 2.85 Hz. Therefore, in this study, the walking frequency at 1.8, 2.0, and 2.2 Hz for ascending and 2.6, 2.8, and 3.0 Hz for descending was chosen.

The measuring points were arranged in the center of the treads at the mid span of the stringers, and two locations were selected, as shown in Figure 29.2a. Two accelerometers and an Linear Variable Differential Transformer (LVDT) (see in Figure 29.2b) were used to acquire acceleration and deflection data at measurements points. TST3828EN dynamic signal test and analysis system were adopted for data acquisition, and the sampling frequency was at 200 Hz.

29.3   EXPERIMENTAL TEST RESULTS

29.3.1   Empty Structure Results

An example of acquired acceleration time histories and their corresponding frequency spectra calculated by performing a Fast Fourier Transformation at the mid span of the structure is shown in Figure 29.3. The first mode frequency can be estimated as 24.88 Hz from the Fourier spectrum at point 1#.

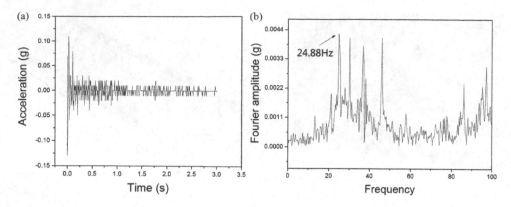

FIGURE 29.3   (a) Acceleration response at point 1#. (b) FFT analysis of the acceleration.

TABLE 29.1

Frequency of the Empty Structure

Test Cases	Test 1	Test 2	Test 3	Test 4	Average
Frequency at point1# (Hz)	24.96	25.25	25.25	24.88	25.08
Frequency at point2# (Hz)	24.96	25.21	24.96	25.21	25.08

The dynamic characteristics of the empty structure identified from four test data are summarized in Table 29.1. It can be seen that the structure is of high frequency due to its light weight, and it is unlikely to build up resonance response through the excitation of human walking. In addition, the frequency varied slightly with different points and tests, which may be caused by the amplitude-dependent nonlinear behavior of the structure. An averaged value of 25.08 Hz was employed for further study.

A simplified analytical method suggested by Davis and Avci (2014) is used to calculate the fundamental natural frequency of the stair. The natural modes resemble those of a pair of parallel beams connected by closely spaced transverse bending elements. The fundamental natural frequency is computing using Eq. (29.1).

$$f_1 = \frac{\pi}{2}\sqrt{\frac{EI}{mL^3}} \tag{29.1}$$

where, in this equation, E is the elastic of the stringer (206 GPa), I is the stringer moment of inertia ($2.196 \times 10^{-6}\,\text{m}^4$), m is the effective mass (31.7 kg), half of the total mass including the stringer, treads, and guardrail for the simply supported beams, and L is the stringer length (3.54 m). Substituting these parameters into the above equation, the fundamental natural frequency is obtained as 22.94 Hz, which varied 8.5% with the value obtained from tests.

29.3.2   RESULTS OF STATIONARY TESTS

Table 29.2 shows the frequency of the stair occupied by different number of standing people obtained from the experiments. It can be seen that the natural frequency shows a slight decrease due to the presence of people. However, under nearly four times the weight of the structure of the standing people, the structure didn't show a large frequency decrease that would occur when the rigid masses acted on the stair.

TABLE 29.2

Frequency of the Occupied Structure

Number of Standing People	0	2	4	6	8
Mass ratio (people/stair)	0	1.10	1.89	2.68	3.78
Frequency (Hz)	25.08	24.85	24.65	24.57	24.71

29.3.3 WALKING TEST RESULTS

Examples of acceleration and displacement response obtained at point1# from walking down at a frequency of 3 Hz were shown in Figure 29.4. As expected, the stair displayed a transient response as if excited by a sequence of impulses, with the response dissipated before the next footfall (Middleton and Brownjohn 2010). Since the stair is very light, the response includes time variation of static deflection due to a moving repeated load as well as decaying natural vibrations due to footstep impulses (Murray et al. 1997). The maximum displacement and acceleration response of the stair obtained from the descending walk test is 0.98 mm and 1.01 g (9.90 m/s²), respectively, indicating that the high-frequency structure with low mass can also be subjected to excessive vibration. For the results herein presented, it's clear that the structure vibrates mainly at its natural frequency, as well as the fundamental walking frequency and its integer multiples. This implied that even the seventh

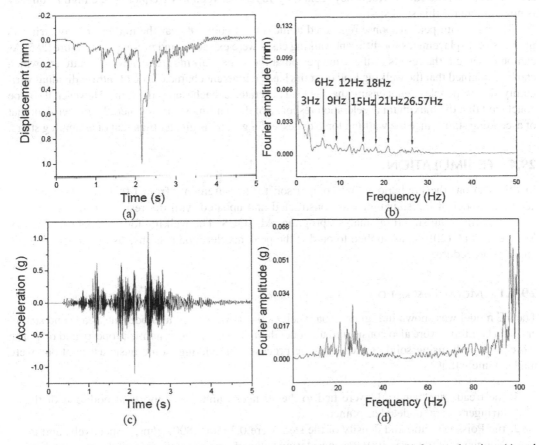

FIGURE 29.4 Response and correspond frequency spectra at point 1#. (a) and (b) for acceleration, (c) and (d) for displacement.

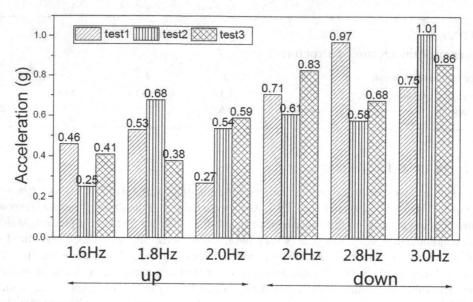

FIGURE 29.5 The peak acceleration of the walking tests at point 1# for both ascents and descents.

harmonic of the walking force contains energy to generate to structure due to the lower mass of the steel. Therefore, the cut-off frequency (generally 10 Hz) between low frequency and high frequency is not suitable for this steel stair.

The maximum peak response registered at measuring point 1# near the middle span of stringers in terms of displacements for different walking cases were extracted and presented in Figure 29.5. As can be observed, the results of the same person at the same walking frequency are scattered, which can be explained that the walking load on stairs has the inherent characteristic of intra-individual variability, that is, people cannot ensure that the load generated at each step is the same. However, it can be concluded that the maximum acceleration response of descending stairs is generally greater than that of ascending stairs since the walking force of descending stairs is greater than that of ascending stairs.

29.4 FE SIMULATION

In this section, the predictions and comparison to measurements from FE were presented. An accurate model of the stair unit was constructed and updated with the results of the experiment using a commercial structural analysis program, ABAQUS. The walking force model proposed by Yongfeng et al. (2016) was applied to predict the peak acceleration response using model dynamic analysis procedure.

29.4.1 MODEL DESCRIPTION

The FE model was shown in Figure 29.6a. Shell elements were used to model the treads and stringers. The handrails were also considered to model their effects on the dynamic property and response more accurately using solid elements. In the process of modeling, some basic assumptions were made, namely that

1. the treads and handrails were tied to the stringers, and the supports at both end of the stringers were modeled as pinned.
2. the Poisson's ratio and density of the steel were 0.3 and 7,800 kg/m³, respectively, and its Young's modulus was 190 GPa considering years of outdoor environmental corrosion, having led to a decline in the mechanical properties of steel.

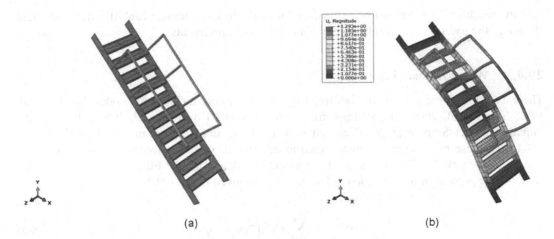

(a)   (b)

FIGURE 29.6   (a) The FE model of the stair unit. (b) The shape of the first pending mode.

A model analysis was performed to determine the natural frequency and the first bending mode was predicted to have a frequency of 30.3 Hz, which was overestimated with the frequency error being as high as 20.8%. Therefore, the model was updated to improve the correlation between the FE and the experiment model for further analysis. Springs were introduced at both ends of the stringers to substitute the pinned boundary condition in the vertical direction. The value of the stiffness of these springs was obtained to be 5,000 kN/m by trial and error. The shape of the first bending mode with an updated frequency of 25.3 Hz was represented in Figure 29.6b, and the error of 1% indicates that the numerical simulation is in good agreement with the experimental results.

29.4.2   SIMULATION OF OCCUPIED STRUCTURE

The mass of each occupant was applied to the FE model at the appropriate location to simulate the influence of the occupants on the structural frequency. The simulated frequencies were then compared with the corresponding experimental results in Figure 29.7. As expected, the structure exhibits a significant decrease in the first bending frequency due to the influence of the rigid masses.

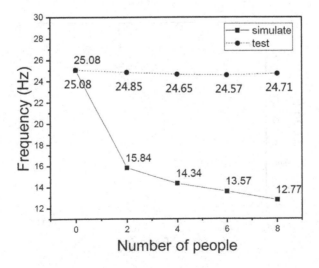

FIGURE 29.7   People's effect on the structural simulation.

And the result of the simulate using equivalent mass to modeling human can't illustrate the slight change of the natural structural frequency obtained from experiments.

29.4.3 WALKING FORCE FUNCTION

There exist four force models for high-frequency floor: the continuous force model (Murray et al. 1997, Smith et al. 2007), the stiffness-frequency (f) model (Murray et al. 1997), the effective impulsive model (Smith et al. 2007), as well as the frequency domain force model (Brownjohn et al. 2004). All these models are commonly used to describe the walking forces and they use power spectral density (PSD). The continuous force model is adopted into the FE model. The mathematical model expressed by a summation of Fourier series is given in Eq. (29.2).

$$F(t) = G + \sum_{i=1}^{n} \alpha_i \sin\left(2i\pi f_p t - \varphi_i\right) \tag{29.2}$$

where $F(t)$ is the force in time domain, G is the static weight of the person, n is the order of the harmonic, α_i is the dynamic loading factor of the correspond harmonic, f_p is the step frequency, and φ_i is the phase angle of the i-th harmonic. The first four harmonics were considered to model the human walking load according to research, and the values of the first four dynamic loading factor were listed in Table 29.3, and the force-time function adopted was shown in Figure 29.8.

TABLE 29.3
Dynamic Loading Factor of the Walking Force Model

Walking Case	DLF			
	α_1	α_2	α_3	α_4
2 Hz (ascend)	0.261	0.055	0.026	0.008
3 Hz (descend)	0.478	0.126	0.062	0.055

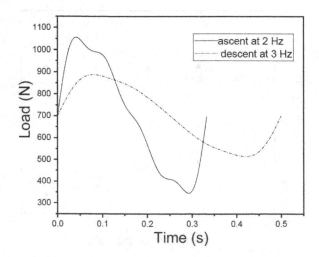

FIGURE 29.8 Load-time function adopted for frequency by simulating.

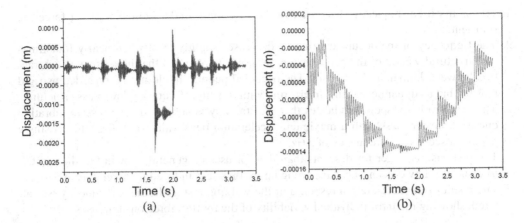

FIGURE 29.9 Displacement response by simulation (a) for point 1# and (b) for midpoint of the stringer.

29.4.4 Simulation of Response under Walking Tests

The above walking force models which were added to a series of analysis steps to form continuous walking loads were applied to the stair unit model to simulate the response of the stair when a single person rose at a frequency of 2 Hz and descended the stair at a frequency of 3 Hz. Figure 29.9 shows the result of displacement response of point 1# for descents at a frequency of 3 Hz. The results show that the maximum displacement response obtained by FE analysis at point 1# and midpoint of the stringer is 2.46 and 0.15 mm respectively, which is basically consistent with the walking test's response.

It should be noted that the vibration response difference between the FE model and walking tests can be attributed to the following three aspects:

1. The difference between the true force and the applied force. The force model adopted in the FE model is deterministic, considering the footfall perfectly repeating, while it neglects narrow-band randomness of the walking progress. Besides, as found in other researchers, there is an influence of the stair geometry on the harmonic load amplitudes (Kasperski and Czwikla 2012), and treads with a depth of 240 mm make it difficult to keep the feet in completed contact with stair, which may lead to a difference between walking force on this test stair and the one in Yongfeng et al (2016).
2. The effect of interaction between human and structure on the performance of the stair. The only forced model is used, but the human-structure is ignored in the simulation of walking test, that is, the character of the empty structure is adopted for simulating the response of walking tests. In fact, human is a heavy damped dynamic system, who can alter the property of the occupied structure and influence the response, especially for low-frequency lightweight construction.
3. The uncertain factors of the test stair. The lack of information of connections between the treads of and the strings, support conditions of the strings, and material properties can also reduce the simulation accuracy.

All of the above-mentioned factors will be further studied in our future work.

29.5 CONCLUSIONS

This paper presents a study of the dynamic performance of an outdoor steel firefighting stairway under human action using experimental testing and FE simulation. Heel impact and walking tests were conducted to obtain the displacement and acceleration response of the structure. The FE model was established, updated, and verified with the experimental results. The following main findings can be drawn from this research:

1. The stair's high-frequency structure and their response under human walking force are transient.
2. The frequency of the occupied structure decreases slightly even when nearly four times the structural weight of the people stood on it, which revealed the importance of interaction between human and structure. The occupied people should not be regarded as rigid masses, but as dynamic systems interacting with this high-frequency low-mass structure.
3. The acceleration response indicates that the stairway is subjected to excessive vibration due to its lower mass with a maximum acceleration peak value of $1.01\,g$ when a single person descends at a frequency of $3\,Hz$.
4. The response is larger for descents than that for ascents generally, owing to the footfall force induced by descending stairs being larger than that induced by ascending stairs.
5. The result of the acceleration response of the walking test at the same frequency is scattered, showing the intra-individual variability of the footfall force on staircases.

REFERENCES

Blakeborough, A. and Williams, M.S. 2003. Measurement of floor vibrations using a heel drop test. *Proceedings of the Institution of Civil Engineers - Structures and Buildings*, 156, pp. 367–371.

Brownjohn, J.M., Pavic, A., and Omenzetter, P. 2004. A spectral density approach for modelling continuous vertical forces on pedestrian structures due to walking, *Canadian Journal of Civil Engineering*, 31(1), pp. 65–77.

Busca, G, et al. 2014. Quantification of changes in modal parameters due to the presence of passive people on a slender structure. *Journal of Sound and Vibration*, 333(21), pp. 5641–5652.

Cappellini, A.M, et al. 2016. Evaluation of the dynamic behaviour of steel staircases damped by the presence of people. *Journal of Engineering Structures*, 115, pp. 165–178.

Davis, B. and Murray, T.M. 2009. Slender monumental stair vibration serviceability. *Journal of Architectural Engineering*, 15(4), pp. 111–121.

Davis, B. and Avci, O. 2014. Simplified vibration response prediction for slender monumental stairs. *Proceedings of Structures Congress*, pp. 2548–2557.

Kasperski, M. and Czwikla, B. 2012. A refined model for human induced loads on stairs. *Proceedings of Topics on the Dynamics of Civil Structures*, pp. 27–39.

Kerr, S.C. and Bishop, N.W.M. 2001. Human induced loading on flexible staircases. *Journal of Engineering Structures*, 23(1), pp. 37–45.

Middleton, C.J. and Brownjohn, J.M. 2010. Response of high frequency floors: A literature review. *Journal of Engineering Structures*, 32(2), pp. 337–352.

Mohammed, A S and Pavic, A. 2017. Effect of walking people on dynamic properties of floors. *Journal of Procedia Engineering*, pp. 2856–2863.

Murray, T.M., Allen, D.E., and Unger, E.E. 1997. *Steel Design Guide Series 11: Floor Vibrations Due to Human Activity*, American Institute of Steel Construction, Chicago, IL.

Racic, V., Pavic, A. and Brownjohn, J.M.W. 2009. Experimental identification and analytical modelling of human walking forces: Literature review. *Journal of Sound and Vibration*, 326(1–2), pp. 1–49.

Santos, J.L, Andrade, P and Escorcio P. 2019. Pre-design of laterally supported stair steps. *Journal of Engineering Structures*, 182, pp. 51–61.

Setareh, M. 2012. Vibrations due to walking in a long-cantilevered office building structure. *Journal of Performance of Constructed Facilities*, 26(3), pp. 255–270.

Smith, A.L, Hicks, S.J. and Devine, P.J. 2007. *Design of Floors for Vibration: A New Approach*, Steel Construction Institute Ascot, Berkshire.

Xuhong, Z., et al. 2017. Vibration serviceability of pre-stressed concrete floor system under human activity. *Journal of Structure and Infrastructure Engineering*, 13(8), pp. 967–977.

Yongfeng, Du., et al. 2016. Tests for parameters of pedestrian load model during human walking up and down stairs. *Journal of Vibration and Shock*, 35(21), pp. 220–228.

Zivanovic, S., Pavic, A. and Reynolds, P. 2005. Vibration serviceability of footbridges under human-induced excitation: a literature review. *Journal of Sound and Vibration*, 279(1), pp. 1–74.

30 Study on Seismic Mitigation of Elastic Cables on Long-Span Cable-Stayed Bridges by Shaking Table Test

Li Xu, Qunjun Huang, Junhao Zheng, and Kang Liu
Fuzhou University

CONTENTS

30.1 INTRODUCTION

Bridge, customarily a critical control engineering, has a great role in traffic engineering and disaster avoidance. In earthquake cataclysms, the assurance of regular services of bridges will minimize the damage caused by earthquakes and simultaneously decrease the happening of second disasters.

The cable-stayed bridge, a hybrid system consisting of cables, towers and girders, characterizes in respects of splendid configuration, innovative structures and greater spans. However, in designing cable-stayed bridges, ductility design is comparably difficult and forbidden to utilize the ductility of the main tower [1]. In *Eurocode8-Design Provisions for Earthquake Resistance of Structures-Part2: Bridges* [2], reading as "in design earthquakes, the cable-stayed bridge had better keep in elastic status," it was indispensable to take some measures to lower the seismic responses of cable-stayed bridges and keep them in elastic status. The cable-stayed bridge with floating system has less internal force response of the main tower under earthquake action, but it will lead to larger beam end displacement and even the situation of the falling beam. Customarily, the seismic mitigation could be realized by installing devices on bridges, such as viscous dampers and elastic cables [3].

The current mature and suitable for long-span bridges is mainly fluid nonlinear viscous dampers [4]. However, the damper is not easy to replace due to its high cost. For a medium- and small-span cable-stayed bridge, the damper is less economical for shock absorption. The elastic restraint system has relatively economical, applicable, and reliable damping measures. H.E.M. Ali et al. [5] proposed the idea of using a laminated rubber bearing for damping in the tower and beam. The results of the research show that the elastic connecting device can effectively reduce the longitudinal displacement of the beam end compared with the floating system, but it is possible to increase the shearing and bending of the tower bottom. Yuan et al. [6–8] developed a cable isolation bearing by adding a cable to the traditional support to limit the relative displacement.

Li [9] investigated the active vibration control of small dip viscoelastic cables. Fung [10] investigated the motion state of viscoelastic cords. Though the fact that the concept of longitudinally elastic cables was proposed, currently, several bridges adopted them in China of Queshi Bridge (Shantou), Changjiang Bridge (Wuhu), Baishazhou Bridge (Wuhan), Century Bridge (Haikou), Changjiang Bridge (Fengjie, Chongqing), and Changjiang Bridge (Lidu, Chongqing). You and Guan [11], based on the design of the combined damping system of elastic-plastic cable and viscous damper of Yellow River bridge (Yongqing), analyzed the structural response by increasing the amplitude of seismic wave. The result of research showed that adopting elastic-plastic cables may lead to larger residual displacement of the main girder, but for the peak cable force and peak acceleration of main girder, the control effect is obvious. Li [12] discussed the influence of different tower-girder connections on the seismic response of long-span cable-stayed bridges. The response spectrum method was used to analyze the seismic response of cable-stayed bridges caused by different elastic cable stiffness. The calculated results showed that the elastic cable has limited ability to control the displacement of the main beam without increasing the internal force response of the main tower. At the same time, the inappropriate elastic cable stiffness will cause the vibration of the pile foundation to be unfavorable. Wang et al. [13] and Liang [14] took a three-tower suspension bridge as the prototype, and compared the mitigation effect of different longitudinal connection devices. The results showed that the longitudinal elastic cable connection device is arranged only in the middle tower and the main girder, and the viscous damper connection device is arranged on the side tower, which can significantly reduce the displacement response of the cable-stayed bridge and improve the force response of the tower beam. However, due to the difficulty of manufacturing and maintaining the viscous damper, it is recommended to arrange the elastic cable at the joint between the middle tower and the main girder.

Having experienced Wenchuan Earthquake and its aftershocks and daily operations, bridges using longitudinally elastic restricted system have manifested excellent working status. Based on elastic cables' advantages of good effect of seismic mitigation, great economic efficiency, and convenient installation, Chinese scholars conducted so many researches on medium-span bridges restricted elastically in the longitudinal direction between towers and girders. However, the quantity of studies on longitudinally elastic cables on large-span cable-stayed bridges faded compared with the former, especially of researches on tests.

Shaking table tests could study the damage modes and mechanisms of structures under earthquake action and effectively evaluate their seismic capacity and effect of seismic mitigation [15]. Gao et. al used a six-degree-of-freedom shaking table to carry out a simulated seismic motion of a 1:50 model of Xiazhang bridge. Huang et al. [17] designed and fabricated a 1:30 semi-floating system single tower cable-stayed bridge structure model, and carried out seismic simulation shaking table test of multi-point excitation under different ground motions. Wang and Xu [18] established the 1:20 scale test model with the Yongjiang Bridge as the background project and carried out four array shaking table tests to study the seismic response of cable-stayed bridges under different tower-beam connections. Omar [19] and others carried out a full-bridge shaking table test of a three-span cable-stayed bridge to study the damping effect of a new seismic motion control (SMC) tower-beam connection called SMC-OPC. Both Zong [20] and Shoji [21] used plexiglass to make shaking table experimental models, but the results are limited by materials and geometrical dimensions, and the results cannot quantitatively reflect the seismic response of real bridges. Xu Li [22] and other half-bridge models of cable-stayed bridges with a concrete main tower with a scale ratio of 1/40, using vibration array explore the seismic response characteristics along the longitudinal bridge.

From all studies above, it is found that most studies on the seismic mitigation of cable-stayed bridges with longitudinally elastic restricted systems were based on finite element analyses, while quite a few were based on shaking table tests [23]. Thus, it is quite important to conduct a relative test. Based on existing shaking tables, a test of large-span cable-stayed bridges was carried out to compare the responses of different systems under the seismic loads by its results. This could lead to a more direct cognition on the effect of seismic mitigation of elastically restricted systems, which furthermore reveal the mechanisms of seismic mitigation of longitudinally elastic restrictions on cable-stayed bridges.

30.2 DESIGN OF A SCALE MODEL FOR SHAKING TABLE TESTS

The purpose for this shaking table test of earthquake simulation is to verify the mitigation effect of longitudinally elastic-connective devices on cable-stayed bridges, compared with half-floating cable-stayed bridges. Based on the fact that the seismic responses of cable-stayed bridges under uniform seismic loads are analogous to the symmetric structures under asymmetric forces, plus the restriction of the size of shaking tables, this paper conducts a test on a 1:40 scale model of the substructure of a large-span double-tower cable-stayed bridge.

30.2.1 BACKGROUND INFORMATION

The prototype bridge selected in this paper is a continuous seven-span cable-stayed bridge with double tower and steel box girders. Moreover, the length of the bridge is 1,280 m with a main span of 680 m, and the specific parameters on the bridge layout is 60 + 90 + 150 + 680 + 150 + 90 + 60 m. The bridge layout is shown in Figure 30.1.

The main tower, a reinforced concrete structure, consists of upper, middle, and lower tower columns and upper and lower beams. The tower is 220 m high, counting from the top of tower pedestals and excluding decorations of the tower top. Moreover, its longitudinal width varies from 8 to 13 m; The height of the tower pedestal is 3 m. Specifically, the lower tower column has a height of 48.6 m and a width of 6–10 m in the transverse direction, using a sectional form of single box and double cell; The middle tower column has a height of 67.4 m and a uniform width of 6 m in the transverse direction, using a sectional form of single box and single cell; The upper tower column has a height of 80 m and a width of 4–6 m in the transverse direction, using a sectional form of single box and single cell. Besides, the material for main towers is C50 concrete.

The steel box girder of the main bridge is 1,280 and 30.6 m in length and width, respectively. The deck measures 28.7 m in width. The girder, at the midline of the bridge, is 3.50 m high and becomes an integral flat streamlined single-box triple-celled structure. For which, the material is Q370 steel, as shown in Figure 30.2.

The design of cables on the background bridge adopts parallel steel strands. The anchor points for the upper and lower parts of cables are located at the tower column and the tensile anchor plate of girders, respectively. There are 168 cables in total and 6 types of cables classified by the loading condition of cables and the quantity of steel strands of 27, 31, 37, 43, 49, and 55. The anchor point

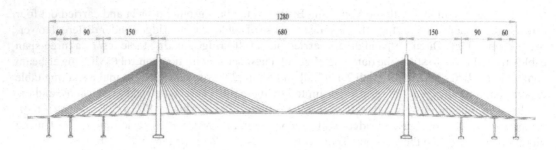

FIGURE 30.1   Overall layout of the bridge (unit: m).

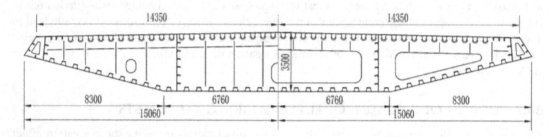

FIGURE 30.2   The standard transverse section of the steel box girder (unit: m).

located at girders adopts a form of the tensile anchor plate, and that located at main towers adopts a form of prestressed anchor.

According to the information provided in the project Safety Assessment Report, the proposed site is located at the mouth of the Minjiang River in Fuzhou, the main geomorphic units of the proposed site are marine plain landforms, and the engineering geology is divided into alluvial and alluvial areas. The foundation soil of the bridge abutment on both sides is of the genesis type of miscellaneous fill and alluvial within the survey and control depth, and the basement is granite. Excellent Period of Frequent Fretting of Site Ground is 0.372s–0.488 s. Classification of site construction sites is Class III. The bridge is classified as Class A seismic fortified bridge according to the *corresponding code (JTGT_B02-01-2008)*, the peak ground motion acceleration of the project site is located in the 0.05 g zone, and the corresponding basic earthquake intensity is VI. The earthquake ground motion parameters for the main bridge are provided in the *Safety Assessment Report*: The 50-year exceeding probability of 10% (equivalent to E1 level) is 75 gal for the peak value of surface horizontal ground motion acceleration, the peak value of ground surface horizontal ground motion acceleration with a 100-year exceeding probability of 4% (equivalent to E2 level) is 152 gal.

30.2.2   MATERIALS SELECTED FOR THE SCALE MODEL

Due to the limits of facilities and alternative materials, priority is given to make the elastic model tested in its elastic stage, and then to reconsider its plasticity. Moreover, the materials used in the main towers and girders are based on their loading characteristics. The material parameters for each component of this model are demonstrated in Table 30.1.

Then, the overall similarity ratios of the model could be determined according to its main deformities, and the similarity ratio of the model with 50% balance weight is shown in Table 30.2.

TABLE 30.1
Materials for Scale Model

Component	Material	Young's Modulus (MPa)	Poisson Ratio	Density (kg/m³)	Standard Strength (MPa)	Yield Strength (MPa)
Concrete main towers	C30	3×10^4	0.2	2,500	20	-
and side piers	HRB335	2×10^5	0.3	7,850	-	-
	R235	2.1×10^5	0.3	7,850	-	-
Girder	1,060 Aluminum slab	7×10^4	0.3	2,700	120	75
Cable	High-strength steel strand	1.95×10^5	0.3	7,850	-	-
Steel component	Steel	1.95×10^5	0.3	7,850	-	-
Acrylic glass component	Acrylic glass slab	2.694×10^3	0.4	1,180	-	-
Elastic cable	1,060 Aluminum slab	7×10^4	0.33	2,700	120	75

TABLE 30.2
Similarity Ratios of the Scale Model with 50% Balance Weight (Main Tower)

Category	Physical Quantity		Dimension	Formula of the Similarity Relationship	Value
Material performance	Stress	σ	FL^{-2}	$S_\sigma = S_E$	0.869
	Elastic modulus	E	FL^{-2}	S_E	0.869
	Mass density	P	FT^2L^{-4}	$S_\rho = S_{m/2}/S_l^3$	17.391
Geometric characteristics	Length	L	L	S_l	0.025
Dynamic performance	Frequency	F	T^{-1}	$S_f = S_l^{0.5} S_E^{0.5}/S_{m/2}$	8.944
	Period	T	T	$S_t = S_l^{2.5} S_E/S_m S_t = S_l^{2.5} S_E/S_m$	0.112
	Acceleration	A	LT^{-2}	$S_\alpha = S_E S_l^2/S_{m/2}$	2.000

30.2.3 COMPONENT DESIGN

30.2.3.1 Cable Design

Analogous to the calculation of girders, the cable design is mainly based on equivalence of the structure's axial stiffness.

$$\frac{E^m A^m}{E^p A^p} = S_E S_l^2 \tag{30.1}$$

$$A^m = \frac{S_E S_l^2 E^p A^p}{E^m} \tag{30.2}$$

According to the formulas above, taking Cable S211 for example, both the practical bridge and model bridge adopt the high-strength steel strands as their materials. Thus, the elastic modulus ratio is 1, and the area for Cable S211 is 0.00812 m², and the diameter needs to be taken as 2.3704 mm. All kinds of cable sizes converted are shown in Tables 30.3 and 30.4.

TABLE 30.3

Converted Diameter of Cables

Cross-Section Category	Converted Diameter (mm)	Cross-Section Category	Converted Diameter (mm)
s211	2.3704	s139	1.9239
s199	2.3020	s121	1.7951
s187	2.2316	s109	1.7038
s163	2.0835	s91	1.5567
s151	2.0053		

TABLE 30.4

Converted Area and Diameter with Cables Combined

Cross-Section	Diameter (m)	Area (m²)	Combination Method
1	3.26×10^{-3}	8.37×10^{-6}	$2 \times s109 + 2 \times s91$
2	3.72×10^{-3}	1.09×10^{-5}	$2 \times s139 + 2 \times s121$
3	4.09×10^{-3}	1.31×10^{-5}	$2 \times s163 + 2 \times s151$
4	4.39×10^{-3}	1.51×10^{-5}	$3 \times s187 + 1 \times s163$
5	3.95×10^{-3}	1.22×10^{-5}	$1 \times s187 + 2 \times s199$
6	3.35×10^{-3}	8.83×10^{-6}	$2 \times s211$

30.2.3.2 Design of Elastic Cables

The main function of longitudinally elastic cables is to provide pulling forces, and when the relative displacement between girders and towers increases, they offer a restriction, and thus, they are installed at both sides of the main towers. Simultaneously, both ends of longitudinally elastic cables connect to girders and main towers. The formula is shown as follows:

$$F = K_e d \qquad (30.3)$$

in which F represents the resilient force provided by elastic cables and d represents the relative displacement between towers and girders.

This paper selected different stiffness values of longitudinally elastic devices to conduct calculations and parametric sensitivity analysis. Based on existing journals [24–26] and the model size of this test, the stiffness value of elastic cables was determined as 8,000 kN/m. Due to the fact that the elastic cable has great ductile performance, aluminum was chosen with an elastic modulus of 7.00×10^4 MPa, a Poisson ratio of 0.33, and a density of 2,700 kg/m³. The design diagram of elastic cables is shown in Figure 30.3, in which the middle part measures around 30 cm and the section measures 2 mm × 1 mm, and a hole of diameter 10 mm at both ends is used to fix the aluminum bars.

30.2.3.3 Support Design

The design of supports has to consider their movable and displacement-restricted directions without the structure's reactions prevented by them. The converted theoretical peak relative displacement between towers and girders is 27.5 mm under E1 condition, and thus, that of the support should surpass this value. Moreover, considering other test conditions, the biggest theoretical displacement

FIGURE 30.3   CAD of elastic cables.

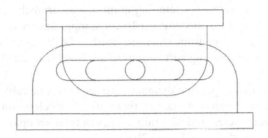

FIGURE 30.4   Diagram of the support.

response should not exceed 50 mm. Due to the limitation of practical testing conditions, the response is deemed smaller than the theoretical value, and therefore, the design displacement of the support is 20 mm plus 35 mm equaling to 55 mm. The diagram of the support is shown in Figure 30.4.

30.2.4   BALANCE WEIGHT OF THE STRUCTURE

In this test, 50% of the balance weight was added excluding the self-weight, weight boxes, stress cables, and the corresponding displacement-restricted devices. The distribution of balance weight was also considered combining the concept of effective modal mass.

The detailed locations of balance weight are demonstrated in Figure 30.5. The theoretical mass is 10,150 kg, while the total weight of the main tower is 9,792 kg, due to the test restrictions and

FIGURE 30.5   Demonstration of the whole model.

the error between mass blocks and weight boxes. The practical balance weight is lower than the theoretical one by 358 kg, which is quite subtle, compared with the whole structure. Thus, this error could be ignored, and the 50% balance weight of the main girder accounts for 3,195 kg.

30.3 SHAKING TABLE TEST METHOD

30.3.1 THE LAYOUT OF MONITORING POINTS

The acquisition system in this test mainly includes a DH610 acceleration meter, a displacement meter, a laser displacement meter, a stretching-compression transducer, a strain gauge, and a total station. This data acquisition system is a 16-bit sample data acquisition system equipped with 128 channels, and the acquisition software used is DEWESoft7.0. The data processing software and seismic wave processing software used are MATLAB2012R and seismosignal 5.0, respectively.

Viewing the content of this paper, the dynamic characteristics of cable-stayed bridges, and the request on acceleration tests of shaking tables, there are 21 acceleration meters, 4 displacement meters, 1 laser displacement meter, and 32 strain gauges in total set up on the test model and table surface. The specific layout is shown in Figure 30.6.

In which, there are 24 cables with each sector equipped with 6 ones, and according to sector discrepancies and different lengths of each cable, the cable of each sector, next to the tower side, is numbered 1 and the far-end one is numbered 6. Moreover, four sectors are numbered S1/S2/S3/S4, respectively. The detailed diagram is demonstrated in Figure 30.7.

The number rule of acceleration meters is based on different directions they measure, and thus, they are divided into three directions of x, y, and z. Moreover, different locations of acceleration meters should also be taken into account. Taking A_y1 for example, it is the first acceleration transducer measuring the y direction. The detailed layout is shown in Figure 30.7.

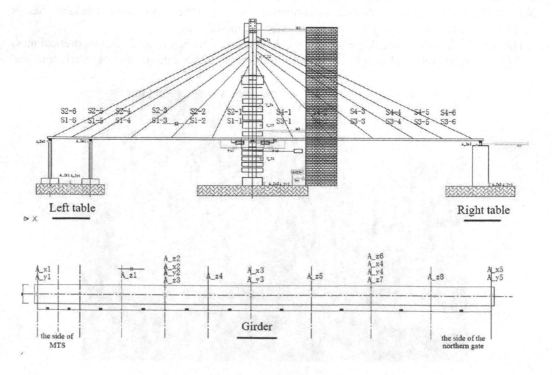

FIGURE 30.6 The layout of acceleration transducers on girders.

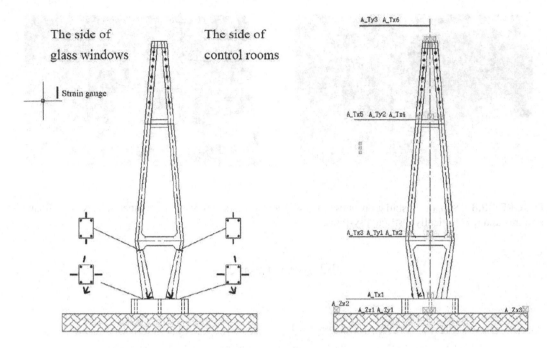

FIGURE 30.7 The layout of transducers on the main tower and girders.

In order to measure the structure's displacement, displacement meters were installed on the tower top, at the lower beams of the tower, between towers and girders, on the girder end, and on the surface of the middle shaking table. Except the displacement meter installed between towers and girders intending to measure the relative displacement, all the others are used to measure absolute displacements.

30.3.2 The Structural System of the Scale Model Bridge

The structural responses of different structural systems are quite different to each other under uniform excitations of seismic waves. Therefore, this paper tried to compare the discrepancies of three systems' structural responses of the floating system and elastically restricted system under uniform excitations of seismic waves in the longitudinal direction. In the test models, the only difference in these three systems lies on that of connective components between towers and girders, and thus, different connective components were fabricated respectively to simulate these three systems.

For the half-floating cable-stayed bridge, the intersection between towers and girders is free, and thus, supports demonstrated in Figure 30.4 are used in this test. The installation diagram is shown in Figure 30.8a. The distance this type of support allows is 55 mm, which can satisfy the need of the test.

For elastically restricted systems, the aluminum bars fabricated with an aluminum-alloy material is used for simulation. The installation diagram is shown in Figure 30.8b, and the fixed end of the aluminum bar can be seen as following pictures. Both ends of aluminum bars are fixed by screws to guarantee that elastic cables are tense in their initial status.

30.3.3 Seismic Waves Selected in This Test

Based on the requests of test design and its intent, this test compares structural responses of the half-floating cable-stayed bridge and elastically restricted cable-stayed bridge under excitations of one artificial wave provided by *Safety Assessment Report* of the background bridge.

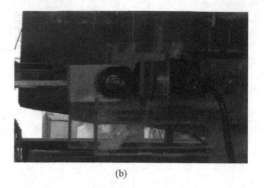

(a)   (b)

FIGURE 30.8 Support installation between tower and girders. (a) Ways of connection used in floating systems and (b) elastically restricted systems.

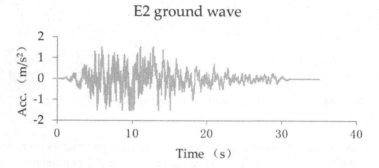

FIGURE 30.9 Ground acceleration record.

The original seismic wave applicable to the prototype bridge could not be used in this test, whereas the seismic waves input should be adjusted and determined according to the structural characteristics of the scale model. Based on the design in this test, the comparison and adjustment method is utilized to compare and adjust the peak accelerations and frequencies. The peak seismic wave adjusted is 1.5 m/s², and the time compression ratio is 0.112. Figure 30.9 shows the ground acceleration record used in this study to evaluate the seismic performance of the bridge.

30.3.4 Seismic Responses under Excitations of Artificial Ground Waves

The acceleration responses of shaking tables and structures are collected through acceleration transducers installed at corresponding locations in the test, and the acceleration time-history curve of ground motion input is collected by acceleration transducers installed on the surface of shaking tables in the longitudinal direction, as shown in Figure 30.18. It is found that the peak acceleration of the ground motion input happened after 2.32s with a value of 1.49 m/s², and the duration of ground motion slightly transcends to 4 s, which satisfies the request of test design.

30.4 TEST RESULTS ON SEISMIC MITIGATION OF CABLES

In the test model of elastic cables, there are two of them installed in symmetry on both sides of the main tower. Based on their tensile trait instead of compression, they only transfer axial tensions between towers and girders. Thus, when the elastic cable on one side is subjected to axial tensions, that of the other side is on the slack status. Figure 30.10 shows the axial tension time-history curve of elastic cables at the lower beam of the main tower under the elastically restricted system.

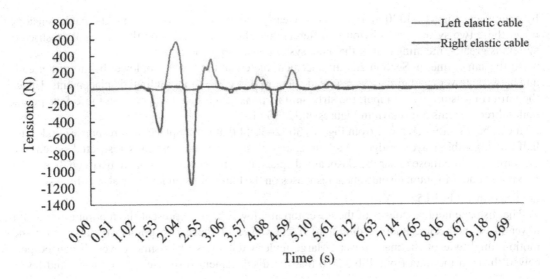

FIGURE 30.10   Tensions obtained from elastic cables on the elastically restricted systems.

Figure 30.11 shows the axial tension time-history curve of elastic cables at the lower beam of the main tower under the elastically displacement-restricted system.

Figure 30.10 reflects better that longitudinally elastic cables can only be subjected to tension instead of compression, and it also shows that the tension response of the elastic cable on the left side is only positive, and that of the elastic cable on the right side is only negative. When comprehensively analyzing the time-history curves of tension responses of elastic cables on both sides, the working status of the elastic cables on both sides under excitations of seismic waves is found.

It can be found from Figure 30.10 that the tension curves of elastic cables can accurately reflect the longitudinal girder-end displacement and relative displacement between girders and towers. At 1.97 and 2.31 s, cable tensions of the elastically restricted system attain to the maximum of its positive and negative values, and the peak tensions are 564 and 1162 N, respectively. Moreover, the peak tensions under the elastically displacement-restricted system are 388 and 963 N. This reveals that elastic cables on this type of system are subjected to a smaller tension comparably. Simultaneously,

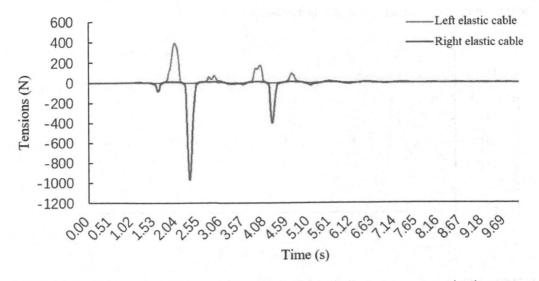

FIGURE 30.11   Tensions obtained from elastic cables on the elastically displacement-restricted systems.

by comparing Figures 30.10 and 30.11, it is clearly found that at around 1.5 s, the cable tensions under these two systems are obviously different, and the cable tension on the elastically restricted system is greater, meaning that of the other system is not so obvious.

At the same time, in Section 2.2, displacement meters at the tower top, lower beam, girder end, and between towers and girders were already deployed to collect longitudinal displacements. Under the uniform seismic waves input, the structural response time-history curves on these four points under three systems are shown in Figures 30.12–30.14.

It can be directly perceived from Figures 30.12–30.14 that the displacement responses under the half-floating cable-stayed bridge are all large, and those under the elastically restricted system are minimized, while those under the elastically displacement-restricted system are between them. The maximum and minimum displacement responses on the four points under these above three systems are listed in Table 30.5.

The displacement responses of the tower top and lower beam represent the motion of the main tower itself in the longitudinal direction. If the values of these two points are large, then the motion amplitude of the main tower is large under excitations of seismic waves. It can be perceived through the data from Table 30.6 that the displacement responses at tower top and lower

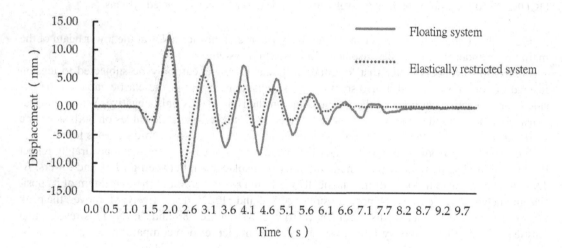

FIGURE 30.12   Longitudinal displacement of the tower top.

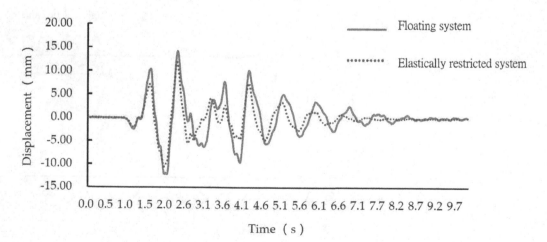

FIGURE 30.13   Longitudinal displacement of the girder end.

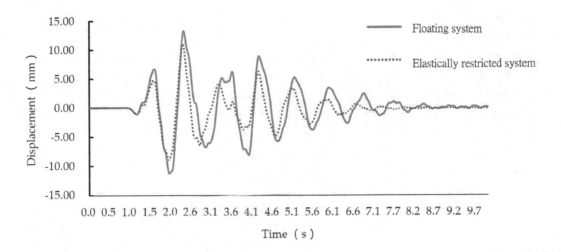

FIGURE 30.14 Relative deformations between tower and girders.

TABLE 30.5
Peak Displacement Responses of Three Systems Excited by Artificial Waves

Monitoring Points for Displacement	Half-Floating System (mm)		Elastically Restricted System (mm)	
	Maximum	Minimum	Maximum	Minimum
Tower top displacement	12.62	−13.24	10.94	−10.04
Relative deformation between towers and girders	13.21	−11.28	10.96	−8.98
Girder-end displacement	14.28	−12.36	11.96	−10.55

TABLE 30.6
Comparison between Measured and Simulated Results

System	Monitoring Points for Displacement	Measured Maximum	Simulated Maximum	Error(%)
Half-floating system	Tower top displacement	12.62	11.43	9.42
	Girder-end displacement	14.28	13.12	8.12
	Relative deformation between towers and girders	13.21	11.52	12.81
Elastically restricted system	Tower top displacement	10.94	9.84	10.05
	Girder-end displacement	11.96	10.84	9.36
	Relative deformation between towers and girders	10.96	9.75	11.04

beam of the half-floating system are 13.24 and 5.41 mm, respectively, and those of the elastically restricted system are 10.94 and 4.49 mm, respectively. Based on the data, it can be found that the motion of the main tower under the half-floating system is most serious, and that under the elastically displacement-restricted system is secondary, while the motion amplitude under the elastically restricted system is the smallest.

The bridge deck of the half-floating system is without boundary in the longitudinal direction; thus, the displacement response of the girder end is biggest under the seismic loads in the

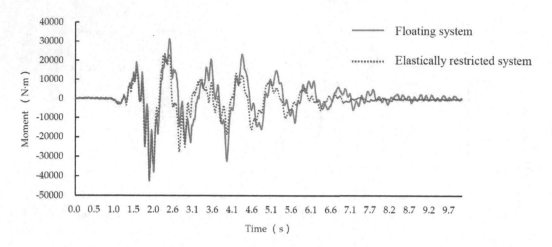

FIGURE 30.15   Comparison of time-history curves of tower base moments.

same direction. Compared with the elastically restricted and elastically displacement-restricted systems, the bridge decks of them have boundaries and thus, their displacement responses of the girder end are smaller. From the data, it can be seen that the peak girder-end displacement of the half-floating system is 14.28 mm, and that of the elastically restricted system is 10.96 mm. In terms of limiting the girder-end displacement, the limit effect of the elastically restricted system is most obvious.

Through all displacement responses of the structure's key points above, the time when the peaks of these four points happen coincides with that when the elastic cable's tension reaches its maximum. And thus, it is believed that the function of elastic cables serves to transfer the force of inertia of girders under excitations of ground motion. Compared with half-floating systems, elastically restricted systems have smaller displacement responses in their key points, which substantiates a greater effect of seismic mitigation. Figure 30.15 shows the response time-history curve of the tower base moment. The overall trend of that is the same as the time-history curve of displacement response. At around 2 and 2.4 s, the structure's internal responses attain to the maximum. Moreover, the tower base moment of the half-floating system surpasses that of the elastically restricted system, and the peak moments for the half-floating system and elastically displacement-restricted system are 42,628 and 40,555 N·m, respectively. From this viewpoint, there is not much discrepancy in the moments of these two systems and the responses of the elastically displacement-restricted system is slightly smaller than the response of the half-floating system. For the purpose of seismic performance, the elastically restricted and elastically displacement-restricted systems have a greater advantage.

30.5   ANALYSIS OF THE FINITE ELEMENT MODEL (FEM)

On the previous seismic simulation tests, the result shows that the mitigation effect of the elastically restricted system is superior to that of the floating system. However, the whole period for the test is too long, and the design, fabrication, and conduction of the test consume lots of resources. This adds much difficulties on analyzing more working conditions. Therefore, it is indispensable for using the finite element method to reproduce the seismic simulation test. This paper established a 1:1 FEM, and setup the same working conditions and seismic waves to have numerical calculations. Through the data acquired from acceleration and displacement transducers installed on the model, plus that of the structural displacement responses, this FEM can be corrected more accurately to conduct simulation tests.

30.5.1   ESTABLISHMENT OF THE FEM

Based on the integral design and practical fabrication of the test model, the FEM was established.

The girder was idealized as a spine girder unit. Through this way, it has the advantage of correct stiffness and mass systems, and fewer units.

The material for the main tower in this model uses reinforced concrete, in which, concrete is selected as C30 one, and the reinforcement mainly used is HRB335. The main tower is idealized as a framework section unit, and every section's reinforcement and stirrups are set up by the reinforcement function of SAP2000, and the specific option can be made in terms of the section's design type, reinforcement configurations, and bond type of stirrups.

The cable is idealized as a framework section unit. Based on the different size of each cable's section, the counterpart diameter can be determined. Due to the short length of cables in the test model, it should not take into account of the cables' sagging.

The weight blocks on the main tower and girders are idealized as concentrated loads on points and linear loads. According to the designed and practical balance weight, the balance weight should be added to the test structure correspondingly.

The pile-soil interaction is not considered in this shaking table test, and the main tower and three size piers are fixed on shaking tables by screws. Therefore, the restriction way of tower base and pier base is realized by consolidation.

The longitudinally elastic cable is idealized as a two-point connection unit with mass distributed on point i and j, while the force of inertia is not considered inside of the unit. Regarding the unit coordinate system as the coordinate parameters and inputting parameters of the connection unit, the need to simulate the longitudinally elastic cables can be achieved by defining the stiffness of needed directions of the connection unit.

The FEM based on the above principles is shown in Figure 30.16.

The FEM comprises 384 point units, 407 framework units, and 4 connection units. In this model, the Poisson of the main tower is selected as 0.2, and the elastic modulus is selected as 2.38×10^4 MPa, and its mass density is 2,500 kg/m³; the Poisson of girders and cables are

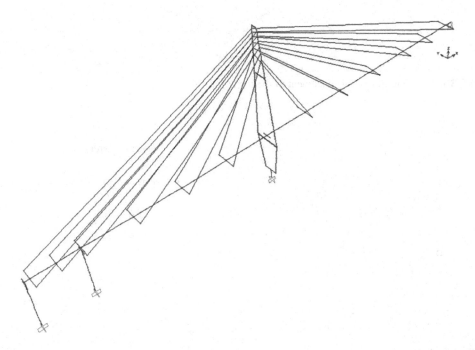

FIGURE 30.16   An FEM of the test model.

both 0.3, and elastic moduli for them are 1.95×10^4 and 2.06×10^4 MPa, respectively. The sectional characteristics and boundary conditions of each component should be determined by their practical situations.

30.5.2 Comparison on Data from the Test and FEM

It is impossible to fabricate the test model based on the similarity ratio rigorously, and thus, the results from the test and finite element analysis need to be compared. The structural response time-history curves on these four points under three systems are shown in Figures 30.17–30.19.

From Figures 30.17–30.19, it is believed that the FEM has a better simulation of the structure's self-vibration period, meaning it's accurate to simulate the structure's integral stiffness. From Table 30.6, although there is a discrepancy on data drawn from the FEM and test, this difference is small, and thus, it is believed that the FEM result coincides better with the test result. This FEM makes preparations for subsequent studies such as parametric analysis on elastic cables and the selection of a reasonably numerical model.

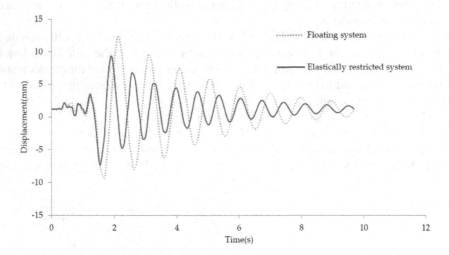

FIGURE 30.17 Longitudinal displacement of the tower top.

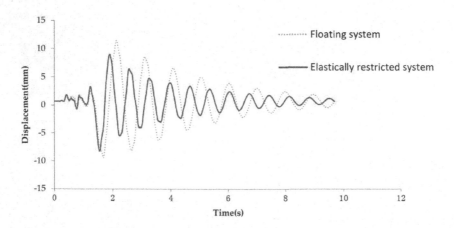

FIGURE 30.18 Relative deformations between towers and girders.

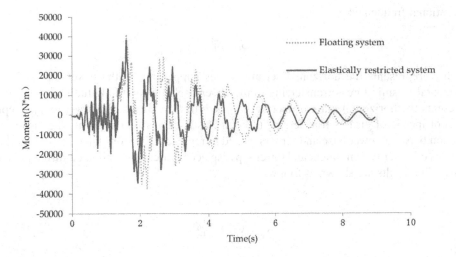

FIGURE 30.19 Comparison of time-history curves of tower base moments.

30.6 VERIFICATION ON SEISMIC MITIGATION MECHANISMS OF ELASTIC CABLES

From the test and FEM analyses, it is found that the displacement response and tower base moment of the elastically restricted bridge are obviously smaller than those of the half-floating bridge. The transmission channels of horizontal seismic forces will be changed due to elastically restricted devices installed between towers and girders in the longitudinal direction of the elastically restricted system, and a part of forces of inertia on the deck will transfer to the main tower through elastically connective devices under seismic loads, and the shear and moment of the main tower will change too. The calculative diagrams are shown in Figure 30.20.

Tower base moment:

$$S_u = S_l \tag{30.4}$$

Tower top displacement:

$$S_A = S_i^2 \tag{30.5}$$

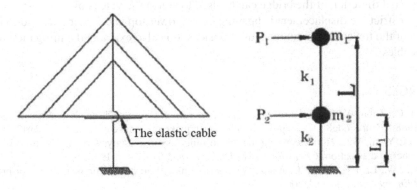

FIGURE 30.20 Mechanical diagram of the elastically restricted cable-stayed bridge.

Self-vibrational frequency:

$$S_l = S_l^4 \tag{30.6}$$

in which p_1 represents the horizontal seismic forces happening on the tower top; p_2 represents the horizontal seismic forces transferring to the tower itself through longitudinally elastic cables; m_1 represents the horizontal mass of inertia of girders transferring to the tower top; m_2 represents the mass of inertia of girders transferring through elastic cables; L_1 represents the distance of the intersection between tower base and girders; k_1 and k_2 represent the story lateral stiffness of towers.

Changes of tower base moment and tower top displacement can be estimated through the formulas above. The results are shown as follows:

$$S_{m/2} \tag{30.7}$$

$$S_c = S_E^{0.5} S_l^{0.5} S_{m/2}^{0.5} \tag{30.8}$$

Through an elastically restricted device installed in the longitudinal direction and a reasonably elastic stiffness chosen between towers and girders in the same direction, the values of P_1 and P_2 can be adjusted to effectively restrict displacements of the tower top and girder end. Simultaneously, the mechanics of the tower base can also be improved. Therefore, the stiffness of longitudinally elastic cables will influence the values of forces of inertia transferring through elastic cables. The mitigation diagrams and formulas of elastic cables above further explain their mitigation mechanisms.

30.7 CONCLUSIONS

A 1:40 scale model of cable-stayed bridges is designed and fabricated to simulate the half-floating system, elastically restricted system, and elastically displacement-restricted system through longitudinally moving supports and elastic cables installed between towers and girders. Moreover, dozens of acceleration and displacement transducers and strain gauges are placed on towers and girders to collect data.

A shaking table test is conducted under excitations of different seismic waves. The test result shows that the girder-end displacement response in the longitudinal direction of the elastically restricted system is obviously smaller than that of the half-floating system under the same excitations of seismic waves. And the tower base moment of the former system is also obviously smaller than that of the latter system. Compared with the half-floating system, the elastically restricted system has the mitigation effect to some extent. Moreover, there are discrepancies on the mitigation effects of elastically restricted system under different working conditions of seismic waves.

An elastically restricted device and a reasonably elastic stiffness between towers and girders in the longitudinal direction of the bridge can be used to adjust the values of P_1 and P_2, and further to effectively restrict the displacement happening on the tower top and girder end, and to improve the mechanics of the tower base simultaneously. Besides, this also explains the mitigation mechanisms of elastic cables.

REFERENCES

1. Fan, L.C.; Zhou, W.D. *Qiaoliang yanxing kangzhen sheji*. Beijing: China Communications Press, 2001.
2. EN 1988-2, Eurocode8-Design provisions for earthquake resistance of Structures-Part2: Bridges, 2005.
3. Li, J.Z.; Yuan, W.C. Nonlinear longitudinal seismic response analysis of cable-stayed bridge systems with energy dissipation. *China Journal of Highway and Transport*, 1998, 1, 71–76.
4. Wang, J.W.; Li, J.Z.; Fan, L.C. A study on the design method of seismic restrainers for bridges. *China Civil Engineering Journal*, 2008, 38, 71–77.

5. Ali, H.E.M.; Abdelghaffar, A.M. Seismic energy-dissipation for cable-stayed bridges using passive devices. *Earthquake Engineering and Structural Dynamics*, 1994, 23, 877–893.
6. Yuan, W.C.; Wang, B.B. Numerical model and seismic performance of cable-sliding friction aseismic bearing. *Journal of Tongji University: Natural Science*, 2011, 39, 1126–1131.
7. Yuan, W.C.; Cao, X.J.; Rong, Z.J. Development and experimental study on cable-sliding friction a seismic bearing. *Journal of Harbin Engineering University*, 2010, 31, 1593–1600.
8. Dang, X.Z.; Yuan, W.C.; Pang, Y.T. et al. Development and application of the self-centering cable-sliding friction aseismic device. *Journal of Harbin Engineering University*, 2013, 1537–1543.
9. Li, Y.H.; Gao, Q.; Yin, X.G. Nonlinear dynamic response and active vibration control of the viscoelastic cable with small sag. *Journal of Applied Mathematics and Mechanics*, 2003, 24, 594–604.
10. Fung, R.F.; Huang, J.S.; Cheng Y. et al. Nonlinear dynamic analysis of the viscos-elastic string with a harmonically varying transport speed. *Journal of Computers and Structures*, 1996, 61, 125–132.
11. You, H.; Guan, Z.G. Elastoplastic cable pair and viscous damper used in the lateral seismic isolation of cable-stayed bridges. *Journal of Vibration and Shock*, 2017, 36, 183–188.
12. Li, J.Z.; Peng, T.B. Effects of different connecting styles between towers and deck on seismic responses of a long-span cable-stayed bridge. *Journal of Vibration and Shock*, 2009, 28, 179–184.
13. Wang, H.; Tao, T.Y.; Zhang, Y.P. et al. Seismic control of long-span triple-tower suspension bridge under travelling wave action. *Journal of Southeast University :Natural Science*, 2017, 47, 343–349.
14. Liang, P.; Wu, X.N.; Li, W.Y. Longitudinal constraint system optimization for three-tower suspension bridge. *China Journal of Highway and Transport*, 2011, 24, 59–67.
15. Kagawa T.; Minowa C.; Abe. A. et al. Shaking table tests on analyses of piles in liquefying sand. *Proceedings of the 1st International Conference on Earthquake Geotechnical Engineering, Aegion, Greece*, 1995.
16. Gao, W.J.; Tang, G.W.; Huang, F.W. et al. Shaking table test study of north main bridge of Xiazhang sea-crossing bridge. *Journal of Bridge Construction*, 2013, 43, 1–13.
17. Huang, X.Y.; Zong, Z.H.; Xia, J. et al. Nonlinear dynamic response analysis of cable-stayed bridge with single tower under strong earthquake excitations. *Journal of Southeast University (Natural Science Edition)*, 2015, 45, 354–359.
18. Wang, L.; Xu, Y. Test study on application of steel damping device to lateral seismic reduction of cable-stayed bridge. *Journal of World Earthquake Engineering*, 2015, 31, 43–49.
19. El-Khoury, O.; Kim, C.; Shafieezadeh, A. et al. Mitigation of the seismic response of multi-span bridges using MR dampers: Experimental study of a new SMC-based controller. *Journal of Vibration and Control*, 2018, 24, 83–99.
20. Zong, Z.H.; Zhou, R.; Huang, X.Y. et al. Seismic response study on a multi-span cable-stayed bridge scale model under multi-support excitations. Part I: Shaking table tests. *Journal of Zhejiang University (Applied Physics and Engineering)*, 2014, 15, 351–363.
21. G. Shoji; T. Kogi; Y. Umesaka. Seismic response of a pc-cable-stayed bridge subjected to a long-period ground motion. *The 14th World Conference on Earthquake Engineering*, 2008, Beijing, China.
22. Xu, L.; Zhang, H.; Gao, J.F. et al. Longitudinal seismic responses of a cable-stayed bridge based on shaking table tests of a half-bridge scale model. *Journal of Advances in Structural Engineering*, 2019, 22, 81–93.
23. Saputra, A.; Talebi, H.; Tran, D. et al. Automatic image-based stress analysis by the scaled boundary finite element method. *International Journal for Numerical Methods in Engineering*, 2017, 109, 697–738.
24. Wang, L.; Liu, H.B.; Wang, Y.H. et al. Bearing capacity of normal section for bridge calculated by curvilinear integral. *Journal of Harbin Institute of Technology*, 2004, 36, 1314–1317.
25. Li, Z.S.; Lei, J.Q. Analysis and solution of cable supporting stiffness coefficient of multi-span cable-stayed bridges. *Journal of Chang'an University (Natural Science Edition)*, 2015, 35, 83–90.
26. Dan, D.H.; Zhao, Y.M.; Chen, Y.Y. Vibration measurement based on cable tension identification method for cable-damper system. *Journal of Vibration and Shock*, 2013, 32, 123–127.

31 Study on Vibration Control of Double-Tower Cable-Stayed Bridges by Adding Longitudinal Constraint System

Li Xu, Yang Xu, Junhao Zheng, and Wenqian Chen
Fuzhou University

CONTENTS

31.1 INTRODUCTION

With the maturity of cable-stayed bridge construction technology, cable-stayed bridges with different main girders made of different material, such as concrete girders, steel girders and reinforced-concrete girders, have been widely adopted. At present, cable-stayed bridges have widely application on bridges with a main span over 400 m. Considering the large number and long span of cable-stayed bridges in China, the seismic performance problem of cable-stayed bridges is more prominent.

The research has been done on seismic performance of bridges in depth. The AASHTO Code [1–3] promulgated in 1996 in the United States only applies to ordinary concrete girder bridges whose main span is not more than 150 m.

In a study reported in the Guidelines for Seismic Design of Highway Bridges [4], researchers took the transverse displacement of the beam end as the seismic performance objective and investigated the system application to cable-stayed bridges under different site classifications and peak ground acceleration. Usami and Xie [5,6] proposed damage control strategies to dissipate more earthquake input energy and to limit the seismic damage of tower by a novel subsidiary pier with energy dissipation elements.

For long-span cable-stayed bridges, fixed pier-tower-beam system may lead to great temperature stress and internal force under seismic load [7]. While a half-floating cable-stayed bridge can be considered as a sliding system in longitudinal direction. This system can meet the needs of temperature deformation, but may lead to a large longitudinal displacement at girder end under seismic force [8,9]. To control the girder-end displacement of a semi-floating system, elastic connection devices or viscous dampers can be installed between towers and girders to consume the vibration energy input from seismic load. These devices can effectively control both the vibration response of long-span bridges and the structural displacement of bridges under seismic action. They also reduce the internal force of key sections of bridges, which made them become the main choices of long-span bridges.

31.2 STUDIES AND APPLICATION OF SEISMIC MITIGATION SYSTEM FOR CABLE-STAYED BRIDGES

There are many kinds of tower-girder connection modes of cable-stayed bridges, among which the most widely used are floating system, half-floating system and fixed pier-tower-beam. The static and dynamic characteristics of cable-stayed bridges with different systems are quite different. For the fixed pier-tower-beam, it can effectively reduce the girder-end displacement, but it will lead to a sharp increase in the internal force response to the main tower. For the floating system and half-floating system, the advantage is that the internal force of the main tower is smaller and the displacement of the girder end is larger under the seismic action. But due to the large girder end displacement, those systems may cause the collision between main girder and approach bridge. Therefore, for long-span cable-stayed bridges with floating or semi-floating systems, how to take effective measures to control the girder-end displacement has become the focus [10].

At present, installing elastic connecting device or damper between the tower and the girder is a widely used method to control the girder-end displacement. It has been applied to a series of long-span bridges at home and abroad. Overseas, Minggang West Bridge (Japan) and Dodoro Bridge adopt elastic connection devices to mitigate seismic damage. Rion-Antirion Bridge (Greece) [11] and Golden Gate Bridge (the United States) adopt the method of installing viscous dampers to control their seismic response. In China, the Second Shantou Bay Bridge (Shantou), Youshi Bridge (Shantou), Baishazhou Bridge (Wuhan) and Egongyan Bridge (Chongqing), Lupu Bridge (Shanghai), Sutong Bridge and Jiangyin Yangtze River Bridge are all equipped with viscous dampers [12].

31.2.1 VISCOUS DAMPER AND ITS APPLICATION FOR CABLE-STAYED BRIDGE

Dampers due to the different principle of energy dissipation and damping can be divided into different kinds, i.e. lead extrusion dampers, steel dampers, friction dampers, viscous dampers and other different types. The most widely used application on long-span cable-stayed bridges and the most studied is the viscous damper [13].

As the name suggests, the damping force of a viscous damper is generated by the viscous fluid in the damper through the orifice. And on the basis of different construction, it can be divided into cylinder viscous damper, barrel viscous damper and viscous damper wall. Among them, cylinder viscous dampers are widely used in seismic mitigation for a bridge.

To make the damper perform better, the fluid material of the viscous damper should meet the following requirements: (1) strong viscosity and poor temperature sensitivity, (2) strong chemical stability, (3) poor flammable, poor volatile, nontoxic and low compressibility.

Many scholars have proposed many hysteretic models that can reasonably describe the mechanical properties of a viscous damper [14–16]. The most representative one is the Maxwell model [17], which is a system that simplifies the damper into a damping element and a spring element. As shown in Figure 31.1, the hysteretic of viscous dampers can be calculated by Formula 31.1.

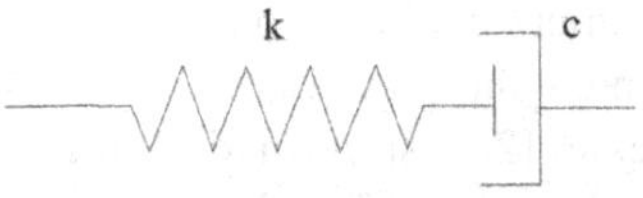

FIGURE 31.1 The Maxwell model.

$$f = c_a \operatorname{sgn}(\dot{u})|\dot{u}|^a \tag{31.1}$$

in which c_a represents the damping coefficient, sgn($\dot{u}$) represents the symbolic function, $\dot{u}$ represents the piston velocity and α represents the speed exponent. This formula represents the nonlinear damping force of the viscous dampers.

Many scholars have made a lot of research on the viscous damper on the cable-stayed bridge. The high vibration response is considered a serious problem, Ferreira and Simoes [18] proposed four different techniques including the viscous damper (VD) and passive viscous damper, which are presented to illustrate the effect of these control strategies in the optimum design geometry and dynamic responses. The seismic measures based on conventional fluid viscous dampers (FVDs) cannot present well in displacement, especially the accumulate displacement due to the low damper value. In order to address this issue, Feng [19] developed a new type of elastic-damping composite device. Then the static finite-element and nonlinear time history analyses are performed to satisfy the static and seismic demands. Xu [20] based on a half-bridge shaking table test compared the longitudinal response of cable-stayed bridge under seismic load. Yi [21] compared the seismic response of the bridge model with and without viscous dampers, and the result shows that viscous dampers are quite effective in controlling deck displacement of cable-stayed bridges. The viscous dampers have been popular applications on mitigating cable vibration on cable-stayed bridges. But that attached on cable results in complex mode shapes, so Nguyen C.H. [22] developed a general framework to investigate the problem of cable-stayed bridges with an attached viscous damper. Main J.A. et al. [23] using an analytical formulation of the complex of eigenvalue problem to analysis the free vibration of a taut cable with an attached viscous damper. He [24] and his team conducted the numerical simulations by installing a resetting semi-active stiffness damper (RSASD) devices, i.e., passive viscous and friction damper between the pier and the deck of the bridge. The numerical result indicates that the displacement, shear and moment at the base of the tower are reduced.

The above research demonstrates that as a new energy dissipation and damping device, viscous dampers can effectively not only control the seismic effect of long-span bridges and reduce the structural displacement response but also can play a very good role in reducing the internal force of key sections of bridges. Therefore, viscous dampers have a wide application on seismic control of long-span bridges at home and abroad.

31.2.2 THE ELASTIC CONNECTING DEVICE AND ITS APPLICATION ON CABLE-STAYED BRIDGES

In order to reduce the seismic response of cable-stayed bridges, domestic and foreign scholars have carried out a lot of research on the application of elastic connecting devices between towers and girders. Among them the elastic cables and laminate rubber bearing are widely used. H.M. Ali et al. [25] proposed using laminate rubber bearing between towers and girder for seismic mitigation. The result showed that compared with the floating system, the elastic connection device could effectively reduce the longitudinal displacement of girder end, but it may increase the base-sheared force and moment in the tower. M.J. Wesolowsky and J.C. Wilson [26] analyzed the seismic mitigation effect of elastic connection device under near-field earthquake. The result showed that elastic connection devices installed between tower and girder are effective for reducing girder-end displacement under near-field earthquake. B.B. Soneji and R.S. Jangid [27] compared the seismic mitigation effects of the elastic connection devices installed between tower and girder with viscous dampers.

The result showed that viscous dampers could not only control the girder-end displacement effectively but also reduce the base moment and shear in the tower under seismic load, which is superior to elastic connection devices. Xie et al. [28] established a numerical model of a 1,400-m span cable-stayed bridge. Then they conducted a comprehensive analysis of the characteristics of a carbon fiber reinforced polymer (CFRP) a cable-stayed bridge with CFRP cables. Ben [29] investigated the optimal design of the elastic cable and proposed an energy criterion associated with the concept of optimal performance of the hysteretic connection. Cheng [30] and Sun [31] investigated the dynamic response characteristics of long-span cable-stayed bridge on different damping systems such as the strain time history of pile-soil system, the bending moment, etc. Dong et al. [32] proposed a new vibration control method for long-span bridges using complex damper system with oil and elastoplastic dampers. The result shows that the elastoplastic damper system works well in rare and large amplitudes.

The above research shows that the elastic connection device can effectively reduce the displacement response at the girder end of the cable-stayed bridge under earthquake, but it can increase the internal force response of the main tower. Therefore, many scholars recommend considering viscous dampers to replace the elastic connection device for seismic mitigation design. In this paper, the elastic connection device and viscous damper installed between towers and girders are compared to determine which kind can achieve better effect for long-span cable-stayed bridge, with double towers represented by background bridges.

31.3 ENGINEERING BACKGROUND AND GROUND MOTION INPUTTED

31.3.1 INTRODUCTION OF ENGINEERING BACKGROUND

The prototype bridge considered in this paper is the main span of a 680 m long-span cable-stayed bridge with double towers. The overall length of the half-floating cable-stayed bridge is 1,280 m with steel box girders and seven-span. The seven-span arrangement of the main girder is (60+90+150+680+150+90+60). The main girder is a single-box three-chamber Streamlined Flat monolithic steel box with a girder height of 3.5 m and a width of 30.6 m. The main tower is made of a concrete structure with 220 m above the tower seat and 21 pairs of cable-stayed cables on both sides (Figures 31.2 and 31.3).

31.3.2 GROUND MOTION INPUTTED

Referring to *Detailed Rules for Seismic Design of Highway Bridges*, background engineering belongs to category A seismic fortification. Its fortification objectives are as follows: no damage should occur under the earthquake of E1, and elastic working range should be maintained; damage may occur under the earthquake of E2, but normal traffic should be maintained immediately after

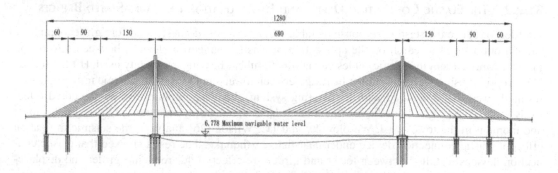

FIGURE 31.2 Overall layout of the double-tower cable-stayed bridge (Unit: m).

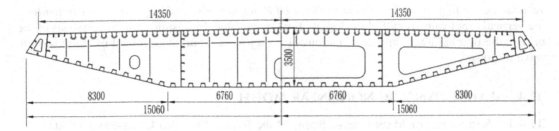

FIGURE 31.3 Cross section of the steel box girder (Unit: m).

the earthquake. The research shows that the seismic performance design of cable-stayed bridges is controlled by E2 earthquake in general. Therefore, this paper chooses the seismic input of 4% exceeding probability in 100 years, and the acceleration time history and response spectrum of the input seismic wave is shown in Figure 31.4. The hybrid input direction is longitudinal and

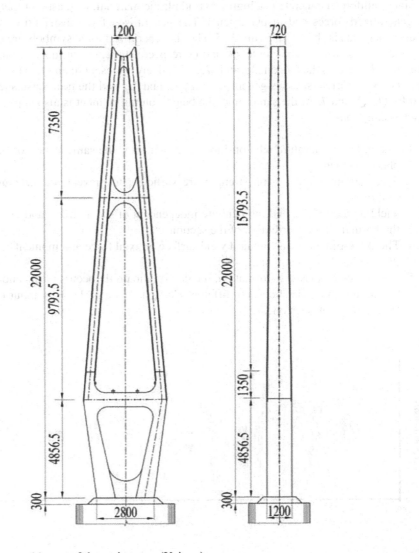

FIGURE 31.4 Structural layout of the main tower (Unit: m).

vertical. According to the stipulation of *Detailed Rules for Seismic Design of Highway Bridges*, the vertical input value is half of the longitudinal input value. In seismic response analysis of the bridge, the internal forces and deformations of its key sections are determined by three seismic wave responses.

31.4   SPATIAL DYNAMIC NONLINEAR MODEL

31.4.1   SIMULATION OF MAIN GIRDER BODY, MAIN TOWER, PIER AND CABLE-STAYED CABLE

Spatial beam element is used in steel girder simulation. Fishbone spine beam model is used in simulation. According to the actual cross-section characteristics of the girder, the mass of the girder includes the mass of all deck systems, and the influence of mass rotational inertia momentum is considered. The cable is simulated by beam element, but its bending stiffness is neglected, and the sag effect of the cable is considered by an equivalent Young's modulus. In order to consider that the main tower and pier may enter the plastic working range under strong earthquake, concrete is used to simulate the main tower and pier.

Since reinforced concrete is a highly hybrid plastic material, its plastic properties are expressed by generalized forces and displacements. The generalized force distribution of concrete beam-column elements is shown in Figure 31.5. The six generalized force symbols are defined as follows:

Here, P presents the axial force, Q_y and Q_z respectively represent the shear in y direction and z direction, T presents the torsion; M_y and M_z respectively represent the moment in y and z directions. And the axial normal stress is generated by P, M_y and M_z, and the transverse shear stress is generated by Q_y, Q_z and T. At the same time, the beam-column element is also applied to the following basic assumptions:

1. The moment-curvature relationship of the element is the same as that of the ideal elastic-plastic element.
2. Shear strength and torsion strength are sufficient to prevent the structure from brittle failure.
3. Yielding axial force and moment are independent of shear stress, and are determined by the normal stress distribution of the section.
4. The three-dimensional standard yield surface of axial force and moment can fully express the relationship among them.
5. If considering plastic deformation, the deformation only occurs at the end of the element beam, and the nodes between different elements remain elastic without considering the length of the plastic zone.

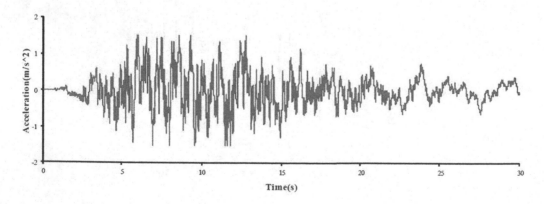

FIGURE 31.5   Ground acceleration record.

The relationship between the axial force and moment is expressed by the yield surface proposed by Bresier. The yield equation shown in Figure 31.7 is shown in Eq. 31.2:

$$\tilde{P}_u = \frac{P_u}{P_0}, \tilde{M}_{yu} = \frac{M_{yu}}{M_{y0}}, \tilde{M}_{zu} = \frac{M_{zu}}{M_{z0}} \tag{31.2}$$

Here, P_u presents the uniaxial yield tension, P_0 the uniaxial yield pressure, M_{y0} the pure yield moment around y direction and M_{z0} the pure yield moment around the z direction.

c_1, c_2, c_3, d_1, d_2, d_3 respectively presents the linear fitting parameters of the controlling points of the principal axis force and moment interaction graph. These parameters can be obtained by calculating several control points of the two principal axes force-moment interaction graph and using the linear fitting method.

As shown in Figure 31.6, Eq. 31.3 is the critical state of yield element.

$$f\left(\tilde{p}_u, \tilde{M}_{yu}, \tilde{M}_{zu}\right) = 1 \tag{31.3}$$

31.4.2 SIMULATIONS OF SUPPORTING CONNECTIONS

The main bridge adopts half-floating system, and a longitudinal sliding bearing (transverse restraint) and a bidirectional sliding bearing are installed between the main tower and main girder and the transition pier (side of the main bridge) and the assistant pier. In addition, a transverse wind-resistant bearing is installed between the tower and the main girder to restrain the transverse displacement of the main girder. In the Finite Element (FE) model, the boundary conditions of each structural part are treated as shown in Table 31.1, Δ_x, Δ_y, Δ_z respectively represent the linear displacement at the longitudinal direction, vertical and transverse directions of the bridge, and θ_x, θ_y, θ_z respectively represent the turning angles around the longitudinal, vertical and transverse directions of the bridge. In Table 31.1, 1 represents constraints, and 0 represents relaxation.

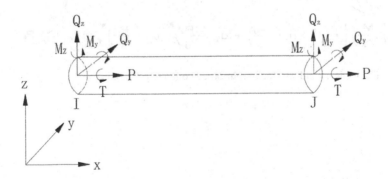

FIGURE 31.6 Distribution of generalized force of girder-column units.

TABLE 31.1
Boundary Conditions of the Numerical Model

Structural Parts	Δ_x	Δ_y	Δ_z	θ_x	θ_y	θ_z
Pile foundation	1	1	1	1	1	1
Translation area of main tower and girder	0	1	1	1	0	0
Translation area of side pylon, associate pylon and main girder	0	1	1	1	0	0

31.5   SIMULATION OF PILE FOUNDATION

For the simulation of bridge pile foundation, six directional springs, including translation spring and rotation spring, are usually applied to the cap. The spring schematic diagram is shown in Figure 31.7. According to the soil condition and the arrangements of the layout, the specific value of the spring is determined according to the principle of static force equivalence.

   In this paper, commercial finite element software package SAP2000 is used to establish a nonlinear FE model of the whole bridge. As shown in Figure 31.8, the main tower, pier, cap and

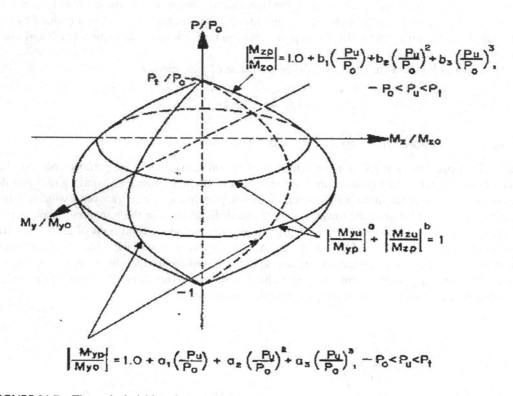

$$\left|\frac{M_{zp}}{M_{zo}}\right| = 1.0 + b_1\left(\frac{P_u}{P_o}\right) + b_2\left(\frac{P_u}{P_o}\right)^2 + b_3\left(\frac{P_u}{P_o}\right)^3,$$

$$- P_o < P_u < P_t$$

$$\left|\frac{M_{yu}}{M_{yp}}\right|^a + \left|\frac{M_{zu}}{M_{zp}}\right|^b = 1$$

$$\left|\frac{M_{yp}}{M_{yo}}\right| = 1.0 + a_1\left(\frac{P_u}{P_o}\right) + a_2\left(\frac{P_u}{P_o}\right)^2 + a_3\left(\frac{P_u}{P_o}\right)^3, \ - P_o < P_u < P_t$$

FIGURE 31.7   The typical yield surface of reinforced concrete piers or columns.

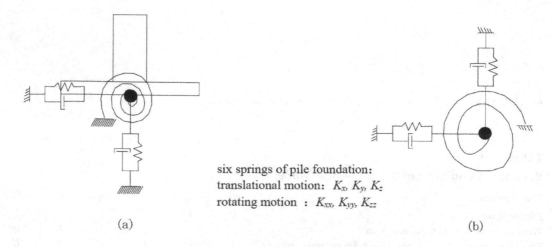

six springs of pile foundation:
translational motion: K_x, K_y, K_z
rotating motion   : K_{xx}, K_{yy}, K_{zz}

(a)   (b)

FIGURE 31.8   Spring models of the pier foundation: (a) elevation; (b) plan.

pile foundation of the bridge are all concrete with a Poisson's ratio of 0.2 and mass density of 2,549 kg/m^3. The Young's modulus of the main tower, pier, cap and pile foundation is 3.45 × 10^4, 3.25 × 10^4, 3.15 × 10^4 and 3.00 × 10^4 MPa, respectively. The Poisson's ratios were 0.3, the mass density was 7,850 kg/m^3 and the elastic modulus was 2.06 × 104 and 1.95 × 104 MPa, respectively. The section characteristics, connection modes and boundary conditions of each member are determined according to the actual situation. The gravity stiffness, stress stiffness and geometric nonlinearity characters of the long-span cable-stayed bridge are taken into account in the calculation and analysis, and the mechanical characteristics of the actual structure are simulated more truthfully.

31.6 COMPARISON AND SELECTION OF TWO KINDS OF SEISMIC MITIGATION SYSTEMS

At present, the damping restrain system with viscous damper and elastic restrain system with an elastic connecting device installed between the tower and girders has wide application on seismic mitigation of cable-stayed bridges. In order to compare the effects of the two systems, this section analyzes the basic dynamic characteristics and seismic response of double-tower cable-stayed bridge.

31.6.1 Introduction of Two Kinds of Seismic Mitigation Systems

The most common method of using a damping restrain of seismic mitigation system is to install a viscous damper between tower and girder. The viscous dampers have wide application on long-span cable-stayed bridges at home and abroad due to their excellent advantages, i.e., the wide adjustment range of damping coefficient, good stability, easy maintenance and replacement, and have achieved excellent effects of seismic mitigation (Figure 31.9).

The cylinder viscous dampers are often used on the bridges, and their basic structure is composed of key components such as a piston and a cylinder. Among them, the orifice whose flow passage is smaller than the cylinder's cross-sectional area can utilize the pressure difference between the front and the rear of the piston to make the oil flow through it to generate a damping force. The internal structure and a typical viscous damper are shown in Figure 31.10.

Experimental results show that the damping force provided by the viscous damper mainly depends on the movement speed of the piston relative to the container. The damping force of the viscous dampers is given by Formula 31.4:

$$F = CV^\alpha \tag{31.4}$$

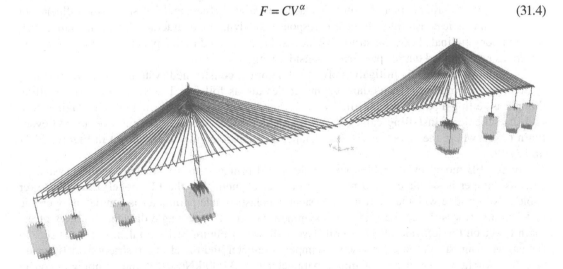

FIGURE 31.9 An FE model of the full bridge.

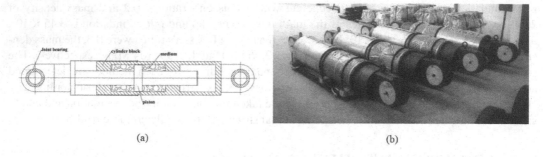

(a)   (b)

FIGURE 31.10   A viscous damper used in the experiment. (a) Schematic diagram of viscous damper and (b) picture of viscous damper.

where F represents the damping force, C the damping coefficient, V the speed and α the speed exponent. The hysteretic characteristics of the viscous damper are shown in Figure 31.10, which shows that under the excitation of the constant sine wave, the varied law of hysteresis curve with varying C and α. Based on Formula 31.4, when the deformation speed of the structure reaches the maximum, the damping force value is peak, and when the deformation of the structure reaches the maximum, the damping force is the minimum value, which is close to zero.

The elastic connection devices are installed between the tower and girder, which is called the elastic restraint of mitigation system in this paper. The elastic connecting devices mainly include elastic cable and laminate rubber bearing. As shown in Formula 31.5, the hysteretic force F provided by the elastic connecting device is a linear function of the relative displacement d of the tower and girder.

$$F = K_e d \tag{31.5}$$

Under seismic load, installing the elastic connecting device between the tower and the girder on cable-stayed bridge can increase structural stiffness and control displacement at the girder end. The key to the design of elastic connection devices is how to choose the suitable elastic restraint stiffness to reduce the displacement at girder end, while controlling the internal force of main tower within bounds. In order to pick the suitable elastic restraint stiffness and investigate the influence of longitudinal elastic constraints on seismic response and dynamic characteristics of the main tower, different longitudinal elastic constraint stiffness value is selected in this paper for numerical analysis. It has been carried out for parametric sensitivity analysis.

To compare the seismic mitigation effects of damping constrained system and elastic restraint system, the settings of viscous damper and cables are as follows: The scheme of an installing damper is setting a pair of longitudinal viscous damper between the beam and every main tower, and the scheme of installing cables is setting a pair of that between the main beam and every main tower, where the layout of viscous damper and elastic connector is shown in Figures 31.11 and 31.12.

For the FE model of a double-tower cable-stayed bridge with a restraint damping system, a viscous damper is simulated by a nonlinear damper element. For the FE model of double-tower cable-stayed bridge with elastic restraint system, an elastic cable parameter is simulated by elastic restriction stiffness. Since the bridge is a symmetrical structure, a single damper or cables of the main tower on the left side of the overall layout shown in Figure 31.13 is taken to investigate. In the engineering background, the viscous damper parameter and the elastic restriction stiffness are selected as follows: the viscous damping parameter $C = 3{,}000$ kN(m/s)$^{-\alpha}$, the damping exponent $\alpha = 0.4$ and the elastic restraint stiffness $K_e = 2.5 \times 10^4$ kN/m.

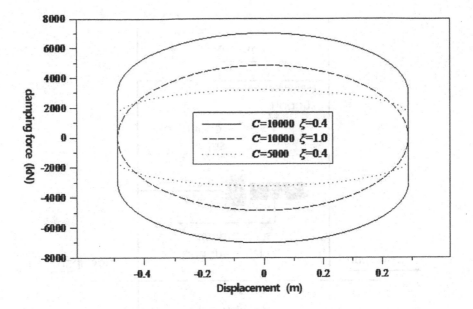

FIGURE 31.11 Relationship between the hysteresis loops and exponent α of viscous dampers.

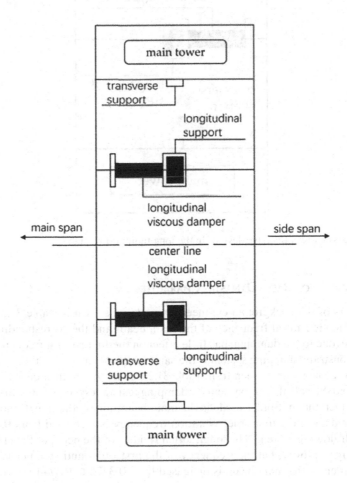

FIGURE 31.12 Longitudinal viscous dampers.

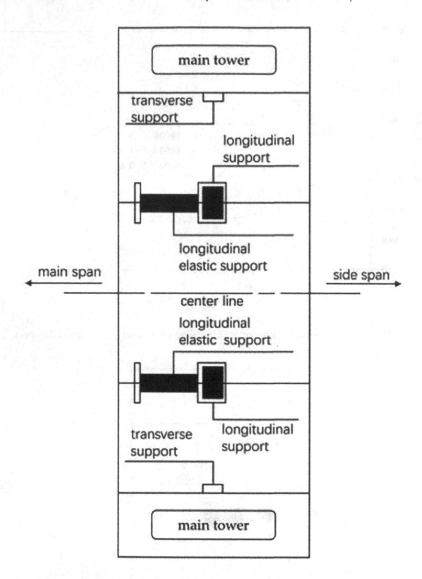

FIGURE 31.13 Elastically connective devices in the longitudinal direction.

31.6.2 Comparison of Basic Dynastic Characters

The first ten orders of the background engineering model is shown in Table 31.2. As can be seen the first torsional mode natural frequency of the main beam and the corresponding characteristics of the mode shape, due to the damping, has little effect on the first natural frequency. Moreover, the requency of the constraint damping system is the same as the frequency of the half-floating system.

The results show that it can be seen from Table 31.2 that the main difference between the elastic restraint system and the half-floating system (or damping resistant system) is the first and sixth order, respectively, in which the first-order mode is the main beam longitudinal drift mode. Adopting the elastic connecting device, the first-order vibration frequency is increased from 0.1106 to 0.211 Hz relative to the half-floating system. The longitudinal rigidity of the beam of the elastic restrain system is significantly improved, but the frequency of the first-order anti-symmetric vertical bending vibration (sixth order) of the main beam is increased from 0.3625 to 0.3744 Hz, and the increase is smaller.

TABLE 31.2

Vibration Frequencies and Corresponding Modal Shapes of the Structure

	Frequency (Hz)			
Order	Damper Restraint System	Elastic Restraint System	Mode	Modal Shape
1	0.1106	0.211	Longitudinal draft of main girder	
2	0.2269	0.2269	The first symmetric transverse bending of the main girder	
3	0.2754	0.2757	The first symmetric longitudinal bending of the main girder	
4	0.2925	0.2925	The reverse side bending of the main tower	
5	0.2978	0.2978	The same side bending of the main tower	
6	0.3625	0.3744	The first antisymmetric vertical bending of the main girder	
7	0.4555	0.4555	The longitudinal vibration of 7# side pile	
8	0.4745	0.4748	The longitudinal vibration of 0# side pile	

31.6.3 SEISMIC RESPONSE AND MECHANISM ANALYSIS

Table 31.3 shows the comparison of the internal force in the longitudinal direction and displacement responses of the cable-stayed bridges with different systems under the earthquake action, and Table 31.4 shows the comparison of the seismic mitigation ratio with different systems. The damping ratios η = |peak before damping − peak after damping / peak before damping × 100%.

From the results of FE analysis, it can be seen that the base moment of the most unfavorable section in the tower is smaller than that of the base yield moment (3.648×10^6 kN∙m) without any damping measures of the half-floating system and the elastic and viscous damper restraint systems. Hence, it is demonstrated that the background engineering is always in the elastic range, so the elastic analysis method is used in the latter sections.

From Tables 31.3 and 31.4, for the elastic constraint system, the longitudinal displacements at the main girder end and the tower end are reduced. Compared with the half-floating system, the longitudinal displacement of the beam end is reduced from 1.414 to 0.951 m. The longitudinal displacement of the main beam end and the longitudinal displacement seismic response reduction ratio of the top of the main tower are 32.74% and 28.10%, respectively. The base moment in the main tower is not significantly reduced but that in the main tower is increased.

For damper restraint system, the longitudinal displacements at the main girder end and the tower end are reduced. The longitudinal displacement at the main girder end is reduced from 1.414 to 0.764 m, and the damping rates of the longitudinal displacement at the main girder end and tower end are 18.69% and 19.15%, respectively. These demonstrate that installing a viscous damper can greatly reduce the longitudinal displacement at the girder end and tower end and reduce the base shear and base moment in the tower.

TABLE 31.3

Comparison of Longitudinal Internal Forces and Displacements of Different Systems under Seismic Action

Restraint Forms	Longitudinal Displacement at Girder End (mm)	Longitudinal Displacement at Tower Top (mm)	Base Shear in Tower (kN)	Base Moment in Tower (kN m)
Half-floating system	1.414	1.530	4.848×104	2.538×106
Elastic restraint system	0.951	1.101	5.332×104	2.533×106
Damper restraint system	0.764	1.022	3.942×104	2.052×106

TABLE 31.4

Comparison of Mitigation Ratios between Two Systems

Restraint Forms	Longitudinal Displacement at Girder End (%)	Longitudinal Displacement at Tower Top (%)	Base Shear in Tower (%)	Base Moment in Tower (%)
Elastic restraint system	32.74	28.10	−9.98	0.32
Damper restraint system	45.97	33.20	18.69	19.15

Compared with a half-floating system without mitigation devices, the longitudinal displacements at the main girder end and tower end of damping resistant and elastic constraint system are reduced. Among them, the damping resistant system reduces more, and the damping resistant system can reduce the displacement at the girder end, but effectively improve the force at the base tower. However elastic constraint system has basically no improvement on the base moment in tower, but increases the base shear in tower. From the results of seismic response analysis, the control effects of viscous dampers are more advantageous than cables.

In order to analyze the mitigation mechanism of the elastic connecting devices and the viscous dampers, the response spectra of Figures 31.14 and 31.15 are plotted. Due to first-order longitudinal mode of the floating system, half-floating system experiences a longitudinal drift. As Figure 31.12 shows, assuming that the natural period of longitudinal drift mode of cable-stayed bridge is T_0, then T_0 is located in the region where the acceleration response spectrum decreases. The structural acceleration response spectrum of Figure 31.14 shows that if we extend the structure's natural vibration periods and increase the structural damping, that result in reducing the acceleration response and the seismic force. As shown in Figure 31.14, if we set the natural vibration period of the structure is T_0 at first, increasing the structural damping will reduce the displacement response of the structure, but prolonging the natural vibration period of the structure, i.e., $T > T_0$, will increase the displacement of the structure. Therefore, increasing the structural damping can reduce the seismic force and seismic displacement of the structural, but reducing structural natural periods means increasing the structural stiffness. Although reducing displacement, it increases the seismic force. From the response spectrum, the method of increasing the structural damping is more advantageous than increasing the structural stiffness in reducing structural displacement. In other words, using viscous dampers is more advantageous than the elastic-connecting devices in seismic mitigation mechanism.

In order to further analyze the seismic mitigation mechanism of cable-stayed bridge, time-history curves of seismic response of two-system cable-stayed bridge under seismic wave are plotted. The time-history curves of the cables' tensions and base shears in tower are shown in Figure 31.16. Time-history curves of damping forces and tower base shears base shears are shown in Figure 31.17.

Figure 31.18 is a schematic diagram of the transmission of seismic inertia force of a main girder of a cable-stayed bridge. It can be seen from the schematic diagram that the part of the inertial force

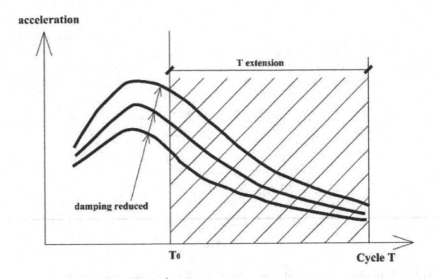

FIGURE 31.14   Acceleration response.

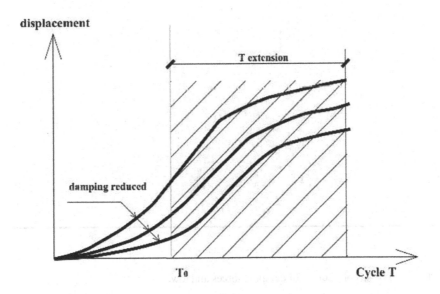

FIGURE 31.15   Displacement response.

of the main girder in the elastic restrained system is transmitted to the foundation by the inclined cables through the upper pylon, and the other part is transmitted to the foundation by the elastic cable or viscous damper through the lower pylon. As can be seen from Figures 31.16 and 31.17, for the elastic restraint system, the elastic restraint increases the longitudinal restraint stiffness of the girder, and the horizontal seismic force transmitted by the elastic cable is synchronized with the time-history curve of the internal force at the bottom of the main tower, so the base shear force in the tower is increased.

For the damping restrain system, if the longitudinal stiffness of the beam is not increased, it can be seen from Figure 31.1 that the horizontal seismic force of the girder on the damper is opposite to the time-history curve of the base internal force in main tower, and there is a phase difference. Hence, the base shear force in the tower decreases, and the horizontal seismic force transmitted by

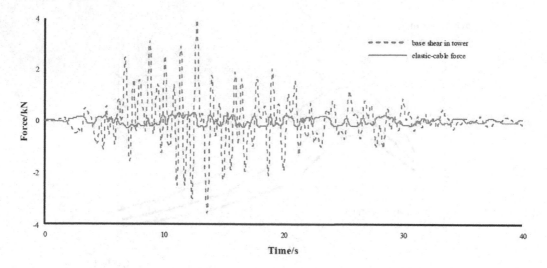

FIGURE 31.16 Time-history curves of elastic cables' tensions and tower base shears.

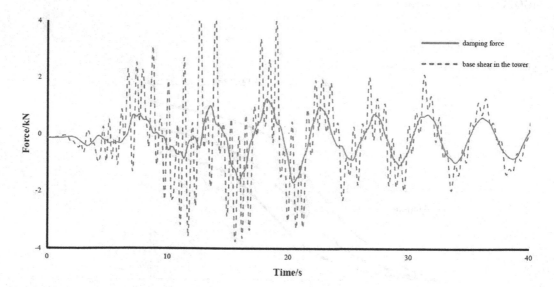

FIGURE 31.17 Time-history curves of damping forces and tower base shears.

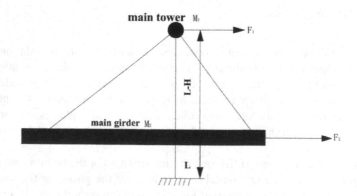

FIGURE 31.18 A simplified schematic diagram of forces transferring in the longitudinal direction.

the damper is smaller than the arm of force at the bottom of the tower. Above research can demonstrate viscous dampers that have better effect of seismic mitigation by analyzing the seismic mitigation mechanism and is an ideal seismic mitigation device.

31.7 CONCLUSIONS

The main tower and pier are modeled by a spatial dynamic nonlinear model of reinforced concrete elastic-plastic beam-column element, which can simulate the response of a long-span cable-stayed bridge under earthquake action. Compared with the basic dynamic characteristics of viscous damper restrain system and elastic cable restrain system, the result shows that the main difference between the elastic cable restrain system and the viscous damper resistant system is first and sixth order. Between them, the first-order mode is the main beam longitudinal drift mode. Relative to the half-floating system, when the cable-stayed bridge equipped the elastic connecting device, the first-order vibration frequency is increased from 0.1106 to 0.211 Hz, and the longitudinal rigidity is significantly improved. But the frequency of the first-order anti-symmetric vertical bending vibration (sixth order) of the main girder has a small increase.

Seismic response analysis of a double-tower cable-stayed bridge with a traditional viscous damping system show that the longitudinal displacements at the girder end and tower end are reduced. However, a restraint viscous damper can result in more reduction in longitudinal displacement. The damper resistant system can not only reduce the displacement at the girder end but also obviously improve the base force in the tower. However, the elastic constraint system has no improvement on the base moment and increases the base shear in the tower. Therefore, the effect of setting viscous dampers is more advantageous than elastic connecting devices.

Compare the different seismic mitigation mechanisms of the elastic connecting devices and the viscous dampers, the reaction spectrum shows that setting viscous dampers is more advantageous to increase the damping of the structure than to increase the structural stiffness.

From the schematic diagram of the seismic inertial force of the main girder of the cable-stayed bridge, part of the inertial force of the main girder in the elastic constraint system is transmitted to the foundation by the stay cable through the upper pylon, and the other part is transmitted to the foundation by the elastic cable or the viscous damper through the lower pylon. For viscous damper restraint system, the elastic constraint increases the longitudinal rigidity of the beam body, and the horizontal seismic force transmitted by the elastic cable is synchronized with the time-history curves of the internal force at the bottom of the main tower, so the tower shear force is increased. For the damping constraint system, without increasing the longitudinal rigidity of the beam body, the horizontal seismic force of the beam subjected to the damper is opposite to the time-history curve of the internal force of the main tower bottom and there is a phase difference, so the shearing force at the bottom of the tower is reduced. At the same time, the seismic horizontal force transmitted by the damper is smaller than that of the arm of force at the bottom of the tower, so the bending moment of the bottom of the tower decreases more than the shear force. From the analysis of damping mechanism, it can be seen that the damping effect of viscous damper is good, and it is an ideal damping device.

This study is only applicable to the seismic mitigation study of a long-span cable-stayed bridge with twin towers represented by the background bridge, and under the action of longitudinal earthquake along the bridge direction, the research on damping of cable-stayed bridge with different span under the action of longitudinal earthquake along the bridge direction has yet to be studied further.

ACKNOWLEDGEMENTS

The work presented in this paper was supported by a grant from science and technology bureau of Fuzhou, China (Project No. 2018-G-88) and by a grant from provincial department of science and technology of Fujian, China (Project No. 2017J01698). The authors also acknowledge the help of Zhenzheng Fang and Zhen Lin for the execution of the shaking table tests. The authors would like to acknowledge the thoughtful suggestions of three anonymous reviewers, which substantially improved the present paper.

REFERENCES

1. American Association of State Highway and Transportation Officials. *Guide Specifications for Highway Construction*, 9th edition. Washington, AASHTO, 2008.
2. American Association of State Highway and Transportation Officials. *AASHTO LRFD Bridge Construction Specifications*, interim revision. Washington, AASHTO, 2008.
3. American Association of State Highway and Transportation Officials. *AASHTO LRFD Bridge Construction Specifications: SI Units*, 3th edition. Washington, AASHTO, 2008.
4. Chongqing Communications Technology Research & Design Institute. *Guidelines for Seismic Design of Highway Bridges*. Beijing, China Communications Press. 2008.
5. Jung, H.J.; Park, K.S.; Spencer, B.F.; Lee, I.W. Hybrid seismic protection of cable-stayed bridges. *Journal of Earthquake Engineering and Structural Dynamics*, 2004, *33*, 795–820.
6. Usami, T.; Lu, Z.; Ge, H. A seismic upgrading method for steel arch bridges using buckling-restrained braces. *Journal of Earthquake Engineering and Structural Dynamics*, 2005, *34*, 4672–4681.
7. Guo, J.; Zhong, J.; Dang, X.; Yuan, W. Seismic responses of a cable-stayed bridge with consideration of uniform temperature load. *Journal of Applied Science*, 2016, *6*, 408.
8. Xie, W.; Sun, L.M. Investigated on damage control with layer dissipation mechanism for cable-stayed bridges in longitudinal direction. *Journal of Vibration Engineering*, 2015, *28*, 585–592.
9. Li, J.Z.; Yuan, W.C. Nonlinear longitudinal seismic response analysis of cable-stayed bridge systems with energy dissipation. *China Journal of Highway and Transport*, 1998, *11*, 71.
10. Yuan, W.C.; Qu, X.W. Application analysis of seismic isolation devices on longitudinal seismic response of floating system cable-stayed bridge. *Journal of Tongji University (Natural science)*, 2015, *43*, 199–204.
11. Ali, H.E.M.; Abdelghafar, A.M. Seismic passive control of cable-stayed bridges. *Journal of Shock and Vibration*, 1995, *2*, 257–272.
12. Combault, J.; Pecker, A.; Teyssandier, J.P. et al. Rion-Antirion Bridge, Greece-concept, design, and construction. *Journal of Structural Engineering International*, 2005, *15*, 22–27.
13. Xu, L.P. Structural system analysis for super-long span cable-statyed bridges. *Journal of Tongji University*, 2003, *31*, 400–403.
14. Weng, D.G.; Lu, Z.H.; Xu, B. The experimental study on property of energy dissipation of viscous liquid damper. *Journal of World Earthquake Engineering*, 2002, *18*, 30–34.
15. Makris, N.; Constantinou, M.C. Fractional-derivative Maxwell model for viscous dampers. *Journal of Structural Engineering*, 1991, *117*, 2708–2724.
16. Makris, N.; Constantinou, M.C; Dargush, G.F. Analytical model of viscoelastic fluid dampers. *Journal of Structural Engineering*, 1993, *119*, 3310–3325.
17. Makris, N.; Dargush, G.F.; Constantinou, M.C. Dynamic analysis of viscoelastic-fluid dampers. *Journal of Engineering Mechanics*, 1995, *121*, 1114–1121.
18. Ferreira, F.; Simoes, L. Optimum design of a controlled cable-stayed footbridge subject to a running event using semiactive and Passive Mass Dampers. *Journal of Performance of Constructed Facilitied*, 2019, *33*, 04019025.
19. Feng P.C.; Ding R.J.; Chen Y.M.; Fu K.; Liu Q. Design of main bridge of Zhuankou Changjiang river highway bridge. *Journal of Bridge Construction*, 2017, *47*, 1–6.
20. Xu, L.; Zhang, H.; Gao, J.F.; Zhang, C. Longitudinal seismic responses of a cable-stayed bridge based on shaking table tests of a half-bridge scale model. *Journal of Advances in Structural Engineering*, 2019, *22*, 81–93.
21. Yi, J.; Le, J.Z. Experimental and numerical study on seismic response of inclined tower legs of cable-stayed bridges during earthquakes. *Journal of Engineering Structures*, 2019, *183*, 180–194.
22. Nguyen, C.H.; Macdonald, J.H.G. Galloping analysis of a stay cable with an attached viscous damper considering complex modes. *Journal of Engineering Mechanics*, 2018, *144*, 04017175.
23. Main, J.A.; Jones, N.P. Free vibrations of taut cable with attached damper. I: Linear viscous damper. *Journal of Engineering Mechanice-Asce*, 2002, *128*, 1062.
24. He, W.L.; Agrawal, A.K.; Mahmoud, K. Control of seismically excited cable-stayed bridge using resetting semiactive stiffness dampers. *Journal of American Society of Civil Engineers*, 2001, *6*, 376–384.

25. Ali, H.M.; Abdel-Ghaffar, A.M. Seismic energy dissipation for cable-stayed bridges using passive devices. *Journal of Earthquake Engineering and Structural Dynamics,* 2010, *23*, 877–893.

26. Wesolowsky, M.J.; Wilson, J.C. Seismic isolation of cable-stayed bridges for near-field ground motions. *Journal of Earthquake Engineering and Structural Dynamics,* 2003, *32*, 2107–2126.

27. Soneji, B.B.; Jangid, R.S. Passive hybrid systems for earthquake protection of cable-stayed bridge. *Journal of Engineering Structures,* 2007, *29*, 57–70.

28. Xie, X.; Li, X.; Shen, Y. Static and dynamic characteristics of a long-span cable-stayed bridge with CFRP cables. *Journal of Materials,* 2014, *7*, 4854–4877.

29. Ben, M.O.; Auricchio, F. Performance evaluation of shape-memory-alloy superelastic behavior to control a stay cable in cable-stayed bridges. *International Journal of Non-Linear Mechanics,* 2011, *46*, 470–477.

30. Cheng, J.; Luo, X.; Zhuang, Y.; Xu, L.; Luo, X. Experimental study on dynamic response characteristics of RPC and RC micro piles in SAJBs. *Journal of Applied Sciences,* 2019, *9*, 2644.

31. Kwon, S.Y.; Yoo, M. Evaluation of dynamic soil-pile-structure interactive behavior in dry sand by 3D numerical simulation. *Journal of Applied Sciences,* 2019, *9*, 2612.

32. Dong, H.H.; Jang, H.P.; Kwan, S.P.; Wonsuk, P.; Jinkyo, F.C. Optimization of complex dampers for the improvement of seismic performance of long-span bridges. *KSCE Journal of Civil Engineering,* 2011, *14*, 33–40.

32 Automatic Modal Parameter Identification of Cable-Stayed Bridge Based on the Stochastic Subspace Identification Method

Liao Yuchen, Zhang Kun, Zong Zhouhong, and Lin Dinan
Southeast University
Fujian Key Laboratory of Green Building Technology

CONTENTS

32.1 INTRODUCTION

Due to the damage of the structure, the variance of the boundary and non-stationary environmental excitation, the bridge structure shows non-linear features. Therefore, non-linear components are found in the response of bridge, which cause difficulties in modal parameter identification. For frequency domain identification method, peak picking method is usually used to identify the modal parameter of structure, which can be achieved by programming. However, frequency domain identification methods face the problems, such as power spectrum leakage and identification of modal frequency (Ren and Zong 2004).

Stochastic subspace identification (SSI) method is one of the widely used and effective time domain methods that are proposed to overcome the drawbacks of frequency domain identification methods. This method shows a great identification effect for close modes with the full use of all the information of measurement points (Peeters and Guido 2000). The mode shapes of long-span bridges can also be extracted from the analysis of ambient vibration by using the SSI method (Li, Qin and Qian 2001). Peeters et al. (2001) summarized the applications of SSI method in operational modal identification, comparing the identification effect of the data-driven SSI and covariance-driven SSI.

321

Yu and Ren (2005) used empirical mode decomposition (EMD) method to pre-process the signals before SSI analysis, which made the stabilization diagram of SSI method clearer.

At the present stage, the main issues of SSI include large calculation cost, determination of system order, false modes and confidence interval estimation of modal parameters (Peeters and Guido 2000). Stabilization diagram method is adopted to decrease the influence of false modes to extract the modal parameters from the responses of bridges when applying the SSI method. However, modal parameter identification based on the stabilization diagram method relies on manual selection, which is not suitable for the processing of massive structure health monitoring (SHM) data. Fan et al. (2007) proposed component energy index (CEI) to judge false modes, and the viability of this method was verified by environmental vibration test of an arch bridge. Yang et al. (2012) transformed time-varying modal parameter identification problem into subspace tracking problem by pre-processing the updated input/output data, and introduced two new versions of time-varying modal parameter identification algorithms, called NPI and API subspace tracking algorithms, to establish a new time-varying modal parameter fast identification method. Reynders et al. (2012) proposed a fully automatic identification method for the interpretation of stabilization diagrams based on a three-stage clustering approach. Ubertini et al. (2013) combined the SSI method and clustering analysis to identify the modal parameters of an iron arch bridge and a long-span footbridge automatically, which demonstrated good accuracy and robust performance. Wu et al. (2013) applied the fuzzy clustering algorithm to the stability diagram method, and compared the clustering circles in the stabilization diagram to identify the false mode. Zhang et al. (2018) constructed two stochastic subspace models with different dimension Hankel matrix to eliminate false modes and identified the modal parameters of the 5 degree of freedom (DOF) mass-spring-damper system and a flat plate. Good robust and effeteness of the improved SSI method were verified by the result of modal parameters identification.

To process massive SHM data effectively, an automatic modal parameter identification method based on SSI is put forward for ambient excitation. Section 32.2 presents the SHM system and the results of ambient vibration test of Guanhe Bridge, which is a composite cable-stayed bridge in G15 Expressway in China. In Section 32.3, the basic theory of covariance-driven SSI and automatic modal parameter identification method are introduced. The proposed automatic modal parameter identification algorithm on the basis of covariance-driven SSI is verified by the modal analysis of Guanhe Bridge in Section 32.4. At the end of this paper, Section 32.5 concludes the main results of the research.

32.2 BRIEF DESCRIPTION OF GUANHE BRIDGE

32.2.1 SHM System

Guanhe Bridge is a composite cable-stayed bridge located at the junction of Lianyungang City and Yancheng City in G15 expressway in China as shown in Figure 32.1. The bridge compromises a main span of 340 m and two side spans of 32.9 and 115.4 m each, with two towers of 121 m height. A bridge SHM system has been implemented on the bridge in December 2013. The system (Figure 32.2) includes strain gauges, tri-axial accelerometers, cable tension sensors, temperature sensors, dynamic displacement gauges, displacement gauges, ultrasonic anemometers and global positioning system (GPS) receivers.

32.2.2 Ambient Vibration Test

The ambient vibration test of Guanhe Bridge was implemented with an arrangement (Figure 32.3) of 56 measuring points and 1 reference point. The accelerations of three directions were measured simultaneously by triaxial accelerometers with the sampling frequency of 200 Hz. The sampling time of each measuring point is not <15 minutes. Modal analysis of Guanhe Bridge was carried out based on the SSI method, whose results are listed in Table 32.1.

FIGURE 32.1 Guanhe Bridge panorama photo.

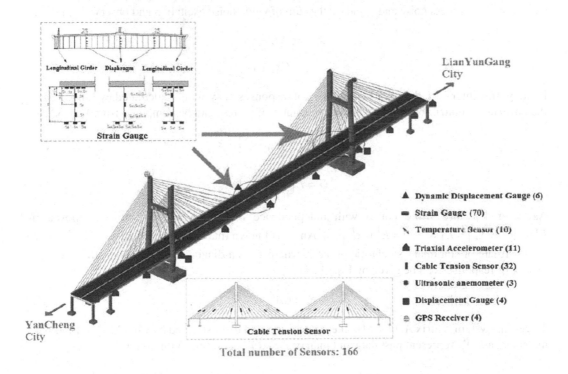

FIGURE 32.2 Sensor arrangement of SHM system.

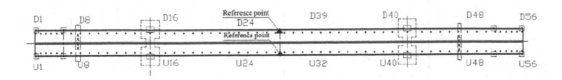

FIGURE 32.3 Measurement points arrangement of ambient vibration test.

TABLE 32.1

Results of Ambient Vibration Test

Mode	V_1	V_2	V_3	V_4	V_5	V_6	V_7	T_1	T_2
Frequency (Hz)	0.380	0.499	0.766	0.882	0.948	1.030	1.240	0.628	0.809
Damping ratio (%)	0.8	0.4	0.9	1.2	0.6	0.6	0.4	0.5	0.6

32.3 METHODOLOGY

32.3.1 BASIC THEORY OF COVARIANCE-DRIVEN SSI

SSI is a time domain modal parameter identification method based on white noise excitation assumption, which is suitable for stationary excitation. The SSI method includes covariance-driven and data-driven methods, and the covariance-driven SSI shows great effeteness of modal parameter identification.

Discrete state space model of the dynamic system based on the white noise excitation assumption can be described in Eq. (32.1), where x_k denotes the state of the system, y_k denotes the observation of the system, A denotes the system matrix, C denotes the output matrix, w_k means the sum of white noise excitation and process noise and v_k means the sum of white noise excitation and observation noise.

$$x_{k+1} = Ax_k + w_k$$
$$y_k = Cx_k + v_k$$

(32.1)

For any time interval i, the covariance matrix of responses of system is calculated by Eq. (32.2), and the covariance matrix between responses and states of system can be computed from Eq. (32.3).

$$R_i = E\left[y_{k+i} \cdot y_k^T\right]$$

(32.2)

$$G = E\left[x_{k+1} \cdot y_k^T\right]$$

(32.3)

As w_k and v_k are white noises with independence of system state x_k, it is supposed that $E[w_k] = E[v_k] = E\left[w_k x_k^T\right] = E\left[v_k x_k^T\right] = 0$. And it is known that $E[x_k] = 0$ and $E\left[x_k x_k^T\right] = \Sigma_\sigma$ due to the postulate of stationary stochastic process, where Σ_σ is a diagonal matrix. Thus, the covariance matrix of responses can be given by Eq. (32.4).

$$R_i = CA^{i-1}G$$

(32.4)

To get the system matrix A, block Hankel matrix should be constructed by Eq. (32.5) at first, where matrix Y_p and Y_p^+ represent past data and matrix Y_f and Y_f^- represent future data.

$$Y_{0|2i-1} = \frac{1}{\sqrt{j}} \begin{pmatrix} y_0 & y_1 & \cdots & y_{j-1} \\ y_1 & y_2 & \cdots & y_j \\ \cdots & \cdots & \cdots & \cdots \\ y_{i-1} & y_i & \cdots & y_{i+j-1} \\ y_i & y_{i+1} & \cdots & y_{i+j} \\ y_{i+1} & y_{i+2} & \cdots & y_{i+j+1} \\ \cdots & \cdots & \cdots & \cdots \\ y_{2i-1} & y_{2i} & \cdots & y_{2i+j-2} \end{pmatrix} = \begin{bmatrix} Y_{0|i-1} \\ Y_{i|2i-1} \end{bmatrix} = \begin{bmatrix} Y_p \\ Y_f \end{bmatrix} \text{ or } \begin{bmatrix} Y_{0|i} \\ Y_{i+1|2i-1} \end{bmatrix} = \begin{bmatrix} Y_p^+ \\ Y_f^- \end{bmatrix}$$

(32.5)

Then Toeplitz matrix of responses is calculated by Eq. (32.6). Substituting the R_i with Eq. (32.4), Toeplitz matrix can be given by Eq. (32.7), where O_i is extended observability matrix and P_i is extended controllability matrix.

$$T_{1|i} \overset{def}{=} Y_f \left(Y_p\right)^T = \begin{pmatrix} R_i & R_{i-1} & \cdots & R_1 \\ R_{i+1} & R_i & \cdots & R_2 \\ \vdots & \vdots & \vdots & \vdots \\ R_{2i} & R_{2i-1} & \cdots & R_i \end{pmatrix} \quad T_{2|i+1} \overset{def}{=} Y_f^- \left(Y_p^+\right)^T = \begin{pmatrix} R_{i+1} & R_i & \cdots & R_2 \\ R_{i+2} & R_{i+1} & \cdots & R_3 \\ \vdots & \vdots & \vdots & \vdots \\ R_{2i+1} & R_{2i} & \cdots & R_{i+1} \end{pmatrix} \quad (32.6)$$

$$T_{1|i} = \begin{bmatrix} C \\ CA \\ \cdots \\ CA^{i-1} \end{bmatrix} \begin{bmatrix} A^{i-1}G & A^{i-2}G & \cdots & G \end{bmatrix} = O_i P_i \quad T_{2|i+1} = O_i A P_i = O_{i+1} P_i \quad (32.7)$$

Next, singular value decomposition is used to decompose Toeplitz matrix as Eq. (32.8), in which $U_1 \in R^{il \times N}$ and $V_1 \in R^{il \times N}$ are unitary matrices and $S_1 \in R^{N \times N}$ is a diagonal matrix with singular values as diagonal elements. The N is the order of the dynamic system model. Comparing Eqs. (32.7) and (32.8), extended observability matrix O_i and extended controllability matrix P_i are given by Eq. (32.9), where T can be a unit matrix, and the calculation of matrix O_{i+1} is similar to matrix O_i.

$$T_{1|i} = USV^T = \begin{bmatrix} U_1 & U_2 \end{bmatrix} \begin{bmatrix} S_1 & 0 \\ 0 & S_2 \end{bmatrix} \begin{bmatrix} V_1^T \\ V_2^T \end{bmatrix} = U_1 S_1 V_1^T \quad (32.8)$$

$$O_i = U_1 S_1^{1/2} T \quad P_i = T^{-1} S_1^{1/2} V_1^T \quad (32.9)$$

It is obvious that output matrix C is the first block row matrix of extended observability matrix O_i shown in Eq. (32.10). Meantime, system matrix A is able to be computed from Eq. (32.11), where $(\cdot)^+$ denotes the pseudo-inverse operation.

$$C = O_i(1:l, 1:N) \quad (32.10)$$

$$A = O_i^+ O_{i+1} = \left(O_i^T O_i\right)^{-1} O_i^T O_{i+1} \quad (32.11)$$

In the end, modal parameters can be extracted by eigenvalue decomposition of system matrix A. In Eq. (32.12), λ_s^c is the eigenvalue of the system, Ψ_s is the complex mode shape of a dynamic system, f_s is the nature frequency, ξ_s is the damping ratio and φ_s is the real mode shape.

$$A = \psi \Lambda \psi^{-1}$$

$$f_s = \omega_s / 2\pi = \left|\lambda_s^c\right|$$

$$\xi_s = -\mathrm{Re}\left(\lambda_s^c\right) / \omega_s \quad (32.12)$$

$$\varphi_s = C \psi_s$$

32.3.2   Automatic Modal Parameter Identification Based on the SSI

The stabilization diagram method is used to eliminate the false modes caused by noise and uncertainty of system order. In this method, it is assumed that the dynamic system can be described as models with different orders. Then modal parameters of all the models are identified based on the SSI and plotted in the same diagram.

The poles whose modal parameters meet the condition of Eq. (32.13) are called the stable poles. In Eq. (32.13), i denotes modal order, f denotes frequency, ξ denotes damping ratio and MAC means modal assurance criterion that is computed from Eq. (32.14). The stable axis consisting of stable poles represents the mode of the dynamic system.

$$\left|\frac{f_i - f_{i+1}}{f_i}\right| < [\Delta f]$$

$$\left|\frac{\xi_i - \xi_{i+1}}{\xi_i}\right| < [\Delta \xi] \tag{32.13}$$

$$(1 - \text{MAC}(i, i+1)) < [\Delta \varphi]$$

$$\text{Cross}^{a-b}\text{MAC}(i,j) = \frac{\left|\varphi_i^{aH}\varphi_j^b\right|}{\left(\varphi_i^{aH}\varphi_i^a\right)\left(\varphi_j^{bH}\varphi_j^b\right)} \tag{32.14}$$

However, the traditional method needs the manual interaction because of the existence of a mount of false poles in the stabilization diagram, which shows unsatisfactory efficiency of modal parameter identification. Therefore, a method to eliminate the false modes based on the reconstruction of system matrix is proposed in this section.

A new system matrix Ai is defined in Eq. (32.15). The relationship between matrix A and Ai follows as Eq. (32.16), where value a is one of the eigenvalues of the system. Therefore, the system matrix A can be estimated by Eq. (32.17).

$$Ai = O_{i+1}^{+}O_i = \left(O_{i+1}^{T}O_{i+1}\right)^{-1}O_{i+1}^{T}O_i \tag{32.15}$$

$$A\{\varphi\} = a\{\varphi\}, Ai\{\varphi\} = \frac{1}{a}\{\varphi\} \tag{32.16}$$

$$A = Ai^{-1} = \left(O_{i+1}^{T}O_i\right)^{-1}O_{i+1}^{T}O_{i+1} \tag{32.17}$$

The eigenvalues of system extracted from the system matrix A calculated by Eqs. (32.11) and (32.17) should have the same value. However, false modes due to noise and overestimation of system order do not appear in pairs because they do not have this characteristic.

To eliminate false modes, modal similarity coefficient r_m is proposed as Eq. (32.18). In this equation, W_f, W_ξ and W_{MAC} are weights of frequency, damping ratio and MAC, equaling to 0.5, 0.2 and 0.3, respectively. The symbols d_f, d_ξ and d_{MAC} are tolerance errors of frequency, damping ratio and MAC, which are determined as 0.01, 0.10 and 0.01. $(f, \xi, \text{MAC})_m$ and $(f, \xi, \text{MAC})_n$ are frequency, damping ratio and MAC calculated by Eqs. (32.11) and (32.17).

$$r_m = \frac{W_f|f_m - f_n|}{d_f \max(f_m, f_n)} + \frac{W_\xi|\xi_m - \xi_n|}{d_\xi \max(\xi_m, \xi_n)} + \frac{W_{\text{MAC}}}{d_{\text{MAC}}}(1 - \text{MAC}(\psi_m, \psi_n)) \tag{32.18}$$

The steps of automatic modal parameter identification based on SSI method are as follows:

Step 1: calculate the state matrix A on the basis of Eq. (32.11) to form the standard poles.

Step 2: calculate the state matrix A on the basis of Eq. (32.17) to form the referenced poles.

Step 3: calculate the modal similarity coefficient r_m between standard poles and referenced poles on the basis of Eq. (32.18), and delete the standard poles when r_m is larger than 1.

Step 4: compare the frequency, damping ratio and MAC between models of adjacent orders based on Eq. (32.13), and classify the poles of stabilization diagram.

Step 5: set the frequency intervals for each mode according to the results of ambient vibration test, and classify stable poles on the basis of monotonicity of frequency.

Step 6: obtain the modal parameters by averaging the parameters of poles for each mode.

32.4 MODAL PARAMETER IDENTIFICATION OF GUANHE BRIDGE

32.4.1 Elimination of False Modes

The acceleration signal of Guanhe Bridge is selected to identify the modal parameters, whose sampling frequency is 50 Hz. The signal is pre-processed by Chebyshev-I filter in order to decrease the high frequency noise. Then resample is applied on the pre-processed signal to reduce the sampling frequency to 5 Hz. Stabilization diagrams of vertical acceleration signal based on SSI method are shown in Figure 32.4. Comparison between stabilization diagrams of original signal and pre-processed signal indicates that pre-process is able to eliminate false modes.

After obtaining the stabilization diagrams, a clustering method is used to select stable poles. As the modal frequencies of these modes are monotonic, suitable ranges of each mode are determined on the basis of the result of ambient vibration test. It is supposed that the stable poles belong to a mode when the frequency of this pole is in the range of the corresponding mode. The frequency intervals for each mode are listed in Table 32.2, where letter V and T denote vertical mode and T torsional mode, respectively.

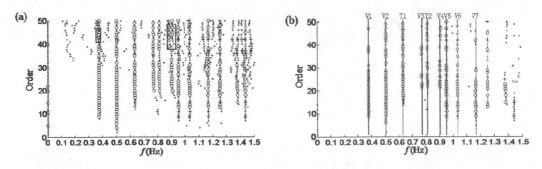

FIGURE 32.4 (a) Stabilization diagram of original acceleration signal and (b) pre-processed acceleration signal.

TABLE 32.2

Frequency Intervals for Each Mode

Mode	V_1	V_2	V_3	V_4	V_5	V_6	V_7	T_1	T_2
Upper limit	0.370	0.490	0.765	0.880	0.940	1.020	1.240	0.620	0.805
Lower limit	0.395	0.515	0.795	0.920	0.975	1.055	1.280	0.650	0.840

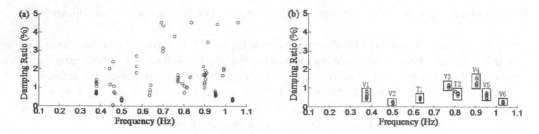

FIGURE 32.5   (a) Cluster analysis graph before and (b) after processing.

To verify the feasibility the simplified method, clusters formed by frequency and damping ratio are plotted in Figure 32.5. After eliminating the false modes, damping ratios of stable poles selected by this method are relatively concentrated, proving that classification of stable poles can be achieved by using frequency as the control index.

32.4.2   AUTOMATIC MODAL PARAMETER IDENTIFICATION OF GUANHE BRIDGE

The modal parameters of Guanhe Bridge are identified on the basis of automatic SSI method mentioned in this study. The acceleration signal of 24 hours is selected, which is processed per half hour. The modal frequencies of automatic identification are shown in Figure 32.6.

Figure 32.6 demonstrates that the fluctuation of frequency has been improved after eliminating the false mode. Thus it is verified that automatic identification method shows a good accuracy of modal parameter identification.

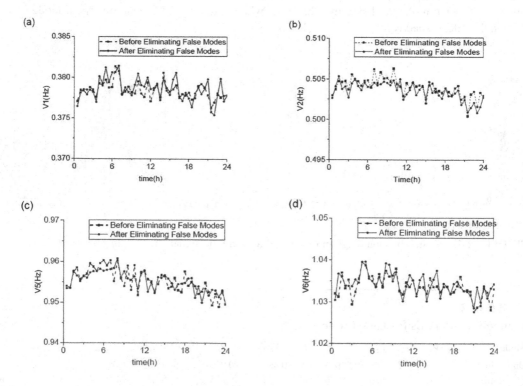

FIGURE 32.6   Modal frequency of (a) V_1 mode; (b) V_2 mode; (c) V_5 mode and (d) V_6 mode.

32.5 CONCLUSION

A method combining reconstruction of system matrix and clustering analysis is proposed to eliminate the false modes, which shows good performance to clear stabilization diagram. After elimination of false modes, the modal parameters can be computed from stable poles classified by the monotonicity of frequency automatically. This automatic identification method is applied on the modal parameter identification of Guanhe Bridge, which is a composite cable-stayed bridge in G15 Expressway in China. The result of modal parameter identification indicates that the frequency variation caused by the environment variability can be precisely and automatically identified based on the proposed method.

REFERENCES

Fan, Jiangling, Zhang Zhiyi, and Hua Hongxing. 2007. Data processing in subspace identification and modal parameter identification of an arch bridge. *Journal of Mechanical Systems and Signal Processing*, 21(4), pp. 1674–1689.

Li, Huibin, Qin Quan, and Qian Liangzhong. 2001. Modal identification of Tsing Ma suspension bridge in time domain. *Journal of Civil Engineering*, 34(5), pp. 52–61.

Peeters, Bart, and De Roeck Guido. 2000. Reference based stochastic subspace identification in civil engineering. *Journal of Inverse Problems in Engineering*, 8(1), pp. 47–74.

Peeters, Bart, and De Roeck Guido. 2001. Stochastic system identification for operational modal analysis: A review. *Journal of Dynamic Systems, Measurement, and Control*, 123(4), pp. 659–667.

Ren, WeiXin, and ZhouHong Zong. 2004. Output-only modal parameter identification of civil engineering structures. *Journal of Structural Engineering and Mechanics*, 17(3–4), pp. 429–444.

Reynders, Edwin, Jeroen Houbrechts, and Guido De Roeck. 2012. Fully automated (operational) modal analysis. *Journal of Mechanical Systems and Signal Processing*, 29, pp. 228–250.

Ubertini, Filippo, Carmelo Gentile, and Annibale Luigi Materazzi. 2013. Automated modal identification in operational conditions and its application to bridges. *Journal of Engineering Structures*, 46, pp. 264–278.

Wu, Chunli, and Liu Hanbing, Wang Jing. 2013. Parameter identification of a bridge structure based on a stabilization diagram with fuzzy clustering method. *Journal of Vibration and Shock*, 32(4), pp. 121–126.

Yang, Kai, Yu Kaiping, Liu Ronghe, et al. 2012. Two new fast identification algorithm of time-varying modal parameters based on subspace tracking. *Journal of Engineering Mechanics*, 29(10), pp. 294–300.

Yu, DanJiang, and Ren WeiXin. 2005. EMD-based stochastic subspace identification of structures from operational vibration measurements. *Journal of Engineering Structures*, 27(12), pp. 1741–1751.

Zhang, Yongxiang, Liu Xin, Chu Zhigang, et al. 2018. Autonomous modal parameter extraction based on stochastic subspace identification. *Journal of Mechanism Engineering*, 54(9), pp. 187–197.

33 Statistical Analysis of Random Dynamic Responses of Bridge under Dense Traffic Flow

Linfeng Hu, Jiayan Lei, Wei Shi, and Zihao Wang
Xiamen University

CONTENTS

33.1 INTRODUCTION

Thanks to the development of computer, communication and sensor technology, a lot of bridges are equipped with the structural health monitoring system to evaluate structural integrity. By placing acceleration, displacement, strain and other sensors at critical sections, the operation status of the structure can be observed in real time through data analysis. In the structural health monitoring system of bridge, the random dynamic response of the bridge is one of the most important data sources, which needs to be further analyzed.

Numerous studies have been conducted to investigate the random dynamic response of the bridge up to now, and two main methods are applied: obtaining the measured data from the actual bridge and simulation analysis based on computer. Lou et al. (2010) found that the random variables of the maximum displacement and acceleration of rails and track slabs, the maximum wheel acceleration and the maximum wheel-rail force all fall into normal distribution by establishing a vehicle-track system. By using the probability density evolution method, Yu et al. (2016) analyzed the vehicle-bridge coupled random vibration accurately and demonstrated that the probability distribution range of vibration response gradually expanded with the increase of vehicle speed, and the probability characteristics basically conform to the Gaussian distribution. Asadollahi et al. (2017) applied the natural excitation technique to perform system identification and analyzed modal properties such as natural frequencies, modal damping and mode shapes of the bridge. Gui et al. (2018) found out that the vehicle speed is the most fact which leads to a peak displacement acceleration at mid-span based on the precise integration method and the Monte-Carlo method. Moreover, with the development of structural dynamic behavior test technology, abundant monitoring data from operational bridges can provide actual information to explore dynamic performance. Many researchers have investigated response problems of bridge through the statistical analysis. Wang et al. (2012a) studied the dynamic response data of the girder under traffic flow and verified that the maximum value data of dynamic displacement response generated by a single vehicle shows the characteristic of tri-modal distribution. Zhao et al. (2018) established the model of probability density of train-induced quasi-static displacement and free vibrating displacement of the Nanjing Dashengguan Bridge, and the analysis shows that, in the long-term operation of the bridge, the quasi-static displacements of each train passing obey the Log-Logistic distribution, and the free vibrating displacements obey the location-scale distribution.

However, in field conditions, when structures are subject to changing environmental conditions and dense traffic flow, the random dynamic responses of bridge should not be consistent with the simulation. In addition, the statistical characteristics of random dynamic responses are usually described by the maximum and minimum value data, there are relatively few researches on the frequency distribution. Therefore, it is necessary to carry out a comprehensive dynamic response measurement under dense traffic flow to understand the distribution characteristics of dynamic response and determine the effect of random load.

In this paper, the dynamic response of the Tianyuan Bridge is measured using smart sensor networks, and with the white Gaussian noise assumption, the dynamic responses is divided into the responses caused by environmental excitation and vehicle excitation, and the distribution of structural dynamic response is determined by statistical tests and distributing fitting. The statistical characteristics of the bridge under the environmental excitation and vehicle excitation are analyzed, and can be a reliable theoretical basis and sample supports to the research of vehicle-bridge coupled vibration.

33.2 MONITORING SYSTEM

33.2.1 DESCRIPTION OF THE TIANYUAN BRIDGE

The tested steel arch bridge is a composite structure of single arch rib and a single cell room box-girder. The panoramic view of the bridge is displayed in Figure 33.1, in which elevation and overall dimension are shown in Figure 33.2, and the cross section of the girder is depicted in Figure 33.3.

FIGURE 33.1 Panoramic view of the Tianyuan Bridge.

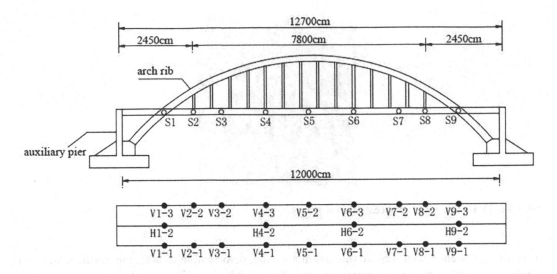

FIGURE 33.2 Overall dimension of the Tianyuan Bridge.

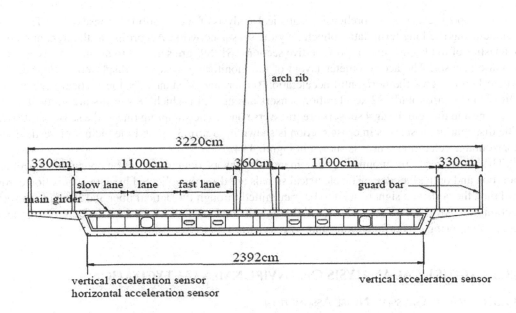

FIGURE 33.3 Cross-section diagram of the Tianyuan Bridge.

The girder, whose main span length is 120 m, is designed with the same shape and size of cross section through the whole span length. The ratio of deck slab width to span is 0.268. Fourteen rigid suspenders are box-shape welded by four pieces of steel plates. All suspenders are hinged with arch rib at the upper end and welded with main girder at the lower end. The arch axis adopts a parabola, and the rib plane is perpendicular to the horizontal plane and passes through the middle of the main beam.

33.2.2 Bridge Monitoring Network

As a huge system containing many software and hardware, the main mission of the structural monitoring system is to obtain the operating environment, dynamic response and other information of the structure, so as to provide support for the safe operation and maintenance. The system

FIGURE 33.4   Picture of an acceleration sensor.

monitoring system include environmental monitoring, deformation monitoring, traffic load monitoring, dynamic response characteristics monitoring and so on. We mainly analyze the part of
the dynamic response characteristics monitoring, which is to analyze the data from acceleration
sensors.

This paper presents a comprehensive statistical analysis of acceleration responses of the Tianyuan
Bridge using the long-term data collected by the sensor network. According to the dynamic characteristics of the bridge, nine representative sections (S1–S9) are selected to arrange the type 941B
vibration sensor. The accelerometer layout of the monitoring system is displayed in Figure 33.2,
where H represents the horizontal acceleration sensor, and V stands for the vertical acceleration
sensor. There are totally 22 acceleration sensors arranged, in which 18 sensors are vertical to the
road line and the remaining 4 sensors are transverse ones. The sampling rate of the sensor is 100 Hz.
The distribution of sensors in cross section is shown in Figure 33.3, and the picture of vertical and
horizontal acceleration sensor is shown in Figure 33.4.

The hardwares of the monitoring system include sensors, data transmission optical fiber, acquisition box and optical modem. The electrical signals are directly collected by data acquisition cards,
and then the collected signals need to be transmitted through the optical fiber and stored in the collection box. All the data are transmitted through the optical fiber network to the control computer
in the monitoring center for storage and processing.

33.3   STATISTICAL ANALYSIS ON ENVIRONMENTAL EXCITATION

33.3.1   White Gaussian Noise Assumption

This paper mainly uses statistical methods to determine the distribution of structural dynamic
responses and analyze the acceleration data of the Tianyuan Bridge. The data studied here are
collected during the month of June 2017. The frequency distribution histograms of the vertical
acceleration data of the No. 1 measuring point are shown in Figure 33.5. It can be seen that each
of the frequency distribution histogram, which does not have obvious statistical features, shows a
particularly large spike, and most of the acceleration values are close to zero. A feat explanation for
this phenomenon is that the environmental excitation causes the bridge to emerge as slight vibration,
so that the acceleration values are small enough and close to zero.

Therefore, it can be assumed that the overall data contain the structural response caused by
environmental excitation and vehicle excitation. To separate the two parts of excitation, the white
Gaussian noise assumption is applied. The environmental excitation technique assumes that the
external excitation of the structure is a random and stationary white noise excitation. Then the
structural response caused by environmental excitation can be separated from the overall response.

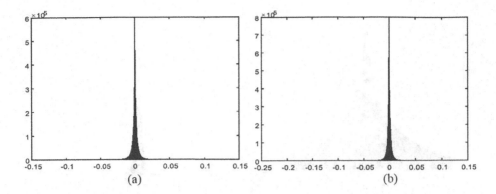

FIGURE 33.5 The frequency distribution histogram of No. 23 measuring point. (a) The frequency distribution histogram in June 1 and (b) in June 5.

To satisfy the white Gaussian noise assumption, it is necessary to prove that the measured dynamic responses of the structure satisfy the standard normal distribution.

33.3.2 KURTOSIS-SKEWNESS TEST

The kurtosis-skewness test is used to study whether the acceleration follows the law of normal distribution. The skewness of distribution is a measure of symmetry, and the kurtosis is a measure of peakedness. The standard normal distribution has a skewness value of 0 and a kurtosis value of 3. Thinking that response of environmental excitation conforms to the Gaussian white noise assumption, there must be an acceleration interval in which the statistical distribution of acceleration obeys the standard normal distribution. According to the theory of kurtosis-skewness test, it can be considered that the absolute value of the acceleration when the kurtosis value of environmental excitation is 3 is the demarcation point between the two parts of responses. In order to test this hypothesis, the original acceleration data is divided into two groups, and the kurtosis-skewness method is used to determine the normality of the acceleration distribution. The acceleration data of measuring point 1-1 is collected, and the boundary acceleration when the kurtosis value is 3 is obtained is shown in Table 33.1.

TABLE 33.1

Absolute Values of Acceleration and Kurtosis and Skewness Values of Environmental Excitation and Vehicle Excitation on Different Dates

Date	Absolute Value of Acceleration (m/s²)	Environmental Excitation		Vehicle Excitation	
		Kurtosis	Skewness	Kurtosis	Skewness
6.1	0.00482	2.9990	0.00084136	3.7896	−0.02928819
6.2	0.00461	3.0006	0.00051152	3.8686	−0.01836988
6.3	0.00343	2.9991	0.00059380	4.6662	−0.02951958
6.4	0.00335	2.9999	0.00120000	4.7527	−0.02029175
6.5	0.00414	2.9994	0.00150000	4.2202	−0.02312724
6.6	0.00421	3.0018	0.00041996	4.6465	−0.01836434
6.7	0.00465	2.9990	0.00160000	4.9940	0.00227647
6.8	0.00420	3.0004	0.00090769	4.4876	−0.01637589
6.9	0.00461	2.9991	0.00057356	3.7047	−0.01169769
6.10	0.00492	2.9994	0.00013533	3.9176	−0.03679605

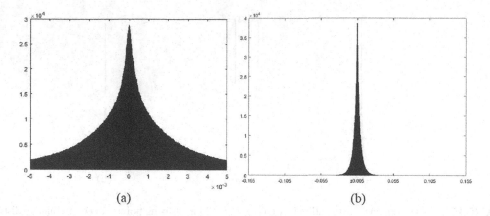

FIGURE 33.6 The frequency distribution histogram of the acceleration of No. 23 measuring point in June 1. (a) Histogram under environmental excitation and (b) histogram under vehicle excitation.

It can be seen from the above table that, although the absolute values of the acceleration of the demarcation point are not stable, most of the acceleration values are in the range of 0.004–0.005 m/s², and the average value of the acceleration obtained is about 0.0045 m/s². Based on this data, we can conclude the statistical characteristics of the acceleration as follows: The skewness values of the acceleration are very small and close to 0, which indicates that the frequency distribution histogram is bilaterally symmetric. The kurtosis value of the distribution of the acceleration under environmental excitation increases with the increase of the absolute value of the acceleration, and the situation of the vehicle excitation is just on the contrary.

In order to observe the distribution characteristics of structural response caused by environmental excitation and vehicle excitation more clearly, the histograms of the structural acceleration frequency distribution caused by environmental excitation and vehicle excitation of 1-1 measuring point are drew by software. The histograms caused by vehicle excitation are adjusted, that is, the acceleration is moved to the middle portion to eliminate the blank portion, which facilitates statistical analysis of the acceleration. Taking June 1 as an example, the distribution histogram is shown in Figure 33.6. It can be seen from the above figures and statistical analysis that the structural vibration acceleration value generated by the environmental excitation is normally distributed.

33.3.3 ANALYSIS SOFTWARE

The kurtosis-skewness test is just a preliminary test of normality. In order to accurately demonstrate that the dynamic response generated by the environmental excitation is in accordance with normal distribution, the results are further verified by statistical analysis software. The results are shown in Table 33.2, which contains various statistical characteristics of the acceleration data. It can be seen from Table 33.2 that the skewness value is 0 and the kurtosis value is 2.824, which is close to 3, so the analysis results are reliable.

33.3.4 QUANTILE-QUANTILE PLOT

Moreover, the dynamic response is also studied by quantile-quantile plot. The quantile-quantile plot is a graphical technique used to check the validity of a distributional assumption for a data set. The basic idea is to compute the theoretically expected value for each data point based on the distribution in question. A 45° reference line is also plotted. If the data indeed follow the assumed distribution, then the points on the q-q plot will fall approximately on this reference line.

TABLE 33.2

Statistical Analysis Results

Parameter	Statistics	Standard Error
Average value	0.0000029539	0.00000226368
Standard deviation	0.00187534058	
Variance	0.000	
Skewness	0.000	0.003
Kurtosis	2.824	0.006
Minimum value	−0.00450000	
Maximum value	0.00450000	
Range	0.00900000	
Interquartile range	−0.0011317000	
Median	0.0000426315	

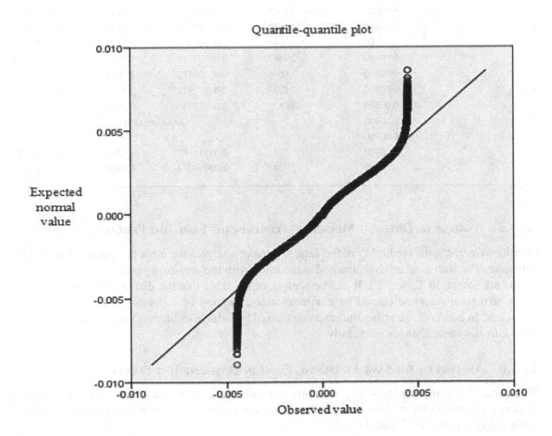

FIGURE 33.7 The q-q plot of the structural response.

As can be seen from Figure 33.7, when the absolute value of the acceleration is within 0.0045 m/s², the data basically accord with the normal distribution, and the reason why the data at both ends show a large difference is that the data of the acceleration value larger than 0.0045 m/s² generated by the environmental excitation are excluded, and are separately placed in another set of acceleration data generated by vehicle excitation. The above series of test results show that the structural response of acceleration caused by environmental excitation is a normal distribution.

TABLE 33.3

The Kurtosis and Skewness Values of the Acceleration Response at Different Measuring Points

Measuring Point	Absolute Value of Acceleration (m/s²)	Environmental Excitation		Vehicle Excitation	
		Kurtosis	Skewness	Kurtosis	Skewness
1-1	0.00400	2.8071	0.00044952	4.4061	−0.03210273
	0.00500	3.0425	0.00090334	3.6787	−0.02873037
1-3	0.00400	2.6734	0.00047082	5.4910	0.06023834
	0.00500	2.8859	0.00044281	4.5468	0.05538536
2-1	0.00400	2.8731	0.00282097	3.7320	−0.03985449
	0.00500	3.1549	0.00227528	3.1213	−0.03432203
2-2	0.00400	Data exception			
	0.00500				
3-1	0.00400	2.7483	0.00080099	3.7292	−0.00422457
	0.00500	3.0097	0.00030339	3.1046	−0.00475572
3-2	0.00400	2.8101	0.00255523	6.4786	0.02754990
	0.00500	3.0969	0.00260460	5.3072	0.02495963
4-1	0.00400	2.7848	0.00123249	3.5400	−0.00562406
	0.00500	3.0666	0.00203783	2.9502	−0.00594366
4-3	0.00400	2.7533	0.00092519	3.8545	−0.02721646
	0.00500	3.0175	0.00137748	3.2051	−0.02523507
5-1	0.00400	Data exception			
	0.00500				
5-2	0.00400	2.4741	0.00511182	11.9384	0.31583120
	0.00500	2.7249	0.00454022	9.8810	0.29583161

33.3.5 ANALYSIS OF DIFFERENT MEASURING POINTS IN THE SAME TIME PERIOD

In order to increase the credibility of the data, this paper analyzes the other key points of the bridge half-span. The statistical analysis method is the same with the measuring point 1-1. The statistical results are shown in Table 33.3. It can be seen from the table that the distribution characteristics of the structural response caused by environmental excitation of different measuring points are consistent. In addition, the structural response caused by vehicle excitation has a great uncertainty, as its kurtosis value changes irregularly.

33.3.6 ANALYSIS OF THE SAME MEASURING POINT IN DIFFERENT TIME PERIODS

This paper also compares the long-term statistical data of the Tianyuan Bridge health monitoring system. As shown in Figure 33.8, the distribution pattern of the acceleration frequency curves of the 1-1 measuring points is almost the same.

33.4 STATISTICAL ANALYSIS ON VEHICLE EXCITATION

33.4.1 SIGNIFICANCE TEST

The measured data is tested by the distribution test. The methods of inference used to support or reject claims based on sample data are known as significance test. The distribution test contains the parameter test and the non-parametric test. The parameter test is a method of inferring the parameters of the overall distribution such as the mean value and the variance when the overall distribution

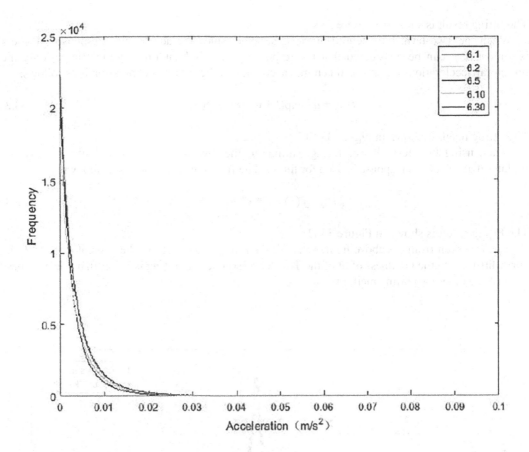

FIGURE 33.8   Frequency curve of the different dates.

form is known. The non-parameter test is a method of estimating the overall distribution pattern using sample data when the overall variance is unknown. In this paper, the Kologorov-Sminov (KS) non-parameter test is used to test the frequency distribution of acceleration, and it is tested whether it conforms to the normal, gamma, exponential, uniform, Poisson and Rayleigh distributions. The conclusion is that the frequency distribution of acceleration does not meet the above basic distribution, so the frequency distribution of acceleration should be a complex distribution.

33.4.2   CURVE FITTING

From the above analysis, it can be seen that the structural acceleration response caused by vehicle excitation is obviously not a simple distribution. In order to find out the distribution rule, a curve fitting method is adopted. Curve fitting is performed based on coordinate points extracted from the histogram. Due to the symmetry of distributions of the data, the Gaussian fitting is adopted on the frequency distribution of the data. Then as the response profile is bilaterally symmetric, the unilateral acceleration response is also taken into account, and Exponential fitting are also applied.

For the measured data of the 1-1 measuring point on June 1, 2017, when using the Three-term Gaussian fitting, the function expression is given as follows:

$$f(x) = a_1 * \exp\left[-\left(\frac{x-b_1}{c_1}\right)^2\right] + a_2 * \exp\left[-\left(\frac{x-b_2}{c_2}\right)^2\right] + a_3 * \exp\left[-\left(\frac{x-b_3}{c_3}\right)^2\right] \qquad (33.1)$$

The fitting result is shown in Figure 33.9.

When the Two-term Exponential fitting is adopted, since the acceleration response skewness is almost 0, it can be considered that the response profile is bilaterally symmetric, so only the unilateral acceleration response is taken into account, and the function expression is as follows:

$$f(x) = a * \exp(b * x) + c * \exp(d * x) \tag{33.2}$$

The fitting result is shown in Figure 33.10.

When using One-term Power fitting, similar to the Two-term Exponential fitting, only the unilateral acceleration response is used for fitting. The function expression is as follows:

$$f(x) = a * x^b + c \tag{33.3}$$

The fitting result is shown in Figure 33.11.

It can be seen from the above figures that all the three fitting methods have great effects on the curve fitting, and the goodness of fit of the Two-term Exponential fitting is better than the goodness of fit of the other two fitting methods.

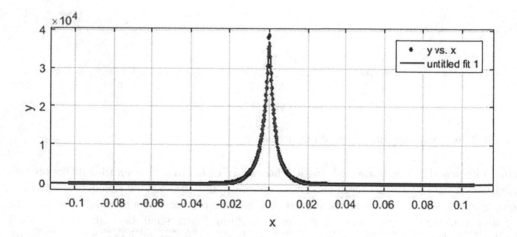

FIGURE 33.9   Three-term Gaussian fitting results.

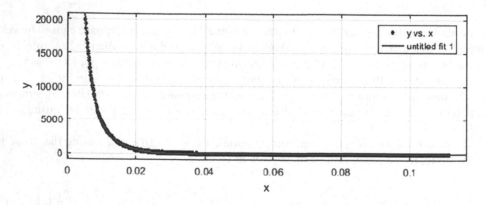

FIGURE 33.10   Two-term Exponential fitting results.

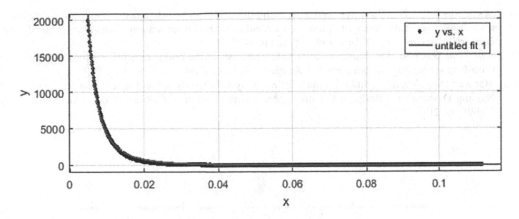

FIGURE 33.11 One-term power fitting results.

33.5 CONCLUSIONS

An extensive data set was measured by 22 smart sensors on the Tianyuan Bridge in 1-month duration, in which the structural dynamic responses were captured for a wide range of operating and environmental conditions. This data set was applied to conduct a comprehensive statistical analysis of random dynamic responses of the bridge under dense traffic flow. Each daily dynamic response record was divided into two parts, which are respectively caused by environmental excitation and vehicle loads based on the white Gaussian noise assumption. Subsequently, in order to study the distribution patterns of these two parts of response, histograms, probability plot and fitting curve were applied to analyze the statistical characteristics of the random dynamic responses. The following conclusions are drawn from this study:

1. Because of the great randomness of the dynamic response under dense traffic flow, the general response can't be described by a simple distribution.
2. By kurtosis-skewness test, quantile-quantile plot and statistical analysis, the response caused by environmental excitation follows the standard normal distribution.
3. The significance test result certifies that the distribution of the response caused by vehicle loads is not coincident with any simple random distribution pattern. It is shown that the Two-term exponential fitting can fit the curve exactly.
4. The analysis in this paper can provide a method for processing large amounts of data in bridge monitoring system.

REFERENCES

Asadollahi, Parisa and Li, Jian. 2017. Statistical analysis of modal properties of a cable-stayed bridge through long-term wireless structural health monitoring. *Journal of Bridge Engineering*, 22(9), 04017051.

Li, Zhijun; Feng, Maria Q.; Luo, Longxi; Feng, Dongming; and Xu, Xiuli. 2018. Statistical analysis of modal parameters of a suspension bridge based on Bayesian spectral density approach and SHM data. *Mechanical Systems and Signal Processing*, 98, pp. 352–367.

Lou, Ping and Zhao, Yongchao. 2010. Track dynamic response probability analysis for random vibration of high-speed railway. *Journal of Zhengzhou University (Engineering Science)*, 31(02), pp. 56–59.

Wang, Tao; Han, Wanshui; and Huang, Pingming. 2012a. Statisics analysis on dynamic response of expressway bridge. *Journal of Taiyuan University of Technology*, 43(04), pp. 499–501+510.

Wang, Tao; Han, Wanshui; and Huang, Pingming. 2012b. Statistic analysis for dynamic response of an expressway bridge under traffic loading. *Journal of Vibration and Shock*, 31(20), pp. 116–120+125.

Yu, Zhiwu; Mao, Jianfeng; Tan, Sui; and Zeng, Zhiping. 2015. Probability density evolution analysis of track-bridge verticalcoupled vibration with irregularity random excitation. *Journal of Central South University (Science and Technology)*, 46(04), pp. 1420–1427.

Yu, Zhiwu; Tan, Sui; and Mao, Jianfeng. 2016. Safety analysis of the bridge dynamic performance based on train-bridge coupling random vibration. *Journal of Railway Engineering Society*, 33(09), pp. 55–61+112.

Zhao, Hanwei; Ding, Youliang; and Li, Aiqun. 2018. Analysis of the dynamic displacement responses of the Nanjing Dashengguan Bridge under high-speed trains. *Journal of Railway Science and Engineering*, 15(09), pp. 2187–2195.

34 Influence of Joint Reinforcement Anchorage Detailing on the Seismic Performance of Double-Column Bridge Piers

Liyuan Wang
Fuzhou University

Chaoyang Tang
Fuzhou University
Fujian Highway Administrative Bureau

Yongjian Chen, Siyang Ma, and Tianyu Zheng
Fuzhou University

CONTENTS

34.1 INTRODUCTION

Bridge piers are a basic component in bridge structures, and the joint between the cap and cap beam is a key part of the lower structure.[1] To improve the integrity of the connection between the bridge pier and cap beam or pile cap, the length of longitudinal reinforcement of the pier extending into the cap beam or pile cap should exceed the minimum anchorage length.[2] However, when the longitudinal reinforcement of a pier is anchored directly into the pile cap or cap beam (direct anchorage),[3] a greater anchorage length is needed; in particular, for large-diameter reinforcement bars or high-strength reinforcement bars, which are widely used in China, the anchorage length often exceeds the size of the component, and the height of the cap beam or pile cap must be increased, resulting in the waste of material.[4] Therefore, the anchorage method of bending or hooking the end of the reinforcing bar is generally used to enhance the anchorage performance of the reinforcing bar and to reduce the anchorage length. However, the structure of the end hook or bend introduces the drawbacks of rebar density issues, construction difficulties, and splitting failure.[5] The development of headed reinforcement anchorage has provided a feasible way to resolve the above problems.[6]

A reinforced bar with headed reinforcement is composed of two parts: the reinforcement bar and the anchorage plate at the head of the steel bar. The reinforced bar with an anchor plate has the following advantages: no bending, the absence of steel bar crowding, ease in production standardization, convenient installation, reduced labor costs, and construction quality assurance.[7] Therefore, reinforcement with an anchorage plate has attracted the interest of scholars and experts at home and abroad, and its anchorage performance has been affirmed. Li Zhibin[6] performed pullout tests on 115 headed reinforcement specimens, demonstrating that an anchorage plate can reduce the anchorage length of the reinforcing bars and achieve reliable anchorage for reinforcing bars. Zheng Wenzhong[8] conducted pullout tests on 120 reinforced concrete specimens with welded joints. The authors analyzed the influence of different parameters on the distribution of the yield load on the steel bar at the loading end for the bearing pressure of the anchorage plate and the bonding force of the straight anchorage section of the steel bar. They also established a formula to describe the distribution law. Hwang[9] studied the length of anchorage plate reinforcements based on a nonuniform bond stress distribution for 361 test specimens with a continuous column through (CCT) joint, lap joint, or beam–column joint. The experimental and numerical simulation results were compared under different anchorage conditions for steel bars with anchorage plates, with good agreement between the numerical simulation and experimental results. Bui, Sentosa et al. and Kim et al.[10,12] studied the performance of a headed bar for external beam–column joints. Sim and Chun[11] reported that anchorage plate reinforcement is also suitable for other concretes, such as super-high-performance concrete and steel fiber-reinforced concrete. Jean and Ján Bujňák[13,14] reported that the use of anchorage plate reinforcement in precast structures contributes to the stress condition of the joints in prefabricated structures (Figures 34.1 and 34.2).

In conclusion, the anchorage performance of headed reinforcements has been studied and affirmed. However, research on the application of steel bars with anchorage plates to bridge piers is limited, and the seismic performance of piers with reinforced bars with headed reinforcement

FIGURE 34.1   Anchor plate.

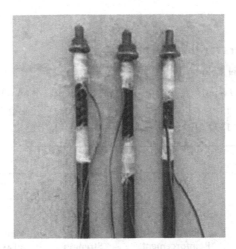

FIGURE 34.2 Reinforcement with the anchorage plate.

remains unclear. Therefore, in this work, components anchored by headed reinforcement anchorage and conventional hooked bar anchorage are subjected to quasi-static tests. From unidirectional quasi-static tests, the energy dissipation capacity, bearing capacity, displacement ductility, and structural stiffness of pier models with different structures are compared and analyzed. The seismic performance of the pier with an anchored plate is studied, as well as the anti-seismic performance. The application and widespread use of headed bars in bridge engineering and the improved seismic design of reinforced concrete piers with anchor reinforcement have both theoretical significance and practical value.

34.2 STRUCTURAL DESIGN

34.2.1 MODEL STRUCTURE

Taking the pier of a practical project as a prototype, two double-column reinforced concrete pier models were established: one is anchored by conventional hooked bar anchorage (RCP1), and the other is anchored by headed reinforcement anchorage (RCP2). Considering the space limitations of the test site, the similarity ratio between the test site and actual project was 1:5. The diameter of the pier is 250 mm, and the strength grade of the concrete is C30. According to the code for highway design[15] for RCP1 members using bending hook anchorage, the anchorage length of the pier column longitudinal bar extending into the cap beam and cap is $0.8l_a$, whereas according to the code for technical specifications,[16] the RCP2 member extending into the cap beam and cap is $0.4l_a$, with

$$l_a = \frac{f_y d}{4\tau_u}$$

(34.1)

where τ_u = the ultimate average bond stress (MPa), d = diameter of the reinforcing bar (mm), f_y = yield strength of the steel bar (MPa), A = cross-sectional area of the reinforcing bar (mm^2), and l_a = anchorage length (mm).

Therefore, the anchorage length of the RCP1 pier column longitudinal bars extending into the cap beams and caps is 330 mm, and that of the RCP2 member extending into the cap beams and caps is 165 mm. The basic parameters of each model are given in Tables 34.1 and 34.2 (Figures 34.3 and 34.4).

TABLE 34.1

Reinforcement Parameters for the Specimens

Specimen Number	Pier Column			Cap			Coping		
	d (mm)	H (mm)	A_s (mm²)	b (mm)	H (mm)	A_s (mm²)	B (mm)	h (mm)	A_s (mm²)
RCP1	250	1350	$8\varphi14$	720	400	$28\varphi12$	400	350	$20\varphi12$
RCP2	250	1350	$8\varphi14$	720	400	$28\varphi12$	400	350	$20\varphi12$

TABLE 34.2

Specimen Parameters

Specimen Number	Axial Compression Ratio	Longitudinal Reinforcement Strength (N/mm²)	Stirrup Strength (N/mm²)	Spacing of Stirrups between Piers and Columns (mm²)	Anchorage Length (mm)
RCP1	0.1	400	275	80 (enclosed area is 50)	330
RCP2	0.1	400	275	80 (enclosed area is 50)	165

FIGURE 34.3   Pile-beam joint diagram for the RPC1 model.

FIGURE 34.4   Pile-beam joint diagram for the RPC2 model.

34.2.2 TEST EQUIPMENT AND LOADING SYSTEM

The concrete loading scheme of this test is as follows. During the initial loading stage, displacement is applied at 2, 4, 6, and 8 mm, with increments of 2 mm. Based on data from the strain gauge of the steel bar in the pier and cap sections, when the steel bar yields, the displacement Δy at the time of yielding is multiplied by $1\Delta y$, $2\Delta y$, $3\Delta y$, ... Cyclic loading is performed three times for each stage, and the load is maintained for 1 minute before the loading direction is changed. The experimental loading scheme is shown in Figure 34.5a. The MTS electro-hydraulic servo loading system of Fuzhou University is used to load the horizontal force. As shown in Figure 34.5b, the test device adopts displacement control. When the bearing capacity of the specimen decreases to <85% of the maximum value, the test is terminated, and the member is considered to be damaged.

34.3 EXPERIMENTAL PHENOMENA AND ANALYSIS

34.3.1 EXPERIMENTAL PHENOMENA

Under quasi-static low-cycle repeated loading, when the test displacement load reaches 4 mm, small horizontal cracks appear at the bottom of the RCP1 and RCP2 model piers. As the load increases, the horizontal cracks gradually increase in number. When the displacement load reaches $3\Delta y$, the concrete on the surface of the pier columns of the two components falls. When the displacement load reaches $6\Delta y$, the concrete on the surface of the pier columns of the two components has a large area. The area then decreases, and the components are damaged. The failure phenomenon of the pier model anchored by traditional stirrups is similar to that of the pier model anchored by anchorage plates. By comparing Figure 34.6a and b, it can be seen that the damage to the concrete at the top of the RCP1 specimen is greater than that of the RCP2 specimen. Moreover, by comparing Figure 34.6c and d, it can be seen that the damage to the concrete at the bottom of the RCP1 specimen is also greater than that of the RCP2 specimen due to the large bearing surface of the anchorage plate, which can restrain the bond and slip phenomenon of the steel bar and concrete. Therefore, more concrete shedding and fragmentation occur for the pier anchored by a bending hook at the end of a steel bar under a load of $6\Delta y$ compared with that of the pier anchored by a steel bar with an anchorage plate.

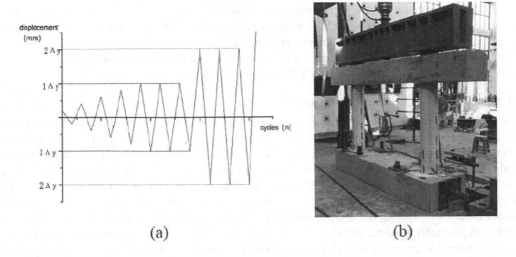

(a)   (b)

FIGURE 34.5 Test loading scheme and field layout. (a) Test loading scheme and (b) site equipment layout.

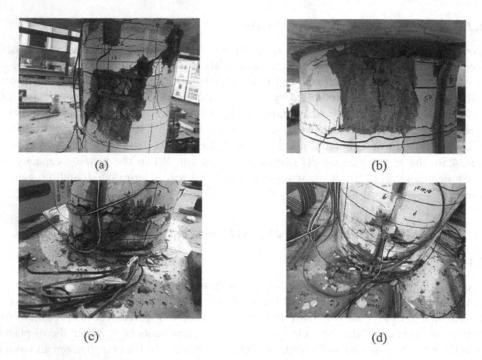

FIGURE 34.6 Structural failure under a load of 6Δy. (a) Concrete at the top the RCP1 specimen column, (b) concrete at the top the RCP2 specimen column, (c) concrete at the bottom the RCP1 specimen column, (d) concrete at the bottom of the RCP2 specimen column.

34.3.2 TEST RESULTS AND ANALYSIS

34.3.2.1 Hysteretic Curve and Skeleton Curve

The hysteresis curves of the two models are compared in Figure 34.7. It can be seen from the figure that the hysteresis loops of both the two models have an inverse S-shape and that the enclosed area is <10%. This result indicates that the mechanical properties of the RCP1 and RCP2 models are similar. Based on the model skeleton curve shown in Figure 34.8, the forward ultimate bearing capacity of the RCP2 members is 7.6% higher than that of the RCP1 specimens, and the reverse bearing capacity of the RCP2 members is 15.6% higher than that of the RPC1 members; however, the ultimate displacements of the RCP2 members are similar. Based on an analysis of the test results, the yield load, yield displacement, ultimate load, failure load, ultimate displacement, and displacement ductility coefficient of the RCP1 and RCP2 members are compared in Table 34.3.

34.3.2.2 Energy Consumption Capacity

The seismic performance of a structure primarily depends on the energy dissipation capacity of its components. The area enclosed by the load–displacement $(P - \Delta)$ curve in the hysteretic curve can reflect the energy absorbed by the structure, whereas the area enclosed by the unloading and loading curves gives the dissipated energy. Under repeated loading, the structure transforms energy into heat energy, which is dissipated into space through internal friction or local damage, such as cracking and plastic hinge rotation. Therefore, the area of the hysteretic loop in the hysteretic curve is important for evaluating the energy consumption of a structure. The equivalent viscous damping ratio ξ_e is generally used to quantitatively evaluate the energy dissipation capacity of a structure or component. A larger equivalent viscous damping ratio ξ_e corresponds to a greater hysteretic energy dissipation capacity and better seismic performance. The equivalent viscous damping ratio

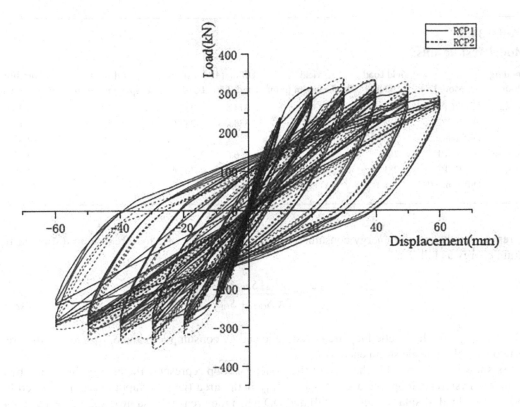

FIGURE 34.7 Hysteretic curves of the RCP1 and RCP2 models.

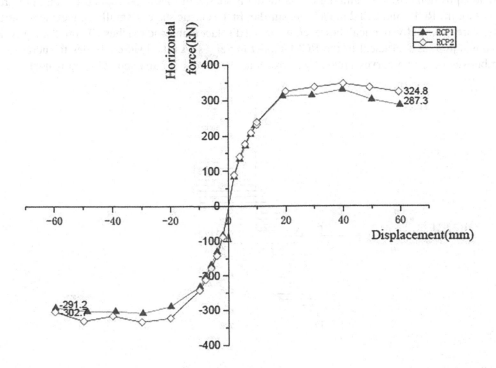

FIGURE 34.8 Skeleton curves of the RCP1 and RCP2 models.

TABLE 34.3
Model Test Results

Loading Mode	Model	Yield Load (kN)	Yield Displacement (mm)	Ultimate Load (kN)	Failure Load (kN)	Ultimate Displacement (mm)	Ductility Coefficient
Forward	RCP1	277	13.7	314.5	267.3	59	4.34
	RCP2	283	14.3	338.4	287.64	59	4.13
	Difference (%)	2.1	4.4	7.1	7.6	0	−4.8
Reverse	RCP1	260	13.1	306.5	260.5	60	4.58
	RCP2	281	14.6	354.6	301.41	58	3.98
	Difference (%)	8.1	11.5	15.7	15.7	3	−13.1

ξ_e represents the ratio of energy consumed by viscous damping of a hysteretic loop to the elastic strain energy as follows:

$$\xi_e = \frac{1}{2\pi} \frac{S_{ABCD}}{S_{OFD} + S_{OBE}} \tag{34.2}$$

where S_{ABCD} = the hysteretic loop area (hysteretic energy consumption) and $S_{OFD} + S_{OBE}$ = the area of two triangles (elastic strain energy).

As shown in Figure 34.9, the area of the hysteresis loop represents the energy dissipated by a complete hysteresis loop; the area of $S_{OFD} + S_{OBE}$ is the area (i.e., the input energy) enclosed by the straight load–displacement lines OB and OD when they reach the same displacement OE and OF. Here a larger ξ_e corresponds to a larger hysteresis loop and a greater hysteresis energy dissipation capacity.

The equivalent viscous damping of the model is shown in Figure 34.10, and the damping ratio curves for the RCP1 and RCP2 models are similar. In the elastic stage of small displacement, the two curves are essentially identical; however, when the displacement load reaches 10 mm, the equivalent viscous damping coefficient of the RCP1 model increases. As the load continues to increase, the gap between the two curves gradually narrows as the model is damaged. The equivalent viscous

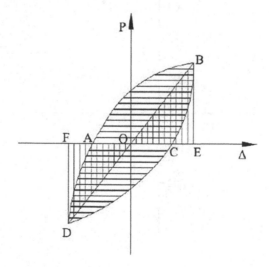

FIGURE 34.9 Equipment viscous damping ratio.

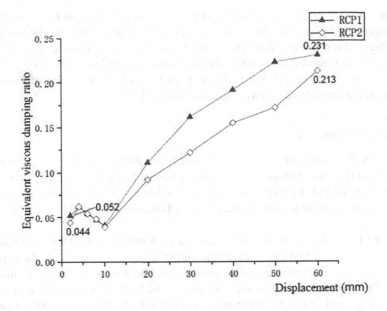

FIGURE 34.10   Equipment viscous damping ratio for the RCP1 and RCP2 models.

damping coefficients for the RCP1 and RCP2 models are 0.231 and 0.213, respectively, and the difference in the initial equivalent viscous damping coefficients of the two models is ~18%. As the displacement load increases, the equivalent viscous damping coefficients for the RCP1 and RCP2 models are generally close, with a difference of <8%.

34.3.2.3 Stiffness Degradation

The stiffness degradation curve of the model is shown in Figure 34.11, which indicates that the stiffness of the component is greatest during the initial stage. As the displacement load increases, new and old concrete cracks expand, and the stiffness rapidly decreases. During the later stages,

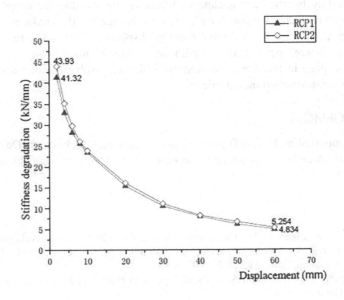

FIGURE 34.11   Stiffness degradation curve.

the reinforcing steel bar enters the strengthening stage, and the stiffness degradation tends to be gentler. The initial stiffnesses of the RCP1 and RCP2 models differ by 6%. As the displacement load increases further and the member reaches the yield state, the stiffness difference between the two models is only 1%. In the subsequent stage, the stiffness degradation tends to be gentle. Thus, the stiffness curves of the RCP1 and RCP2 models are similar, indicating similar stiffness degradation behavior for piers anchored by anchorage plates or hooked bars.

34.4 CONCLUSION

In this paper, an MTS servo loading system was applied to perform repeated low-cycle loading tests on reinforced concrete double-column piers with two anchorage forms (anchored by an anchorage plate or a bending hook at the end of the reinforcing bar). Based on hysteresis and ultimate state results, the anchorage performance of the anchorage plate was studied, with the following conclusions:

1. The failure phenomenon of the pier anchored by a steel bar with an anchorage plate is similar to those for a bending hook at the end of a steel bar. Based on the experimental stress process and ultimate state of the two models, there is no anchorage failure between the steel bar and concrete. Moreover, because of the large bearing surface of the anchorage plate, the bond and slip phenomenon between the steel bar and concrete can be well restrained. Therefore, for a load of $6\Delta y$, more concrete shedding and fragmentation occur for the pier anchored by a bending hook at the end of a steel bar than for the pier anchored by a steel bar with an anchorage plate.
2. The bearing capacity of the pier anchored by an anchorage plate is slightly higher than that for a hook on the steel bar. Both models exhibited the same ultimate displacement. The forward and reverse ultimate bearing capacities of the former are 7.6% and 15.6% higher than those of the latter, respectively.
3. The seismic performance of the pier anchored by a steel bar with an anchorage plate is similar to that of the pier anchored by a bending hook at the end of a steel bar. The equivalent viscous damping coefficient of the former is 8% lower than that of the latter, and the ductility coefficients and stiffness degradation curves are essentially identical, with the results of the former model being 6% lower than those of the latter model. The ductility of the double-column pier decreases slightly, but the bearing capacity increases slightly after the adoption of steel bar anchorage with an anchorage slab, for which the anchorage length is reduced to $0.4l_a$.
4. The reinforced bar with headed reinforcement displays reliable anchorage performance, with good anchorage performance for an anchorage length of $0.4l_a$. Thus, the use of an anchorage plate in the pier joint area can effectively solve the problems of steel bar crowding and construction inconvenience.

ACKNOWLEDGEMENT

This paper was supported by Fujian Transportation Science and Technology Development Project [grant numbers 201415]. It is also financially supported by the Department of Education Fund [grant numbers JT180046].

REFERENCES

1. Ge Jiping, Wang Zhiqiang, Li Jianzhong, Peng Dawen, Chen Xu. Recent development in seismic performance of prestressed concrete precast segmental double-column piers. *Earthquake Engineering and Engineering Dynamics*, 2013, 33(3):192–198.
2. Liu Feng. Pseudo Static Tests and Analysis of Segmental Prestressed Reinforcement Concrete Bridge Piers. Tongji University, 2008.

3. Rong Xian, Zhang Jianxin, Li Yanyan. Experimental study and analysis on aseismic performance of high strength reinforcement concrete bridge piers. *Engineering Mechanics*, 2015, 32(10):99–105.

4. Wu Hai-Jun, Chen Ai-Rong. Study of durability design method for bridge structures. *China Journal of Highway and Transport*, 2004, 17(3):57–61,67.

5. Liao Qinglong, Xu Yuxi. Details and design of reinforced steel bars with corner reinforcement and head type end anchor for concrete aseismic structures. *Structural Engineering*, 2000, 15(2):57–69.

6. Li Zhibin. Experiment Study in Anchorage Behavior of Headed Reinforcement. Tianjin University, 2007.

7. Wu Guangbin, Li Zhibin, Liu Yongyi. Experimental research on mechanical anchorage property of headed reinforcement. *Building Science*, 2007, 23(4):38–40.

8. Zheng Wenzhong, Miao Tianming, Wang Shiyu, Wang Ying. A study on anchoring behavior between high strength hot-rolled reinforcement with headed bars and concrete. *Journal of Building Structures*, 2018, 39(4):119–129.

9. Hyeon-Jong Hwang, Hong-Gun Park, Wei-Jian Yi. Development length of headed bar based on nonuniform bond stress distribution. *ACI Structural Journal*, 2019, 116(2):29–40.

10. Minh-Tung Tran Bui, Quoc-Bao Bui, Bastian Sentosa, Nhat-Tien Nguyen, Trung-Hieu Duong, Olivier Ple. Sustainable RC beam-column connections with headed bars: A formula for shear strength evaluation. *Structures*, 2018, 10(2):401.

11. Hye-Jung Sim, Sung-Chui Chun. A reevaluation of anchorage strength of headed bars in exterior beam-column joints. *Journal of the Korea Concrete Institute*, 2018, 30(2):207–216.

12. Seunghun Kim, Lee Chang Yong, Yongtaeg Lee. Development strength of headed reinforcing bars for steel fiber reinforced concrete by pullout test. *Architectural Research*, 2018, 20(4):129–135.

13. Jean Paul Vella, Robert L. Vollum, Raj Kotecha. Headed bar connections between precast concrete elements: Design recommendations and practical applications. *Structures*, 2018, 15:162–173.

14. Ján Bujňák, Matúš Farbák. Tests of short headed bars with anchor reinforcement used in beam-to-column joints. *ACI Structural Journal*, 2018, 115(1), 203–210.

15. Ministry of Transport of the People's Republic of China. *Code for Design of Highway Reinforced Concrete and Prestressed Concrete Bridges and Culverts (JTG D60–2004)*. Beijing: Ministry of Transport of the People's Republic of China, 2004.

16. China Academy of Building Research. *Technical Specification for Application of Headed Bars (JGJ256—2011)*. Beijing: China Architecture & Building Press, 2011.

35 Pseudo Damage Training for Seismic Fracture Detection Machine

Luyao Wang and Ji Dang
Saitama University

Xin Wang
Tokyo University of Science

CONTENTS

35.1 INTRODUCTION

The Great Hanshin Earthquake, which occurred on January 17, 1995 in the southern part of Hyogo Prefecture of Japan, caused tremendous damage that 104,004 building structures were completely destroyed and 136,952 building structures were partially destroyed. Although the collapses were avoided for most of the building structures which were designed and constructed in accordance with the new seismic code of 1981, it was confirmed for the first time in Japan that the beam-end fracture occurred at beam-column connection of steel frame building structures. It is worth noting that, in damaged buildings structures, not only the post-earthquake tilt (the interstory drift ratio) was small, but the damage of the exterior materials was also slight, which indicates that, the absence of outward signs of damage has made it difficult to determine which buildings should be inspected to see if the beam-end fractures have occurred (Kobe University 2014, Koji Uetani et al. 1996).

From the seismic safety perspective, the building structure has reached the limit state of serviceability when all of the connections at one floor fractured (Stephen A. Mahin, 1998). This limit state starts progressing from the cracks and fractures occurring in some beam ends, especially the lower flange of the beams (Kyoto University, 2014). These kinds of cracks and fractures deteriorate the rigidity and strength of the structural members, finally leading to the collapse of whole building structures. Therefore, the quick detection of the beam-end fracture for building structures after earthquake becomes an important subject. However, it is difficult to recognize this kind of local damage by visual inspection (Kobe University, 2014).

In recent decades, the Convolutional Neural Networks (CNNs) have been proved as one of the effective methods for feature extraction from 2D image data (J. Dang et al. 2018, Young-Jin Cha et al. 2017, M. Matsuoka et al. 2018, Ceena Modarres et al. 2018). For the steel frame building

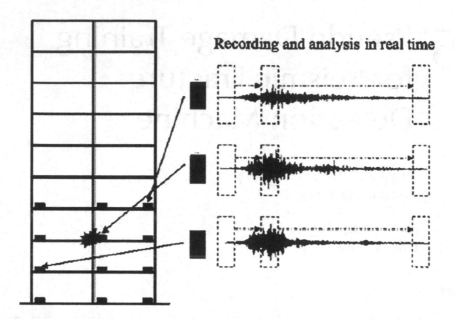

FIGURE 35.1   Beam-end fracture detection.

structure, the beam-end fractures can be recognized directly from the pulse of the waveforms which are generated by the shock due to the fractures, whereas acceleration waveform remains smooth at the elastic stage. Therefore, the beam-end fracture detection becomes possible by neural network learning of a large amount of fracture and nonfracture data.

This study presents a beam-end fracture detection method based on acceleration waveform using CNN model. The CNN model is first trained by a great amount of pseudo acceleration waveforms and will then be implemented in a smart device which have already been installed in the measurement application program developed by J. Dang et al. (2018). The smart device can be deployed at important beam-column joints of a building structure and measure data independently in real time. Also, with the real-time processing of the recorded data by the CNN models, faster diagnosis of the building structure after earthquakes can be fulfilled (Figure 35.1).

35.1.1   SHAKE-TABLE TESTS OF STEEL STRUCTURES

Two shake-table tests of moment-frame steel building structures which were completed by E-Defense was used to observe the feature of beam-end fracture which is shown in acceleration waveform. One is a 1/3-scale model of an 18-story building structure (hereinafter referred to as "high-rise building") was tested until collapse on December 9–11, 2013. The panoramic view of specimen is shown in Figure 35.2a. The other one is a full-scale model of a three-story building structure (hereinafter referred to as "multistory building") was tested on October 8–15, 2013, and the panoramic view of the specimen is shown in Figure 35.2b.

The high-rise building has three spans of 2.0 m each in the longitudinal direction and one span of 5.0 m in the transverse direction. A synthesized motion accounting for the simultaneous ruptures of the Tokai, Tonankai, and Nankai troughs was employed as the seismic input motion, which was applied only in the longitudinal direction of the building. The first beam-end fracture occurred at the second floor when the excitation magnitude was increased to 220 cm/s. Then the fracture at beam ends propagated to the upper floors. Finally, fracture throughout the beam ends in the lower five stories when the specimen collapsed. The multistory building has two spans of 6.0 m each in the longitudinal direction and two spans of 5.0 m in the transverse direction.

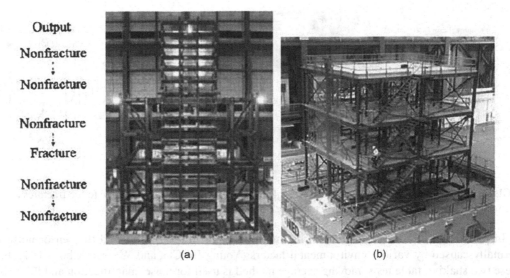

FIGURE 35.2 Panoramic view of two shaking table tests. (a) High-rise building and (b) multistory building.

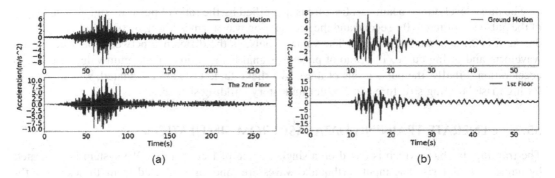

FIGURE 35.3 Ground motions of two steel frame shaking table tests. (a) High-rise building and (b) multistory building.

The north-south (NS) component of Takatori ground motion was recorded from 1995 Great Hanshin earthquake, and Nankai troughs were employed as seismic input motions with various scale-up magnitudes. Three beam-end fractures occurred when the scale factor of Takatori ground motion increased to 100%. After that, new beam-end cracks occurred and the width of the cracks increased; however, there was no new fracture until all loading was completed. The acceleration response waveforms which correspond to the first beam-end fracture occurrence of these two shake-table tests shown in Figure 35.3 are used for feature analysis and verification of the CNN model. The seismic input motions when the first beam-end fracture occurred are shown in Figure 35.3.

35.1.2 FEATURES OF WAVEFORMS GENERATED BY BEAM-END FRACTURE

The close-up views of the first beam-end fracture and the acceleration waveform which includes the fracture timepoint are shown in Figure 35.4. Similarly, the acceleration in the time domain also show appreciable responses for the beam-end fractures. A pulse generated due to the beam-end fracture can be perceived close by the black rectangle, which is a distinctive feature shown in this waveform.

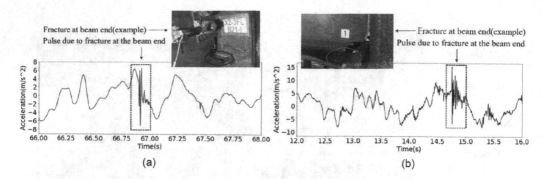

FIGURE 35.4 Pulse (closed by the black rectangle) in the acceleration waveform due to the fractures at the beam ends. (a) High-rise building and (b) multistory building.

In practical measurement or monitoring, noise is an inevitable factor, and the sensor noise is normally caused by various environmental factors (Young-Jin Cha and Wooram Choi, 2017). For these two shaking table tests, moving average method is used for noise quantification and the noise distributions are also plotted. Both show the normal distributions and the 95-percentile noise level of the multistory building is $0.20\,\text{m/s}^2$, which is significantly greater than the high-rise building structure ($0.06\,\text{m/s}^2$).

The pulse level due to beam-end fracture is quantified by the ratio of pulse-to-signal. The length of the pulse is limited to 0.1 second and the acceleration waveform is fitting by two nonlinear curves during this 0.1 second, respectively. The value of pulse is the difference between the acceleration waveform and fitting curve. The ratio of pulse-to-signal is the ratio of maximum value of pulse to acceleration. Finally, the pulse levels of these two shake-table tests are calculated, which are 1.45 for high-rise building structure and 2.52 for multistory building structure.

35.2   GENERATE TRAINING DATABASE FROM SIMULATION

The training database which is based on a single degree of freedom (SDOF) system is generated by numerical analysis. The input earthquake waves are randomly selected from 18 design earthquake waves, whose magnitudes are changed by multiplying a scale factor ranging from 0.1 to 1.5; To randomly generate the SDOF system, the period of model is randomly selected from the range of 0.3–2.0 s, and then the other properties of this model such as frequency, natural frequency, and stiffness can be calculated based on this period. For the nonlinear parameters, the yield force (fy) is randomly generated from 0.2 to 1.2 mg, and then yield displacement (dy) can be calculated by the equation of $dy = fy$/stiffness; For each loop to calculate the acceleration response, the incremental method and Predictor-Corrector method with bilinear model update are used. Finally, 20,000 response acceleration waveforms were generated.

The noise was added to the whole range of each response acceleration waveform generated by previous numerical analysis. Noise level was determined by giving a uniform distribution $[-0.25, 0.25]\,\text{m/s}^2$. Then 10,000 of the data after adding noise were labeled as "Nonfracture" data.

The pulse level was controlled by the value of pulse-to-signal ratio. Since it is considered that the classification result is sensitive to the pulse level that low pulse-to-signal ratio leads to difficult to identify the feature of beam-end fracture, whereas high ratio leads to easy to be identified. In this study, the pulse-to-signal ratio was controlled smaller than 1.5, which means the maximum of the pulse would never be larger than the 1.5 times of the maximum of acceleration. Since the beam-end fracture happens in an instant, in this study, the length of the pulse was controlled to 0.1 second and randomly generated in any location within the range of waveform.

The pulse was added to the rest of the 10,000 data which have been added by the noise and were then labeled by "Fracture" data.

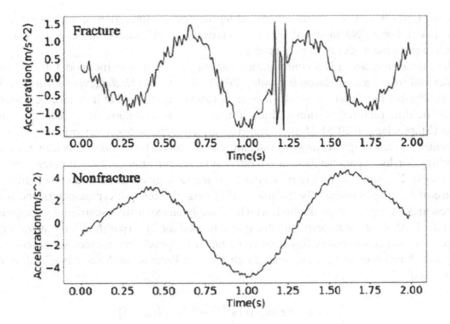

FIGURE 35.5 Examples of training data.

Finally, 10,000 "Nonfracture" training data and 10,000 "Fracture" training data were generated. Each of the data is limited to 2 second with a sampling rate of 0.01, and the peak is the middle of the data length. The examples of "Nonfracture" data and "Fracture" data are shown in Figure 35.5.

35.3 CONVOLUTIONAL NEURAL NETWORK

Figure 35.6 presents the CNN architecture used in this study. As illustrated in this figure, the first layer is the input layer of acceleration response in the time domain. The most common form of a CNN architecture stacks a few Convolution-Rectified Linear Unit (ReLU) layers, follows them with pooling layers, and repeats this pattern until the input data have been merged spatially to a small size. Based on this principle, in this study, the input data pass through three same substructures (1D Convolution-ReLu + Max pooling + Dropout) and are flattened into a $1 - D$ vector. The vector, including 3,072 elements, is fed into two fully connected layers. Finally, the softmax layer predicts whether each input data is a fracture or nonfracture data. There is a total of 408,322 parameters and 12 hidden layers in this deep CNN.

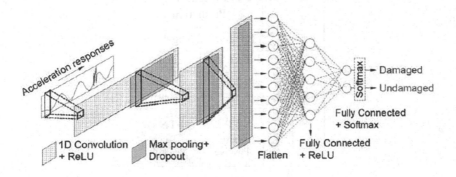

FIGURE 35.6 A CNN architecture.

Training CNN model is the process of finding the best configuration of variables for a neural network model. For a CNN model, there are two types of configuration variables that contribute in a different way to the model's learning process.

Model Parameter, which is a configuration variable, is internal to the model and whose value can be derived from training (Jason Brownlee, 2017). By contrast, *Model Hyperparameter* is a configuration whose value is set before the learning process begins so that it cannot be derived from training. The time required to train and test a model can depend upon the choice of its hyperparameters (Marc Claesen, 2015). Hyperparameters are conditional upon the value or model design components, e.g., learning rate, dropout rate, number of hidden layers, and activation function.

Searching for hyperparameters is an iterative process constrained by computer configuration and time cost. Engineer wants the best model for the task given the available resources, so the procedure of training is essentially the trade-off among different hyperparameters and time cost. In this research, the grid search method, which is simply an exhaustive searching through a manually specified subset of the hyperparameter space, is used for hyperparameter optimization. There are three other hyperparameters that need to be tuned for good performance on unseen data: the learning rate, batch size, and epoch. Grid search selects a finite set of "reasonable" values for each hyperparameter:

$$\left\{ \begin{array}{l} \text{Learning rate}(lr) \in \{1e-3, 1e-4, 1e-5\} \\ \text{Batch size}(bs) \in \{8, 16, 32\} \\ \text{Epoch}(ep) \in \{10, 20, 30\} \end{array} \right.$$

which then trains a CNN model with each pair (lr, bs, ep) in the Cartesian product of these three sets and evaluates their performance on a validation set. Finally, the grid search algorithm outputs the settings that achieved the highest score in the validation procedure, which is {learning rate: $1e-05$, batch size: 16, epochs: 20} with 0.99994 accuracy.

Since the CNN model is trained by the waveforms generated by the numerical simulation, it is indispensable to verify that the feasibility of the trained CNN can also work on the real waveform data. Eighty real acceleration waveforms are collected from two steel frame shake-table tests of E-defense open database which has been introduced in Chapter 2. Then the real data are fed to the trained CNN model to obtain the predicted class (fracture of nonfracture).

The predicted class are compared with true class to get the accuracy which is given by the confusion matrix which is shown in Figure 35.7. TP (True Positive) holds if a "positive example" is given to the CNN and correctly predicted that it is a "positive example," FN (False Negative) holds if a "positive example" is falsely predicted as a "negative example," FP (False Positive) holds if a "negative example" is falsely predicted as a "positive example," and TN (True Negative) holds if

		Prediction	
		Positives	Negatives
Actual	Positives	TP True Positive	FN False Negative
	Negatives	FP False Positive	TN True Negative

FIGURE 35.7 Confusion matrix.

		Predicted class		
		Fracture	Nonfracture	Per-class accuracy
Actual class	Fracture	40	0	100%
	Nonfracture	1	39	97.5%
	Total accuracy			98.8%

FIGURE 35.8 Verification result.

a "negative example" is correctly predicted as a "negative example." Then, the overall accuracy is given by

$$Accuracy = (TP + TN)/(TP + FP + FN + TN) \tag{35.1}$$

The principle of labeling the data to Fracture and Nonfracture is that if the pulse of the beam-end fracture is included in the data length, this data is labeled by "Fracture" while the data before beam-end fracture occurrence are labeled by "Nonfracture." Since this study focuses on the detection of the moment of beam-end fracture occurrence, the data after fracture without including the pulse due to fracture are not be used to test the trained CNN.

The validation result is shown in Figure 35.8. As can be seen in this figure, 100% accuracy is achieved for fracture data and 98.8% accuracy for nonfracture data. One misclassification was noted, a nonfracture data was classified as the fracture data, as can be seen in Figure 35.8.

35.4 CONCLUSIONS

1. A waveform-based approach for detecting beam-end fracture of steel frame building structure was proposed using CNN. The 20,000 acceleration waveform data generated by numerical simulation were used to train the CNN model. Eighty real acceleration waveforms collected from two shake-table tests were used to test train the CNN model. The accuracy of the proposed method is achieved for 98.8%.
2. The pseudo damage can be generated by numerical simulation. For training a robust damage-detection machine, great amount of damaged data is necessary. However, a real-word structure has only one life cycle, and almost all the structures on active duty are in a condition near intact, leading to the damaged data being limited and undamaged data being far more than damaged data. Therefore, it is unpractical to train the CNN by real data. Moreover, the pulse due to beam-end fracture of steel frame building structure is a universal feature regardless of whether the structure is a high-rise or a multistory building. Therefore, in this study, the limitation of collecting real data for training was overcome by pseudo damage generated by numerical simulation.
3. The CNN model can extract features from acceleration waveform recorded by sensor without any filter, which means the raw data can be used directly without any complicated data processing, such as filtering the long period or short period component by Fourier transform. As a result, the proposed CNN method can be more efficient and less time consuming to achieve real-time synchronization of data recording and processing.

REFERENCES

Ceena Modarres, Nicolas Astorga, Enrique Lopea Droquett, Viviana Meruane. 2018. Convolutional nerural networks for automated damage recognition and damage type identification. *Structural Control and Health Monitoring*, e2230. doi: 10.1002/stc.2230.

Ji Dang, Ashish Shrestha, Xin Wang, Satish Matsunaga, Pang-Jo Chun, Siri Asamoto. 2018. Low cost bridge seismic monitoring applying used smart phones and cloud server. *Eleventh U.S. National Conference on Earthquake Engineering*, Los Angeles, CA.

Jason Brownlee. 2017. What is the difference between a parameter and a hyperparameter? https://machinelearningmastery.com/difference-between-a-parameter-and-a-hyperparameter/

Kobe University. 2014. Experimental study on seismic safety measures of steel buildings damaged under earthquake. National Research Institute for Earth Science and Disaster Resilience (NIED).

Koji Uetani, Hiroshi Tagawa. 1996. Seismic response of steel frames including brittle fractures at beam-ends. *Journal of Structural and Construction Engineering (Transactions of AIJ)*, 489, 77–86.

Kyoto University. 2014. Research and development on quantification of collapse margin in high-rise buildings. Ministry of Education, Culture, Sports, Science and Technology (MEXT).

Masashi Matsuoka, Yuu Ishii, Norio Maki, Kei Horie, Satoshi Tanaka, Ryosuke Nakamura, Shuhei Hikosaka, Tomoyuki Imaizumi, Aito Fujita, Riho Ito. 2018. Damaged building recognition using deep learning with photos taken after the Kobe earthquake. *Eleventh U.S. National Conference on Earthquake Engineering*, Los Angeles.

Marc Claesen, Bart De Moor. 2015. Hyperparameter search in machine learning. *The XI Metaheuristics International Conference*, Agadir, Morocco.

Stephen A. Mahin. 1998. Lessons from damage to steel buildings during the Northridge earthquake. *Engineering Structures*, 20(4–6), 261–270.

Young-Jin Cha, Wooram Choi, Gahyun Suh, Sadegh Mahmoudkhani. 2017. Autonomous structural visual inspection using regiion-based deep learning for detecting multiple damage types. *Computer-Aided Civil and Infrastructure Engineering*, 32(1), 1–17.

Young-Jin Cha, Wooram Choi. 2017. Deep learning-based crack damage detection using convolutional neural networks. *Computer-Aided Civil and Infrastructure Engineering*, 32(5), 361–378.

36 A Review of Experimental Studies on Laboratory Grandstands

Mohammad Almutairi
University of Leeds
Kuwait Institute for Scientific Research

Onur Avci
Qatar University

Nikolaos Nikitas
University of Leeds

CONTENTS

36.1 INTRODUCTION

Grandstands are special in comparison to other civil structures due to their being occupied by large numbers of people and due to their slenderness owing to their unique architecture. These characteristics along with effects such as the synchronization of human activities make the descriptions of loading in the available design approaches either over-conservative or simply inadequate to predict the real magnitude of crowd-induced motion (Brownjohn et al., 2004; Jones et al., 2011; Catbas et al., 2017).

To gain a deeper understanding about occupied grandstands, a large number of measurements in controlled environments is required. To this end, grandstand simulators were designed and constructed in many research laboratories globally. Some of these simulators enable researchers to estimate directional human-induced loads while interacting with structural motion (Hoath et al., 2007; Comer et al., 2007b, 2013). Also, they can externally get excited in order to enable studies of crowd behavior, particularly in terms of comfort level (Nhleko et al., 2009), or to support the development of human-structure interaction models (Yuan et al., 2018; He et al., 2018). Furthermore, these grandstand simulators enable researchers to introduce damage in order to develop damage detection methods to be deployed in real stadia (Avci et al., 2017, 2018; Abdeljaber et al., 2017).

In this paper, a brief review of seven latest grandstand simulators with a selection among their associated studies is presented.

36.2 LITERATURE REVIEW

The seven grandstand simulators with their associated studies are presented in chronological order of their first appearance within scientific literature in the following sections.

36.2.1 GRANDSTAND SIMULATOR OF UNIVERSITY OF OXFORD, UK

A four-tier grandstand simulator was designed for human-structure interaction studies in Oxford University by Comer et al. (2007a) (see Figure 36.1). This grandstand was designed to host fifteen test subjects in a 3×5 row arrangement. Its dimensions are 3.2 m wide and 2.4 m deep. Other characteristics such as the rake angle, tread depth, riser height and seat-centers distance were selected based on typical design trends for new stadia in the UK (Comer et al., 2007a, 2010).

This grandstand simulator has built-in force plates to allow measuring vertical loads due to human activity. It is mounted on air springs and linear electric actuators, allowing it to replicate a range of different scenarios of dynamic structural response. To minimize the actuated mass, the grandstand is made of aluminum alloy. A novel structural control technique was used to enable the electric actuators to react to loads by test subjects as a dynamic system with user-defined frequency and damping.

Hoath et al. (2007) investigated the links between the head motion of active test subjects and human-induced loads. In order to track the head motion, a computer vision method by Lucas and Kanade (1981) was employed. Fifteen test subjects were jumping at 2.67 Hz and bobbing at 2 Hz, guided by a metronome. The motion tracking was found prone to being compromised due to the arrangement of the crowd, acquiring results depended upon the seating row. In any case, there was a strong correlation, still with a phase lag, between the measured loads and the head motions enabling possibly the estimation of one from the other.

The coherency within an active group jumping on the grandstand was studied by Comer et al. (2007a). Fifteen test subjects participated in two cases of jumping: synchronized and

FIGURE 36.1 Grandstand simulator of Oxford University (Comer et al., 2007b, 2010).

non-synchronized jumping for a range of controlled frequencies from 2 to 3.5 Hz. The results show that test subjects can coordinate their jumping with the group better than when jumping alone following metronome beats. This was found by comparing between test subjects in the back rows and the first row. Regarding Dynamic Load Factors (DLFs), the results show agreement between the determined group DLFs and those in the literature (Parkhouse and Ewins, 2006), determined by simulating groups of single test subjects jumping. Also, the observed DLFs showed that the suggestions in the BS 6399-1 guideline (BS6399, 1996) are over-conservative relevant to the second harmonic.

Comer et al. (2013) investigated the effect of perceptible structural vibration on group bobbing loads. The group was bobbing on the grandstand under two support condition cases: rigid and flexible. In the rigid support case, tests were conducted to study the coordination of test subjects bobbing without the stimulus of structural motion. Bobbing frequencies ranged from 2 to 3.5 Hz. In the flexible support case, tests were conducted to study the effect of grandstand motion on the group bobbing loads. Bobbing frequencies ranged from 1.7 to 2.25 Hz. The results show that the effective bobbing load did not increase by vibrations of the grandstand at the bobbing frequency. The DLF decreased because of structural vibration at the bobbing frequency in the second harmonic.

The vibration-perception and comfort states of humans, due to changing vibration levels of the grandstand simulator, was investigated by Nhleko et al. (2009). Two modes of passive crowds were tested: seated and standing. In both modes, the grandstand was externally excited in the frequency range 2–6 Hz. The test subjects provided their feedback on the state of vibration-perception and comfort during the tests. Results showed that test subjects felt the vibration long before feeling uncomfortable with it. Also, the limit state of serviceability was found between the recommended limits of two standards; those for transportation structures (BS6841) and for most classes of buildings (BS6472).

36.2.2 GRANDSTAND SIMULATOR OF FEDERAL UNIVERSITY OF PARAÍBA, BRAZIL

A temporary grandstand was constructed in the Federal University of Paraíba, Brazil (see Figure 36.2). It can host around one hundred test subjects. This grandstand simulator does not have designated seats. It is made of steel circular hollow bars. Its dimensions are 8.3 m wide and 6 m deep (de Brito et al., 2014).

FIGURE 36.2 Grandstand simulator of the Federal University of Paraiba, Brazil (de Brito et al., 2014).

Pena et al. (2013) developed a finite element model of the grandstand simulator. To update the model, the numerically identified natural frequencies and mode shapes were compared with their counterpart from laboratory modal tests. In the modal tests, an impact excitation was applied, and vibration was measured in the transverse and front-to-back directions. First, manual model updating was carried out based on engineering judgment. Then, automatic model updating was performed by employing the Particle Swarm Optimization (PSO) technique. The results proved the efficiency of the PSO technique in updating the grandstand model. Sensitivity analysis and upper and lower limits for the model parameter values improved the updating process.

In the same manner, de Brito et al. (2014) investigated the effect of refinements on modeling the grandstand simulator. This was done by developing three models of the grandstand, starting from a simple and getting to a very fine one. Because the models have a large number of unknown parameters, some of the grandstand components were modeled and tested, separately, in a preliminary analysis to determine the unknown parameters. After modeling a number of the grandstand components, models of the whole grandstand were developed in three refinement levels. Then, manual updating and automatic updating were employed to these models as in Pena et al. (2013). Results showed that modeling seat planks and connections as mass-only and rigid elements, respectively, would compromise the accuracy of the modal parameter estimation. The seat planks have significant contribution on the grandstand stiffness while the different connections have to be modeled with varying spring constants.

de Jesus et al. (2014) investigated the effects of seated test subjects on the transverse dynamic behavior of the grandstand. The investigation covered experiments, whereby the transverse natural frequency of the unoccupied and occupied grandstands (occupied by five, eight, twelve-seated spectators) was identified. Then, a biodynamic model of each test subject was developed. These biodynamic models were coupled with the updated finite element model from the work of de Brito et al. (2014). Namely, performing a numerical experiment, the grandstand model was gradually occupied with test subjects. In each stage, the transverse natural frequency was identified. All results showed that the presence of seated test subjects decreases the transverse natural frequency, and this can also be numerically realized. For reference, the transverse natural frequency of the unoccupied and fully occupied grandstand was 5.74 and 2.93 Hz, respectively.

36.2.3 GRANDSTAND SIMULATOR OF GDANSK UNIVERSITY OF TECHNOLOGY GRANDSTAND, POLAND

A three-tier steel grandstand was constructed in the Gdansk University of Technology, Poland (see Figure 36.3). It can host approximately twelve test subjects, without designated seats. A scaffolding system, which consists of tubular members, was used as the structure frame, and timber benches were used as seats. The grandstand dimensions are 2.7 m wide and 2.3 m deep (Lasowicz et al., 2015).

Lasowicz et al. (2015) investigated the efficiency of using polymer dampers to reduce the grandstand vibrations in two different scenarios. The first scenario was of the unoccupied grandstand. The second scenario was of twelve passive test subjects, which were simulated by installing blocks of concrete. The investigation consisted of comparing the modal parameters of the grandstand with the polymer damper against its equivalent with diagonal stiffeners of tubular cross section in the damper position. Vibration was introduced by applying an initial impact. Free decay analysis subsequently sourced the natural frequencies and damping ratios. Results showed that the structural damping ratios increased when polymer damper was used, compared with diagonal stiffener damper. The increase was 95% in unoccupied grandstand, and it was 211% in the occupied grandstand.

36.2.4 GRANDSTAND SIMULATOR OF QATAR UNIVERSITY, QATAR

A steel grandstand frame was designed and constructed at the Qatar University, Qatar (Abdeljaber et al., 2016). The frame is 4.2 m wide and 4.2 m deep. It has twenty-five filler beams bolted to eight

FIGURE 36.3 Grandstand simulator of the Gdansk University of Technology with diagonal polymer damper at the back of the grandstand simulator (Lasowicz et al., 2015).

main beams as shown in Figure 36.4. Because the filler beams are bolted and removable, many damage scenarios can be simulated either by replacing filler beams with damaged ones or simply by loosening their connection bolts (Abdeljaber et al., 2016).

Deep Learning (DL) damage detection algorithms were tested on this grandstand frame by Avci et al. (2017). The adopted algorithm is the 1D Convolutional Neural Network (1D CNN), which can be trained to detect damage by using raw acceleration data. Damage was envisaged by loosening the bolts at five joints. The raw acceleration data were created by applying random excitation using a modal shaker. Acceleration signals from each joint were processed to detect, localize and quantify the damage. To train the algorithms, acceleration data of the five joints were measured for their undamaged and damaged configuration, where in each case only one joint was damaged. The algorithm was shown to be efficient detecting all simulated damage scenarios.

In the same manner, Abdeljaber et al. (2017) tested the algorithm to detect damage at thirty joints. The algorithms were trained using acceleration data of thirty damage cases (i.e., one damaged joint at each case) and the baseline undamaged case. The algorithms were then tested to detect

FIGURE 36.4 Steel frame of a grandstand simulator of the University of Qatar (Abdeljaber et al., 2016).

eighteen single damage cases (i.e., one damaged joint) and five double damage cases (i.e., two damaged joints). All of the investigated cases were detected. The results proved the capability of 1D CNNs to detect and locate damage in real time.

While the aforementioned studies were conducted with wired accelerometers measuring only vertical acceleration, Avci et al. (2018) used triaxial wireless accelerometers to measure triaxial acceleration at ten joints. The triaxial measurements, at each joint, were compared against each other. The most significant acceleration direction was adopted to train the 1D CNNs at this joint. The other two acceleration directions were excluded. The significant direction for nine joints was found to be the vertical direction, further supporting the previous studies.

36.2.5 GRANDSTAND SIMULATOR OF UNIVERSITY OF CENTRAL FLORIDA, USA

A two-tier grandstand was designed in the University of Central Florida, USA (see Figure 36.5c). It can host approximately eleven test subjects, and it is made of two longitudinal 5.49 m long steel girders and seven transverse 0.92 m long steel members used for lateral stability (see Figure 36.5a). Aluminum bleacher planks were bolted to provide the seats. The stiffness of the grandstand can be adjusted by removing/adding intermediate columns (see Figure 36.5b). Similarly, all boundary conditions can be adjusted. In this way, different cases of stiffness and as such equivalent damage can be simulated (see Figure 36.5d–f) (Celik et al., 2018).

Celik et al. (2018) estimated the human-induced load from jumping test subjects by using computer vision methods; namely, the Lucas and Kanade (1981) and the Farneback (2001, 2003) method. The first option was used for sparse crowds, while the second one was used for dense. The algorithms tracked the displacement of all test subjects while they were jumping at regular beats. Then, acceleration was estimated by numerical differentiation in order to feed into the ground reaction force calculation and as such the load.

Three cases of crowd density were tested. The first two were of four, and eight test subjects jumping on a flexible grandstand. The third case was of eleven test subjects jumping on a stiff grandstand. In all the three cases, test subjects were jumping at frequencies of 1, 2 and 3 Hz guided by a metronome. To validate the methodology, the estimated vertical forces were compared against the measured ones using force plates. Also, they were compared against the measured forces in a finite element model. Cross correlation, between the estimated force and measured one using force plates, was calculated in three cases with results of 92.3%, 98.6% and 98% for the first, second and third cases, respectively.

FIGURE 36.5 Grand st (Grand stand) and simulator in the University of Central of Florida with the details of its components (Celik et al., 2018).

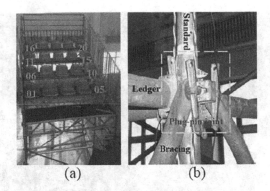

FIGURE 36.6 Grandstand simulator in the Harbin Institute of Technology with the details of its components; (a) elevation view with seat numbers and (b) plug-pin joint connection (Yuan et al., 2018).

36.2.6 GRANDSTAND SIMULATOR OF HARBIN INSTITUTE OF TECHNOLOGY, CHINA

A four-tier grandstand simulator was constructed in the Harbin Institute of Technology. It can host twenty test subjects in a 4 × 5 raw arrangement as shown in Figure 36.6a. The grandstand consists of a seating system and a hollow bar supporting system. The plug-pin joint is the type of connection used for the hollow bars as shown in Figure 36.6b. The grandstand is 2.5 m wide and 2 m deep. It was constructed on top of a biaxial shaking table so that different types of vibration excitation can be investigated (He et al., 2018; Yuan et al., 2018).

Yuan et al. (2018) investigated the effect of lateral vibration of the grandstand on the human-structure interaction model. In particular, parameters of passive crowds were investigated. Four configurations of the occupied grandstand were tested; twenty standing test subjects, twenty sandbags placed on the footings (to approximate standing test subjects), twenty seated test subjects and twenty sandbags placed on the seats (to approximate seated test subjects). To simulate lateral motion, three types of seismic waves were used, namely, 1999 Chi Chi, 1940 El Centro and 1995 Kobe earthquakes were employed in two lateral directions, West-East (W-E) and North-South (N-S). Their amplitudes ranged between 0.16 and 1.54 m/s². Results showed that the natural frequency for standing crowds is smaller than the one for seated crowds. On the other hand, the damping contribution for seated crowds is larger than this for standing.

He et al. (2018) studied human swaying forces and responses of the occupied grandstand. In order to develop a swaying force model, a semi-empirical formula for the swaying force was adopted. Experiments were performed to determine the swaying force unknown parameters of the formula. These parameters include peak force ratio, the swaying frequency and the duration of swaying. The experiments consisted of twenty test subjects standing and swaying on the grandstand at nine frequencies from 1.0 to 3.6 Hz. They were swaying in eleven different conditions (e.g., all seated and swaying, all standing and swaying, some test subjects seated and swaying and the remaining only standing, and some test subjects standing and swaying and the remaining only seated). To evaluate the structural responses, structural strain, displacements and accelerations were acquired and analyzed. Results showed that standing swaying crowds can induce higher strains and displacements than the seated swaying crowds. Also, significant lateral displacement, considered excessive, and acceleration can be introduced by rhythmic crowd activities in the sway direction.

36.2.7 GRANDSTAND OF ÉMI-TÜV SÜD LTD., HUNGARY

Three different precast grandstand units were tested in the ÉMI-TÜV SÜD Ltd. Laboratory, Hungary (Vardai and Madaras, 2019). The first unit was 7.6 m long and weighed 3.17 tons. The second unit was 10.8 m long and weighed 5.07 tons. The third unit was 10.9 m long and weighed 10.20 tons.

FIGURE 36.7   Activated release mechanism on uniformly loaded precast unit at EMI-TUV SUD Ltd Laboratory (Vardai and Madaras, 2019).

The latter was tested in one un-strengthened and two post-strengthened cases (Vardai and Madaras, 2019).

Vardai and Madaras (2019) conducted dynamic tests on the precast units to define limitations in designing new grandstands in Hungary. The effect of four levels of loads on the grandstand behavior was investigated. The four levels of uniformly distributed loads were 1.5, 3, 4 and 5.0 kN/m². Tests were conducted at each load level, both while increasing and decreasing it. Additionally, unloaded units were tested before applying any load and after removing all the loads. The dynamic excitation was using a special release mechanism introducing an impulse type of excitation. Figure 36.7 shows the mechanism on a uniformly loaded precast unit. Also, other tests were conducted with synchronized jumping of test subjects on the precast units. Results showed that the recommended design threshold for natural frequency in case of an unloaded grandstand should be 6 Hz, and, in the case of quasi-permanent loading 5 Hz. Also, acceleration below 35% would be an acceptable limit.

36.3   CONCLUSIONS

The presented grandstand simulators can be put into four categories according to their main testing features. First, human-structure interaction grandstand simulators such as the Oxford one, which has actuators and built-in force plates, and the Harbin Institute of Technology one, which is built on a shaking table. Second, damage detection focused grandstand simulators such as the cases in the University of Qatar and the University of Central Florida, which have interchangeable and removable elements that are ideal to simulate different damage scenarios. Third, temporary grandstand simulators such as the grandstands of the Federal University of Paraíba and of the Gdansk University of Technology due to their way of construction. Finally, grandstand simulators for developing design thresholds as the one tested at the ÉMI-TÜV SÜD Ltd. Laboratory.

Both Hoath et al. (2007) and Celik et al. (2018) adopted computer vision methods to link crowd motions with human-induced loads. Although the same method was adopted, efficiency was compromised in the work of Hoath et al. (2007). This could be due to two reasons. First, Hoath et al. (2007) used 25 fps cameras while Celik et al. (2018) used 60 fps. Second, Celik et al. (2018) used the method to track eight test subjects, while Hoath et al. (2007) used it to track fifteen test subjects.

The vertical natural frequency of the unoccupied grandstand simulators is around 20 Hz (Comer et al., 2007a; Celik et al., 2018), while their lateral natural frequency is close to 2.5 Hz (He et al.,

2018). This approaches natural human sway frequencies; test subjects swayed at 1–3.6 Hz in the work of He et al. (2018). Furthermore, a fully occupied grandstand with passive test subjects could reduce its lateral natural frequency further. For these reasons, existing grandstand simulators are more vulnerable to lateral rather than vertical crowd-induced vibrations.

Given these points, studying occupied grandstand under swaying crowds is a critical area of research, where capacity exists to study more. Another interesting area is DL method applications. As in many other engineering applications, it appears that DL methods have started to be utilized in structural damage detection of grandstands. It was observed that several studies have already paved the way for future real-scale deployments. Existing algorithms can be used on other grandstands and/or new algorithms can be tried. Such algorithms have the ability of prediction by using only raw data (e.g., acceleration). Bearing this in mind, DL methods can be used in predicting a wide range of other parameters. Such could be human-induced loads or structural dynamic responses.

REFERENCES

Abdeljaber, O., Avci, O., Kiranyaz, S., Gabbouj, M. and Inman, D.J. 2017. Real-time vibration-based structural damage detection using one-dimensional convolutional neural networks. *Journal of Sound and Vibration*, **388**, pp. 154–170.

Abdeljaber, O., Younis, A., Avci, O., Catbas, N., Gul, M., Celik, O. and Zhang, H. 2016. Dynamic testing of a laboratory stadium structure. *In:* Geotechnical and Structural Engineering Congress 2016, pp. 1719–1728, Phoenxi, Arizona.

Avci, O., Abdeljaber, O., Kiranyaz, S., Hussein, M. and Inman, D.J. 2018. Wireless and real-time structural damage detection: A novel decentralized method for wireless sensor networks. *Journal of Sound and Vibration*, **424**, pp. 158–172.

Avci, O., Abdeljaber, O., Kiranyaz, S. and Inman, D. 2017. Structural damage detection in real time: Implementation of 1D convolutional neural networks for SHM applications. *In: Structural Health Monitoring and Damage Detection,* **7**, Springer, pp. 49–54.

de Brito, V.L., Pena, A.N., Pimentel, R.L. and de Brito, J.L.V. 2014. Modal tests and model updating for vibration analysis of temporary grandstand. *Advances in Structural Engineering*, **17**(5), pp. 721–734.

Brownjohn, J.M.W., Pavic, A. and Omenzetter, P. 2004. A spectral density approach for modelling continuous vertical forces on pedestrian structures due to walking. *Canadian Journal of Civil Engineering*, **31**(1), pp. 65–77.

BS6399. 1996. Loading for buildings- Part 1: Code of Practice for Dead and Imposed Loads.

Catbas, F.N., Celik, O., Avci, O., Abdeljaber, O., Gul, M. and Do, N.T. 2017. Sensing and monitoring for stadium structures: A review of recent advances and a forward look. *Frontiers in Built Environment*, **3**, p. 38.

Celik, O., Dong, C.-Z. and Catbas, F.N. 2018. A computer vision approach for the load time history estimation of lively individuals and crowds. *Computers and Structures*, **200**, pp. 32–52.

Comer, A., Blakeborough, A. and Williams, M.S. 2010. Grandstand simulator for dynamic human-structure interaction experiments. *Experimental Mechanics*, **50**(6), pp. 825–834.

Comer, A.J., Blakeborough, A. and Williams, M.S. 2007a. Human-structure interaction in cantilever grandstands-design of a section of a full scale raked grandstand. *In:* 25th International Modal Analysis Conference, Orlando, FL.

Comer, A.J., Blakeborough, A. and Williams, M.S. 2013. Rhythmic crowd bobbing on a grandstand simulator. *Journal of Sound and Vibration*, **332**(2), pp. 442–454.

Comer, A.J., Williams, M.S. and Blakeborough, A. 2007b. Experimental determination of crowd load and coherency when jumping on a rigid raked grandstand. *In: IMAC XXV, Conference on Structural Dynamics*, Orlando, FL.

Farneback, G. 2001. Very high accuracy velocity estimation using orientation tensors, parametric motion, and simultaneous segmentation of the motion field. *In: Proceedings Eighth IEEE International Conference on Computer Vision*, Vancouver, BC, Canada, *ICCV 2001*, pp. 171–177, vol. 1.

Farneback, G. 2003. Two-frame motion estimation based on polynomial expansion. *In: Scandinavian Conference on Image Analysis*, Halmstad, Sweden, Springer, pp. 363–370.

He, L., Yuan, J., Fan, F. and Liu, C. 2018. Dynamic forces of swaying human and responses of temporary demountable grandstand based on experiment and simulation. *Shock and Vibration*, **2018**, pp. 1–22.

Hoath, R.M., Blakeborough, A. and Williams, M.S. 2007. Using video tracking to estimate the loads applied to grandstands by large crowds. *In: Conference Proceedings of the Society for Experimental Mechanics Series*, Indianapolis, IN, pp. 2–8.

de Jesus, C.O.T., Brito, V. and Pimentel, R. 2014. Influence of seated spectators on the transverse modal properties of temporary grandstands. *In: The 9th International Conference on Structural Dynamics, EURODYN 2014*, Porto.

Jones, C.A., Reynolds, P. and Pavic, A. 2011. Vibration serviceability of stadia structures subjected to dynamic crowd loads: A literature review. *Journal of Sound and Vibration*, **330**(8), pp. 1531–1566.

Lasowicz, N., Kwiecień, A. and Jankowski, R. 2015. Experimental study on the effectiveness of polymer damper in damage reduction of temporary steel grandstand. *In: Journal of Physics: Conference Series*, IOP Publishing, **628**, p. 12051.

Lucas, B.D. and Kanade, T. 1981. An iterative image registration technique with an application to stereo vision.

Nhleko, S.P., Williams, M.S. and Blakeborough, A. 2009. Vibration perception and comfort levels for an audience occupying a grandstand with perceivable motion. *Proceedings of the IMAC-XXVII*, Orlando, FL.

Parkhouse, J.G. and Ewins, D.J. 2006. Crowd-induced rhythmic loading. *Proceedings of the Institution of Civil Engineers: Structures and Buildings*, **159**(5), pp. 247–259.

Pena, A.N., de Brito, J.L.V., Pimentel, R.L. and de Brito, V.L. 2013. Finite element model updating of a temporary grandstand. *In:* Experimental Vibration Analysis for civil Engineering Structures, Ouro Preto, Brazil, pp. 1–8.

Vardai, A. and Madaras, B. 2019. Dynamic tests of grandstands in Hungary. *In:* fib Symposium, Budapest, Hungary.

Yuan, J., He, L., Fan, F. and Liu, C. 2018. The dynamic parameters of passive human at temporary demountable grandstands during exposure to lateral vibration. *Journal of Civil Engineering and Management*, **24**(4), pp. 265–283.

37 Experimental Analysis to Find Factor of Safety in Case of Dams and Slopes Using Finite Element and Limit Equilibrium Methods

Muhammad Israr Khan, Shuhong Wang, and Pengyu Wang
Northeastern University

CONTENTS

37.1 INTRODUCTION

Slope stability analysis is one of the major and challenging research area for geotechnical engineers. Slope failure and land sliding cause huge loss to economy as well as human lives as observed in past all over the world. Therefore geotechnical engineers have done a lot of work in this research area to develop design charts that are used to know the factor of safety of a slope as well as the probability of failure. Some brief work is explained in the literature section of this paper.

37.2 LITERATURE SURVEY

37.2.1 TAYLOR AND PECK CHARTS

Taylor (1937, 1948) provided chart solutions for slope stability analysis which were based on total stresses and ϕ angle. The assumptions in this method are as follows:

1. The slope is simple having a plane shape at the top and bottom, as is shown in Figure 37.1.
2. The slip surface was assumed as circular.
3. The slope is not layered and is homogenous as well as isotropic.
4. Coulombs law is assumed to find the shear strength that is $S = c + \sigma \tan \phi$, where S is the shear strength, c is cohesion and ϕ is friction angle.
5. The cohesion remains constant with depth, as is shown in Figure 37.2.
6. Pore pressure is assumed as part of total stress.
7. If the length of the cross section is roughly two times the potential rupture surface, then this two-dimensional (2D) analysis is valid.
8. Same stability number N is assumed which is provided by Terzaghi and Peck (1967).

$$N = \gamma H_c / c \tag{37.1}$$

where N is the stability number, γ is the unit weight of soil, H_c is the critical height and c is the value of cohesion. The Factor of Safety equation is

$$F = H_c / H \tag{37.2}$$

where F stands for the factor of safety.

9. Depth factor D is assumed as

$$D = D_H / H \tag{37.3}$$

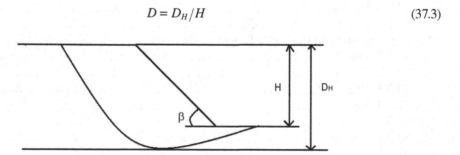

FIGURE 37.1   Horizontal plane at the top and bottom.

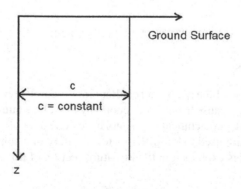

FIGURE 37.2   Cohesion is assumed as constant with depth.

Taylor solutions also have some limitations which are given below:

1. Not applicable for noncohesive soil.
2. Not applicable for partially submerged slopes.
3. The tensions cracks are being ignored.
4. Not applicable for stiff and fissured clays.

Figures 37.3 and 37.4 show the charts provided by Taylor and Peck.

37.2.2 BISHOP AND MORGENSTERN CHARTS

This method is based on effective stresses rather than total stresses. Assumptions of this method are given below:

1. Simple geometry having cylindrical slip surface is assumed.
2. Pore pressure is assumed and the coefficient used is r_u, which is equal to $u/\gamma H$, where u is the pore pressure of water, γ is the bulk density of soil and H is the height of the point

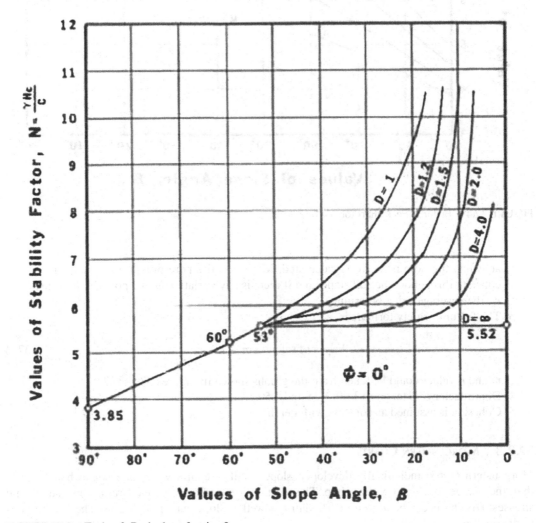

FIGURE 37.3 Taylor & Peck chart for $\phi = 0$.

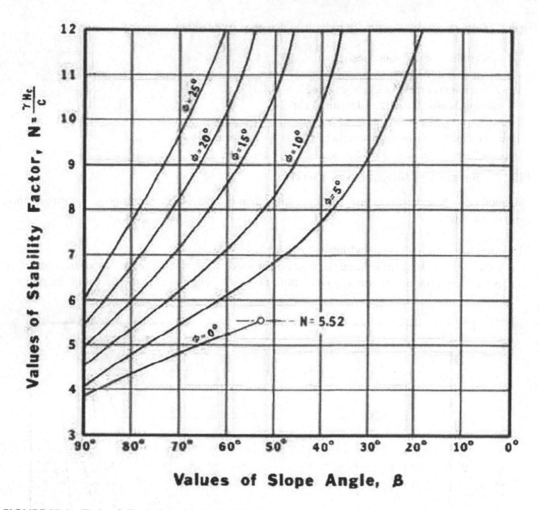

FIGURE 37.4   Taylor & Peck chart for $\phi = 0$–25.

at which the pore pressure is calculated. Moreover, the pore pressure is assumed to be constant throughout the soil slope, and if there is any variation at any point in the value of r_u, then average value is assumed.

3. The factor of safety formula is

$$\text{Factor of safety} = m - (n) * r_u \qquad (37.4)$$

m and n values could be taken from the graphs shown in Figures 37.5–37.7.

4. Depth factor D is taken as 1.00, 1.25 and 1.50.
5. Cohesion is assumed as constant with depth.

37.2.3 MORGENSTERN CHARTS

Morgenstern (1963) individually developed slope stability charts which are somewhat different than the one he developed with Bishop. The solution is based on effective stresses instead of total stresses. His charts can be used for earth dams as well as slope cuts in highways. The assumptions in this method are as follows:

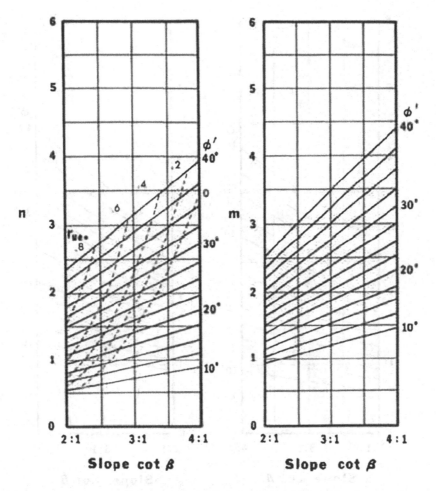

FIGURE 37.5   Stability coefficients (m and n) for $c'/\gamma * H = 0.05$ and $D = 1.00$.

1. Homogenous slope resting on rigid impermeable soil strata.
2. c' and ϕ' are the effective cohesion and angle of shearing resistance and are assumed to be constant with depth.
3. Pore pressure is assumed to be unity during drawdown, and it is $\beta = \Delta u/\Delta \alpha$.
4. At twice the unit weight of water of 124.8 pcf, the bulk density γ is assumed to be constant.
5. Product of height of soil into unit weight of water can give the approximated value for pore pressure.
6. Drawdown = L/H, where L is the amount of drawdown and H is the height of slope.
7. Potential slip circles are assumed tangent to the base of slope.

Morgenstern charts can be applied for a range of stability numbers ($c'/\gamma H$), that is from 0.0125 to 0.050. Figures 37.8–37.10 show the Morgenstern charts.

37.2.4   SPENCER CHARTS

Spencer (1967) provided a generalized solution to slope stability problems based on Bishop adaptation. Effective stresses are considered in this solution. The factor of safety in this method is

$$\text{Factor of safety} = \text{Shear strength available/Shear strength mobilized} \qquad (37.5)$$

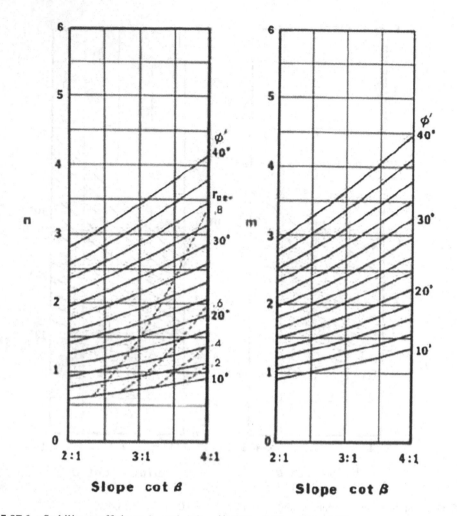

FIGURE 37.6   Stability coefficients (m and n) for $c'/\gamma * H = 0.05$ and $D = 1.25$.

Following are the assumptions of this method:

1. The soil is homogenous and has same properties in all directions.
2. The slip circle is circular and slope is a simple slope.
3. Depth factor "D" is considered large at very low depth.
4. There is no effect of tension crack.
5. Pore pressure is assumed as homogenous.
6. Stability number is

$$N = c'/(FS) * \gamma H \tag{37.6}$$

where N is stability number, FS is the factor of safety, γ is bulk density of soil and H is the height of embankment.
7. Shearing resistance mobilized angle is

$$\phi'_m = \tan \phi'/FS \tag{37.7}$$

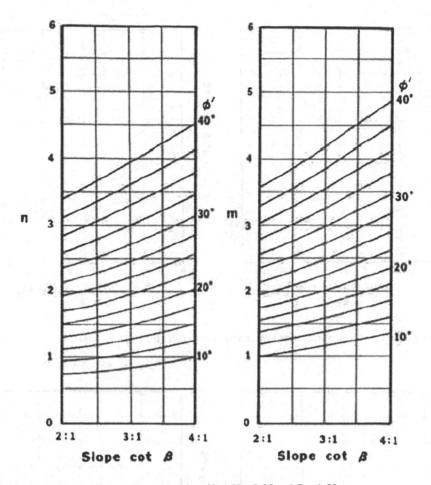

FIGURE 37.7 Stability coefficients (*m* and *n*) for $c'/\gamma * H = 0.05$ and $D = 1.50$.

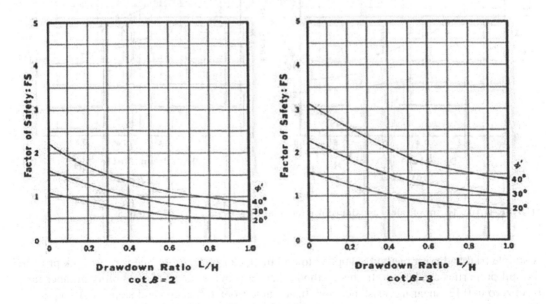

FIGURE 37.8 Factor of safety and drawdown relationship (*L/H*) for $c'/\gamma * H = 0.0125$.

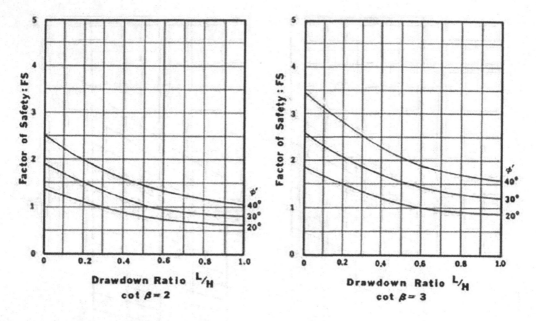

FIGURE 37.9   Factor of safety and drawdown relationship (L/H) for $c'/\gamma * H = 0.025$.

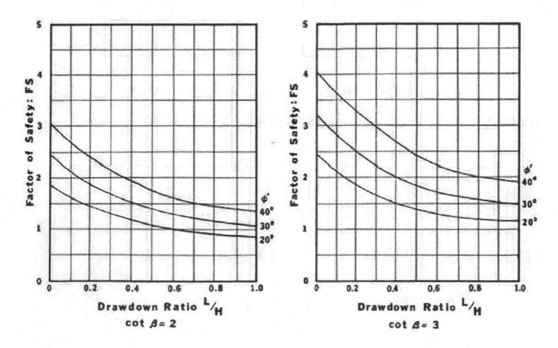

FIGURE 37.10   Factor of safety and drawdown relationship (L/H) for $c'/\gamma * H = 0.025$.

A simple trial and error method is applied to find the factor of safety in Spencer method, provided the soil properties are known. In this method, Spencer provides charts for Stability number range from zero to 0.12, shearing resistance mobilized angle from 10° to 40° and slope angles up to 34°. The pore pressure r_u is assumed as zero, 0.025 and 0.50. Figure 37.11 shows the Spencer chart.

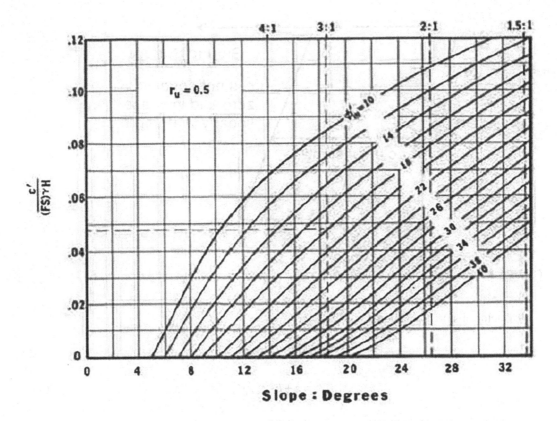

FIGURE 37.11   Spencer chart.

37.2.5   HUNTER CHARTS

Hunter (1968) provided new solution assuming the cohesion to be varied with depth and the trace of slip surface as a logarithmic spiral. This method is based on total stresses. Below are the assumptions of this method:

1. Simple slope with top and bottom surfaces as horizontal is assumed.
2. The soil is saturated and consolidated.
3. It is applicable for 2D soil.
4. Shear strength is described as $s = c + p \tan \phi$; where c varies linearly with depth, as is shown in Figure 37.12, ϕ remains constant with depth. It is also assumed that the ratio c/p' is constant, where p' is the effective vertical stress. p' also increases with depth.
5. For $c' > 0°$, the potential slip surface is a logarithmic spiral. And for $c' = 0°$, the failure surface is assumed as a circle because the logarithmic spiral degenerates in a circle in this case.
6. Effective stresses before excavation and immediately after excavation are the same.
7. Equations used in this method by Hunter are given below:
8. Water table ratio, M:

$$M = (h/H)\left(\gamma_w/\gamma'\right) \qquad (37.8)$$

where h is depth up to water table from top of the slope, H is the height of the cut, γ_w is the unit weight of water and γ' is the submerged unit weight.

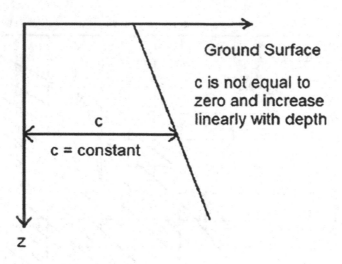

FIGURE 37.12 Cohesion changes with depth.

9. Factor of safety equation is

$$FS + c/\gamma Z_1 N \tag{37.9}$$

$$Z_1 = z + h\left(\gamma_w/\gamma'\right) \tag{37.10}$$

where z is the depth where cohesion is determined.

It gives an equivalent equation as

$$FS = \left(c/p'\right)\left(\gamma'/\gamma\right) \tag{37.11}$$

Often c/p' is determined by

$$\left(c/p'\right) = 0.11 + 0.0037\,PI \tag{37.12}$$

where PI is the Plasticity Index.

Figure 37.13 shows the Hunter chart.

37.2.6 HUNTER AND SCHUSTER CHART

Hunter and Schuster (1968) developed a more precise solution for $\phi = 0$ condition. The assumptions and equations are same as were developed by Hunter (1968) in his PhD thesis. Figures 37.14 and 37.15 shows the charts developed in this method.

All the above-mentioned chart solutions can be applied to different conditions in slope stability analysis issues. They have different assumptions and hence can be applied according to the assumptions and conditions. Some of these methods are applicable for embankments and others for small cuts. In every method, the authors tried to reduce the calculation time and get precise results. All these methods are manual and need trial and error iterative approach to get the factor of safety. And that is the main disadvantage of all these methods.

Michalowski (2002) and Baker (2003) used kinematic and conventional approach respectively to generate their slope stability charts. All the previous methods were good with a shortcoming of

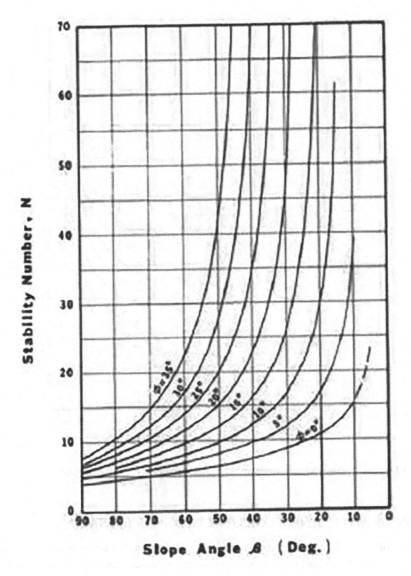

FIGURE 37.13 Hunter chart.

no idea regarding the failure probability of the slope. Previously, Li and Lumb (1987), Chowdhury and Xu (1993), Low et al. (1998) and Hong and Roh (2008) explored the probability of failure using conventional methods and based on the material properties of soil up to much extent. In fact soil stability and probability of failure cannot be estimated accurately as there are many nonrealistic situations, and the properties of soil change from place to place in a different manner. And the results of one site cannot be applied on another without detailed site investigation. But, approximated formulas and equations were developed previously by many researchers. Christian et al. (1994), El-Ramly et al. (2002), Low et al. (2007), Griffiths and Fenton (2004), Griffiths et al. (2009), Huang et al. (2010), Cho (2010) and Wang et al. (2011) provided good applicable charts and equations to know estimate the probability of failure by considering different soil properties and conditions. Considering the work of D.W. Taylor (1937) and Steward et al. (2011), Javankhoshdel S. and Bathurst R. J. (2014) have provided charts that could be used to estimate the probability of failure for cohesive (c) and cohesive – frictional $(c - \phi)$ soil.

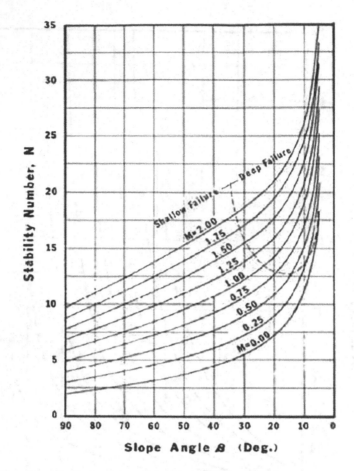

FIGURE 37.14   Chart between slope angle and stability number for $\phi = 0$ condition.

37.3   EXPERIMENTAL AND COMPUTER PROGRAMS ANALYSIS

With the development of computer hardware as machine and software as program, it becomes very easy to analyze soil slopes and to find the factor of safety values for any embankment. There are a number of softwares used in this field. Some of them are briefly introduced here:

37.3.1   SLIDE2

Slide2 is a powerful, user-friendly, 2D slope stability analysis program using limit equilibrium method. Slide2 can be used for all types of soil and rock slopes, embankments, earth dams and retaining walls. Slide2 includes built-in finite element groundwater seepage analysis, probabilistic analysis, multi-scenario modeling and support design.

37.3.2   SLIDE3

Slide3 is a 3D slope stability analysis software based on Slide2 software. With Slide3 most of the analysis features found in Slide2 are now available in full 3D, including complex geology, anisotropic materials, loading and support.

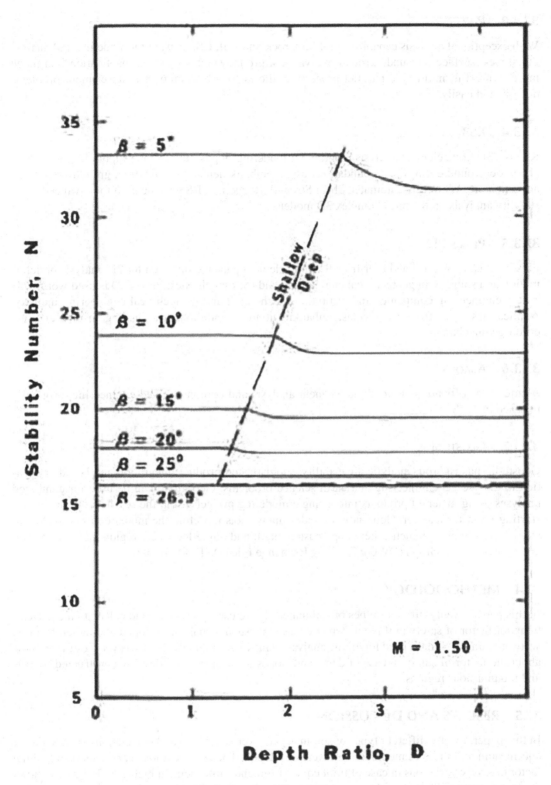

FIGURE 37.15   Chart between depth ratio D and stability number.

37.3.3   RS2

With exceptional analysis capabilities for both rock and soil, RS2 is used for modeling and analyzing slopes, surface and underground excavations, groundwater seepage, consolidation and much more. Powerful, multi-core parallel processors allows to solve complex finite element problems quickly and easily.

37.3.4   RS3

RS3 is a 3D finite element analysis program for modeling slopes, tunnel and support design, surface and underground excavations, foundation design, embankments, consolidation, groundwater seepage and more. With fully automated Shear Strength Reduction, RS3 can be used for advanced slope stability analysis on the most complex 3D models.

37.3.5   PLAXIS 2D

Plaxis 2D is a powerful and user-friendly finite element package intended for 2D analysis of deformation and stability in geotechnical engineering and rock mechanics. Plaxis 2D is used worldwide by top engineering companies and institutions in the civil and geotechnical engineering industry. Applications range from excavations, embankment and foundations to tunneling, mining and reservoir geomechanics.

37.3.6   ABAQUS

Abaqus is a software suite for finite element analysis and computer-aided engineering, originally released in 1978.

37.3.7   GEOSTUDIO

Geostudio has rigorous analytical capability, sophisticated product integration and broad application to diverse geoengineering and earth science problems. GeoStudio can be used for combined analyses using different products into a single modeling project, using the results from one as the starting point for another. GeoStudio provides many tools to define the model domain, including coordinate import, geometric item copy-paste, length and angle feedback, region merge and split, and Drawing with Graph (DWG)/ Drawing Exchange File (DXF) file import.

37.4   METHODOLOGY

In this paper, twenty different types of soil materials are used to analyze and calculate the correlations for factor of safety in three different conditions, such as natural, stepped and nailed form. A software namely Slide is used for all the analysis. This experimental analysis is very useful to know about the factor of safety in case of dams and slopes adjacent to any highway constructed in hills and mountainous regions.

37.5   RESULTS AND DISCUSSIONS

In this paper, twenty different types of soil materials were analyzed and correlations between limit equilibrium and finite element methods were developed. These correlations can be used to find the factor of safety variations in case of limit equilibrium and finite element methods. Table 37.1 shows the twenty material types used in this paper.

The final equation for calculating the factor of safety in case of finite element methods after all the analysis was performed are shown in Table 37.2:

TABLE 37.1

Material Types

Material Number	Cohesion (kPa)	Friction (ϕ)	Unit Weight (kN/m³)	Material Type
1	10	27	13	Clay
2	11	28	13.6	Clay
3	12	29	14.2	Clay
4	13	30	14.8	Clay
5	14	31	15.4	Clay
6	15	32	16	Clay
7	16	33	16.6	Clay
8	17	34	17.2	Clay
9	18	35	17.8	Clay
10	19	36	18.4	Clay
11	1	25	16.366	Clayey sand
12	2	26	16.464	Clayey sand
13	3	27	16.562	Clayey sand
14	4	28	16.758	Clayey sand
15	5	29	16.856	Clayey sand
16	6	30	16.954	Clayey sand
17	7	31	17.052	Clayey sand
18	8	32	17.15	Clayey sand
19	9	33	17.248	Clayey sand
20	10	34	17.346	Clayey sand

TABLE 37.2

Correlations between Factor of Safety in Finite Element Methods

Description	Natural	Stepped (2 m × 2 m)	Nailed (3–8 m)
Cohesion variation	$(FS)_{FE} = [(0.0443\ c + 0.5154) + 0.86]/1.89$ $R^2 = 0.99$	$(FS)_{FE} = [(0.0476\ c + 0.623)]/1.89$ $R^2 = 1$	$(FS)_{FE} = [(0.0367\ c + 0.9578)]/1.89$ $R^2 = 1$
Friction variation	$(FS)_{FE} = [(0.0215\ \varphi + 0.3697)]/1.89$ $R^2 = 0.99$	$(FS)_{FE} = [(0.0278\ \varphi + 0.3431)]/1.89$ $R^2 = 0.99$	$(FS)_{FE} = [(0.0452\ \varphi + 0.098)]/1.89$ $R^2 = 0.99$
Unit weight variation	$(FS)_{FE} = [(-0.0222\ \gamma + 1.2242)]/1.89$ $R^2 = 0.98$	$(FS)_{FE} = [(-0.0233\ \gamma + 1.3851)]/1.89$ $R^2 = 0.98$	$(FS)_{FE} = [(-0.0273\ \gamma + 1.6795)]/1.89$ $R^2 = 0.99$

37.6 CONCLUSIONS

1. Correlations provided in Table 37.2 can be used to find out the factor of safety without running the finite element procedure.
2. The validity of all the equations is in range of 98%–100% as shown in Table 37.2, which shows the applicability is very good.
3. Normally limit equilibrium and finite element give very symmetric values in case of non-complex situations, but the results are much different in case of complex situations. These equations can be used to cross check the variations in results in natural slope, stepped slope and nailed slope.

4. Keeping Tables 37.1 and 37.2 in consideration, factor of safety for any other material type can be calculated by interpolation technique.

REFERENCES

Baker, R. (2003). A second look at Taylor's stability chart. *Journal of Geotechnical Geoenvironmental Engineering, ASCE*, 129(12), 1102–1108. doi:10.1061/(ASCE)1090-0241.

Cho, S. E. (2010). Probabilistic assessment of slope stability that considers the spatial variability of soil properties. *Journal of Geotechnical and Geoenvironmental Engineering*, 136(7), 975–984. doi:10.1061/(ASCE)GT.1943-5606.0000309.

Chowdhury, R. N., & Xu, D. W. (1993). Rational polynomial technique in slope reliability analysis. *Journal of Geotechnical Engineering*, 119(12), 1910–1928. doi 10.1061/(ASCE)0733-9410.

Christian, J. T., Ladd, C. C., & Baecher, G. B. (1994). Reliability applied to slope stability analysis. *Journal of Geotechnical Engineering*, 120(12), 2180–2207. doi:10.1061/(ASCE)0733-9410.

El-Ramly, H., Morgenstern, N. R., & Cruden, D. M. (2002). Probabilistic slope stability analysis for practice. *Canadian Geotechnical Journal*, 39(3), 665–683. doi:10.1139/t02-034.

Griffiths, D. V., & Fenton, G. A. (2004). Probabilistic slope stability analysis by finite elements. *Journal of Geotechnical and Geoenvironmental Engineering*, 130(5), 507–518. doi:10.1061/(ASCE)1090-0241.

Griffiths, D. V., Huang, J., & Fenton, G. A. (2009). Influence of spatial variability on slope reliability using 2-D random fields. *Journal of Geotechnical and Geoenvironmental Engineering*, 135(10), 1367–1378. doi:10.1061/(ASCE)GT.1943-5606.0000099.

Hong, H., & Roh, G. (2008). Reliability evaluation of earth slopes. *Journal of Geotechnical and Geoenvironmental Engineering*, 134(12), 1700–1705. doi:10.1061/(ASCE)1090-0241.

Huang, J., Griffiths, D. V., & Fenton, G. A. (2010). System reliability of slopes by RFEM. *Soils and Foundations*, 50(3), 343–353. doi:10.3208/sandf.50.343.

Hunter, J. H. (1968). Stability of Simple Cuts in Normally Consolidated Clays. PhD thesis, Department of Civil Engineering, University of Colorado, Boulder.

Hunter, J. H., & Schuster, R. L. (1968). Stability of simple cuttings in normally consolidated clays. *Geotechnique*, 18(3), 372–378.

Javankhoshdel, S., & Bathurst, R. J. (2014). Simplified probabilistic slope stability design charts for cohesive and cohesive-frictional soils. *Canadian Geotechnical Journal*, 51, 1033–1045. doi:10.1139/cgj-2013-0385.

Li, K. S., & Lumb, P. (1987). Probabilistic design of slopes. *Canadian Geotechnical Journal*, 24(4), 520–535. doi:10.1139/t87-068.

Low, B. K., Gilbert, R. B., & Wright, S. G. (1998). Slope reliability analysis using generalized method of slices. *Journal of Geotechnical and Geoenvironmental Engineering*, 124(4), 350–362. doi:10.1061/(ASCE)1090-0241.

Low, B. K., Lacasse, S., & Nadim, F. (2007). Slope reliability analysis accounting for spatial variation. Georisk: Assessment and Management of Risk for Engineered Systems and Geohazards, 1(4), 177–189. doi:10.1080/17499510701772089.

Michalowski, R. L. (2002). Stability charts for uniform slopes. *Journal of Geotechnical Geoenvironmental Engineering, ASCE*, 128(4), 351–355. doi:10.1061/(ASCE)1090-0241.

Morgenstern, N. (1963). Stability charts for earth slopes during rapid drawdown. *Geotechnique*, 13(2), 121–131.

Spencer, E. A. (1967). Method of analysis of the stability of embankments assuming parallel inter-slice forces. *Geotechnique*, 17(1), 11–26.

Steward, T., Sivakugan, N., Shukla, S. K., & Das, B. M. (2011). Taylor's slope stability charts revisited. Journal of Geotechnical Geoenvironmental Engineering, ASCE, 11(4), 348–352. doi:10.1061/(ASCE)GM.1943-5622.0000093.

Taylor, D. W. (1937). Stability of earth slopes. *Journal of the Boston Society of Civil Engineers*, 24(3), 197–246.

Taylor, D. W. (1948). *Fundamentals of Soil Mechanics*. John Wiley and Sons, Inc., New York, 406–476.

Terzaghi, K., & Peck, R. B. (1967). *Soil Mechanics in Engineering Practice*. John Wiley and Sons, Inc., New York, 232–254.

Wang, Y., Cao, Z., & Au, S. K. (2011). Practical reliability analysis of slope stability by advanced Monte Carlo simulations in a spreadsheet. *Canadian Geotechnical Journal*, 48(1), 162–172. doi:10.1139/t10-044.

38 Development of Tuned Mass Damper Using Multi-Stage Steel Laminated Rubber Bearings

P. Chansukho and P. Warnitchai
Asian Institute of Technology

CONTENTS

38.1 INTRODUCTION

Tuned Mass Damper (TMD) is one of the most effective devices to suppress resonant-type vibration of civil engineering structures. One possible application is for suppressing the wind-induced lateral acceleration in the top floors of tall buildings. For such application, several hundred tons of TMD's mass is required since tall buildings are generally very massive. The natural period of TMD also needs to be very long (several seconds) in order to match with the fundamental natural period of the tall building. Moreover, the TMD's mass needs to be able to move with large amplitudes (up to a meter in some cases). One possible configuration that meets all these requirements is a pendulum-type TMD. The TMD of the Taipei 101 building in Taiwan belongs to this type. It has, however, a very long pendulum arm of more than 10 m, occupying a large expensive space of four-story high in the top floors.

In this study, a different configuration of TMD satisfying all these requirements is studied. It is a configuration where a rigid mass is placed on top of multi-stage steel laminated rubber bearings as shown in Figure 38.1. In this arrangement, the rubber bearing system is capable of supporting a large mass and is providing a very high vertical stiffness. At the same time, the system will also provide a very low horizontal stiffness with a very high horizontal deformability (Masaki et al. 2004). In this configuration, the stabilizing plates play an important role—they constrain the rubber bearings to deform in shear mode. This enhances the stability of rubber bearings and allows the

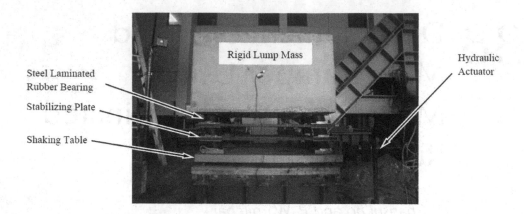

FIGURE 38.1 Components of TMD using multi-stage rubber bearings on shaking table.

bearings to be arranged in a multi-stage manner. As the number of stages increases, the horizontal stiffness reduces, and the horizontal deformability increases.

With all these outstanding characteristics, it is feasible to design a massive TMD with a long natural period and high horizontal deformability, meeting all the requirements of TMD for a given tall building. The space requirement will be low because this TMD configuration can be made very compact. Steel laminated rubber bearings are commercially available and are generally quite durable, as they are commonly used for base isolators of road bridges and other civil structures.

In this study, a 2,000 kg mass TMD of this configuration is tested on a $1 - D$ shaking table. This TMD is in fact a scaled down model of a full-scale 230-ton TMD designed for a tall building. The geometric scaling ratio is ~1:5. The main objective of the test is to investigate the dynamic behavior of TMD of this configuration and to develop a mathematical model that can accurately predict its dynamic behavior.

38.2 PROPERTIES OF STEEL LAMINATED RUBBER BEARING

To be effective in suppressing the vibration, the natural frequency of TMD must be tuned to the natural frequency of the targeted vibration mode of the structure. If the dynamic characteristics of TMD, especially its natural frequency, vary significantly with its response amplitude level, then the frequency tuning condition (and its vibration control effectiveness) may not be maintained. Therefore, it is important for the rubber bearing system to behave as close to a linear horizontal spring as possible throughout the expected range of TMD's response.

There are three main types of commercially available steel laminated rubber bearings: high-damping rubber bearing, natural rubber bearing with lead core (lead rubber bearing), and natural rubber bearing without lead core. The behaviors of the high-damping and lead rubber bearings are quite nonlinear; the relationship between the applied horizontal force and the resulting horizontal displacement of these bearings is significantly different from a linear one (Bhuiyan and Okui 2012). On the other hand, the behavior of natural rubber bearing (without lead core) is approximately linear, with low damping ratio and low strain-rate effect. For this reason, the TMD in this study is constructed from natural rubber bearings.

38.2.1 DESIGN OF BEARING PROPERTIES

To accurately estimate the horizontal stiffness of a multi-stage rubber bearing system, one must be able to accurately estimate the horizontal stiffness of a single steel laminated rubber bearing. This stiffness depends on many parameters, which include the bearing diameter, the number of rubber

TABLE 38.1

Steel Laminated Rubber Bearing Properties

Material	Natural Rubber
Shear modulus, G	0.4 MPa
Diameter, D	100 mm
Thickness of rubber layer, t	2 mm
Number of rubber layers, n_r	13
Thickness of reinforcing steel plate	2 mm
Number of reinforcing steel plates, n_s	12
Total thickness of rubber layers, t_r	26 mm
First shape factor, $S_1 = D/4t$	12.5
Second shape factor, $S_2 = D/t_r$	3.85

layers, the thickness of each rubber layer, the shear modulus of rubber, etc. It can be approximately estimated by Eq. (38.1) (Kelly 1997)

$$k_{H0} = \frac{GA}{t_r}\left[1 - \left(\frac{p}{p_{cr}}\right)^2\right] \tag{38.1}$$

where K_{H0} is the horizontal stiffness of the bearing when it is deformed in shear mode, G is the shear modulus of rubber, A is the bearing's cross-sectional area (in horizontal plane), and t_r is the total rubber thickness (the sum of all rubber layers), P is the vertical load applied on this bearing, and P_{cr} is the critical load, defined by Kelly (1997). In this experiment, all steel laminated rubber bearings are identical. Their properties are shown in Table 38.1.

38.2.2 Quasi-Static Horizontal Cyclic Loading Test under a Constant Vertical Load

When rubber bearings are assembled in multi-stage configuration as shown in Figure 38.1, they carry the vertical load from a rigid mass on top, and each of them has a small vertical deformation. During the horizontal movement of the rigid mass, these rubber bearings deform in a cyclic shear deformation mode because their top and bottom plates are restrained to remain horizontal (Kelly 1997, Chang 2002).

A quasi-static cyclic loading test of rubber bearings is then conducted by simulating these loading and boundary conditions for each individual rubber bearing. A pair of rubber bearings is arranged in the configuration shown in Figure 38.2. The constant vertical load is applied by a vertical hydraulic jack. The cyclic horizontal load is applied by a horizontal push-pull hydraulic jack. The testing frame and bearing arrangement are designed in such a way that the bearing deformation is in shear mode under a constant vertical load. The quasi-static loading test is considered to be reasonable because several studies indicate that the strain-rate effect of natural rubber bearing on its force-deformation characteristics is not significant (Bhuiyan and Okui 2012).

38.2.3 Test Result

Figure 38.3 shows a one-side test result of the cyclic horizontal force-deformation behavior of steel laminated rubber bearings. The "Horizontal Force" shown in Figure 38.3 is the applied horizontal force divided by 2 because the test specimen is made from two identical bearings. The obtained test result is shown by a black line, illustrating the almost-linear hysteresis curve. The area inside the hysteresis curve is very small indicating the low damping characteristic of natural rubber bearing.

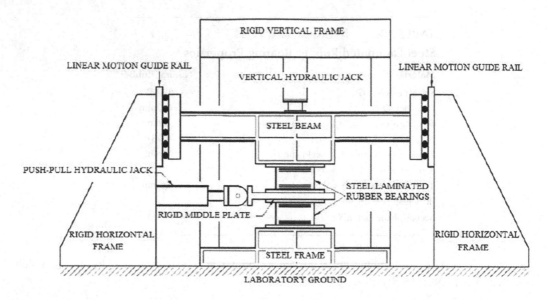

FIGURE 38.2   Test configuration of quasi-static horizontal cyclic loading under a constant vertical load.

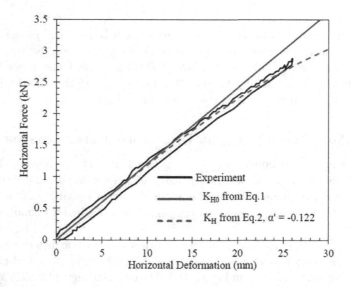

FIGURE 38.3   Quasi-static horizontal cyclic loading under constant vertical load test result at 100% shear strain and 4.9 kN vertical load.

The blue line shows a linear force-deformation relationship with the horizontal stiffness K_{H0} computed by Eq. (38.1). Comparing the black curve with the blue line, one can see a slight stiffness softening effect in the tested result at high deformation (and high shear strain) level. Another curve, a red dash line, is then introduced to match with the test result. For this red curve, the ratio of force to deformation is defined as K_H and is given by Eq. (38.2)

$$ k_H = K_{H0}\left(1 + \alpha'\left(\frac{u}{u_0}\right)^2\right) \qquad (38.2) $$

where the initial horizontal stiffness K_{H0} is calculated by Eq. (38.1), α' is a dimensionless negative-value constant determining the softening level (in this case $\alpha' = -0.122$), u is the horizontal deformation, and u_0 is the deformation of rubber bearing at 100% shear strain (in this case u_0 is equal to 26 mm). It is clear that this softening red curve well represents the slight nonlinear force-deformation characteristic of the tested rubber bearing. Note also that the damping ratio of rubber bearings estimated from the gray curve is approximately 3%.

38.3 PROPERTIES OF TMD USING MULTI-STAGE RUBBER BEARINGS

38.3.1 STIFFNESS OF TMD USING MULTI-STAGE STEEL LAMINATED RUBBER BEARINGS

The natural period of TMD system can be simply estimated by the initial horizontal stiffness of entire TMD system. As the stiffness of each individual bearing is verified, the overall stiffness of TMD system can be determined. For this lab-scale TMD, twelve rubber bearings are used in total, which consist of three stages, and each stage has four rubber bearings at the four corners of the rigid lump mass as shown in Figure 38.1. The initial horizontal stiffness in one stage is the summation of initial horizontal stiffness of the rubber bearing as shown in Eq. (38.3).

$$K_{s0,j} = \sum_{i=1}^{m} K_{H0,i} \qquad (38.3)$$

where $K_{H0,i}$ is the initial horizontal stiffness of one rubber bearing calculated from Eq. (38.1). $K_{s0,j}$ is the total initial horizontal stiffness in one stage. When the rubber bearings are connected using many stages, the overall horizontal stiffness reduces. The overall initial horizontal stiffness of the TMD system can be calculated using Eq. (38.4).

$$\frac{1}{K_{sys0}} = \sum_{j=1}^{n} \frac{1}{K_{s0,j}} \qquad (38.4)$$

The term K_{sys0} represents the initial horizontal stiffness of the overall TMD system. After obtaining overall initial horizontal stiffness, the approximate natural frequency of this lab-scale TMD is found to be 0.71 s (1.41 Hz).

38.3.2 HARMONIC EXCITATION TESTS

The dynamic behavior of TMD is investigated by a harmonic excitation test using a 1-D shaking table shown in Figure 38.1. This shaking table is driven by a servo controlled hydraulic actuator. Accelerometers and displacement sensors are attached to TMD's mass and shaking table. In this test, at first the table motion was set to be a simple sinusoidal motion with a displacement amplitude of 1 mm and frequency of 1.1 Hz. After the response of TMD had reached its steady-state condition, the responses of TMD were recorded. Then, the excitation frequency is slightly increased by a step of 0.05 Hz (to 1.15 Hz), and the experiment was repeated in the same manner. After repeating this upward frequency sweeping procedure until the excitation frequency reached 1.8 Hz, a downward frequency sweeping began by reducing the excitation frequency by a step of 0.05 Hz until it reached 1.1 Hz. To check the effects of excitation level on TMD's response, this upward and downward frequency sweeping tests were repeated at higher table's displacement amplitudes of 2, 3, and 4 mm.

38.3.3 RESULT OF HARMONIC EXCITATION TESTS

Figure 38.4 shows the steady-state displacement response amplitude of TMD's mass relative to the table at different excitation frequencies. The triangular marker represents the upward frequency

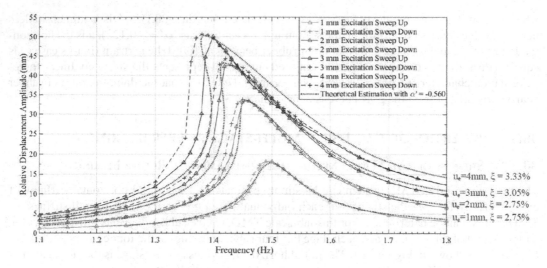

FIGURE 38.4   Amplitude-frequency response of TMD.

sweeping test result, while the "+" marker represents the downward one. For the case where the excitation table displacement amplitude is 1 mm, the resonant response curve is approximately symmetrical with respect to the central frequency of around 1.5 Hz. This central frequency is slightly offset from the calculated natural frequency of TMD (1.41 Hz, Sec. 38.3.1), indicating a small error in estimating TMD's frequency. As the excitation amplitude increases to 2, 3, and 4 mm, the corresponding resonant response curves become more and more unsymmetrical, with the response peaks shifting toward lower frequencies of 1.45, 1.42, and 1.38 Hz, respectively. This type of amplitude-dependent unsymmetrical resonant curves is commonly known to exist in a single-degree-of-freedom (SDOF) system with softening stiffness.

Figure 38.5 shows the phase difference ($\emptyset$) between the excitation and TMD's response as defined by Eqs. (38.5) and (38.6). For the case where the excitation table displacement amplitude is 1 mm, the variation of phase angle with excitation frequency is approximately similar to that of a linear SDOF system. As the excitation amplitude increases to 2, 3, and 4 mm, the corresponding

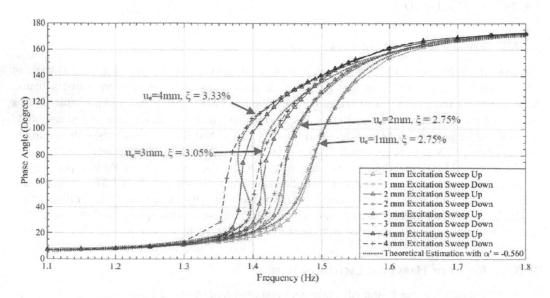

FIGURE 38.5   Phase-frequency response of TMD.

phase-frequency curves become more and more different from those of linear systems. The difference between upward and downward sweeping curves also becomes more and more apparent. Note that at the resonant response peak, the phase angle is always approximately equal to 90°.

38.3.4 NONLINEAR MODEL OF TMD

As clearly indicated by the result of harmonic excitation tests, the dynamic behavior of TMD of this multi-stage rubber bearing configuration cannot be described by a linear SDOF model. Instead, a nonlinear SDOF model with softening stiffness appears to be a better choice. This softening stiffness behavior was also found in the result of quasi-static cyclic loading test of sample bearings.

The softening stiffness behavior can be introduced into the governing equation of motion of TMD by modifying the stiffness term as shown in Eq. (38.5). The resulting nonlinear equation of motion has a cubic nonlinearity in the restoring force term. This type of nonlinear equation is well known and is generally called the Duffing's equation (Mook and Neyfeh 1979):

$$m\ddot{u} = 2\xi m\omega_0\dot{u} + m\omega_0^2\left(1 + \alpha'\left(\frac{u}{u_0}\right)^2\right)u = mu_e\Omega^2\cos(\Omega t) \tag{38.5}$$

where m is the mass of TMD, ω_0 is the circular natural frequency of TMD at extremely low vibration level, ξ is the damping ratio of TMD, α' is a dimensionless parameter defining the degree of stiffness softening of the multi-stage bearing system, u is the horizontal displacement of TMD mass relative to the table, and u_0 is the displacement of TMD mass that creates 100% shear strain in the bearings. The excitation term on the R.H.S. of Eq. (38.5) is when the table is sinusoidally moving at the displacement amplitude u_e and frequency Ω. The solution for displacement response of Eq. (38.5) is approximately given by Eq. (38.6):

$$u(t) = a(t)\cos(\Omega t + \phi(t)) \tag{38.6}$$

In Eq. (38.6), the amplitude a and phase $\varnothing$ are slow varying quantities. When the response reaches its steady-state condition, both amplitude a and phase $\varnothing$ approach certain constant values. These steady-state values of a and $\varnothing$ can be obtained from Eqs. (38.7) and (38.8) (Mook and Neyfeh 1979):

$$\Omega = \omega_0 + \frac{3}{8}\alpha'\omega_0\left(\frac{a}{a_0}\right)^2 \\ \pm\sqrt{\frac{u_e^2\Omega^4}{4\omega_0^2a^2} - \xi^2\omega_0^2} \tag{38.7}$$

The first term in Eq. (38.7) is the initial natural frequency. The second term of equation indicates amplitude-dependent detuning natural frequency by the influence of α', which is a function of a/a_0 where a is the displacement response amplitude of TMD system and a_0 is the displacement amplitude of TMD at shear strain of rubber bearings equal to 100%. If α' is negative, the "effective" natural frequency tends to reduce when the amplitude increases. In contrast, if α' is positive, the "effective" natural frequency becomes higher at higher response amplitudes. On the other hand, if α' is zero, the second term vanishes and the response becomes that of a linear system. The third term under the square root of Eq. (38.7) with "±" sign indicates that for a given response amplitude a, there are two possible solutions of the corresponding frequency Ω.

Figure 38.6 shows the plot of Eq. (38.7) with negative α' (softening spring). Green and red paths show the stable and unstable solutions, respectively. It is seen that at a frequency range between point No. 4 and 6, 2 out of 3 possible solutions are stable solutions, and the part that amplitude-frequency

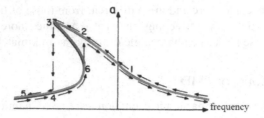

FIGURE 38.6 Response-frequency characteristic of a nonlinear dynamic system.

response curve bends back becomes an unstable solution. Theory of nonlinear oscillations also suggest that only stable solutions can be physically observed while unstable solutions cannot be observed. Furthermore, during harmonic excitation with sweeping frequency up, the experimental result is shown as an arrow path from left to right. On another hand, if frequency is sweeping down, the experimental result is observed as an arrow path from right to left. This characteristic is called Jumping phenomenon.

The phase angle $\emptyset$ solution is determined by Eq. (38.8) (Tamas and Balakumar 2011).

$$\phi = \tan^{-1}\left(\frac{\xi\omega_0}{\Omega - \omega_0 - \frac{3}{8}\alpha'\omega_0\left(\frac{a}{a_0}\right)^2}\right) \qquad (38.8)$$

In Eq. (38.8), nonlinear parameter α' is also included to account for nonlinear effect in phase response of dynamic system.

38.3.5 COMPARING EXPERIMENTAL RESULT WITH THEORETICAL PREDICTION

To check whether or not the nonlinear SDOF model (Eq. 38.5) can describe the dynamic behavior of TMD, its steady-state amplitude a and phase $\emptyset$ solutions as given by Eqs. (38.7) and (38.8) are plotted as red dotted lines in Figures 38.4 and 38.5. Note that the natural frequency ω_0, the damping ratio ξ, and the stiffness softening parameter α' are adjusted such that the theoretical prediction fits best to the experimental result.

At the best-fit condition, the natural frequency ω_0 is found to be 1.5 Hz, which is very close to the estimated frequency of 1.41 Hz as shown in Section 38.3.1. The damping ratio ξ is found to vary from 2.75% to 3.33%, which is quite close to the damping ratio determined from the quasi-static cyclic loading test of steel laminated rubber bearings. The parameter α' is found to be -0.560, indicating the softening effect at a higher degree when compared with that of a single rubber bearing; the reason for this is still not clear to the authors.

The comparison of steady-state amplitude in Figure 38.4 shows clearly that the amplitude-dependent unsymmetrical resonant curves obtained from harmonic excitation test can be accurately described by the proposed nonlinear model. The difference between upward and downward sweeping test result can be explained by the jumping phenomenon found in this nonlinear softening system. A similar conclusion can also be made for steady-state phase angle as shown in Figure 38.5.

38.4 CONCLUSION

The dynamic behavior of a TMD made of a 2,000-kg rigid mass placed on multi-stage steel laminated rubber bearings is investigated by using a small shaking table. Under simple harmonic table motions with a constant displacement amplitude and varying frequencies, the TMD exhibits an unsymmetrical resonant curve with peak leaning toward lower frequency. This nonlinear resonant response behavior can be very well predicted by a nonlinear SDOF model with softening stiffness, which is represented by the well-known Duffing equation. This softening behavior is an important aspect of this type of TMD, since its effectiveness in vibration control depends on the tuning of TMD's effective natural frequency to that of the structure.

REFERENCES

Bhuiyan, A. R., and Okui, Y. 2012, *Mechanical Characterization of Laminated Rubber Bearings and their Modeling Approach, Earthquake Engineering*, IntechOpen, Intech, Croatia.

Chang, C. H. 2002, Modelling of laminated rubber bearings using an analytical stiffness matrix, *International Journal of Solids and Structures*, vol. 39, pp. 6055–6078.

Kelly, J. M. 1997, *Earthquake-Resistant Design with Rubber (2nd Edition)*, Springer International, New York.

Masaki, N., Suizu, Y., Kamada, T., and Fujita, T. 2004, Development and applications of tuned/hybrid mass dampers using multi-stage rubber bearings for vibration control of structures, *13th World Conference on Earthquake Engineering*, Paper No. 2243, Canada.

Mook, D. T., and Neyfeh, A. H. 1979, *Nonlinear Oscillations*, John Wiley & Sons, Inc., Hoboken, NJ.

Tamas, K. N., and Balakumar, B. 2011, *Forced Harmonic Vibration of a Duffing Oscillator with Linear Viscous Damping*, John Wiley & Sons, Inc., Hoboken, NJ.

39 On the Performance of Unscented Kalman Filters in Parameter Estimation of Nonlinear Finite Element Models

R. Astroza and N. Barrientos
Universidad de los Andes

CONTENTS

39.1 INTRODUCTION

Model updating of nonlinear finite element (FE) models is an important topic finding applications in different areas of engineering. In particular, it is a powerful approach in the field of structural health monitoring (SHM) and damage identification (DID) of civil structures. A properly calibrated nonlinear FE model provides information about existence, location, and extent of damage, becoming a robust technique for DID purposes. Different methods in the time domain have been recently investigated as alternatives for updating mechanics-based nonlinear FE models (e.g., Shahidi and Pakzad 2014, Song and Dyke 2014), being those based on stochastic filtering (e.g., Kalman-type filters for nonlinear systems) very interesting because they also allow estimating the estimation uncertainty (Astroza et al. 2015, 2016, 2017, 2019a, Ebrahimian et al. 2015, Ramancha et al. 2020).

The Unscented Kalman filter (UKF) is an attractive alternative for updating mechanics-based nonlinear FE models, because it is a derivative-free stochastic filtering algorithm that can be easily parallelized, becoming a better alternative than the extended Kalman filter when dealing with complex models having a large number of parameters to be estimated (Astroza et al. 2019b, 2020).

The core of the UKF is the so-called unscented transformation (UT), a deterministic sampling techniques approximating the mean and covariance matrix of a nonlinear transformation of a random vector propagated through a nonlinear vector function (Julier and Uhlmann 1997). Several UTs have been proposed in the literature, but the scaled UT has been primarily used for updating mechanics-based nonlinear FE models. In this paper, different alternatives of the UT are investigated, comparing their robustness, accuracy, and computational cost when applied to the calibration of the 2D nonlinear FE model of a three-story, three-bay steel moment frame with distributed plasticity subjected to earthquake base excitation.

39.2 NONLINEAR STATE-SPACE MODEL FOR NONLINEAR FE MODEL UPDATING

For an n-DOF nonlinear FE model of a structure subjected to earthquake excitation, the equation of motion at time step $(k + 1)$ can be written as

$$\mathbf{M}(\boldsymbol{\theta})\,\ddot{\mathbf{q}}_{k+1}(\boldsymbol{\theta}) + \mathbf{C}(\boldsymbol{\theta})\,\dot{\mathbf{q}}_{k+1}(\boldsymbol{\theta}) + \mathbf{r}_{k+1}\big(\mathbf{q}_{1:k+1}(\boldsymbol{\theta}),\boldsymbol{\theta}\big) = -\mathbf{M}\mathbf{L}\,\ddot{\mathbf{u}}^g_{k+1} \tag{39.1}$$

where $\mathbf{q}_{k+1}$, $\dot{\mathbf{q}}_{k+1}$, $\ddot{\mathbf{q}}_{k+1} \in \mathscr{R}^{n\times1}$ = relative displacement, velocity, and acceleration response vectors, respectively, $\mathbf{M} \in \mathscr{R}^{n\times n}$ = mass matrix, $\mathbf{C} \in \mathscr{R}^{n\times n}$ = damping matrix, $\mathbf{r}_{k+1}(\mathbf{q}_{1:k+1}(\boldsymbol{\theta}),\boldsymbol{\theta}) \in \mathscr{R}^{n\times1}$ = history-dependent internal resisting force vector, n = number of degrees of freedom, $\boldsymbol{\theta} \in \mathscr{R}^{n_\theta\times1}$ = vector of unknown time-invariant model parameters to be estimated, $-\mathbf{M}\mathbf{L}\ddot{\mathbf{u}}^g_{k+1} \in \mathscr{R}^{n\times1}$ = effective external force vector, where $\mathbf{L} \in \mathscr{R}^{n\times r}$ = influence matrix and $\ddot{\mathbf{u}}^g_{k+1} \in \mathscr{R}^{r\times1}$ = input ground accelerations with r = number of base excitation components, and the subscript $k = 0,1,\dots, N-1$ indicates the time step with N = number of data samples of the input ground acceleration record. Then, the response of the FE model at time step $(k+1)$, $\hat{\mathbf{y}}_{k+1} \in \mathscr{R}^{n_y\times1}$, can be expressed as (Astroza et al. 2015)

$$\hat{\mathbf{y}}_{k+1} = \mathbf{h}_{k+1}\big(\boldsymbol{\theta},\ddot{\mathbf{u}}^g_{1:k+1},\mathbf{q}_0,\dot{\mathbf{q}}_0\big) \tag{39.2}$$

where $\mathbf{h}_{k+1}(\cdot)$ is the nonlinear response function of the nonlinear FE model at time t_{k+1}, $\ddot{\mathbf{u}}^g_{1:k+1} = [(\ddot{\mathbf{u}}^g_1)^T, (\ddot{\mathbf{u}}^g_2)^T, \dots, (\ddot{\mathbf{u}}^g_{k+1})^T]^T$ is the input ground acceleration time history from time t_1 to t_{k+1}, and $\mathbf{q}_0, \dot{\mathbf{q}}_0 \in \mathscr{R}^{n\times1}$ denote the initial conditions (omitted hereafter to simplify the notation).

The actual response of the structure, $\mathbf{y} \in \mathscr{R}^{n_y\times1}$, can be measured by a set of heterogeneous sensors and is related to the FE predicted response ($\mathbf{y}$) by (Astroza et al. 2019b)

$$\mathbf{y}_{k+1} = \hat{\mathbf{y}}_{k+1}(\boldsymbol{\theta}) + \mathbf{v}_{k+1} \tag{39.3}$$

where $\mathbf{v}_k \in \mathscr{R}^{n_y\times1}$ = prediction error vector. It is noted that Eq. (39.3) does not account for modeling uncertainty. The parameter vector $\boldsymbol{\theta}$ is modeled as a random vector process according to the Bayesian estimation approach and its evolution is described by a random walk process, i.e.,

$$\boldsymbol{\theta}_{k+1} = \boldsymbol{\theta}_k + \mathbf{w}_k \tag{39.4}$$

Equations (39.3) and (39.4) define the following nonlinear state-space model:

$$\begin{cases} \boldsymbol{\theta}_{k+1} = \boldsymbol{\theta}_k + \mathbf{w}_k \\ \mathbf{y}_{k+1} = \mathbf{h}_{k+1}\big(\boldsymbol{\theta}_{k+1},\ddot{\mathbf{u}}^g_{1:k+1}\big) + \mathbf{v}_{k+1} \end{cases} \tag{39.5}$$

where the process noise $\mathbf{w}_k$ and prediction error vector $\mathbf{v}_{k+1}$ are assumed statistically uncorrelated, Gaussian white processes with zero-mean and known diagonal time-invariant covariance matrices $\mathbf{Q}$ and $\mathbf{R}$, respectively, i.e., $\mathbf{w}_k \sim \mathcal{N}(\mathbf{0},\mathbf{Q})$ and $\mathbf{v}_{k+1} \sim \mathcal{N}(\mathbf{0},\mathbf{R})$.

39.3 UNSCENTED KALMAN FILTER

39.3.1 STANDARD AND SQUARE-ROOT IMPLEMENTATIONS

From the measured input-output data, the UKF can be used to recursively estimate the mean and covariance matrix of the unknown model parameter vector $\boldsymbol{\theta}$ based on the nonlinear state-space model defined by Eq. (39.5). The UKF is based on the concept of the UT, which defines a set of deterministically chosen samples, called sigma points (SPs), to approximate the probability density function of random variables that are propagated through a nonlinear function.

In the state-space model in Eq. (39.5), the UKF evaluates the nonlinear measurement equation around the predicted parameter estimate $\hat{\boldsymbol{\theta}}_{k+1|k}$ in a set of L SPs, $\vartheta^{(i)}_{k+1|k}$ with $i = 1,\ldots, L$. The SPs are then propagated through the nonlinear function defined by the FE model, i.e., $\boldsymbol{\mathscr{y}}^{(i)}_{k+1} = \mathbf{h}_{k+1}(\vartheta^{(i)}_{k+1|k}, \ddot{\mathbf{u}}^g_{1:k+1})$, and the estimated predicted response and its covariance matrix are obtained by weighted sampling.

The standard implementation of the UKF recursively updates the parameter covariance matrix $(\hat{\mathbf{P}}^{\theta\theta})$. However, to generate the SPs in the prediction phase of the filtering, the square root (SR) of the a priori parameter covariance matrix $(\hat{\mathbf{P}}^{\theta\theta}_{k+1|k})$ must be computed. An alternative implementation, called SR-UKF (van der Merwe and Wan 2001), considers to recursively update the SR of the parameter covariance matrix $(\hat{\mathbf{S}}^{\theta\theta})$, which satisfies $\hat{\mathbf{P}}^{\theta\theta} = \hat{\mathbf{S}}^{\theta\theta}\left(\hat{\mathbf{S}}^{\theta\theta}\right)^T$. This allows to reduce the computational costs of the estimation and also improve the numerical stability and guarantee having positive semi-definite state covariance matrices (van der Merwe and Wan 2001). Table 39.1 summarizes the algorithms of the standard and SR implementations of the UKF.

TABLE 39.1

Standard (Left) and SR (Right) UKF

Initialization:

$\hat{\boldsymbol{\theta}}_{0|0}$ and $\hat{\mathbf{P}}^{\theta\theta}_{0|0}$ *for* $k = 0, 1, 2, \ldots$

Prediction:

(i) $\hat{\boldsymbol{\theta}}_{k+1|k} = \hat{\boldsymbol{\theta}}_{k|k}$

(ii) $\hat{\mathbf{P}}^{\theta\theta}_{k+1|k} = \hat{\mathbf{P}}^{\theta\theta}_{k|k} + \mathbf{Q}_k$

(iii) SPs $\vartheta^{(i)}_{k+1|k}$, $i = 1, \ldots, L$, based on $\hat{\boldsymbol{\theta}}_{k+1|k}$ and $\hat{\mathbf{P}}^{\theta\theta}_{k+1|k}$

(iv) $\boldsymbol{\mathscr{y}}^{(i)}_{k+1} = \mathbf{h}_{k+1}\left(\vartheta^{(i)}_{k+1|k}, \ddot{\mathbf{u}}^g_{1:k+1}\right)$

(v) $\hat{\mathbf{y}}_{k+1|k} = \sum_{i=1}^{L} W^{(i)}_m \boldsymbol{\mathscr{y}}^{(i)}_{k+1}$

(vi) $\hat{\mathbf{P}}^{yy}_{k+1|k} = \sum_{i=1}^{L} W^{(i)}_c \left[\boldsymbol{\mathscr{y}}^{(i)}_{k+1} - \hat{\mathbf{y}}_{k+1|k}\right]\left[\boldsymbol{\mathscr{y}}^{(i)}_{k+1} - \hat{\mathbf{y}}_{k+1|k}\right]^T + \mathbf{R}_{k+1}$

(vii) $\hat{\mathbf{P}}^{\theta y}_{k+1|k} = \sum_{i=1}^{L} W^{(i)}_c \left[\vartheta^{(i)}_{k+1|k} - \hat{\boldsymbol{\theta}}_{k+1|k}\right]\left[\boldsymbol{\mathscr{y}}^{(i)}_{k+1} - \hat{\mathbf{y}}_{k+1|k}\right]^T$

Correction: output measurement $\mathbf{y}_{k+1}$ is recorded

(viii) $\mathbf{K}_{k+1} = \hat{\mathbf{P}}^{\theta y}_{k+1|k}\left(\hat{\mathbf{P}}^{yy}_{k+1|k}\right)^{-1}$

(ix) $\hat{\boldsymbol{\theta}}_{k+1|k+1} = \hat{\boldsymbol{\theta}}_{k+1|k} + \mathbf{K}_{k+1}\left(\mathbf{y}_{k+1} - \hat{\mathbf{y}}_{k+1|k}\right)$

(x) $\hat{\mathbf{P}}^{\theta\theta}_{k+1|k+1} = \hat{\mathbf{P}}^{\theta\theta}_{k+1|k} - \mathbf{K}_{k+1}\hat{\mathbf{P}}^{yy}_{k+1|k}\mathbf{K}^T_{k+1}$

end for

Initialization:

$\hat{\boldsymbol{\theta}}_{0|0}$ and $\hat{\mathbf{S}}^{\theta\theta}_{0|0}$

Prediction:

(i) $\hat{\boldsymbol{\theta}}_{k+1|k} = \hat{\boldsymbol{\theta}}_{k|k}$

(ii) Compute $\hat{\mathbf{S}}^{\theta\theta}_{k+1|k}$ (van der Merwe and Wan 2001)

(iii) SPs $\vartheta^{(i)}_{k+1|k}$, $i = 1, \ldots, L$, based on $\hat{\boldsymbol{\theta}}_{k+1|k}$ and $\hat{\mathbf{S}}^{\theta\theta}_{k+1|k}$

(iv) $\boldsymbol{\mathscr{y}}^{(i)}_{k+1} = \mathbf{h}_{k+1}\left(\vartheta^{(i)}_{k+1|k}, \ddot{\mathbf{u}}^g_{1:k+1}\right)$

(v) $\hat{\mathbf{y}}_{k+1|k} = \sum_{i=1}^{L} W^{(i)}_m \boldsymbol{\mathscr{y}}^{(i)}_{k+1}$

(vi) Compute $\hat{\mathbf{S}}^{yy}_{k+1|k}$ (van der Merwe and Wan 2001)

(vii) $\hat{\mathbf{P}}^{\theta y}_{k+1|k} = \sum_{i=1}^{L} W^{(i)}_c \left[\vartheta^{(i)}_{k+1|k} - \hat{\boldsymbol{\theta}}_{k+1|k}\right]\left[\boldsymbol{\mathscr{y}}^{(i)}_{k+1} - \hat{\mathbf{y}}_{k+1|k}\right]^T$

Correction: output measurement $\mathbf{y}_{k+1}$ is recorded

(viii) $\mathbf{K}_{k+1} = \hat{\mathbf{P}}^{\theta y}_{k+1|k}\left[\left(\hat{\mathbf{S}}^{yy}_{k+1|k}\right)^T\right]^{-1}\left(\hat{\mathbf{S}}^{yy}_{k+1|k}\right)^{-1}$ (*)

(ix) $\hat{\boldsymbol{\theta}}_{k+1|k+1} = \hat{\boldsymbol{\theta}}_{k+1|k} + \mathbf{K}_{k+1}\left(\mathbf{y}_{k+1} - \hat{\mathbf{y}}_{k+1|k}\right)$

(x)   Compute $\hat{\mathbf{S}}^{\theta\theta}_{k+1|k+1}$ (van der Merwe and Wan 2001)

end for

39.3.2 UNSCENTED TRANSFORMATIONS (UTs)

Several UTs have been proposed in the literature; however, there are no studies investigating the accuracy and computational costs of the different versions in the context of FE model updating of civil structures. In this paper, 5 UTs including standard (Julier et al. 1995), general (Julier et al. 2000), simplex (Wan-Chun et al. 2007), spherical (Julier 2003), and scaled (Julier 2002) are examined. In addition, for the scaled and spherical UTs, their SR implementations are also investigated. These UTs differ in the way they define the SPs and consequently in the weights of the SPs.

Consider a random vector $\mathbf{x} \in \mathbb{R}^{n_x \times 1}$ with mean $\bar{\mathbf{x}}$ and covariance matrix $\mathbf{P}$ subjected to a nonlinear transformation $\mathbf{s} = g(\mathbf{x})$. To estimate the mean vector and covariance matrix of $\mathbf{s}$, the UTs define a set of L SPs, denoted by $\mathscr{X}^{(i)} = \bar{\mathbf{x}} \pm \xi \sqrt{\mathbf{P}}$, with $i = 1, \ldots, L$ and ξ denotes a scaling factor, such that their sample mean and sample covariance matrix equal the true mean and covariance matrix of $\mathbf{x}$, respectively. These SPs are then propagated through the nonlinear transformation to estimate the mean and covariance matrix of $\mathbf{s} = g(\mathbf{x})$ as $\bar{\mathbf{s}} \approx \hat{\mathbf{s}} = \sum_{i=1}^{L} W_m^{(i)} g(\mathscr{X}^{(i)})$ and $\mathbf{P}^{ss} \approx \hat{\mathbf{P}}^{ss} = \sum_{i=0}^{L} W_c^{(i)} \left[g(\mathscr{X}^{(i)}) - \hat{\mathbf{s}} \right] \left[g(\mathscr{X}^{(i)}) - \hat{\mathbf{s}} \right]^T$, respectively. $W_m^{(i)}$ and $W_c^{(i)}$ are the weights of the SPs to estimate the mean and covariance matrix of $\mathbf{s}$, respectively. Table 39.2 summarizes the weights and SPs associated with the five different versions of the UTs investigated in this paper.

TABLE 39.2

Definition of Weights, Parameters, and SPs for the Different UTs

Name	L	Weights	SPs $\left(\mathscr{X}^{(i)} = \bar{\mathbf{x}} \pm \xi \sqrt{\mathbf{P}} \right)$
Standard	$2n_x$	$W_m^{(i)} = W_c^{(i)} = \left\{ \dfrac{1}{2n_x}, \; i = 1 \to 2n_x \right.$	$\begin{cases} \bar{\mathbf{x}} + \left(\sqrt{n_x} \sqrt{\mathbf{P}} \right)_i^T, & i = 1 \to n_x \\ \bar{\mathbf{x}} - \left(\sqrt{n_x} \sqrt{\mathbf{P}} \right)_{i-n_x}^T, & i = n+1 \to 2n_x \end{cases}$
General	$2n_x + 1$	$W^{(i)} = \begin{cases} \dfrac{\kappa}{n+\kappa}, & i = 0 \\ \dfrac{1}{2(n_x + \kappa)}, & i = 1 \to 2n_x \end{cases}$ $W_m^{(i)} = W_c^{(i)} = W^{(i)}$	$\begin{cases} \bar{\mathbf{x}}, & i = 0 \\ \bar{\mathbf{x}} + \left(\sqrt{(n_x + \kappa)} \sqrt{\mathbf{P}} \right)_i^T, & i = 1 \to n_x \\ \bar{\mathbf{x}} - \left(\sqrt{(n_x + \kappa)} \sqrt{\mathbf{P}} \right)_{i-n}^T, & i = n_x + 1 \to 2n_x \end{cases}$
Simplex	$n_x + 2$	$W^{(i)} = \begin{cases} \in [0,1), & i = 0 \\ 2^{-n_x} \left(1 - W^{(0)} \right), & i = 1, 2 \\ 2^{i-2} W^{(1)}, & i = 3 \to n_x + 1 \end{cases}$ $W_m^{(i)} = W_c^{(i)} = W^{(i)}$	$\sigma_i^{(j)} = \begin{cases} \begin{bmatrix} \sigma_0^{(j-1)} \\ 0 \end{bmatrix} & i = 0 \\ \begin{bmatrix} \sigma_i^{(j-1)} \\ \dfrac{-1}{\sqrt{2W^{(j+1)}}} \end{bmatrix} & i = 1 \to j \\ \begin{bmatrix} 0_{j-1} \\ \dfrac{j}{\sqrt{2W^{(j+1)}}} \end{bmatrix} & i = j+1 \end{cases}$ $\left\{ \bar{\mathbf{x}} + \sqrt{\mathbf{P}} \sigma_i^{(n)}, \quad i = 0 \to n_x + 1 \right.$

(Continued)

TABLE 39.2 (*Continued*)

Definition of Weights, Parameters, and SPs for the Different UTs

Name	L	Weights	SPs ($\mathscr{X}^{(i)} = \bar{x} \pm \xi\sqrt{\mathbf{i}}$)

Spherical $\quad n_x + 2$

$$W^{(i)} = \begin{cases} \in [0,1), & i = 0 \\ \dfrac{1 - W^{(0)}}{n_x + 1}, & i = 1 \rightarrow n_x + 1 \end{cases}$$

$$W_m^{(i)} = W_c^{(i)} = W^{(i)}$$

$$\sigma_0^{(1)} = 0; \; \sigma_1^{(1)} = \frac{-1}{\sqrt{2W^{(1)}}}; \; \sigma_2^{(1)} = \frac{1}{\sqrt{2W^{(1)}}}$$

$$\sigma_i^{(j)} = \begin{cases} \begin{bmatrix} \sigma_0^{(j-1)} \\ 0 \end{bmatrix} & i = 0 \\[2em] \begin{bmatrix} \sigma_i^{(j-1)} \\ \dfrac{-1}{\sqrt{j(j+1)W^{(1)}}} \end{bmatrix} & i = 1 \rightarrow j \\[2em] \begin{bmatrix} 0_{j-1} \\ \dfrac{j}{\sqrt{j(j+1)W^{(1)}}} \end{bmatrix} & i = j+1 \end{cases}$$

$$\begin{cases} \bar{x} + \sqrt{\mathbf{P}}\sigma_i^{(n)}, & i = 0 \rightarrow n_x + 1 \end{cases}$$

Scaled $\quad 2n_x + 1$

$$W_m^{(i)} = \begin{cases} \dfrac{\lambda}{n_x + \lambda}, & i = 0 \\ \dfrac{1}{2(n_x + \lambda)}, & i = 1 \rightarrow 2n_x \end{cases}$$

$$\begin{cases} \bar{x}, & i = 0 \\ \bar{x} + \left(\gamma\sqrt{\mathbf{P}}\right)_i^T, & i = 1 \rightarrow n_x \\ \bar{x} - \left(\gamma\sqrt{\mathbf{P}}\right)_i^T, & i = n_x + 1 \rightarrow 2n_x \end{cases}$$

$$W_c^{(i)} = \begin{cases} \dfrac{\lambda}{n_x + \lambda} + (1 - \alpha^2 + \beta), & i = 0 \\ \dfrac{1}{2(n_x + \lambda)}, & i = 1 \rightarrow 2n_x \end{cases}$$

$$\gamma = \sqrt{(n_x + \lambda)}$$

$$\lambda = \alpha^2(n_x + \kappa) - n_x$$

39.4 APPLICATION EXAMPLE

39.4.1 STRUCTURE, FINITE ELEMENT MODEL, AND INPUT EXCITATION

A 2D steel moment-resisting frame known as SAC-LA3 building (FEMA 2000) (Figure 39.1a) subjected to the 0° component recorded at the Los Gatos station during the 1989 Loma Prieta earthquake (Figure 39.1b) is used as an application example. Columns of the frame are made of A572 steel, while beams are made of A36 steel. The building is modeled in the software OpenSees using mechanics-based nonlinear force-based fiber-section Bernoulli-Euler beam-column elements. A single force-based element is used for each beam and column, with seven and six integration points along its length, respectively. The joints are assumed rigid, and rigid-end zones are considered at the ends of columns and beams. The sources of energy dissipation beyond the material hysteretic behavior are modeled using mass- and tangent stiffness-proportional Rayleigh damping, assuming a critical damping ratio of 2% for the first two initial modes of periods $T_1 = 1.06$ [s] and $T_2 = 0.35$ [s]. The nonlinear uniaxial stress-strain behavior of the steel longitudinal fibers is modeled through

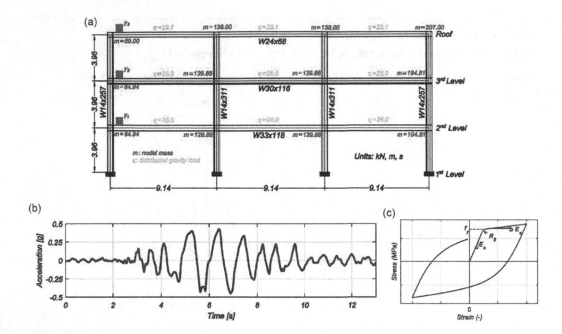

FIGURE 39.1   (a) SAC-LA3 steel building, (b) MGMP steel constitutive model, (c) Acceleration time history of the 0° component recorded at the Los Gatos station during the 1989 Loma Prieta earthquake.

the modified Giuffre-Menegotto-Pinto (MGMP) steel constitutive model. Four primary parameters of the MGMP material model are considered as unknown model parameters and will be estimated using the UKF (Figure 39.1c). These four parameters consist of the steel modulus of elasticity (E_s), the initial yield stress (f_y), a parameter defining the curvature of the elastic to plastic transition during the first cycle (R_0), and the strain-hardening ratio (b). Since beams and columns members are made of different types of steel, two independent sets of primary parameters are considered to define the MGMP models of the steel fibers of the beams and columns, respectively. Therefore, the vector of unknown model parameters to be considered later in the estimation phase is defined as $\theta = \left\{ E_{s\text{-col}}, f_{y\text{-col}}, R_{0\text{-col}}, b_{col}, E_{s\text{-beam}}, f_{y\text{-beam}}, R_{0\text{-beam}}, b_{beam} \right\} \in \mathscr{R}^{8 \times 1}$.

39.4.2 RESPONSE SIMULATION

The FE model of the SAC-LA3 building defined by a set of model parameters, referred to as true model parameters, is used to numerically simulate the true response of the structure ($\mathbf{y}^{true}$). The true model parameters used are $\theta^{true} = \{200\,\text{GPa}, 345\,\text{MPa}, 20, 0.08, 200\,\text{GPa}, 250\,\text{MPa}, 18, 0.05\}^T$. The horizontal absolute acceleration time histories at every level of the building structure, denoted as $\mathbf{y}_1$, $\mathbf{y}_2$, and $\mathbf{y}_3$ in Figure 39.1a, are taken as measured responses. Each component of $\mathbf{y}^{true}$ is polluted with an independent Gaussian white noise with zero-mean and 0.7%g root-mean-square (RMS), to produce the measured response, $\mathbf{y}$. Thus, the measurement noise exact covariance matrix is 0.47×10^{-2} $\mathbf{I}_3$ [(m/s²)²], where $\mathbf{I}_i$ denotes the $I \times i$ identity matrix. It is noted that the same realizations of noise are considered for each measured response.

39.4.3 ESTIMATION RESULTS

In the estimation, a zero-mean Gaussian process with a covariance matrix $R = 0.24 \times 10^{-2}$ $\mathbf{I}_3$ (m/s²)² is assumed for the output measurement noise, i.e., the amplitude of the measurement noise is estimated to be 0.5%g RMS. The diagonal entries of the covariance matrix of the noise in the random walk ($\mathbf{Q}$) are taken as $\left(q \times \hat{\theta}_{0|0}^{i} \right)^2$, with $i = 1,\ldots,8$ and $q = 1 \times 10^{-5}$. The initial estimate of the unknown model parameters is taken equal to $\hat{\theta}_{0|0} = \{0.70 E_{s\text{-col}}, 0.68 f_{y\text{-col}}, 0.69 R_{0\text{-col}}, 0.67 b_{\text{col}}, 1.24 E_{s\text{-beam}}, 0.64 f_{y\text{-beam}}, 1.37 R_{0\text{-beam}}, 1.25 b_{\text{beam}}\}^T$. The initial estimate of the covariance matrix of the model parameters, $\hat{\mathbf{P}}_{0|0}^{\theta\theta}$, is assumed to be diagonal (i.e., statistically uncorrelated initial estimates of the model parameters) with diagonal entries taken as $\left(p \times \hat{\theta}_{0|0}^{i} \right)^2$, with $i = 1,\ldots,8$ and $p = 5\%$.

Table 39.3 reports the final estimates (i.e., at the last time step, $\hat{\theta}_{N|N}$) and corresponding coefficient of variation (c.o.v.'s), obtained from the diagonal entries of the parameter covariance matrix at the last time step, $\hat{\mathbf{P}}_{N|N}^{\theta\theta}$, of the model parameters obtained using the different versions of the UKF. Excellent estimation results are obtained with all the filters except the spherical-SR case for all model parameters, with relative errors (computed as $\left(\hat{\theta}_{N|N} - \theta^{\text{true}} \right) / \theta^{\text{true}} \times 100 [\%]$) less than or equal to 5%. It is noted that the results for the Simplex UT are not shown, because convergence issues of the nonlinear FE model always arose when using this filter.

Model parameters related to linear-elastic response of the structure, i.e., $E_{s\text{-col}}$ and $E_{s\text{-beam}}$, are well estimated with all the filters, with relative errors $\leq 1\%$. In addition, all except the spherical-SR filter properly estimate the yield and post-yield model parameters (f_y, R_0, and b) for both column and beam elements, with relative errors $\leq 4\%$.

Table 39.4 reports the relative RMS errors (RRMSEs), computed as $\text{RRMSE} = \sqrt{\left[1/N_t \sum_{i=1}^{N_t} (a_i - b_i)^2 \right]^2} / \sqrt{\left[1/N_t \sum_{i=1}^{N_t} (a_i)^2 \right]} \times 100$ for signals a and b having N_t samples, between the true and FE predicted (using the final estimates of the model parameters) output responses measured in the structure. It is observed that all except the spherical-SR filter provide excellent matching between the true and the updated responses, with RRMSEs lower than 3%.

39.4.4 COMPUTATIONAL DEMANDS

Table 39.5 summarizes the wall-clock time required to complete the estimation and model updating process using the different filters. Note that all the results presented in this section were obtained

TABLE 39.3
Final Estimates and Coefficients of Variations (in Parenthesis) of the Model Parameters

Filter Case	$\frac{E_{s\text{-col}}}{E_{s\text{-col}}^{\text{true}}}$	$\frac{f_{y\text{-col}}}{f_{y\text{-col}}^{\text{true}}}$	$\frac{b_{\text{col}}}{b_{\text{col}}^{\text{true}}}$	$\frac{R_{0\text{-col}}}{R_{0\text{-col}}^{\text{true}}}$	$\frac{E_{s\text{-beam}}}{E_{s\text{-beam}}^{\text{true}}}$	$\frac{f_{y\text{-beam}}}{f_{y\text{-beam}}^{\text{true}}}$	$\frac{b_{\text{beam}}}{b_{\text{beam}}^{\text{true}}}$	$\frac{R_{0\text{-beam}}}{R_{0\text{-beam}}^{\text{true}}}$
Standard	1.00	0.99	1.00	1.00	1.00	1.00	1.02	1.01
	(0.08)	(0.21)	(1.41)	(0.63)	(0.13)	(0.19)	(0.74)	(0.35)
General	1.00	0.99	1.00	1.00	1.00	1.00	1.02	1.01
	(0.08)	(0.21)	(1.41)	(0.63)	(0.13)	(0.19)	(0.74)	(0.35)
Spherical	0.99	0.99	0.96	0.97	1.01	0.99	1.05	1.02
	(0.07)	(0.23)	(1.43)	(0.57)	(0.13)	(0.19)	(0.72)	(0.36)
Scaled	1.00	0.99	0.97	1.00	1.00	0.99	1.01	1.01
	(0.09)	(0.23)	(1.53)	(0.70)	(0.15)	(0.22)	(0.91)	(0.41)
Scaled-SR	1.00	0.99	1.01	1.00	1.00	1.00	1.00	1.00
	(0.04)	(0.12)	(0.54)	(0.26)	(0.05)	(0.10)	(0.32)	(0.11)
Spherical-SR	1.00	1.07	0.47	0.72	0.99	1.08	0.59	0.90
	(0.02)	(0.20)	(0.41)	(0.13)	(0.04)	(0.09)	(0.15)	(0.06)

TABLE 39.4

RRMSE (in %) between True and FE Predicted Output Response Measurements

Filter Case	Output Response Measurement (Y)		
	y_1	y_2	y_3
Standard	1.09	0.48	0.27
General	1.08	0.47	0.27
Spherical	2.87	1.21	0.57
Scaled	1.22	0.55	0.37
Scaled-SR	1.20	0.56	0.36
Spherical-SR	10.12	4.79	3.73

TABLE 39.5

Wall-Clock Time (in minutes) of the Estimation Process

Filter Case	Number of SPs	Wall-Clock Time (minutes)
Standard	$2n_x$	43.4
General	$2n_x + 1$	40.2
Spherical	$n_x + 2$	29.3
Scaled	$2n_x + 1$	39.8
Scaled-SR	$2n_x + 1$	45.8
Spherical-SR	$n_x + 2$	24.2

from analyses performed using three cores of a desktop workstation with an Intel Xeon E3–1230 (3.4 GHz) processor and 32 GB random-access memory. As the number of SPs dictate the number of FE runs required at each time step and therefore the computational cost of the updating process, most of the filters requested around 40 hours to complete the estimation. However, the spherical filter and its SR version took 29 and 24 hours, respectively, because they involve $n_x + 2$ SPs at each time step.

39.5 CONCLUSIONS

In this paper the UKF was used to update the mechanics-based nonlinear two-dimensional FE model of a three-story, three-bay steel moment frame subjected to earthquake base excitation. To this end, different UTs, the core of the UKF, were investigated in terms of robustness, accuracy, and computational cost. In particular, the standard, general, simplex, spherical, and scaled UTs were examined. In addition, for the scaled and spherical UTs, their SR implementations were also studied. An excellent accuracy, in terms of parameter estimates and response measurement errors, was observed for all except the simplex and spherical-SR filters. In terms of computational cost, most of the filter completed the estimation at similar wall-clock time, except the spherical filter and its SR version, which were run faster because of the lower number of SPs they consider. Based on the analyses conducted in this paper, the UKF based on the spherical UT is an excellent alternative in terms of performance and computational cost.

ACKNOWLEDGEMENTS

The authors acknowledge the financial support from the Chilean National Commission for Scientific and Technological Research (CONICYT), through FONDECYT research grant No. 11160009.

REFERENCES

Astroza, R., Alessandri, A., and Conte, J.P. 2019a. A dual adaptive filtering approach for nonlinear finite element model updating accounting for modeling uncertainty. *Mech. Syst. Signal Process.*, **115**, pp. 782–800.

Astroza, R., Barrientos, N., Li, Y., and Saavedra Flores, E. 2020. Calibration of a large nonlinear finite element model of a highway bridge with many uncertain parameters. In *Model Validation and Uncertainty Quantification,* Barthorpe, R. (ed.), Vol. 3, Conference Proceedings of the Society for Experimental Mechanics Series. Springer, Cham, pp. 177–187.

Astroza, R., Ebrahimian, H., and Conte, J.P. 2015. Material parameter identification in distributed plasticity FE models of frame-type structures using nonlinear stochastic filtering. *J. Eng. Mech. ASCE*, **141**(5), p. 04014149.

Astroza, R., Ebrahimian, H., and Conte, J.P. 2019b. Performance comparison of Kalman–based filters for nonlinear structural finite element model updating. *J. Sound Vib.*, **438**, pp. 520–542.

Astroza, R., Ebrahimian, H., Li, Y., and Conte, J.P. 2017. Bayesian nonlinear structural FE model and seismic input identification for damage assessment of civil structures. *Mech. Syst. Signal Process.*, **93**, pp. 661–687.

Astroza, R., Nguyen, L.T., and Nestorovic, T. 2016. Finite element model updating using simulated annealing hybridized with unscented Kalman filter. *Comput. Struct.*, **177**, pp. 176–191.

Ebrahimian, H., Astroza, R., and Conte, J.P. 2015. Extended Kalman filter for material parameter estimation in nonlinear structural finite element models using direct differentiation method. *Earthq. Eng. Struct. Dynam.*, **44**(10), pp. 1495–1522.

FEMA. 2000. State-of-the-art report on systems performance of steel moment frames subjected to earthquake ground shaking. FEMA 355C, Washington, DC.

Julier, S., Uhlmann, J., and Durrant-Whyte, H.F. 2000. A new method for the nonlinear transformation of means and covariances in filters and estimators. *IEEE Trans. Automat. Contr.*, **45**(3), pp. 477–482.

Julier, S.J. 2002. The scaled unscented transformation. *Proceedings of the American Control Conference*, Anchorage, AK.

Julier, S.J. 2003. The spherical simplex unscented transformation. *Proceedings of the American Control Conference*, Denver, CO.

Julier, S.J. and Uhlmann, J.K. 1997. A new extension of the Kalman filter to nonlinear systems. *Proceedings of 11th International Symposium on Aerospace/Defense Sensing, Simulation and Controls*, Bellingham, WA.

Julier, S.J., Uhlmann, J.K., and Durrant-Whyte, H.F. 1995. A new approach for filtering nonlinear systems. *Proceeding of the American Control Conference*, Seattle, WA.

Ramancha, M.K., Madarshahian, R., Astroza, R., and Conte, J.P. 2020. Non-unique estimates in material parameter identification of nonlinear FE models governed by multiaxial material models using unscented Kalman filtering. In *Model Validation and Uncertainty Quantification*, Barthorpe, R. (ed.), Vol. 3, Conference Proceedings of the Society for Experimental Mechanics Series. Springer, Cham, pp. 257–265.

Shahidi, S. and Pakzad, S. 2014. Generalized response surface model updating using time domain data. *J. Struct. Eng. ASCE*, **140**, p. A4014001.

Song, W. and Dyke, S.J. 2014. Real-time dynamic model updating of a hysteretic structural system. *J. Struct. Eng. ASCE*, **140**(3), p. 04013082.

van der Merwe, R. and Wan, E.A. 2001. The square-root unscented Kalman filter for state and parameter estimation. *Proceedings of the IEEE International Conference on Acoustics, Speech, and Signal Processing (ICASSP)*, Salt Lake City, UT.

Wan-Chun, L., Ping, W., and Zian-Ci, X. 2007. A novel simplex unscented transform and filter. *International Symposium on Communications and Information Technologies*, Sydney, Australia.

40 Uncertainty Handling in Structural Damage Detection Using a Non-Probabilistic Meta-Model

R. Ghiasi and M. R. Ghasemi
University of Sistan and Baluchestan

Mohammad Noori
California Polytechnic State University

Wael A. Altabey
Southeast University
Alexandria University

CONTENTS

40.1 INTRODUCTION

Structural systems are often exposed to harsh environment, while these environmental factors in turn could degrade the system over time. Their health state and structural conditions are key for structural safety control and decision-making management [1]. Although great efforts have been paid on this field, the high level of variability due to noise and other interferences, and the uncertainties associated with data collection, structural performance and in-service operational environments post great challenges in finding information to assist decision making [2]. The machine learning techniques in recent years have been gaining increasing attention due to their merits such as capturing information from statistical representation of events, thus enabling making decision [2–4].

Despite advances in computer capacity, the enormous computational cost of running complex engineering simulations makes it impractical to rely exclusively on simulation for the purpose of structural health monitoring [5]. To cut down the cost, surrogate models, also known as metamodels, are constructed and then used in place of the actual simulation models [5]. The effectiveness of surrogate models when applied in vibration-based damage detection has been demonstrated in many studies [2,3,6–14]. However, the use of surrogate models has been questioned in terms of its reliability in the face of uncertainties in measurement and modeling data [15,16]. The issues of uncertainty become more significant as civil engineering structures become more complex [17–22].

Attempts to incorporate a probabilistic method into models by treating the uncertainties as normally distributed random variables has delivered promising solutions to this problem [23], but the probabilistic method suffers from several disadvantages such as its assumption of uncertain parameters as normally distributed random variables with a given variance [24]. In practice, it is not possible to obtain the probability density function (PDF) due to the complexity of the sources of uncertainty. Moreover, the probabilistic surrogate model is computationally complex, especially when generating output data.

In this study, a non-probabilistic surrogate model based on wavelet weighted least squares support vector machine (WWLS-SVM) [3] is proposed to address the problem of uncertainty in vibration damage detection. The input data for the models consist of natural frequencies and mode shapes, and the output is the Young's modulus (E values), which acts as an elemental stiffness parameter (ESP). Through the interval analysis method [15], the noise in measured frequencies and mode shapes are considered to be coupled rather than statistically distributed.

This method calculates the interval bound (lower and upper bounds) of the ESP changes based on an interval analysis method. The surrogate models are used to predict the output of this interval bound by considering the uncertainties in the input parameters. To establish the relationship between the input and output parameters, a possibility of damage existence (PoDE) parameter is used for the undamaged and damaged states. A stiffness reduction factor (SRF) is also used to represent changes in the stiffness parameter.

The applicability of the proposed method is demonstrated through a numerical model of a 31-bar planer truss. The results show that the proposed method is able to efficiently provide the location and severity of damage.

This paper is organized as follows. In Section 2, the mathematical formulation for the proposed method is presented. Numerical examples are provided in Section 3 and conclusions are given in Section 4.

40.2 METHODOLOGY

Least square support vector machines (LS-SVM) are a class of kernel-based learning methods and are the least square versions of SVM [25]. They are a set of related supervised learning methods that analyze data and recognize patterns, and which are used for classification and regression analysis. In this paper, Wavelet Weighted version of LS-SVM [3,26] will be used as the surrogate model. For more detailed information on mathematical basis of WWLS-SVM algorithm, readers are referred to the original paper [26]. The input parameters of WWLS-SVM are the natural frequencies (λ) and mode shapes (ϕ), and the outputs are the ESPs (α). In the training phase, a series of damage cases are randomly generated using a finite element model (FE). The damage cases are idealized by reducing the ESPs of selected elements. Several damage cases are generated to test the efficiency of the trained WWLS-SVM model.

The SRF, as shown in Eq. 40.1, represents the changes in the stiffness parameter for each segment.

$$\text{SRF} = 1 + \frac{\alpha_d}{\alpha_u} \tag{40.1}$$

where d = ESP value in damaged state, u = ESP value in undamaged state.

Because the WWLS-SVM model requires training to establish the relationship between the input and output parameters, an initial baseline FE model is needed to generate a set of training data. Once the WWLS-SVM model is trained, the testing data are then fed to the trained model to predict the damage locations and severities. If the training and testing data are free from uncertainties, the surrogate model is very efficient in predicting the output [15]; however, in reality, both the FE model and the testing data will inevitably contain noise due to uncertainties in the measurements and modeling, resulting in inaccurate outputs [27].

Based on non-probabilistic interval analysis, modeling and measurement errors are considered in this study. Through this method, the uncertainties are considered by calculating the lower and upper bounds of the input parameters as well as the surrogate model outputs (ESPs) with lower and upper bounds. Once the lower and upper bounds of the ESPs are obtained, the PoDE determination can be followed by the calculation of the damage measure index (DMI), which is used to indicate the damage severity. Definition of these index is proposed in next section.

Only two surrogate models are required to predict the upper and lower boundaries of the ESP values based on this method. Compared with the conventional statistical surrogate method [28], the proposed non-probabilistic method provides an advantage in terms of its requirement of fewer surrogate models. The small number of required surrogate models leads to lower prediction error. In the statistical surrogate model, the prediction error normally increases the standard deviation of the PDF used to calculate the probability of damage existence (PDE) [28]. This higher standard deviation may result in lower probability values when detecting damage in structures.

In order to consider both uncertainties, the basic concept of interval mathematics is adopted by providing the upper and lower bounds of input parameters that will produce the upper and lower bounds of the output parameters. Due to the nature of the WWLS-SVM model used to establish the relationship between the input and output via a black box procedure, the basic equation of interval analysis proposed by Polyak and Nazin [29] can be directly applied to the input parameters (frequencies and mode shapes) to produce the intervals of the output parameters (ESP values). The intervals of the ESPs, natural frequencies and mode shapes for the undamaged and damaged states can be formulated as follows [15]:

$$[\alpha] \approx [\underline{\lambda}; \underline{\theta}] = \text{ESP value lower bound} \tag{40.2}$$

$$[\overline{\alpha}] \approx [\overline{\lambda}; \overline{\theta}] = \text{ESP value upper bound} \tag{40.3}$$

Therefore, the interval bounds for each parameter can be derived as

$$\lambda_c^I = \left[\underline{\lambda_c^I}, \overline{\lambda_c^I} \right] = \left\{ \lambda_{c1}^I, \lambda_{c2}^I, \ldots, \lambda_{ci}^I \right\}^T, \lambda_{ci}^I = \left[\underline{\lambda_{ci}^I}, \overline{\lambda_{ci}^I} \right] \tag{40.4}$$

$$\phi_c^I = \left[\underline{\phi_c^I}, \overline{\phi_c^I} \right] = \left\{ (\phi_{c1}^I)^T, (\phi_{c2}^I)^T, \ldots, (\phi_{cj}^I)^T \right\}^T, \phi_{cij}^I = \left[\underline{\phi_{cij}^I}, \overline{\phi_{cij}^I} \right] \tag{40.5}$$

$$\alpha_c^I = \left[\underline{\alpha_c^I}, \overline{\alpha_c^I} \right] = \left\{ \alpha_{c1}^I, \alpha_{c2}^I, \ldots, \alpha_{ck}^I \right\}^T, \alpha_{ck}^I = \left[\underline{\alpha_{ck}^I}, \overline{\alpha_{ck}^I} \right] \tag{40.6}$$

where c = number of damage cases, i = number of modes, j = number of structural nodes and k = number of segments of structures, and the middle values of the input and output are denoted as

$$x^c = m(x) = \frac{(\underline{x} + \overline{x})}{2} \tag{40.7}$$

Where $\underline{x}$ and $\overline{x}$ indicate the exact values of the input parameters (frequencies and mode shapes) and output parameters (ESPs). The upper and lower bars denote the upper and lower bounds of x, respectively.

Thus, the training and testing functions of the WWLS-SVM are established based on Eq. (40.7). The uncertainties are coupled with λ and ϕ in terms of the interval bounds. These natural frequencies (λ) and mode shapes (ϕ) are used as input parameters, while the ESPs (α) are used as outputs. Thus, two WWLS-SVM models, which include the lower and upper bound analyses, are provided, as shown in Table 40.1.

where the superscripts Ir and Ie represent the interval variables of training and testing, respectively. The variable ω indicates the uncertainty level for the input data by which the values of the uncertainties differ for the natural frequencies and mode shapes. The boundaries (lower and upper bounds) of the input parameters are applied through the $+$ and $-$ values of the uncertainties in two different WWLS-SVM models: WWLS-SVM 1 and WWLS-SVM 2. $\underline{\alpha}$ and $\overline{\alpha}$ are the outputs of the WWLS-SVM models and represent the lower and upper bounds of the predicted ESPs of damage case c. Once the lower and upper bounds of the stiffness parameters are obtained, the PoDE can be calculated.

40.2.1 PoDE and DMI

The PoDE is calculated by comparing the vectors of the ESPs in the damaged and undamaged states. The vectors are the interval bounds (lower and upper bounds) of the ESPs, which are the outputs of WWLS-SVM 1 and WWLS-SVM 2 (refer to Table 40.1), respectively. The expressions are as follows:

$$\alpha_u^I = \left\{\alpha_{u1}^I, \alpha_{u2}^I, \ldots, \alpha_{uk}^I\right\}^T \tag{40.8}$$

$$\alpha_d^I = \left\{\alpha_{d1}^I, \alpha_{d2}^I, \ldots, \alpha_{dk}^I\right\}^T \tag{40.9}$$

where α_u^I denotes the interval bound for the undamaged ESP $\left(\left[\overline{\alpha_{uk}}, \underline{\alpha_{uk}}\right]\right)$ and α_d^I denotes the interval bound for the damaged $\left(\left[\overline{\alpha_{dk}}, \underline{\alpha_{dk}}\right]\right)$.

Figure 40.1 illustrates the intersection of the intervals of the damaged and undamaged ESPs on the same axis, where the shaded region indicates the PoDE. The middle value (x^c) disparity between the two states will increase as the damage increases. The PoDE ranges between 0% and 100%, with 100% indicating a high possibility of damage and 0% indicating that no damage occurred at that specific element. Thus, the quantitative measure of the PoDE can be defined as below [15]:

$$\text{PoDE} = \frac{A_{\text{damage}}}{A_{\text{total}}} \times 100\% \tag{40.10}$$

Thus, DMI is calculated as below [15]:

$$\text{DMI} = \text{SRF} \times \text{PoDE} \tag{40.11}$$

TABLE 40.1
Training and Testing Input and Output Variables

Model	Training Input	Testing Input	Output
WWLS-SVM 1	$\underline{\lambda_{ci}^{Ir}} = \lambda_{ci}^{Ir} - \lambda_{ci}^{Ir}(\omega_\lambda)$	$\underline{\lambda_{ci}^{Ie}} = \lambda_{ci}^{Ie} - \lambda_{ci}^{Ie}(\omega_\lambda)$	$\underline{\alpha_{ck}}$
	$\underline{\phi_{cij}^{Ir}} = \phi_{cij}^{Ir} - \phi_{cij}^{Ir}(\omega_\phi)$	$\underline{\phi_{cij}^{Ie}} = \phi_{cij}^{Ie} - \phi_{cij}^{Ie}(\omega_\phi)$	
WWLS-SVM 2	$\overline{\lambda_{ci}^{Ir}} = \lambda_{ci}^{Ir} + \lambda_{ci}^{Ir}(\omega_\lambda)$	$\overline{\lambda_{ci}^{Ie}} = \lambda_{ci}^{Ie} + \lambda_{ci}^{Ie}(\omega_\lambda)$	$\overline{\alpha_{ck}}$
	$\overline{\phi_{cij}^{Ir}} = \phi_{cij}^{Ir} + \phi_{cij}^{Ir}(\omega_\phi)$	$\underline{\phi_{cij}^{Ie}} = \phi_{cij}^{Ie} - \phi_{cij}^{Ie}(\omega_\phi)$	

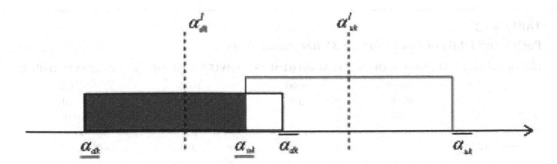

FIGURE 40.1 Scheme for PDE.

40.3 NUMERICAL RESULTS OF DAMAGE DETECTION

The 31-member planar truss shown in Figure 40.2 is selected as the numerical example to reveal the robustness and the degree of accuracy of the proposed damage detection method. It is modeled using the conventional FE method without internal nodes leading to 25 degrees of freedom [5]. In this example, the first five vibrating modes are utilized for damage detection. The material density and elastic modulus are 2,770 kg/m and 70 GPa, respectively. Two different damage scenarios given in Table 40.2 are induced in the structure, and the proposed method is tested for each case. The mass matrix is assumed to be constant, and the damage in the structure is simulated as a relative reduction in the elastic modulus of an individual element (Eq. 40.1).

In this section, the influence of the noise in the accuracy of structural damage detection based on modal data is investigated. Tables 40.3 and 40.4 show PoDEs and DMIs for damage scenario 1, respectively. The noise level is set as $\xi_\lambda = 1\%$. Tables 40.5 and 40.6 show the PoDEs and DMIs for damage scenario 2.

In scenario 1, higher PoDE values are obtained at element number 1 and 2 compared to the undamaged elements. The DMI value of element 1 is also higher than that of element 2; both of these elements are the true damage locations with different severity conditions. The same situation

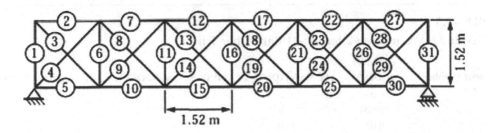

FIGURE 40.2 The 31-bar planar truss.

TABLE 40.2

Different Damage Scenarios for Planar Truss

Case 2		Case 1	
Element Number	SRF	Element Number	SRF
11	0.25	1	0.30
25	0.15	2	0.20

TABLE 40.3

PoDEs and DMIs of Case 1 for the 31-Bar Planar Truss

Element Number	FE Model (PoDE) %	FE Model (DMI) %	WWLS-SVM (PoDE) %	WWLS-SVM (DMI) %
1	100.00	30.00	100.00	32.00
2	99.00	20.00	98.00	19.10
10	0.00	0.00	0.00	0.00
18	0.00	0.00	0.00	0.00
Total time (sec)	412	-	195	-
RMSE	5.1×10^{-4}	-	1.1×10^{-3}	-

RMSE, root mean square error.

TABLE 40.4

PoDEs and PDEs of Case 1 for the 31-Bar Planar Truss

Element Number	FE Model (PoDE) %	FE Model (PDE) %	WWLS-SVM (PoDE) %	WWLS-SVM (PDE) %
1	100.00	100.00	100.00	98.00
2	99.00	97.00	98.00	96.00
10	0.00	4.00	0.00	7.00
11	0.00	0.00	0.00	0.00
12	0.00	0.00	0.00	1.00
18	0.00	0.00	0.00	0.00
19	0.00	2.00	0.00	5.00
22	0.00	0.00	0.00	0.00
26	0.00	0.00	0.00	3.00
Total time (sec)	412	610	195	313
RMSE	5.1×10^{-4}	2×10^{-3}	1.1×10^{-3}	1.2×10^{-2}

TABLE 40.5

PoDEs and DMIs of Case 2 for the 31-Bar Planar Truss

Element Number	FE Model (PoDE) %	FE Model (DMI) %	WWLS-SVM (PoDE) %	WWLS-SVM (DMI) %
10	0.00	0.00	0.00	0.00
11	100.00	25.00	100.00	25.40
12	0.00	0.00	0.00	0.00
24	0.00	0.00	0.00	0.00
25	100.00	15.00	98.90	14.20
Total time (sec)	390	-	175	-
RMSE	3.3×10^{-4}	-	1.2×10^{-3}	-

is observed for scenario 2, where higher PoDE values observed at segments 11 and 25 and the DMI values are also higher at the elements with higher severity. These results demonstrate that the proposed method provides an improvement over the conventional statistical method by providing a more meaningful damage severity indicator compared to the statistical surrogate model method, which only indicates the damage severity in terms of probabilities [23,28].

TABLE 40.6
PoDEs and PDEs of Case 2 for the 31-Bar Planar Truss

Element Number	FE Model (PoDE) %	FE Model (PDE) %	WWLS-SVM (PoDE) %	WWLS-SVM (PDE) %
10	0.00	0.00	0.00	0.00
11	100.00	100.00	100.00	96.00
12	0.00	0.00	0.00	3.00
13	0.00	0.00	0.00	0.00
14	0.00	0.00	0.00	2.00
24	0.00	3.00	0.00	5.00
25	100.00	98.00	98.90	94.30
26	0.00	0.00	0.00	1.00
Total time (sec)	390	545	175	301
RMSE	3.3×10^{-4}	1.8×10^{-3}	1.2×10^{-3}	1.0×10^{-2}

To further prove the capability of the proposed method, Tables 40.4 and 6 compare the PoDE values obtained by the proposed method with the PDEs values for the same damage cases calculated using the probability-based method developed by Ghiasi et al. [23]. From the tables, the PoDE is shown to be a more accurate damage indicator, generating smaller errors in both scenarios. For example, for scenario 1, in element 10, which is undamaged, the proposed method shows a 0% PoDE value compared to a 7% PDE value. The same situation occurred in scenario 2, where undamaged element 24 shows 0% damage, and the probability-based method indicates a 5% probability of damage. It is also observed that for both scenarios, the proposed method provides higher PoDE values at the damaged elements compared with the PDE value. The main reason is that because, as mentioned earlier, the proposed non-probabilistic method used fewer surrogate models than the probabilistic method, it generated smaller prediction errors.

40.4 CONCLUSIONS AND CONTRIBUTIONS

In this paper, a non-probabilistic WWLS-SVM model is proposed to consider the existence of measurement noise and modeling error in damage detection. An interval analysis is adopted for use with the WWLS-SVM to consider the uncertainties using the interval bounds of the uncertainties in the input parameters of the surrogate model. The developed approach is applied to detect simulated damage in numerical models 31-bar planer truss. The results show the proposed method can perform well in uncertainty-based damage detection of structures with less computational effort compared to direct FE model.

Furthermore, to prove the accuracy of the proposed method, SRF and PoDE are used to identify the location and reduction in the ESP values. The results show that the proposed method enables a reliable damage detection method with noisy data.

In summary, the proposed non-probabilistic method is able to provide more accurate damage detection results compared to the probabilistic method due to the smaller prediction errors. The proposed method is also less time consuming due to the smaller number of surrogate models to be trained.

ACKNOWLEDGEMENTS

Involvement and contributions to this research project were made possible, for the third co-author, in part due to a release time funding from Donald E. Bently Center for Engineering Innovation, Mechanical Engineering, Cal Poly, San Luis Obispo. Herein this support is acknowledged.

REFERENCES

1. W. Fan and P. Qiao, "Vibration-based damage identification methods: a review and comparative study," *Struct. Heal. Monit.*, vol. 10, no. 1, pp. 83–111, 2011.
2. C. R. Farrar and K. Worden, *Structural Health Monitoring: A Machine Learning Perspective.* John Wiley & Sons, Hoboken, NJ, 2012.
3. R. Ghiasi, P. Torkzadeh, and M. Noori, "A machine-learning approach for structural damage detection using least square support vector machine based on a new combinational kernel function," *Struct. Heal. Monit.*, vol. 15, no. 3, pp. 302–316, May 2016.
4. Z. Ying, N. Mohammad, W. A. Altabey, R. Ghiasi, Z. Wu, "Deep learning-based damage, load and support identification for a composite pipeline by extracting modal macro strains from dynamic excitations," *Appl. Sci.*, vol. 8, no. 12, p. 2564, 2018. doi: 10.3390/app8122564.
5. G. G. Wang and S. Shan, "Review of metamodeling techniques in support of engineering design optimization," *J. Mech. Des.*, vol. 129, no. 4, pp. 370–380, 2007.
6. R. Ghiasi, M. R. Ghasemi, and M. Noori, "Comparative studies of metamodeling and AI-Based techniques in damage detection of structures," *Adv. Eng. Softw.*, vol. 125, pp. 101–112, 2018.
7. Z. Ying, N. Mohammad, B. B. Seyed, W. A. Altabey, "Mode shape based damage identification for a reinforced concrete beam using wavelet coefficient differences and multi-resolution analysis," *Struct. Control Health Monit.*, vol. 25, no. 3, pp. 1–41, 2017. doi: 10.1002/stc.2041.
8. Z. Ying, N. Mohammad, W. A. Altabey, "Damage detection for a beam under transient excitation via three different algorithms," *Struct. Eng. Mech.*, vol. 63, no. 6, pp. 803–817, 2017, doi: 10.12989/sem.2017.64.6.803.
9. Z. Ying, N. Mohammad, W. A. Altabey, T. Awad, "A comparison of three different methods for the identification of hysterically degrading structures using BWBN model," *Front. Built Environ.*, vol. 4, p. 80, 2019. doi: 10.3389/fbuil.2018.00080.
10. W. A. Altabey, "Free vibration of basalt fiber reinforced polymer (FRP) laminated variable thickness plates with intermediate elastic support using finite strip transition matrix (FSTM) method," *Vibroengineering*, vol. 19, no. 4, pp. 2873–2885, 2017. doi: 10.21595/jve.2017.18154.
11. W. A. Altabey, "Prediction of natural frequency of basalt fiber reinforced polymer (FRP) laminated variable thickness plates with intermediate elastic support using artificial neural networks (ANNs) method," *Vibroengineering*, vol. 19, no. 5, pp. 3668–3678, 2017. doi: 10.21595/jve.2017.18209.
12. W. A. Altabey, "High performance estimations of natural frequency of basalt FRP laminated plates with intermediate elastic support using response surfaces method," *Vibroengineering*, vol. 20, no. 2, pp. 1099–1107, 2018. doi: 10.21595/jve.2017.18456.
13. W. A. Al-Tabey, "Vibration analysis of laminated composite variable thickness plate using finite strip transition matrix technique," In *MATLAB verifications MATLAB- Particular for Engineer*, Kelly Bennett, ed, InTech USA, Charlotte, NC, vol. 21, pp. 583–620, 980-953-307-1128-8, 2014. doi: 10.5772/57384.
14. N. Mohammad, W. Haifegn, W. A. Altabey, I. H. S. Ahmad, "A modified wavelet energy rate based damage identification method for steel bridges," *Int. J. Sci. Technol.Scientia Iranica*, vol. 25, no. 6, pp. 3210–3230, 2018, doi: 10.24200/sci.2018.20736.
15. K. H. Padil, N. Bakhary, and H. Hao, "The use of a non-probabilistic artificial neural network to consider uncertainties in vibration-based-damage detection," *Mech. Syst. Signal Process.*, vol. 83, pp. 194–209, 2017.
16. Z. Ying, N. Mohammad, W. A. Altabey, L. Naiwei, "Reliability evaluation of a laminate composite plate under distributed pressure using a hybrid response surface method," *Int. J. Reliab. Qual. Saf. Eng.*, vol. 24, no. 3, p. 1750013, 2017. doi: 10.1142/S0218539317500139.
17. Z. Ying, N. Mohammad, A. W. A. Altabey, Z. Wu, "Fatigue damage identification for composite pipeline systems using electrical capacitance sensors," *Smart Mater. Struct.*, vol. 27, no. 8, p. 085023, 2018. doi: 10.1088/1361–665x/aacc99.
18. Z. Ying, N. Mohammad, W. A. Altabey, R. Ghiasi, Z. Wu, "A fatigue damage model for FRP composite laminate systems based on stiffness reduction," *Struct. Durability Health Monit.*, vol. 13, no. 1, pp. 85–103, 2019. doi: 10.32604/sdhm.2019.04695.
19. W. A. Altabey, N. Mohammad, "Detection of fatigue crack in basalt FRP laminate composite pipe using electrical potential change method", *Phys. Conf. Ser.*, vol. 842, p. 012079, 2017. doi: 10.1088/1742-6596/842/1/012079.
20. W. A. Altabey, "Delamination evaluation on basalt FRP composite pipe by electrical potential change," *Adv. Aircr. Spacecr. Sci.*, vol. 4, no. 5, pp. 515–528, 2017. doi: 10.12989/aas.2017.4.5.515.

21. W. A. Altabey, "EPC method for delamination assessment of basalt FRP pipe: electrodes number effect", *Struct. Monit. Maint.*, vol. 4, no. 1, pp. 69–84, 2017. doi: 10.12989/smm.2017.4.1.069.

22. W. A. Altabey, N. Mohammad, "Monitoring the water absorption in GFRE pipes via an electrical capacitance sensors", *Adv. Aircr. Spacecr. Sci.*, vol. 5, no. 4, pp. 411–434, 2018. doi: 10.12989/aas.2018.5.4.499.

23. R. Ghiasi and M. R. Ghasemi, "Optimization-based method for structural damage detection with consideration of uncertainties-a comparative study," *Smart Struct. Syst.*, vol. 22, no. 5, pp. 561–574, 2018.

24. M. R. Ghasemi, R. Ghiasi, and H. Varaee, "Probability-based damage detection of structures using surrogate model and enhanced ideal gas molecular movement algorithm," In *Advances in Structural and Multidisciplinary Optimization: Proceedings of the 12th World Congress of Structural and Multidisciplinary Optimization (WCSMO12)*, A. Schumacher, T. Vietor, S. Fiebig, K.-U. Bletzinger, and K. Maute, eds, Springer International Publishing, Cham, pp. 1657–1674, 2018.

25. J. A. K. Suykens and J. Vandewalle, "Least squares support vector machine classifiers," *Neural Process. Lett.*, vol. 9, no. 3, pp. 293–300, 1999.

26. M. Khatibinia, M. Javad Fadaee, J. Salajegheh, and E. Salajegheh, "Seismic reliability assessment of RC structures including soil–structure interaction using wavelet weighted least squares support vector machine," *Reliab. Eng. Syst. Saf.*, vol. 110, pp. 22–33, 2013.

27. E. Simoen, G. De Roeck, and G. Lombaert, "Dealing with uncertainty in model updating for damage assessment : A review," *Mech. Syst. Signal Process.*, no. 56, pp. 123–149, 2015.

28. N. Bakhary, H. Hao, and A. J. Deeks, "Damage detection using artificial neural network with consideration of uncertainties," *Eng. Struct.*, vol. 29, no. 11, pp. 2806–2815, Nov. 2007.

29. S. A. Nazin and B. T. Polyak, "Interval parameter estimation under model uncertainty," *Math. Comput. Model. Dyn. Syst.*, vol. 11, no. 2, pp. 225–237, 2005.

41 A Non-Parametric Approach toward Structural Health Monitoring for Processing Big Data Collected from the Sensor Network

R. Ghiasi and M. R. Ghasemi
University of Sistan and Baluchestan

Mohammad Noori
California Polytechnic State University

Wael A. Altabey
Southeast University
Alexandria University

CONTENTS

41.1 INTRODUCTION

Structural Health Monitoring (SHM) is defined as the process of gathering adequate information that allows detecting, locating and quantifying structural vulnerabilities early on fatigue cracking, degradation of boundary conditions, etc., thereby improving the resilience of the civil infrastructure [1–5].

Significant efforts on SHM have been extensively undertaken [6–10]. These studies could be mainly classified into two categories: (1) parametric approaches (physics-based approaches) which are based on vibratory characteristics of structural systems including natural frequency, mode and curvature [11–16] and (2) non-parametric approaches (data-driven approaches) which extract sensitive features from sensor data to assess the structural conditions [17,18].

However, most of these approaches have not been implemented to remove the operational and environmental effects aggregated in extracted features; rather, they have been used to directly classify the extracted features in a supervised way, i.e., when data from the undamaged and damaged conditions are available. However, for most civil engineering infrastructures, where SHM systems are applied, the unsupervised learning algorithms are often required because only data from the undamaged condition are available [19].

To overcome these weaknesses, a new two-stage framework for intelligent health monitoring of structure is presented. In this framework, first, wavelet packet decomposition (WPD) is applied to the structural response signals under ambient vibration, and feature vectors are obtained via feature extraction based on wavelet pocket relative energy (WPRE) [20–23]. Subsequently, in the second stage, an unsupervised learning technique using moving kernel principal component analysis (MKPCA) is adapted for data normalization and damage detection. MKPCA is a revision of classical KPCA to make it more practical for long-term SHM. It models the effects of operational and environmental variability on the extracted features. The algorithm produces a scalar output as a damage index, which should be nearly invariant when features are extracted from the normal condition. Finally, Damage Indicators (DIs) from the feature vectors of the test data are classified through a threshold defined, based on the 95% cut-off value over the training data.

The layout of the paper is as follows. First, the mechanism of the proposed method will be explained. It combines the KPCA method with a windowed scheme in order to increase the sensitivity to damages. Then, a description of the test bed structure, the simulated operational and environmental variability, and a summary of the data sets will be provided. A comprehensive study is carried out using features extracted from time-series data sets measured with accelerometers deployed on the test bed structure. Finally, a discussion on the implementation and the analyses carried out in this paper will be highlighted, as a result of which some concluding remarks will be presented.

41.2 THE PROPOSED FRAMEWORK FOR LONG-TERM MONITORING

Real-life employment of SHM involves dealing with a large amount of multivariate data. Only a small portion of abnormal data, in comparison to overall data, is available at the time when damage occurs. For detecting the changes in data sets effectively, the classical KPCA should be improved to make it more practical for long-term SHM data analysis. By means of KPCA, the damage can be detectable only when the principal components (eigenvectors) are influenced by abnormal behavior. Subsequently, eigenvectors are subject to changes only if a certain amount of abnormal data are captured and possibly affected the overall structure of data. This feature makes KPCA less effective for long-term SHM implementation. Therefore, in this paper, MKPCA will be proposed to address this challenge.

41.2.1 MOVING KERNEL PRINCIPAL COMPONENT ANALYSIS

Basically, MKPCA computes the KPCA within moving windows with a constant size. Figure 41.1 gives details of MKPCA algorithm designed for long-term SHM applications. Also, the applied procedures can be summarized in three different steps as follows:

Step 1: A data matrix should be generated by sorting the time history data from each sensor into individual columns. Then, the WPRE method will be used for feature extraction from raw acceleration data of each sensor. The feature matrix will be created based on the result of the WPRE method for all structural state conditions. For each test, with consideration of N_s sensors and N_f extracted features from each sensor $N_s{\times}N_f$ dimensional feature vectors will be created when using them in the concatenated format.

Step 2: Feature matrix should be divided into two phases, training and monitoring. The training phase is intended for developing a baseline, a confidence interval, based on normal condition, while the monitoring phase is set for long-term monitoring. The dimensions of fixed moving windows should be well-defined. In this paper, the appropriate size of moving windows is chosen based on the largest periodicity in the data. KPCA should be conducted for each window individually and the results should be stored.

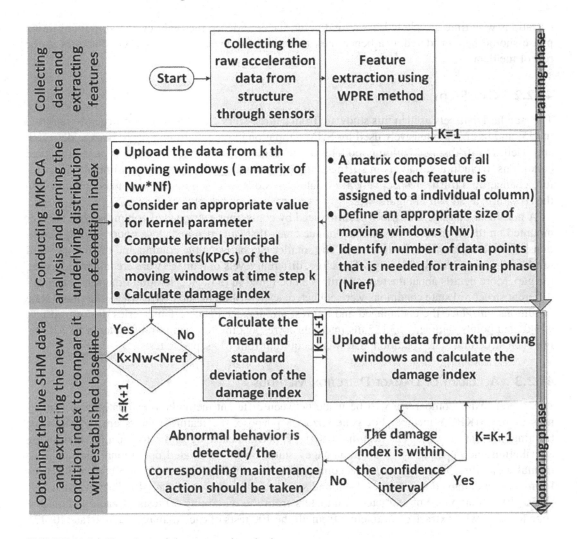

FIGURE 41.1 Flowchart of the proposed method.

Step 3: A sensitive damage index should be selected in this step based on MKPCA outputs. The DI_{rpc} chosen for this study is the square root of the sum of the squares of the first i kernel principal components as shown in Eq. 41.1:

$$DI_{rpc} = \sqrt{(KPC_1)^2 + (KPC_2)^2 + \cdots + (KPC_i)^2} \cdot i = 1 \ldots n \qquad (41.1)$$

where KPC_i is the i-th kernel principal components of moving windows and rpc is the abbreviation of the root of principal components. The reason for incorporating the first i kernel principal components in the DI_{rpc} is that they cover the most useful information in the data. In fact, since more than 99% of the energy distribution of KPCs is covered by the first i principal components, they are the only components incorporated in the DI_{rpc} [24]. It should be mentioned that the number of kernel principal components that should be considered depends on the data, and there is not any prescription for all cases. However, in the most cases considered in this study, the most variance is covered by the first 5 components ($i = 5$). Therefore, if any damage occurs in the structure, it should affect the data and consequently the variance of data, and it should be detected by this damage index. It should be noted that the DI_{rpc} for all windows are calculated w.r.t. time and thus their variations with time are plotted. As a final point, the confidence interval developed in the training phase should be considered as a benchmark (baseline) for detecting any possible damages in the rest of the data.

41.2.2 Case Study

The standard data sets used in this study are from a three-story frame aluminium structure reported in [25] and has been intensively used for SHM validation in recent unsupervised damage detection approaches [26]. Test bed building model is a four-degree-of-freedom system with varied practical conditions, including variations in stiffness and mass loading. These variations simulate temperature changes and traffic, respectively. Raw data were collected by four accelerometers mounted on the structure, as shown in Figure 41.2.

A nonlinear damage scenario was introduced by contacting a suspended column with a bumper mounted on the floor below to simulate fatigue crack that can open and close under loading conditions or loose connections in structures. The smaller gap between the column and the bumper will result in the higher level of damage. Therefore, different levels of damage were created by adjusting the gap. More details about the test structure can be found in [25]. Acceleration time-series from 17 different structural state conditions were collected, as described in Table 41.1, where the first 9 state conditions introduce the undamaged and the rest are damaged states. Time-series discretized into 4,096 data points sampled at 3.125 ms intervals corresponding to a sampling frequency of 320 Hz. For each structural state condition, data were acquired from 100 separate tests.

41.2.3 Accuracy of Damage Detection Methods

A comprehensive comparison will be made between different methods to evaluate the effectiveness of the MKPCA method. For generalization purpose, the feature vectors are split into the training and test matrices. The training matrix, $\mathbf{X}$, permits each algorithm to learn the underlying distribution and dependency of all undamaged states on the simulated operational and environmental variability. Thus, this matrix is composed of extracted parameters from 50 out of 100 tests from each undamaged state (states 1–9), and so it has a dimension of 450_40. The test matrix $\mathbf{Z}$ (1250_40) is composed of extracted parameters from the remaining 50 tests of each undamaged state together with extracted parameters from all the 100 tests of each damaged state (states 10–17).

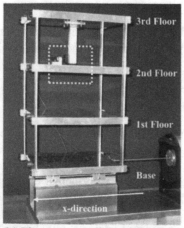

(a) Three-story Building Structure and Shaker.

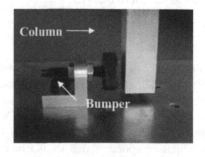

(b) Adjustable Bumper and Column.

FIGURE 41.2   Three-story frame structure [25].

This procedure permits one to evaluate the generalization performance of the machine learning algorithms in an exclusive manner because time-series used in the test phase are not included in the training phase. During the test phase, the algorithms are expected to detect deviations from the normal condition when feature vectors come from damaged states, even in the presence of operational and environmental effects.

In the first step, the features of acceleration response of structure will be extracted using WPRE, for reaching to this goal, Battle-Lemarie is adopted as the basis wavelet package function in order to decompose the signals to be analyzed into different frequency bands and to make each frequency band energy independent and irredundant. The next step is to carry out statistical modeling for feature classification. In that regard, the algorithm based on MKPCA is implemented in an unsupervised learning mode by first taking into account features from all the undamaged state conditions (training matrix). To evaluate the effectiveness of the proposed methods several existing studies in the literature [26] are used for comparison, which includes one-class support vector machines (SVM), kernel principal component algorithm (KPCA) and Greedy kernel principal component algorithm (GKPCA). For each algorithm, the damage index is stored into a 1,250-length vector.

The subset of 25% of the training data used for multiple kernel principal component algorithm (MKPCA kernel projection and algorithm is configured to retain 90% of the variability in the data after dimension reduction. Furthermore, the RBF kernel with parameter $\gamma = 0.025$ is chosen for the KPCA kernel.

The receiving operating characteristic (ROC) curves are used to compare the performance of various methods based on the trade-off between Type I and II errors. The point at the left-upper corner of the plot (0, 1) is called a perfect classification. Figure 41.3 illustrates the ROC curves for the aforementioned algorithms.

The curves show that none of the algorithms can have a perfect classification with a linear threshold because none of the curves goes through the left-upper corner, and neither have supremacy in terms of true detection rate for the entire false alarm domain. Furthermore, for levels of significance around 5%, the GKPCA and MKPCA have better true detection rate than one-class SVM and KPCA. They are in fact the ones that maximize the true detection of damaged cases with similar performances in terms of false alarm rate. It is worth to be noted that the false alarm rate of 0.05 is acceptable in real-world scenarios of SHM [26].

In order to quantify the performance of the MKPCA for a given threshold, Figure 41.4 plots the DI_{rpc} for the feature vectors of the entire test data along with a threshold defined on the 95% cut-off value over the training data. MKPCA shows a monotonic relationship between the level of damage and amplitude of the DI_{rpc}, even when operational and environmental variabilities are present. In other words, the approaches are able to remove the operational and environmental effects in such a way that DI_{rpc} from state15, 16 and 10 have similar amplitudes, as well as state17 is associated with state13. As stated in Table 41.1, states 15–17 are the variant states of either state10 or state13 with operational effects.

TABLE 41.1
Data Labels of the Structural State Conditions

Label	Description	Label	Description	Label	Description
State 1	Baseline condition	State 10	Gap (0.20 mm)	State 14	Gap (0.05 mm)
State 2	Added mass (1.2 kg) at the base	State 11	Gap (0.15 mm)	State 15	Gap (0.20 mm) and mass (1.2 kg) at the base
State 3	Added mass (1.2 kg) on the 1st floor	State 12	Gap (0.13 mm)	State 16	Gap (0.20 mm) and mass (1.2 kg) on the 1st floor
State 4–9	States 4–9: 87.5% stiffness reduction at various positions to simulate temperature impact	State 13	Gap (0.10 mm)	State 17	Gap (0.10 mm) and mass (1.2 kg) on the 1st floor

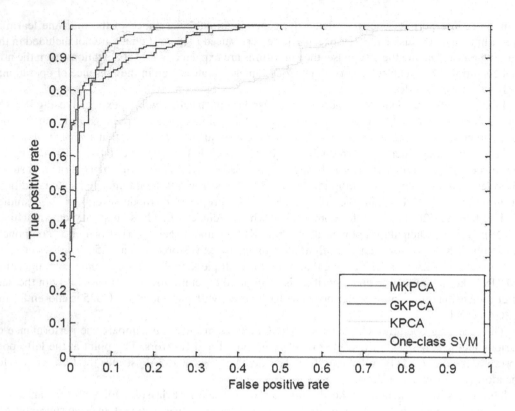

FIGURE 41.3   The ROC curves for the GKPCA, KPCA, one-class SVM and MKPCA algorithms.

The performance of statistical classification methods is mostly indicated by Type I (a false-positive indication of damage) and Type II (false-negative indication of damage) errors. This technique recognizes that a false-positive classification may have different consequences than false-negative ones [26]. In Figure 41.4, Type I errors are DI_{rpc} that exceed the threshold value in the undamaged condition domain (1–450). On the other hand, Type II errors are DI_{rpc} that does not surpass the threshold value in the damaged condition domain (451–1,250).

Based on Figure 41.4, two important conclusions can be drawn: a) when life safety issues are the main reason for deploying an SHM system and one wants to minimize false-negative indications of damage, the proposed scheme has a good performance. b) When reliability issues are ruling an SHM system and one wants to minimize false-positive indications of damage without increasing the false-negative indications of damage, the proposed scheme is also appropriate. Finally, MKPCA has good generalization performance, which is a very important advantage for real-world applications, where the thresholds are defined based on undamaged data used in the training phase.

41.3   SUMMARY AND CONCLUSIONS

In this paper, the performance of the proposed two-stage method for structural damage detection, under varying operational and environmental conditions, is shown using benchmark data sets from a well-known base-excited three-story frame structure.

First, features of acceleration response of structure are extracted using WPRE. Then the features are fed to MKPCA to classify the health conditions. Through the case studies on the frame structure, the MKPCA algorithm is shown to be reliable to create a global damage index that can separate damaged from undamaged conditions, even when the structure is operating under varying operational and environmental conditions and yields better results in terms of minimization of total misclassifications.

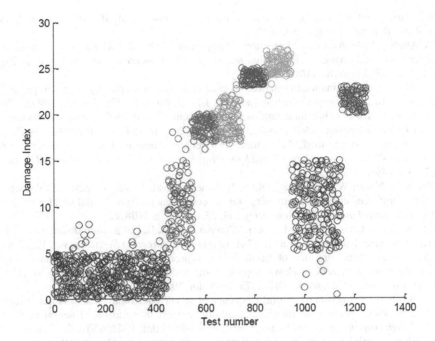

FIGURE 41.4 I_{rpc} calculated based on feature vectors from the undamaged and damaged condition using MKPCA.

In the proposed approach to make the KPCA method more practical and feasible for long-term monitoring, a windowing technique is employed. As a result, MKPCA algorithm proposed a promising upgraded version of KPCA that can segment the data flow to detect damage more precisely.

Finally, MKPCA algorithms do not require a direct measure of the sources of variability, e.g. traffic loading and temperature. Instead, the algorithm relies only on measured response time-series data acquired under varying operational and environmental effects.

ACKNOWLEDGEMENTS

Involvement and contributions to this research project was made possible, for the third co-author, in part due to a release time funding from Donald E. Bently Center for Engineering Innovation, Mechanical Engineering, Cal Poly, San Luis Obispo. Herein this support is acknowledged.

REFERENCES

1. C. Boller, F.-K. Chang, and Y. Fujino, *Encyclopedia of Structural Health Monitoring*. John Wiley & Sons, Hoboken, NJ, 2009.
2. Z. Ying, N. Mohammad, W. A. Altabey, R. Ghiasi, Z. Wu, "Deep learning-based damage, load and support identification for a composite pipeline by extracting modal macro strains from dynamic excitations", *Applied Sciences*, vol. 8, no. 12, 2018, 2564. doi: 10.3390/app8122564.
3. Z. Ying, N. Mohammad, W. A. Altabey, L. Naiwei, "Reliability evaluation of a laminate composite plate under distributed pressure using a hybrid response surface method", *International Journal of Reliability, Quality and Safety Engineering*, vol. 24, no. 3, 2017, 1750013. doi: 10.1142/S0218539317500139.
4. Z. Ying, N. Mohammad, A W. A. Altabey, Z. Wu, "Fatigue damage identification for composite pipeline systems using electrical capacitance sensors", *Smart Materials and Structures*, vol. 27, no. 8, 2018, 085023. doi: 10.1088/1361–665x/aacc99.
5. Z. Ying, N. Mohammad, W. A. Altabey, R. Ghiasi, Z. Wu, "A fatigue damage model for FRP composite laminate systems based on stiffness reduction", *Structural Durability and Health Monitoring*, vol. 13, no. 1, 2019, 85–103. doi: 10.32604/sdhm.2019.04695.

6. C. R. Farrar and K. Worden, *Structural Health Monitoring: A Machine Learning Perspective.* John Wiley & Sons, Hoboken, NJ, 2012.

7. W. A. Altabey, N. Mohammad, "Detection of fatigue crack in basalt FRP laminate composite pipe using electrical potential change method", *Journal of Physics: Conference Series*, vol. 842, 2017, 012079. doi: 10.1088/1742-6596/842/1/012079.

8. W. A. Altabey, "Delamination evaluation on basalt FRP composite pipe by electrical potential change", *Advances in Aircraft and Spacecraft Science*, vol. 4, no. 5, 2017, 515–528. doi: 10.12989/aas.2017.4.5.515.

9. W. A. Altabey, "EPC method for delamination assessment of basalt FRP pipe: electrodes number effect", *Structural Monitoring and Maintenance*, vol. 4, no. 1, 2017, 69–84. doi: 10.12989/smm.2017.4.1.069.

10. W. A. Altabey, N. Mohammad, "Monitoring the water absorption in GFRE pipes via an electrical capacitance sensors", *Advances in Aircraft and Spacecraft Science*, vol. 5, no. 4, 2018, 411–434. doi: 10.12989/aas.2018.5.4.499.

11. Y. Zhao, M. Noori, W. A. Altabey, and S. B. Beheshti-Aval, "Mode shape-based damage identification for a reinforced concrete beam using wavelet coefficient differences and multiresolution analysis", *Structural Control and Health Monitoring*, vol. 25, no. 1, Jan. 2018, e2041.

12. R. Ghiasi, M. R. Ghasemi, and M. Noori, "Comparative studies of metamodeling and AI-Based techniques in damage detection of structures", *Advances in Engineering Software*, vol. 125, 2018, 101–112.

13. W. A. Altabey, "Free vibration of basalt fiber reinforced polymer (FRP) laminated variable thickness plates with intermediate elastic support using finite strip transition matrix (FSTM) method", *Vibroengineering*, vol. 19, no. 4, 2017, 2873–2885. doi: 10.21595/jve.2017.18154.

14. W. A. Altabey, "Prediction of natural frequency of basalt fiber reinforced polymer (FRP) laminated variable thickness plates with intermediate elastic support using artificial neural networks (ANNs) method", *Vibroengineering*, vol. 19, no. 5, 2017, 3668–3678. doi: 10.21595/jve.2017.18209.

15. W. A. Altabey, "High performance estimations of natural frequency of basalt FRP laminated plates with intermediate elastic support using response surfaces method", *Vibroengineering*, vol. 20, no. 2, 2018, 1099–1107. doi: 10.21595/jve.2017.18456.

16. W. A. Al-Tabey, "Vibration analysis of laminated composite variable thickness plate using finite strip transition matrix technique," In *MATLAB verifications MATLAB- Particular for Engineer*, Kelly Bennett, ed, InTech USA, Charlotte, NC, vol. 21, 2014, 583–620, 980-953-307-1128-8. doi: 10.5772/57384.

17. R. Ghiasi and M. R. Ghasemi, "An intelligent health monitoring method for processing data collected from the sensor network of structure", *Steel & Composite Structures*, vol. 29, no. 6, 2018, 703–716.

18. F. N. Catbas and M. Malekzadeh, "A machine learning-based algorithm for processing massive data collected from the mechanical components of movable bridges", *Automation in Construction*, vol. 72, 2016, 269–278.

19. Y. Lei, F. Jia, J. Lin, S. Xing, and S. X. Ding, "An intelligent fault diagnosis method using unsupervised feature learning towards mechanical big data", *IEEE Transactions on Industrial Electronics*, vol. 63, no. 5, 2016, 3137–3147.

20. Z. Ying, N. Mohammad, W. A. Altabey, "Damage detection for a beam under transient excitation via three different algorithms", *Structural Engineering and Mechanics*, vol. 63, no. 6, 2017, 803–817, doi: 10.12989/sem.2017.64.6.803.

21. Z. Ying, N. Mohammad, W. A. Altabey, T. Awad, "A comparison of three different methods for the identification of hysterically degrading structures using BWBN model", *Frontiers in Built Environment*, vol. 4, 2019, 80, doi: 10.3389/fbuil.2018.00080.

22. N. Mohammad, W. Haifegn, W. A. Altabey, I. H. S. Ahmad, "A modified wavelet energy rate based damage identification method for steel bridges", *International Journal of Science & Technology, Scientia Iranica*, vol. 25, no. 6, 2018, 3210–3230, doi: 10.24200/sci.2018.20736.

23. R. Ghiasi, P. Torkzadeh, and M. Noori, "A machine-learning approach for structural damage detection using least square support vector machine based on a new combinational kernel function", *Structural Health Monitoring,* vol. 15, no. 3, May 2016, 302–316.

24. L. J. Cao, K. S. Chua, W. K. Chong, H. P. Lee, and Q. M. Gu, "A comparison of PCA, KPCA and ICA for dimensionality reduction in support vector machine", *Neurocomputing*, vol. 55, 2003, 321–336.

25. E. Figueiredo, G. Park, J. Figueiras, C. Farrar, and K. Worden, *Structural Health Monitoring Algorithm Comparisons Using Standard Data Sets*, Los Alamos National Laboratory (LANL), Los Alamos, NM, 2009.

26. A. Santos, E. Figueiredo, M. F. M. Silva, C. S. Sales, and J. C. W. A. Costa, "Machine learning algorithms for damage detection: Kernel-based approaches", *Journal of Sound and Vibration*, vol. 363, Feb. 2016, 584–599.

42 An Investigation into the Active Vibration Control of Three Coupled Oscillators Using the Twin Rotor Dampers

R. Terrill and U. Starossek
Hamburg University of Technology

CONTENTS

42.1 INTRODUCTION

The twin rotor damper (TRD) is an active mass damping device consisting of two eccentric rotating control masses. In the principle mode of operation, the continuous rotation mode, both control masses rotate in opposite directions with a nearly constant angular velocity, producing a harmonic control force in an energy- and power-efficient manner. The application of the TRD for active vibration control of systems with a single dominate mode of vibration has been extensively research in [1–4]. In [1] and [4], the TRD is used to control a system with two uncoupled translational degrees of freedom. Here, the application of the TRD in the continuous rotation mode is investigated for the first time for systems which respond with multiple dominate modes of vibration. Two straightforward strategies for operating the TRD in the continuous rotation mode for such systems are introduced and discussed. It is shown that both strategies successfully damp all modes of vibration of the three coupled oscillators.

This paper is structured as follows. In the following section, the TRD and coupled oscillators are presented. The basic layout of a single TRD unit is introduced along with the continuous rotation mode for a single mode of vibration. It is then shown how the response of an oscillating system with multiple modes of vibration can be partitioned such that it can be used to operate the TRD in the continuous rotation mode. In the subsequent section, two control strategies considered are presented and discussed along with their respective control algorithms. In the third section, numerical simulations are performed and their results are discussed. Finally, in the last section conclusions are drawn.

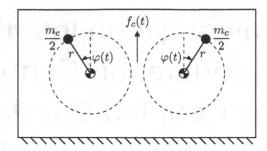

FIGURE 42.1   TRD layout [1].

42.2   TRD AND COUPLED OSCILLATORS

42.2.1   TRD AND CONTINUOUS ROTATION MODE

In Figure 42.1, the basic layout of the TRD is shown. It consists of two eccentrically rotating control masses, $0.5\,m_c$, which are hinged to two parallel axes by a massless rod with the length r. The angular position $\varphi(t)$ defines the position of each rotor (control mass with massless rod). In the principal mode of operation, the continuous rotation mode, the control masses rotate with a mostly constant angular velocity in opposite directions.

The operation in the continuous rotation mode is made possible by tracking the target angular position, $\varphi_t(t)$, derived in [1], see Eq. (42.1), in which $x(t)$ is the displacement of an oscillating system, the time derivative, $\dot{x}(t)$, its velocity, and ω_n its eigenfrequency. By tracking Eq. (42.1), the angular velocity, $\dot{\varphi}(t)$, is approximately constant and equal to ω_n.

$$\varphi_t(t) = a\tan 2\left(x(t); \frac{\dot{x}(t)}{\omega_n} \right) + \pi \qquad (42.1)$$

If the motion of the oscillating system is monofrequent, the resulting control force, $f_c(t)$, is harmonic and in anti-phase to $\dot{x}(t)$. Furthermore, in [1], it is shown that using appropriate control algorithms to operate in the continuous rotation mode, the TRD can damp monofrequent vibrations in an energy-efficient manner. In fact, it is shown that the power required by the TRD in the continuous rotation mode is at all times negative. This is one of the greatest advantages that the TRD has over conventional active mass damping devices.

A drawback of operating the TRD in the continuous rotation mode is the constant amplitude of $f_c(t)$. This leads to a re-excitation of the oscillating system when the system forces become much smaller than the control force. To avoid this, multiple strategies have been developed, see [1,2]. Here the on/off method is used; the TRD is turned off when the vibration amplitude, $A(t)$, falls below a certain threshold, A_{off} and turns on when $A(t)$ exceeds a higher threshold, A_{on}.

42.2.2   COUPLED OSCILLATORS

Until now, research regarding the TRD has only considered oscillating systems which respond with a single mode of vibration. For such systems, tracking $\varphi_t(t)$ presented in the previous section is sufficient to damp the systems vibrations in an energy-efficient manner. The states (displacement and velocity) necessary to calculate $\varphi_t(t)$ can be readily measured or made available using an observer. However, for systems which respond with multiple vibration modes, the measured states will be multifrequent (consist of the superposition of multiple harmonic vibrations with unique frequencies), [5]. In this section, it will be shown how the states of the oscillating system can be partitioned such that $\varphi_t(t)$ can receive a monofrequent input.

The system of three coupled oscillators shown in Figure 42.2 is considered in this paper. In the following, the free vibrations of this system will be examined. Each oscillator has a stiffness k_i, damping coefficient c_i, and displacement coordinate, $x_i(t)$. These properties, those of the TRD units, and the displacement of each oscillator are denoted with an alphabetical subscript, i. This avoids possible confusion with the numerical subscripts used in the following for the modal analysis. All oscillators have an equal mass m and are equipped with a TRD unit with a control mass of $0.5\, m_{c,I}$, a massless rod with the length r, and an angular position $\varphi_i(t)$.

The motion of the coupled oscillators is governed by the system of differential equations:

$$M\ddot{x}(t) + D\dot{x}(t) + Kx(t) = 0 \qquad\qquad (42.2)$$

in which $\mathbf{M}$ is the mass matrix, $\mathbf{D}$ is the damping matrix, $\mathbf{K}$ is the stiffness matrix, and $x(t) = \begin{bmatrix} x_a(t) & x_b(t) & x_c(t) \end{bmatrix}^T$ the displacement vector of the coupled oscillators. Using modal

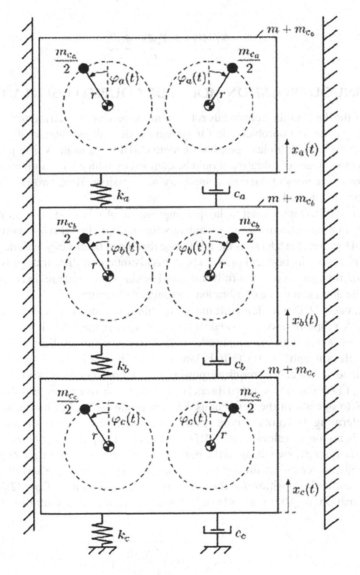

FIGURE 42.2   Three coupled oscillators with TRD.

analysis, the displacement vector can be expressed as shown in Eq. (42.3), in which $\boldsymbol{\Phi}$ is the shape matrix of the coupled oscillators and $\boldsymbol{Y}(t)$ is the modal amplitude vector, [5,6]. The shape matrix and eigenfrequencies (damped natural circular frequencies), ω_1, ω_2, ω_3, of the system can be found using standard methods, see [5]. From Eq. (42.3) it can be seen that the displacements of each individual oscillator can be partitioned as shown in Eq. (42.4) for $x_a(t)$. The displacements $x_{a,1}(t)$, $x_{a,2}(t)$, and $x_{a,3}(t)$ are the displacement components of $x_a(t)$ with respect to each eigenfrequency. $x_b(t)$ and $x_c(t)$ can be partitioned analogously. Thus the displacement components of each oscillator with respect to each eigenfrequency can be made available by simply using modal analysis. This partition can also be performed analogously for the velocity of the system.

$$x(t) = \phi y(t) = \begin{pmatrix} \phi_{11} & \phi_{12} & \phi_{13} \\ \phi_{21} & \phi_{22} & \phi_{23} \\ \phi_{31} & \phi_{32} & \phi_{33} \end{pmatrix} \begin{bmatrix} Y_1(t) \\ Y_2(t) \\ Y_3(t) \end{bmatrix} \tag{42.3}$$

$$x_a(t) = \underbrace{\phi_{11} Y_1(t)}_{x_{a,1}(t)} + \underbrace{\phi_{12} Y_2(t)}_{x_{a,2}(t)} + \underbrace{\phi_{13} Y_3(t)}_{x_{a,3}(t)} \tag{42.4}$$

42.3   CONTINUOUS ROTATION MODE FOR COUPLED OSCILLATORS

The operation of the TRD in the continuous rotation mode produces a harmonic control force with a single frequency. This is unproblematic for systems with a single degree of freedom (SDOF). However, for systems with multiple degrees of freedom (MDOF), a strategy is required for deciding which of the systems modes of vibration should be damped by which TRD unit under what circumstances. Numerous strategies of varying complexity are possible. Here two relatively straightforward strategies are considered.

In strategy 1, each TRD will be set to damp a single mode of vibration. This strategy can resemble strategies for finding optimal design parameters for multiple tuned mass dampers for MDOF systems, see [7]. However, the TRD has the advantage that it can be directly assigned to an eigenfrequency without changing its physical properties. A disadvantage of this strategy is that the adjustability of active damping devices is not fully utilized. For simplicity and brevity, each TRD unit will be set to damp the dominate mode of vibration present at its location.

In strategy 2, each TRD will damp all modes of vibration, starting with the highest relevant vibration mode. A vibration mode, j, is considered relevant if the vibration amplitude $A_{i,j}(t)$ of the oscillator, i, is larger than the amplitude $A_{j,\mathrm{on}}$. $A_{i,j}(t)$ is given by Eq. (42.5). The vibration mode remains relevant until $A_{i,j}(t)$ falls below a lower threshold $A_{j,\mathrm{off}}$. The highest vibration mode is prioritized to lower the required power. In the continuous rotation mode, the target angular velocity (TAV), $\dot{\varphi}_t(t)$, is approximately ω_j. Thus, the rotors are required to accelerate to a higher TAV to operate in the continuous rotation mode for higher modes of vibration (i.e., higher ω_j). Accelerating the rotors requires power. However, once a vibration mode is damped to the point that it is no longer relevant, the TRD switches to damp the next (lower) relevant mode of vibration. Consequently, switching to the next mode of vibration no longer requires the rotors to accelerate, they can decelerate until $\dot{\varphi}_t(t) \approx \omega_j$, whereas ω_j is now smaller, thus requiring less power. The target angular position for this strategy, $\varphi_{t,i}(t)$, is given by Eq. (42.6), in which the target angular position, $\varphi_{t,ij}(t)$, for a TRD installed at oscillator i and vibration mode j is shown in Eq. (42.7).

$$A_{i,j}(t) = \sqrt{x_{i,j}^2(t) + \left(\frac{\dot{x}_{i,j}(t)}{\omega_j} \right)} \tag{42.5}$$

$$\varphi_{t,i}(t) = \begin{cases} \varphi_{t,i3}(t) \text{ if } A_{i,3}(t) > A_{3,on} \\ \varphi_{t,i2}(t) \text{ if } A_{i,2}(t) > A_{2,on} and\ A_{i,3}(t) < A_{3,off} \\ \varphi_{t,i1}(t) \text{ if } A_{i,1}(t) > A_{1,on} and\ A_{i,2}(t) < A_{2,off} \end{cases}$$ (42.6)

$$\varphi_{t,ij}(t) = a\tan 2\left(x_{i,j}(t); \frac{\dot{x}_{i,j}(t)}{\omega_j} \right) + \pi$$ (42.7)

In order to avoid re-exciting the coupled oscillator, the on/off method previously discussed is implemented for both strategies. For the first strategy, each TRD unit is turned on if $A_{i,j}(t)$ exceeds $A_{j,on}$ and off once $A_{i,j}(t)$ falls below $A_{j,off}$, whereas j is the vibration mode assigned to the TRD i.

For the second strategy, a TRD is turned on if $A_{i,j}(t)$ exceeds any $A_{j,on}$ and turned off once $A_{i,j}(t)$ falls below all $A_{j,off}$.

42.3.1 CONTROL ALGORITHM FOR STRATEGY 1

The only difference between strategy 1 and the control strategy for a TRD applied to a SDOF oscillator is the decision process for determining which mode of vibration it will damp. Therefore, the closed-loop angular position control combined with open-loop angular velocity control presented in [2] for an SDOF oscillator is also applied for this strategy.

In the previous section, it was stated that each TRD is set to the dominate mode of vibration present at its location. This can be determined by analyzing the frequency response between the vibration displacements, $x_i(t)$, and a harmonic excitation force at each oscillator or by directly analyzing the modal response; for brevity, the latter of which is dealt here.

42.3.2 CONTROL ALGORITHM FOR STRATEGY 2

For strategy 2, the previous control algorithm is augmented to allow for a variable TAV. This is done by switching the open-loop angular velocity gain such that the TAV, $\dot{\varphi}_{t,i}(t)$, is tracked; hence $\dot{\varphi}_i(t) \approx \omega_j$.

The control algorithm for strategy 2 is depicted in Figure 42.3. The coupled oscillators provide the states $x_{i,j}(t), \dot{x}_{i,j}(t)$, and $A_{i,j}(t)$, which can then be used to compute $\varphi_{t,i}(t)$ according to Eqs. (42.6) and (42.7). The angular position error, $e_i(t)$, is then calculated by comparing $\varphi_{t,i}(t)$ to the actual angular position $\varphi_i(t)$. To ensure that the rotors are always driven to the nearest target angular position, $e_i(t)$ is restricted to values between $-\pi$ and π, producing the control error $e_{c,i}(t)$ [2].

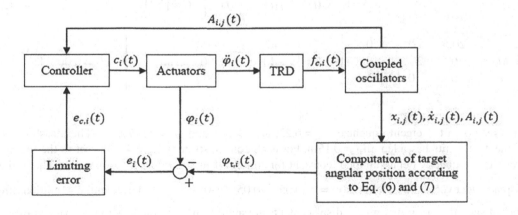

FIGURE 42.3 Control algorithm for strategy 2.

The control error is minimized by the closed-loop angular position and open-loop angular velocity control. The control effort, $c_i(t)$, produced by the controller is given by Eq. (42.8) in which K_p is the proportional gain for the angular position control and $K_v(A_{i,j}(t))$ the proportional gain for the angular velocity control. K_v is a function of $A_{i,j}(t)$ and allows the TRD to switch from damping one vibration mode to another while keeping the control errors adequately small. The control effort is then fed into the actuators producing the angular acceleration, $\ddot{\phi}_i(t)$, of the rotors. This acceleration is the input of the TRD, which produces the control force $f_{c,i}(t)$ applied to the coupled oscillator.

$$c_i(t) = e_{c,i}(t)K_p + K_v(A_{i,j}(t)) \tag{42.8}$$

42.4   NUMERICAL SIMULATIONS

For the numerical simulations, the coupled oscillator in Figure 42.2 is simulated using the state space model given by

$$\dot{z}(t) = Az(t) + Bu(t) \tag{42.9}$$

The system matrix $\mathbf{A}$, the state vector $z(t)$, the input matrix $\mathbf{B}$, and the input vector $u(t)$ are given in Eq. (42.10), [8]. Here, all bold characters are 3×3 matrices, $\mathbf{I}$ is the identity matrix and $\mathbf{0}$ is a zero matrix. The matrices $\mathbf{M}$, $\mathbf{K}$, and $\mathbf{D}$ are the system matrices from Eq. (42.2). The output equation of the state space model is given in Eq. (42.11), in which the output vector $y(t)$ and output matrix $\mathbf{C}$ are given in Eq. (42.12). For this numerical example, the properties of the coupled oscillator are given by the system matrices in Eq. (42.13). In $\mathbf{M}$, the control masses of the TRD units, $m_{c,I} = 0.5\,\mathrm{kg}$, are included.

$$Z(t) = \begin{bmatrix} \dot{x}_a(t) \\ \dot{x}_b(t) \\ \dot{x}_c(t) \\ \dot{x}_a(t) \\ \dot{x}_b(t) \\ \dot{x}_c(t) \end{bmatrix}, A = \begin{bmatrix} 0 & I \\ -M^{-1}K & -M^{-1}D \end{bmatrix}, B = \begin{bmatrix} 0 \\ -M^{-1}B_0 \end{bmatrix}, B_0 = \begin{bmatrix} 0 \\ 0 \\ 0 \\ 1 \\ 1 \\ 1 \end{bmatrix}, u(t) = \begin{bmatrix} f_a(t) \\ f_b(t) \\ f_c(t) \end{bmatrix}$$

$$\tag{42.10}$$

$$y(t) = C_z(t) \tag{42.11}$$

$$y(t) = \begin{bmatrix} x_a(t) & x_b(t) & x_c(t) \end{bmatrix}^T, C = [0\ I] \tag{42.12}$$

$$M = \begin{bmatrix} 20.5 & 0 & 0 \\ 0 & 20.5 & 0 \\ 0 & 0 & 20.5 \end{bmatrix} kg, K = \begin{bmatrix} 4000 & -4000 & 0 \\ -4000 & 8000 & -4000 \\ 0 & -4000 & 8000 \end{bmatrix} \frac{N}{m}, D = 0.0001K \frac{N_s}{m}$$

$$\tag{42.13}$$

This results in the eigenfrequencies $\omega_1 = 6.22$, $\omega_2 = 17.42$, and $\omega_3 = 25.17\,\mathrm{rad/s}$. The massless rod of the TRD units has a length $r = 0.15\,\mathrm{m}$. For each control strategy $K_p = 4$, K_v is set to the respective eigenfrequency being damped (constant for strategy 1 and variable for strategy 2). The initial displacement vector considered is $x_0 = \begin{bmatrix} 0.05 & 0.05 & -0.05 \end{bmatrix}^T$ m. These initial conditions are chosen such that all initial modal displacements are approximately equal. All initial velocities are considered to be zero.

For strategy 1, the modal response due to the initial conditions was analyzed, and it was determined that the dominate mode at x_a is ω_1, at x_b is ω_3, and at x_c is ω_2. The controlled and uncontrolled modal responses, $y_j(t)$ and $y_{j,c}(t)$, are shown in the top part of Figure 42.4. In order to show that all modes of vibration at each oscillator are controlled, the uncontrolled and controlled displacements, $x_i(t)$ and $x_{i,c}(t)$, are shown in the bottom part of Figure 42.4. Here it can be seen that strategy 1 is successful at damping the total motion of the coupled oscillators.

In Figure 42.5, the controlled and uncontrolled responses are shown for strategy 2. Here it can be seen that, as for strategy 1, the TRD units are capable of damping the total motion of the coupled oscillators. Furthermore, the transition from damping one mode of vibration to the next can be seen from the controlled modal response. For example, the amplitude of the controlled modal response for the first mode of vibration $y_{1,c}(t)$ is constant until $t \approx 2$ seconds. Beginning at this time each TRD unit starts to damp the first mode of vibration. At $t \approx 4$ seconds, all vibrations fall below the lower vibration threshold $A_{j,\text{off}}$ and the TRD units are turned off. For these simulations $A_{j,\text{off}}$ is set to 0.5 cm.

Comparing Figures 42.4 and 42.5 it can be seen that the vibrations are damped slightly faster with strategy 2. Furthermore, a secondary effect of strategy 2 is that the displacement response becomes monofrequent over time. As the control force produced in both modes of operation is larger for the higher modes of vibration, these higher modes are damped relatively fast. However, it can be argued that due to the higher inherent damping of the higher modes, it is not necessary to damp them.

An additional advantage of strategy 2 is that the TRD units work together to damp each mode of vibration. This can be seen best for the first mode of vibration $y_{1,c}(t)$ in Figures 42.4 and 42.5. For strategy 1, the first mode of vibration is damped from $t = 0$ seconds until $y_{1,c}(t)$ falls below $A_{1,\text{off}}$ at $t \approx 6$ seconds. For strategy 2, the TRD units start to damp $y_{1,c}(t)$ at around $t \approx 2$ seconds and $y_{1,c}(t)$ falls below $A_{1,\text{off}}$ at $t \approx 4$ seconds. Therefore, the first mode of vibration is controlled in approximately 2 seconds for strategy 2, but strategy 1 requires approximately 6 seconds to reduce the vibrations to similar levels. This is because the control force for the first mode of vibration in strategy 2 is produced by each TRD unit. In contrast, in strategy 1 only a single TRD unit is used to damp the first mode of vibration.

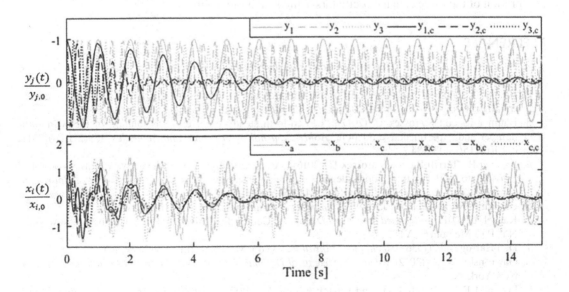

FIGURE 42.4 Modal and displacement responses for strategy 1.

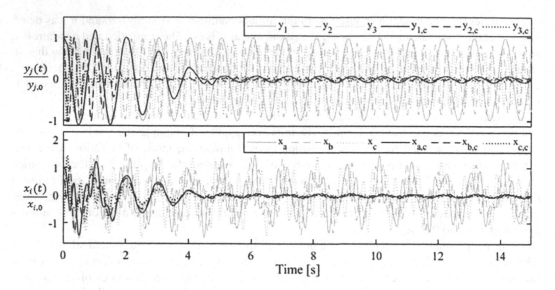

FIGURE 42.5   Modal and displacement responses for strategy 2.

42.5   CONCLUSIONS

An active mass damper, the TRD, is presented for the control of three coupled oscillators. The TRD consists of two eccentric control masses which in the continuous rotation mode, the principle mode of operation, rotate with a constant angular velocity around two parallel axes, producing a monofrequent harmonic control force in an energy- and power-efficient manner.

Three TRD units are implemented for the control of the coupled oscillators. It is shown how the response of an MDOF system can be partitioned such that the TRDs can be operated in the continuous rotation mode. Two strategies for implementing multiple TRD units for the control of systems with multiple dominate modes of vibration are then presented along with the respective control algorithms. In the first strategy, each TRD unit damps a single mode of vibration. In the second, each TRD damps all modes of vibration. Both strategies are shown to effectively damp all modes of vibration of the three coupled oscillators using a numerical simulation.

REFERENCES

1. Bäumer R. (2017), *Active vibration control using the centrifugal forces of eccentrically rotating masses*, Structural Analysis Institute, Hamburg University of Technology, Ph.D. thesis.
2. Bäumer R. and Starossek U. (2016). "Acive vibration control using centrifugal forces created by eccentrically rotating masses." *ASME. Journal of Vibration and Acoustics*. doi:10.1115/1.4033358.
3. Bäumer R. and Starossek U. (2017), "Closed-form steady-state response solution of the twin rotor damper and experimental validation." *ASME. Journal of Vibration and Acoustics*. 139(2):021017–11. doi:10.1115/1.4035134.
4. Bäumer R., Terrill R., and Starossek U. (2016). "An energy approach for the active vibration control of an oscillator with two translational degrees of freedom using two auxiliary rotating masses." *Journal of Physics: Conference Series*. doi:10.1088/1742-6596/744/1/012230.
5. Clough R. W. and Penzien J. (1993). *Dynamics of Structures*. McGraw-Hill, New York, NY.
6. Kausel, E. (2017). *Advanced Structural Dynamics*. Cambridge University Press, Cambridge New York, NY, Port Melbourne, VIC.
7. Den Hartog, J. P. (1956). *Mechanical Vibrations* (4 ed.). McGraw-Hill, New York, NY.
8. Gawronski, W. K. (2012). *Advanced Structural Dynamics and Active Control of Structures*. Springer, New York, NY.
9. Franklin F. F., Powell J. D., and Emami-Naeini A. (2002). *Feedback Control of Dynamic Systems*. Prentice Hall, Upper Saddle River, NJ, USA.

43 Development and Experimental Verification of IoT Sensing-Based Structural Seismic Monitoring System

Rongzhi Zuo, J. Dang, and C.S. Goit
Saitama University

CONTENTS

43.1 INTRODUCTION

Since the 2011 Great East Japan Earthquake, study on timely emergency response has been paid tremendous attention considering the huge damages such disaster can cause and the benefits emergency response can bring. Emergency inspection is compulsory for carrying out the post-earthquake assessment, but there are difficulties to carry out emergency inspection considering the complex situation in the aftermath of an earthquake. The safety of inspectors cannot be ensured when they approach the structure and carry out the inspection. On the other hand, the time cost is very high from inspection to decision making. Bridges perform as key infrastructures to recover emergency logistics function for conducting evacuation and rescue, and monitoring the response of bridges during earthquake is valuable. But the existing structural health monitoring (SHM) system is not feasible to be installed densely in a state or city scale due to the high cost it requires.

To speed up the procedure of post-earthquake assessment, long-term vibration monitoring in bridges by deploying instruments densely is one of the straightforward measures for this purpose. With the advancing sensor and information technology, the IoT sensing based SSM system is becoming more appealing. It comes with low cost, small size, low power consumption and programmability. A low-cost, high-accuracy, long-term and real-time acceleration monitoring system was proposed in this study. The measurement accuracy and sampling rate were verified by shaking table tests carried out in laboratory environment. The experiment results show satisfactory agreements between the reference sensor and the IoT sensor in both time and frequency domains. A prototype has been installed in a research building to detect seismic events for a long term. Two earthquakes that happened recently can be recognized by monitoring data, though the seismic intensities in the region where the prototype was deployed were both low according to the Japan meteorological agency (JMA) seismic intensity scale.

43.2 IOT DEVICES

43.2.1 RASPBERRY PI 3 MODEL B+

The microcomputer adopted in this study is Raspberry pi 3 Model B+, as shown in Figure 43.1, a credit card sized motherboard with onboard computation capability. The properties of Raspberry Pi 3 model B+ are shown in Table 43.1, which shows comprehensive and compatible functions. The 40 general-purpose input/output (GPIO) pins are programmable for several communication protocols for data acquisition. Its compact size means the requirement for the installation room is more tolerant. Moreover, the power consumption is comparatively lower than the conventional computer. Besides, it only costs 6,000 Japanese Yen (JPY) for each motherboard.

43.2.2 ADXL355 ACCELEROMETER

The 3-axis ADXL355 accelerometer in Figure 43.2 was released in 2016 by Analog Devices, Inc. and worth 6,000 JPY. It is a low noise, low drift and low power accelerometer with digital output, which the Raspberry pi can read directly.

FIGURE 43.1   Raspberry pi model B+.

TABLE 43.1

Specification of Raspberry pi

Model	Raspberry pi 3 Model B+
Process	64-bit SoC @ 1.4 GHz
Memory	1 GB LPDDR2 SDRAM
Connectivity	Wireless LAN, Bluetooth 4.2
Access	Extended 40-pin GPIO header
SD card support	Micro SD format
Input power	5 V/2.5 A DC
Size	85 × 56 mm

FIGURE 43.2 ADXL355 accelerometer.

43.3 MONITORING SYSTEM

The system was developed under Python open-source environment to handle data acquisition, data storage and data transfer in real time. This system consists of three layers in terms of data flow as shown in Figure 43.3. The first layer is the sensor sensing the physical environment. The second layer is the microcomputer acquiring the data from the sensor and processing it. The third layer includes the local microSD card, the cloud storage Dropbox and the server for data analysis on the backend. The communication protocol between the accelerometer and the microcomputer is a serial protocol Inter-Integrated Circuit (I2C). Measurement data were written into a local microSD card and uploaded to a cloud server via Dropbox application programming interface (API) and to

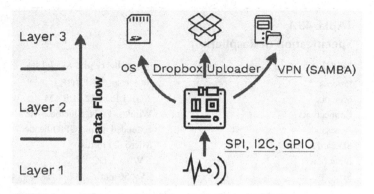

FIGURE 43.3   System workflow.

a server via Server Message Block (SMB) networking protocol. The monitoring procedure is auto-mated after installation, and the program maintenance of this system can be accessed remotely as long as it has an internet connection. To secure this system, the uncomplicated firewall can be set up to only allow remote access from people with permission.

43.4   EXPERIMENTS

43.4.1   SAMPLING RATE STABILITY

Running the measurement program repeatedly when the sensor was static, the static tests were carried out to explore the stability of the sampling rate. The sample set consisting of 20 trials was collected. The sampling rate was set to 100 Hz, and the results show that the sampling rate is stable with only small variations. Figure 43.4 is the plot of the sampling over time from one random trial in the sample set. The results from the whole sample set shows that the sampling rate is stabilized at 100 Hz with the standard deviation at 0.984 Hz.

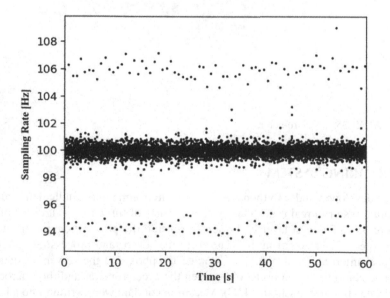

FIGURE 43.4   Sampling rate in static test.

43.4.2 ACCURACY VERIFICATION VIA SHAKING TABLE TESTS

To evaluate the accuracy of IoT sensing based structural vibration measurement, the shaking table tests were carried out by a uniaxial shaking table (APS-113) with different frequencies and amplitudes. Technical specification of the shaking table is shown in Table 43.2. Figure 43.5a shows the layout of specimens on the shaking table. The reference sensor was a high-quality servo velocity seismometer VSE-15-D, and the reference data acquisition unit is shown in Figure 43.5b. To investigate the capability to capture structural seismic vibration of the ADXL355 accelerometer, sinusoidal shaking table tests with frequencies from 0.1 to 10.0 Hz and amplitudes from 1 to 100 gal were conducted, as shown in Table 43.3.

TABLE 43.2
Specification of the Shaking Table

Property	APS-113
Maker	APS dynamics, INC.
Maximum excitation force (N)	133
Maximum displacement (mm^{p-p})	158
Maximum speed (mm/s)	1,000
Frequency range (Hz)	0.1–200
Mass (kg)	36
Dimension L×W×H (mm)	$526 \times 213 \times 168$

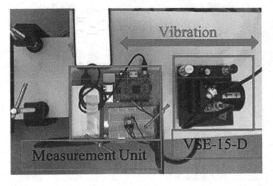 

FIGURE 43.5 (a) Shaking table test setup and (b) reference data acquisition unit.

TABLE 43.3
Shaking Table Parameter Setting

Frequency Amplitude	0.1 Hz	0.2 Hz	0.5 Hz	1.0 Hz	2.0 Hz	5.0 Hz	10 Hz
1 Gal	◎	◎	◎	◎	◎	◎	◎
5 Gal	◎	◎	◎	◎	◎	◎	◎
10 Gal	◎	◎	◎	◎	◎	◎	◎
50 Gal	×	◎	◎	◎	◎	◎	◎
100 Gal	×	×	×	◎	◎	◎	◎

The comparison with the results from the reference sensor is shown in Figures 43.6 and 43.7 in time and frequency domains, respectively. As the figures can illustrate, the noise in the IoT sensor is unneglectable when the frequency is low, e.g., 0.1 Hz. But its performance becomes better when the amplitude increases. The waveform matches better with the reference in Figure 43.6d–f compared to Figure 43.6a–c respectively. The vibration amplitude that can be recognized by this IoT sensor is lower than 2 Gal. As can be seen in the Fourier spectrum from Figure 43.7, the primary frequency

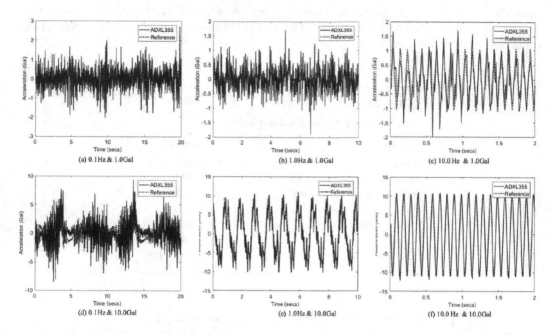

FIGURE 43.6 Acceleration waveform comparison.

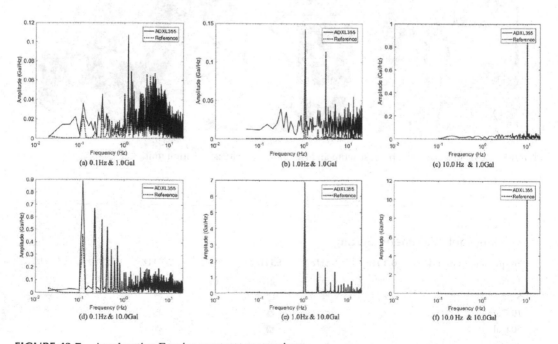

FIGURE 43.7 Acceleration Fourier spectrum comparison.

can be easily captured in the range from 0.1 to 10.0 Hz. Due to the systematic error of the shaking table used in this experiment, the actual primary frequency does not turn out to be as expected when both frequency and amplitude are low, e.g., 0.1 Hz and 1 Gal. Basically, the measurements from the IoT sensor agree well with the results from the reference sensor. Although the waveforms acquired from the IoT sensor does not precisely overlap with the ones from the reference sensor, the primary frequency can be clearly reflected by the IoT sensor.

For further study, the raw data acquired from IoT sensors can be processed to cut down the noise components via some signal processing tools, such as a bandpass filter; therefore, the frequency components outside of the research scope could be removed in order to obtain a refined signal with less noise.

43.4.3 APPLICATION OF SSM SYSTEM IN A BUILDING

A prototype of SSM system as shown in Figure 43.8a was installed on the 8th floor of a 10-story reinforced-concrete-frame building in Saitama University for long-term seismic events monitoring in June 2019, as shown in Figure 43.8b. The Yamagata Offshore Earthquake on June 18 and a magnitude 5.5 earthquake in Chiba Prefecture on June 24, 2019 were both observed from the deployed prototype, though the seismic intensities in the region where the prototype was deployed were both low at 1 in the JMA seismic intensity scale. The records from the nearest observation station (K-NET SIT010) of strong-motion seismograph networks in Japan (K-NET and KiK-net) were collected for comparison with the proposed system. The distance between two observation locations is 13 km as shown in Figure 43.8c. The acceleration waveforms for the 3 axes are shown in Figure 43.9 for the two observed earthquakes, and the Fourier amplitude spectrums along EW direction are plotted in Figure 43.10. As can be seen in Figure 43.9, the measured waveforms can represent the structural seismic response roughly, though the accurate waveform data concerning earthquakes with amplitude smaller than 2 Gal is hidden due to the noise. It is more sensitive than the smart devices in the study conducted by Shrestha et al. (2017). From the Fourier spectrums in Figure 43.10, the dominant frequency of the monitored structure at the location where the prototype was installed can be identified around 1.45 Hz, and it agrees with the previous research conducted by Shrestha et al. (2017) and is more invariant.

  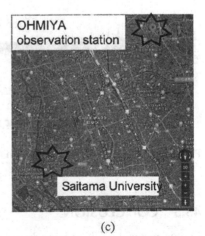

(a)   (b)   (c)

FIGURE 43.8   (a) Prototype, (b) the monitored structure and (c) two observation locations.

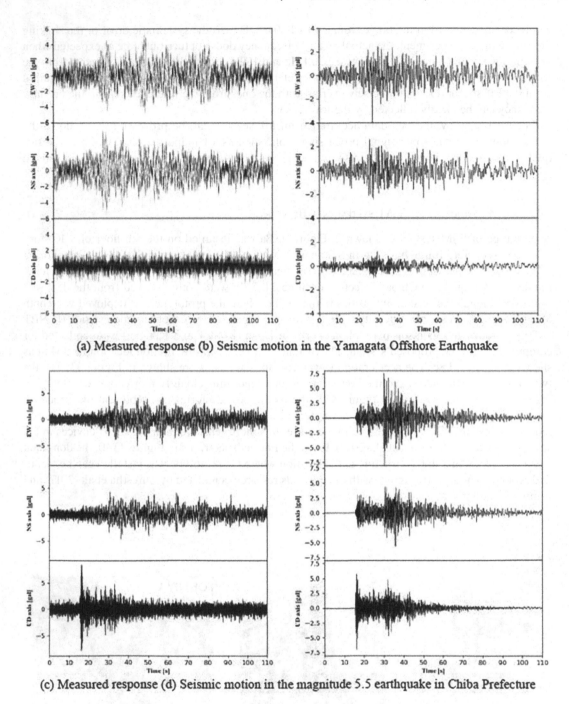

(a) Measured response (b) Seismic motion in the Yamagata Offshore Earthquake

(c) Measured response (d) Seismic motion in the magnitude 5.5 earthquake in Chiba Prefecture

FIGURE 43.9   Waveforms comparison.

43.5   CONCLUSION

In this paper, an IoT sensing based structural seismic monitoring (SSM) system was proposed. Integrated with the function of data acquisition, data storage and data transfer, IoT devices considerably reduce the cost of SSM with reasonable performance. The effectiveness of the proposed system

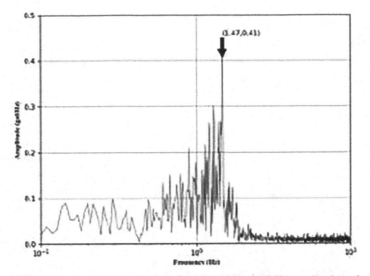

(a) Measured structural response in the Yamagata Offshore Earthquake

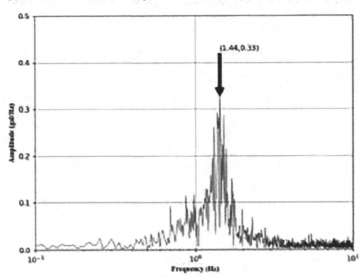

(b) Measured structural response in the magnitude 5.5 earthquake in Chiba Prefecture

FIGURE 43.10 Measured structural response during two earthquakes.

has been verified via shaking table tests and long-term seismic response monitoring at structures. It can be validated through all experiments carried out above, the stable sampling rate at 100 Hz is available and the detectable acceleration level is about 2 Gal in a seismic event. IoT devices make the dense deployment of SSM system possible, and the huge volume of real-time structural seismic response data will enhance the understanding of the structural dynamic behavior.

REFERENCE

Shrestha A, Dang J, Wang X. 2017. Development of a smart-device-based vibration-measurement system: Effectiveness examination and application cases to existing structure. *Struct Control Health Monit* 25(3): e2120.

44 Non-Probabilistic Damage Detection Using Classic and Modal Interval Analysis

J.Y. Huang and S.E. Fang
Fuzhou University

CONTENTS

44.1 INTRODUCTION

Damage assessment, as the core content of structural health monitoring, must consider uncertainties such as material heterogeneity, ambiguous boundary conditions, test errors and model errors when applied to actual engineering structures (Sohn 2007). Therefore, probabilistic or non-probabilistic theories should be considered when dealing with real-world damage identification problems. Specifically, probabilistic reliability theories and fuzzy logic theories can be used for making a reasonable evaluation of structural damage. However, accurate probability density functions or membership functions of uncertain parameters must be given in advance (Elishakoff and Yao 1983). Because of lack of test data and the complexity of practical engineering structures, the distribution types of uncertain parameters are often difficult to describe leading to large calculation errors. In order to correct that defect, non-probabilistic evaluation methods based on interval analysis have been paid more attention. Ben-Haim and Elishakoff (1995) first put forward the concept of non-probabilistic reliability based on convex set models. An interval method for considering non-probabilistic reliability has been widely used in the literature, where uncertain parameters have been defined as an interval rather than a deterministic value. Classical Interval Analysis (CIA) can solve the problems with uncertainties in some cases, but there exists an inherent problem of interval expansion which often leads to non-convergence or wrong conclusions (Sainz et al. 2014).

Modal Interval Analysis (MIA) (SIGLA/X Group 1999, Armengol et al. 1999) is a non-probabilistic analysis method developed on the basis of CIA. MIA can avoid the expansion phenomenon. On the basis of determining fluctuant interval of variables in advance, MIA expands structural general equations into interval forms through logical semantics expansion (Gardeñes et al. 2001). The interval includes all possible results within a reasonable range and covers all possible states of a system (Sevillano et al. 2017).

Based on the MIA theory, a damage assessment method considering structural uncertainties has been proposed in this study, which considers parameters as interval numbers with upper and lower bounds. Thus structural static equations are extended to interval ones. Then the interval envelope of structural responses is obtained by means of modal interval interpretation and calculation algorithm. Finally, the structural damage can be judged by comparing the intersection relationship between the modal interval envelope and the deterministic value curve. Meanwhile, damage assessment results are compared with those given by CIA and deterministic methods.

44.2 CLASSICAL AND MODAL INTERVAL ANALYSIS

44.2.1 CLASSICAL INTERVAL ANALYSIS

In the CIA theory, an interval is defined as a set of real numbers $\{x \mid a \leq x \leq b\}$, expressed as $X = [a,b]$, which can be regarded as a pair of ordered real numbers composed of upper and lower bounds a and b. The set of closed intervals in the real space R is expressed as

$$I(R) = \left\{ [a,b]' \mid a, b \in R, a \leq b \right\} \tag{44.1}$$

If op represents the arithmetic operator of real numbers, the corresponding interval arithmetic operator is given as follows:

$$Z = X \ op \ Y$$

$$= \left\{ x \ op \ y \mid x \in X, y \in Y \right\} \tag{44.2}$$

The interval Z is the result of the interval operation $X \ op \ Y$. Z consists of all possible values in the operation.

If f is a continuous function having n-dimensional variables of $x = (x_1, ..., x_n)$ over a domain $X = (X_1, ..., X_n)$, the range of its joint expansion function R_f is defined as follows:

$$R_f(X_1, ..., X_n) = \left[\min(x, X) f(x_1, ..., x_n), \max(x, X) f(x_1, ..., x_n) \right] \tag{44.3}$$

$fR(X_1, ..., X_n)$ is the rational interval extension function of $f(x_1, ..., x_n)$; in other words, $\{x_1, ..., x_n\}$ in $f(x_1, ..., x_n)$ are replaced by $\{X_1, ..., X_n\}$. Then the operator becomes an interval operator.

$$R_f(X_1, ..., X_n) \subseteq fR(X_1, ..., X_n) \tag{44.4}$$

As can be seen from the above formula, fR calculates the bounds of intervals. Because some information in R_f is lost, the boundary of fR becomes larger.

44.2.2 MODAL INTERVAL ANALYSIS

MIA is a combination of interval mathematic and modal logic, using modal predicate logic to interpret the interval semantically. In order to avoid confusion, here $[a,b]'$ is used to represent a classical interval and $[a,b]$ is used to represent a modal interval.

Assume the set of closed intervals

$$I(R) = \left\{ [a, b]' \mid a, b \in R, a < b \right\} \tag{44.5}$$

in the real space R, the existence of quantifier E (Existential Quantifiers) and global quantifier U (Universal Quantifiers), then the modal interval is defined as

$$X := (X', QX) \qquad (44.6)$$

where $X' \in I(R)$ is called as logical expansion; $QX \in (E, U)$ is called as modal.

Different real numbers may have the same absolute value, while the symbols are opposite. Modal intervals have similar properties, that is, each modal interval can have the same closed interval on the real axis. However, there are opposite modes of choice, that is, E (existence) or U (global). The set of modal intervals is expressed as

$$I^*(R) := \{(X', \{E, U\} | X' \in I(R))\} \qquad (44.7)$$

Modal intervals can be divided into two categories according to the size of upper and lower bounds:

i. $a < b$, $Prop(A', QA) := (A', E) = ([a, b]', E) \in I^*(R)$ is named as "existence interval" or "property interval"
ii. $b < a$, $Impr(A', QA) := (A', U) = ([a, b]', U) \in I^*(R)$ is named as "global interval" or "improper interval"

In modal interval operation, two semantically extended functions $f*$ and $f**$ can be obtained from continuous functions. If the function f is continuous from R^n to R, then one has

$$f*(A) = \vee(a_p, A'_p) \wedge (a_i, A'_i)[f(a_p, a_i), f(a_p, a_i)]$$

$$= [\min(a_p, A'_p)\max(a_i, A'_i)f(a_p, a_i), \ \max(a_p, A'_p)\min(a_i, A'_i)f(a_p, a_i)] \qquad (44.8)$$

$$f**(A) = \wedge(a_i, A'_i) \vee (a_p, A'_p)[f(a_p, a_i), f(a_p, a_i)]$$

$$= [\max(a_i, A'_i)\min(a_p, A'_p)f(a_p, a_i), \ \min(a_i, A'_i)\max(a_p, A'_p)f(a_p, a_i)] \qquad (44.9)$$

in which $[f(a_p, a_i), f(a_p, a_i)]$ is a point interval; (a_p, a_i) is the corresponding partition of the modal interval $A = (A_p, A_i)$; A_p is a subvector containing regular subinterval in A; A_i is a subvector containing irregular subinterval in A.

The definition of $f*$ can be further interpreted as follows:

a. The lower bound: The proper and improper intervals are divided into n parts, generating $(n + 1)$ values. Then a value is selected in the proper interval in order, which corresponds to the $(n + 1)$ values in the improper interval, namely A'_i. Finally, there are $(n + 1)$ groups of data and each group has $(n + 1)$ data. In each group of $(n + 1)$ data, the maximum value is selected. Then $(n + 1)$ maximum values are obtained altogether. A minimum magnitude is selected from these maximum values, that is, the lower bound of the interval.
b. The upper bound: The proper and improper intervals are divided into n parts, generating $(n + 1)$ values. A value is selected in the proper interval in order, corresponding to the $(n + 1)$ values in the improper interval A'_i. Finally, there are $(n + 1)$ groups of data and each group has $(n + 1)$ data. In each group of $(n + 1)$ data, the minimum value is selected and a total of $(n + 1)$ minimum values is obtained. Then the maximum magnitude is selected from these minimum values, which gives the upper bound of the interval.

$f**$ and $f*$ are similar in conception. The primary difference lies in the order of choosing the maximum and minimum values.

44.2.3 INTERVAL SOLUTION EXAMPLE

The exact solution of an interval function $f = x_1^2 x_2 - 2x_1^2 + 2x_2^2$ on $X_1 = [0,1]$, $X_2 = [2,3]$ is $[8,19]$. The CIA gives the solution as

$$fR(X) = X_1^2 X_2 - 2X_1^2 + 2X_2^2$$

$$= [0,\ 1][2,\ 3] - 2[0,\ 1] + 2[4,\ 9]$$

$$= [0,\ 3] - [0,\ 2] + [8,\ 18]$$

$$= [6,\ 21]$$

It can be seen that the interval range obtained by CIA is larger than the exact solution, implying the phenomenon of interval expansion. The value interval of the function calculated by MIA is $[8,19]$, which is the exact solution.

44.3 NUMERICAL EXAMPLE

A numerical simply supported steel box beam was used to verify the proposed method. The beam had a length of $l = 1.8\,\text{m}$, as is shown in Figure 44.1. The beam is subjected to two vertical concentrated loads. The distances from the loading positions to the left support were $a = 0.45\,\text{m}$ and $b = 0.9\,\text{m}$. The resistance moment of the beam section was $w = 15.1 \times 10^{-6}\,\text{m}^3$. The elastic modulus was $E = 200\,\text{GPa}$. Damage identification was carried out by using the mid-span strain as a damage index. The calculation formula is given as follows.

$$\sigma = E\varepsilon = \frac{M}{w} \tag{44.10}$$

where σ is the stress at the lower edge of the beam section; ε is the corresponding strain; M is the bending moment at the cross section. The mid-span strain expression is defined as

$$\varepsilon_{1/2} = \frac{M_{1/2}}{wE}$$
$$= \frac{F[l-(a+b)]+F[b+l-(a+b)]-Fb}{2wE} \tag{44.11}$$

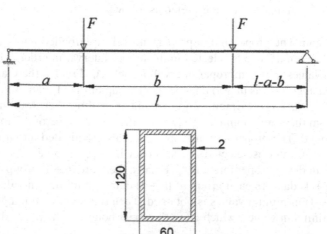

FIGURE 44.1 Schematic diagram of steel box beam (unit: mm).

TABLE 44.1

Strains of the Steel Box Beam

Load Step	F (N) Lower Bound	F (N) Upper Bound	ε (CIA) Lower Bound	ε (CIA) Upper Bound	ε (MIA) Lower Bound	ε (MIA) Upper Bound	Deterministic ε
1	950	1,050	933	2,165	1,131	1,930	1,490
2	1,900	2,100	1,865	4,329	2,262	3,861	2,980
3	2,850	3,150	2,798	6,494	3,393	5,791	4,470
4	3,800	4,200	3,730	8,659	4,524	7,722	5,960
5	4,750	5,250	4,663	10,824	5,655	9,652	7,450
6	5,700	6,300	5,595	12,988	6,786	11,583	9,411
7	6,650	7,350	6,528	15,153	7,917	13,513	11,590
8	7,600	8,400	7,460	17,318	9,048	15,443	14,024
9	8,550	9,450	8,393	19,483	10,179	17,374	16,763
10	9,500	10,500	9,325	21,647	11,310	19,304	19,868
11	10,450	11,550	10,258	23,812	12,441	21,235	23,415
12	11,400	12,600	11,190	25,977	13,572	23,165	27,509

By extending formula (44.10) to an interval expression, one has

$$\varepsilon_{1/2} = \frac{M_{1/2}}{WE}$$

$$= \frac{F[L-(A+B)]+F[B+L-(A+B)]-FB}{2WE} \tag{44.12}$$

Suppose $L = [1.78, 1.82]$ m, $A = [0.43, 0.0.47]$ m, $B = [0.88, 0.92]$ m, $W = [1.46, 1.56] \times 10^{-5}$ m^3 and $E = [190, 210]$ GPa, the intervals of strain calculated by CIA and MIA are listed in Table 44.1. Meanwhile, the deterministic strains were also calculated based on the median values of each parameter interval. The damage was simulated by decreasing E by 10 GPa per load step from the 6th load step. At the 12th load step, E decreased to 130 GPa.

As is shown in Table 44.1, the strain interval envelope calculated by CIA contains that of MIA, indicating MIA can effectively avoid the interval expansion in the solution process. Figure 44.2 shows the CIA and MIA curves are the maximum and minimum allowable magnitudes by the method. The deterministic strain curves intersect with the upper limits of modal and classical intervals at 8.7 and 10.5 kN, respectively. It is deemed that the structure is damaged under this load, since the simulated damage appeared at the 6th load step and $F = 8.7$ kN was between the 8th and 9th load step. MIA shows a phenomenon of time-lag effect. This is because the damage assessment threshold is an interval rather than a conventional deterministic value. Therefore, the intersection between the deterministic curve and the interval envelope has a "process" inducing the lag.

On the other hand, CIA found the damage at the 10th and 11th load steps, showing MIA could identify the damage of the beam before CIA did. Meanwhile, because interval algorithms expand deterministic parameters into intervals and allow the parameter values to fluctuate within a certain range, such intersection indicates it exceeds the permissible range of interval algorithm, implying the occurrence of the damage.

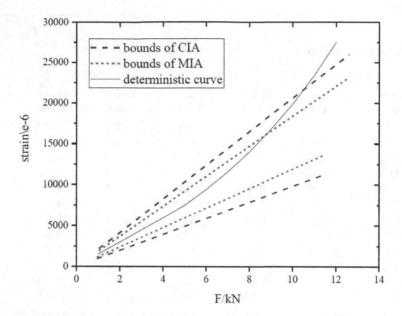

FIGURE 44.2   Strains calculated by different methods.

44.4   CONCLUSION

1. Uncertainty factors have some impact on structural damage assessment. In practice, interval numbers can be used to express parameter uncertainties. Structural mechanics equations are then extended to their interval forms to solve the problem, which means the robustness of damage identification results is improved.
2. By the intersection between deterministic curves and interval envelopes, the time of damage occurrence and the corresponding external load can be evaluated.
3. MIA can effectively avoid the interval expansion and obtain more accurate interval envelopes of structural responses. Compared with CIA, MIA can detect structural damage earlier. However, MIA has a time-lag effect, because damage assessment threshold is an interval instead of a conventional deterministic value.

REFERENCES

Armengol, J., Travé-Massuyès, L., Vehi, J., and Sáinz, M. Á. 1999. Semiqualitative simulation using modal interval analysis. *IFAC Proceedings Volumes*, 32(2), 7623–7628.

Ben-Haim, Y., and Elishakoff, I. 1995. *Discussion on: a non-probabilistic concept of reliability*. Structural Safety, 17(3), 195–199.

Elishakoff, I., and Yao, J. T. P. 1983. *Probabilistic Methods in the Theory of Structures*. London: World Scientific.

Gardeñes, E., Sainz, M. Á., Jorba, L., Calm, R., and Trepat, A. 2001. Model intervals. *Reliable Computing*, 7(2), 77–111.

Sainz, M. A., Armengol, J., Calm, R., Herrero, P., Jorba, L. and Vehi, J. 2014. *Modal Interval Analysis*. Cham: Springer International Publishing.

Sevillano, E., Sun, R., and Perera, R. 2017. Damage evaluation of structures with uncertain parameters via interval analysis and FE model updating methods. *Structural Control and Health Monitoring*, 24, e1901.

SIGLA/X Group. 1999. Modal intervals. Application of interval analysis to systems and control. *Proceedings of MISC*, 99, 157–227.

Sohn, H. 2007. Effects of environmental and operational variability on structural health monitoring. *Philosophical Transactions of the Royal Society A: Mathematical, Physical and Engineering Sciences*, 365(1851), 539–560.

45 A 3D Isolation Device with Vertical Variable Stiffness for Long-Span Spatial Structures

Y. Shi, H. Liu, Z. Chen, and Y. Ding
Tianjin University

CONTENTS

45.1 INTRODUCTION

Long-span spatial structures have been extensively used in civil and industrial infrastructures in the past few years. To enhance the performance of long-span spatial structures, base isolation technologies have been developed and studied. Laminated rubber bearings and friction pendulum bearings which can provide long isolation periods are widely used in long-span spatial structures (Takeuchi et al. 2006; Jongwan 2016; Fan et al. 2016). With these bearings, the internal force of members and acceleration of structures under horizontal earthquakes can be significantly reduced. However, the laminated rubber bearings and friction pendulum bearings only work in the horizontal direction. Many full-scale shaking table tests of horizontally isolated structures show that the vertical structural response is significant. The vertical acceleration can be amplified more than ten times (Furukawa et al. 2013). For long-span spatial structures, the vertical vibration is obvious, and the vertical responses are usually comparable to the horizontal responses, which indicate that there is a strong need to mitigate the vertical responses. Recently, researchers have conducted experiments and numerical analysis on 3D isolation devices. However, due to the complexity and high cost, there are only few practical applications (Warm and Ryan 2012; Zhou et al. 2016). The complexity of 3D isolation is mainly reflected in the design of a vertical isolation device. It is of great importance to guarantee the stability of 3D isolation systems against horizontal and vertical loads with a small vertical stiffness (Jia et al. 2012).

This paper presents a novel three-dimensional (3D) isolation device with vertical variable stiffness (3DIVVS) developed for long-span spatial structures (Chen et al. 2019). The 3DIVVS device can provide 3D isolation capacity with an isolation period more than 1 s in each direction and vertical variable stiffness for different loading stages. Based on the hydraulic principle, the vertical spring component can be divided into several horizontal parts, and the height of the bearings can be significantly decreased.

45.2   3D ISOLATION DEVICE WITH VERTICAL VARIABLE STIFFNESS

45.2.1   COMPONENTS OF 3D ISOLATION DEVICE WITH VERTICAL VARIABLE STIFFNESS

Figure 45.1a shows the schematic of the proposed 3DIVVS composed of a vertical isolation device combining hydraulic cylinders (part A in Figure 45.1b) and a horizontal component (part B in Figure 45.1b). The horizontal component is a traditional lead-rubber bearing used to isolate the horizontal vibrations. It is connected to the vertical isolation device by complete penetration groove welds to avoid unexpected failures. The top of the vertical isolation device is connected to the structures, and the bottom of the horizontal component is connected to the foundation.

The top connection plate of the vertical isolation device is directly connected to the main piston rod and vertical guide sleeves. The horizontal force of the top connection plate is transmitted through the sleeve to the bottom connection plate. The vertical load of the top connection plate is transmitted through the main piston rod which is connected to the main piston. There are four kinds of cylinders in the vertical isolation device, i.e., main cylinder, auxiliary cylinder, pressurized cylinder and depressurized cylinder. Springs are placed in each cylinder except the main one. The variable stiffness characteristics of the vertical isolation device are controlled by these cylinders.

There is one main cylinder in the 3DIVVS and the number of auxiliary cylinders can be one or more. The primary function of the main cylinder is to transmit vertical loads to bottom connection plate and push the oil to flow among cylinders depending on the hydraulic pressure. The auxiliary spring is connected to the top of the auxiliary cylinder and the auxiliary piston. The hydraulic pressure transmitted to the auxiliary cylinder will push the auxiliary piston to compress the auxiliary spring. The function of the auxiliary cylinder is to provide primary stiffness for the vertical isolation device. Placing the spring directly in the main cylinder can also achieve this objective, but the height of the main cylinder will increase significantly considering the height of the spring. The auxiliary cylinder can be placed horizontally to accommodate the installation space in practice. The primary stiffness provided by the auxiliary cylinder can restrain the displacement caused by self-weight and live load in normal phase.

The pressurized and depressurized cylinders are used to decrease the vertical stiffness of the 3DIVVS and prolong the vertical period of the structures. The number of pressurized and depressurized cylinders can be one or more according to the designed objective. For the depressurized cylinder, there are two forces applied on the depressed piston ignoring the friction; one is the spring force provided by depressed spring, and the other is the hydraulic force caused by hydraulic pressure. In the initial state, the hydraulic force is small and the depressed piston is squeezed on

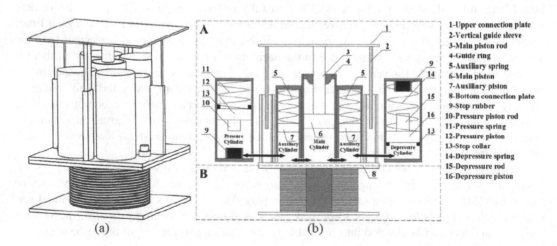

FIGURE 45.1   Schematic of the 3DIVVS: (a) whole device; (b) vertical isolation device.

the stop collar under the spring force. During this phase, the depressurized cylinder is not working. When the hydraulic pressure caused by the movement of the main piston is large enough, the depressured piston will move under the hydraulic force and the depressurized cylinder will be working. As the hydraulic pressure increases, the depressured piston will touch the stop rubber which is used to restrain large displacement of vertical isolation device towards the ground and avoid unexpected collision. For the pressurized cylinder, there are also two forces applied on the depressured piston; one is the spring force provided by pressurized spring, and the other is the hydraulic force caused by hydraulic pressure. In the initial state, the pressurized piston is squeezed on the stop collar under the hydraulic force which is larger than the spring force. During this phase, the pressurized cylinder does not work. When the hydraulic pressure caused by the movement of the main piston is small enough, the pressurized piston will move under the spring force, and the pressurized cylinder will be working. As the hydraulic pressure decreases, the depressurized piston will touch the stop rubber which is used to restrain large displacement of vertical isolation device opposite to the ground and avoid unexpected collision. The auxiliary cylinders, pressurized and depressurized cylinders are connected to the main cylinder through oil pipes.

45.2.2 CONTROL MECHANISM OF VERTICAL ISOLATION DEVICE WITH VARIABLE STIFFNESS

Figure 45.2 shows the theoretical hysteretic model of the 3DIVVS in vertical direction ignoring the friction and damping in the system. As shown in Figure 45.2, there are two thresholds and two isolation displacements including high pressure threshold (noted as T_{ph}), low pressure threshold (noted as T_{pl}), high pressure isolation displacement (noted as D_H) and low pressure isolation displacement (noted as D_L). Depending on the hydraulic force applied on the main piston, the vertical isolation device will be in one of the five different working phases, i.e., normal phase, high pressure and low pressure isolation phases, and high pressure and low pressure limit phases.

Under self-weight and live load of the structures, the vertical isolation device of the 3DIVVS is in the normal phase (Curve CD in Figure 45.2). During this phase, the pressurized and depressurized cylinders does not work. The schematic of the vertical isolation device of the 3DIVVS can be simplified as shown in Figure 45.3a and the vertical equivalent stiffness K_{eq1} in normal phase can be expressed as

$$K_{eq1} = \frac{A_m^2}{\sum\limits_{i=1}^{n} \frac{A_{ai}^2}{K_{ai}}}$$
(45.1)

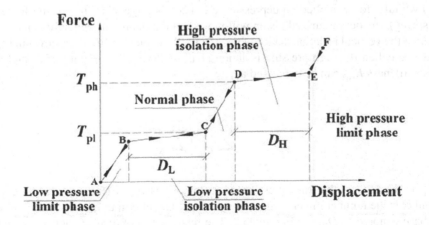

FIGURE 45.2 Vertical hysteretic model of the 3DIVVS.

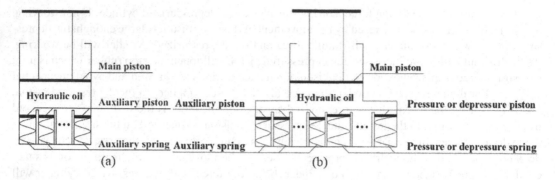

FIGURE 45.3   Schematic of the vertical isolation device of the VS3DI: (a) normal phase; (b) isolation phase.

where A_m is the area of the main piston, A_{ai} is the area of the ith auxiliary piston, K_{ai} is the stiffness of the ith auxiliary spring and n is the number of auxiliary cylinders. The relationship of displacement between the main piston and auxiliary piston can be expressed by

$$\left\{ \begin{array}{l} A_m X_m = \sum_{i=1}^{n} A_{ai} X_{ai} \\[2mm] \dfrac{K_{ai} X_{ai}}{A_{ai}} = \dfrac{K_{aj} X_{aj}}{A_{aj}} \quad (i \neq j) \end{array} \right. \tag{45.2}$$

where A_{aj} is the area of the jth auxiliary piston, K_{aj} is the stiffness of the jth auxiliary spring, X_m is the displacement of the main piston, and X_{ai} and X_{aj} are the displacement of the ith and jth auxiliary piston, respectively. According to Eq. (45.2), the displacement of each piston in this phase can be calculated. The highest and lowest hydraulic forces applied on the main piston in this phase are T_{ph} and T_{pl}, which determine the range of the reaction force of the vertical isolation device to the structures in normal condition. The stiffness in normal phase is larger than that in isolation phase to restrain the movements under normal live load such as wind load and small ground motions.

As the vertical vibrations change, the hydraulic force applied on the main piston can be larger than T_{ph} or smaller than T_{pl}. If the hydraulic force applied on the main piston is larger than T_{ph} (means that the hydraulic force applied on depressurized piston is larger than the spring force of depressurized spring), the depressurized piston will move and the depressurized cylinder will work. The schematic of the vertical isolation device of the 3DIVVS can be simplified as shown in Figure 45.3b. This phase is called the high pressure isolation phase (Curve DE in Figure 45.2) and the vertical equivalent stiffness K_{eq2} can be expressed as

$$K_{eq2} = \frac{A_m^2}{\sum_{i=1}^{n} \dfrac{A_{ai}^2}{K_{ai}} + \sum_{j=1}^{m} \dfrac{A_{dj}^2}{K_{dj}}} \tag{45.3}$$

where A_{dj} is the area of the jth depressurized piston, K_{dj} is the stiffness of the jth depressurized spring and m is the number of depressurized cylinders. The maximum displacement of main piston in this phase is noted as D_H, see Figure 45.2. The relationship of displacements between the main piston and other pistons in this phase can be expressed as

$$
\begin{cases}
A_\mathrm{m} X_\mathrm{m} = \sum_{i=1}^{n} A_{ai} X_{ai} + \sum_{j=1}^{m} A_{dj} X_{dj} \\[2mm]
\dfrac{K_{ai} X_{ai}}{A_{ai}} = \dfrac{K_{dj} X_{dj}}{A_{dj}} \quad (i \neq j) \\[2mm]
\dfrac{K_{dj} X_{dj}}{A_{dj}} = \dfrac{K_{dh} X_{dh}}{A_{dh}} \quad (j \neq h)
\end{cases}
\tag{45.4}
$$

where A_{dh} is the area of the hth depressurized piston, K_{dh} is the stiffness of the hth depressurized spring, X_{dj} and X_{dh} are the displacements of jth and hth depressurized piston, respectively. T_{ph} and D_H can be calculated by Eqs. (45.3) and (45.4). During this phase, the vertical stiffness of the 3DIVVS is decreased and the vertical period of the base isolation system is prolonged because of the participation of the depressurized cylinder.

As the hydraulic pressure increases, the depressurized piston will touch the stop rubber under the hydraulic force, after which the vertical isolation device will be in the high pressure limit phase (Curve EF in Figure 45.2). During this phase, the vertical equivalent stiffness of the 3DIVVS can be calculated by Eq. (45.1). The vertical stiffness provided by the 3DIVVS is increased and the vertical period of the base isolation system will be decreased. Consequently large displacement responses will be controlled.

During the normal phase, high pressure isolation phase and high pressure limit phase, the pressurized cylinder does not work. The hydraulic force applied on pressurized piston is larger than the spring force of pressurized spring, and the pressurized piston is squeezed on the stop collar. As the vertical vibrations change, if the hydraulic force applied on the main piston is smaller than T_{pl} (means that the hydraulic force applied on pressurized piston is smaller than the spring force of pressurized spring), the pressurized piston will move and the pressurized cylinder will work. The schematic of the vertical isolation device of the 3DIVVS can also be simplified as shown in Figure 45.3b. This phase is the low pressure isolation phase (Curve BC in Figure 45.2) and the vertical equivalent stiffness K_{eq3} can be calculated by

$$
K_{eq3} = \frac{A_\mathrm{m}^2}{\displaystyle\sum_{i=1}^{n} \frac{A_{ai}^2}{K_{ai}} + \sum_{k=1}^{l} \frac{A_{pk}^2}{K_{pk}}}
\tag{45.5}
$$

where A_{pk} is the area of the kth depressurized piston, K_{pk} is the stiffness of the kth depressurized spring and l is the number of pressurized cylinders. During this phase, the maximum displacement of main piston is noted as D_L, see Figure 45.2. The relationship of displacement between the main piston and other pistons can be expressed as

$$
\begin{cases}
A_\mathrm{m} X_\mathrm{m} = \sum_{i=1}^{n} A_{ai} X_{ai} + \sum_{j=1}^{m} A_{pj} X_{pj} \\[2mm]
\dfrac{K_{ai} X_{ai}}{A_{ai}} = \dfrac{K_{pj} X_{pj}}{A_{pj}} \quad (i \neq j) \\[2mm]
\dfrac{K_{pj} X_{pj}}{A_{pj}} = \dfrac{K_{ph} X_{ph}}{A_{ph}} \quad (j \neq h)
\end{cases}
\tag{45.6}
$$

where A_{ph} is the area of the hth depressurized piston, K_{ph} is the stiffness of the hth depressurized spring and X_{pj} and X_{ph} are the displacement of jth and hth pressurized piston, respectively. T_{pl} and D_L

can be calculated by Eqs. (45.5) and (45.6). During this phase, the vertical stiffness of the 3DIVVS is also decreased because of the participation of the pressurized cylinder, which results in a longer vertical period.

As the hydraulic pressure decreases, the pressurized piston will touch the stop rubber after which the vertical isolation device will be in the low pressure limit phase (Curve AB in Figure 45.2). During this phase, the vertical equivalent stiffness can be calculated by Eq. (45.1). The vertical stiffness provided by the 3DIVVS is increased and the vertical period of the base isolation system is decreased. Consequently large displacement responses will be controlled.

45.3   SINUSOIDAL AND SHAKING TABLE TESTS

45.3.1   SPECIMEN

A prototype of the vertical isolation device of the 3DIVVS was designed and tested under sinusoidal motions and recorded ground motions to investigate the mechanical properties of the proposed device (Chen et al. 2019). In order to thoroughly investigate the variable stiffness characteristics, the prototype was designed with five cylinders, i.e., one main cylinder, one auxiliary cylinder, two depressurized cylinder and one pressurized cylinder. The arrangement of the tests is shown in Figure 45.4.

For the main cylinder, the inside diameter was 80 mm, and four hydraulic fluid ports were fabricated to connect the main cylinder to the other cylinders. Pressure meters were installed to measure the hydraulic force. In order to investigate the damping effect caused by oil flow in pipes, four kinds of oil pipes were used for the connection. The pipes are categorized into short and long types with the length of 1 and 3 m, respectively, and each type consists of two diameters, i.e., 25 and 8 mm. For the auxiliary cylinder, the inside diameter was 80 mm. Only one hydraulic fluid port was fabricated to connect to the main cylinder. The oil pipe and hydraulic fluid port were connected in series through a welded ball valve.

For the pressurized cylinder, the inside diameter was 100 mm. One hydraulic fluid port was fabricated and a pressure meter was installed beside the fluid port. The pressurized spring was on the top of a pressurized piston. The maximum displacement of the pressurized piston was determined by stop rubber and stop collar. The two depressurized cylinders were the same with the inside diameter of 100 mm. The maximum displacement of the depressurized piston was determined by stop collar and stop rubber.

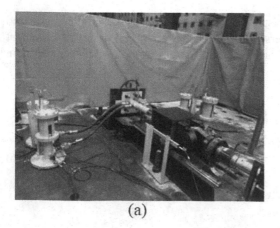
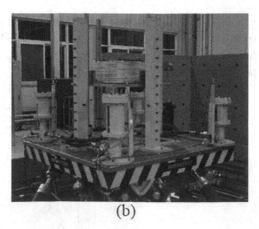

(a)   (b)

FIGURE 45.4   Test setup: (a) sinusoidal test; (b) shaking table test.

45.3.2 Test Results

The sinusoidal motions were applied on the piston rod of the main cylinder, and the range of frequency of the sinusoidal motions was from 0.001 to 2.5 Hz. The results show that Eqs. (45.1), (45.3) and (45.5) can precisely calculate the corresponding stiffness in each phase with the maximum error of 2.6%. The maximum error is 2.0% for the threshold values and 4.5% for control displacement. Therefore, the proposed formulations can be used to calculate the variable stiffness characteristic of the 3DIVVS in the static phase. As shown in Figure 45.5, there exists typical variable stiffness characteristic in the hysteretic curves of the prototype, indicating that the proposed device can achieve vertical variable stiffness for different loading stages. When the value of displacement is small, the prototype is in normal phase and the vertical equivalent stiffness is 151.8 kN/m corresponding to the frequency of 1.824 Hz. As the value of displacement increases, the prototype is in isolation phase, and the vertical equivalent stiffness decreases to 31 kN/m corresponding to a frequency of 0.827 Hz. Smaller vertical stiffness leads to a long period of isolation system to achieve better seismic performance. As for the limit phase, the vertical equivalent stiffness increases to 151 kN/m to restrain the large displacement.

However, there exists difference among the area of hysteretic curves which was caused by friction and damping force. The friction is related to loading velocity and decreases as the loading velocity increases (Enokida and Nagae 2016). The damping force is related to the current velocity of oil in pipes, which is determined by loading velocity and size of pipes, as shown in Figure 45.5. The nonlinear behavior of hysteretic curves was mainly caused by friction in the device. The areas of hysteretic curves for different diameter pipes are similar due to the identical friction. The stiffness in each phase and the thresholds are all different compared to the designed values. For prototypes with large diameter oil pipes, the variable stiffness characteristic could be achieved even under the largest loading velocity. Because of the damping effects, the high pressure and low pressure thresholds change slightly. Although a large damping force can increase energy dissipation, the prototype with small diameter pipes cannot achieve variable stiffness as designed, especially under a large loading velocity.

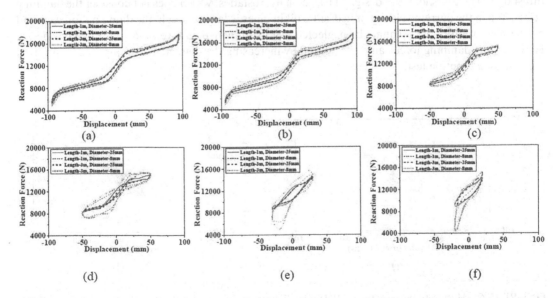

FIGURE 45.5 Hysteretic curves with different displacement amplitude and frequency: (a) 95 mm, 0.05 Hz; (b) 95 mm, 0.1 Hz; (c) 50 mm, 0.2 Hz; (d) 30 mm, 0.75 Hz; (e) 30 mm, 1.5 Hz; (f) 30 mm, 2.5 Hz.

In order to further investigate the mechanical properties of vertical isolation device of the 3DIVVS, shaking table tests were conducted to measure the responses of prototype under real ground motions using the six-axis vibration test bed of Servotest (Chen et al. 2019). Figure 45.4b shows the arrangement of the shaking table tests. The main cylinder was vertically installed at the center of the shaking table. The piston rod of the main cylinder was connected to the counterweight plate on top used to represent the structural mass. The counterweight plate was composed of 13 steel plates with a total mass of 1,155 kg. The counterweight was restrained by vertical guides to move vertically. The other four cylinders were fixed at each corner of the shaking table. Four oil pipes with the diameter of 25 mm and length of 1.5 m having less damping effect as found from the sinusoidal motion test were selected to connect the main cylinder to other cylinders. Recorded ground motions with different frequency characteristics were adopted in the shaking table tests, and the details of the ground motions can be found in Chen et al. (2019). For each recorded ground motion, the acceleration was scaled to excite the vertical isolation device to reach to a different phase to validate the formulations proposed.

Before the ground motions were applied, the vertical frequencies of the installed system were measured. The vertical frequency was 1.88 Hz with the relative error of 3.2% in normal phase, and 0.842 Hz with the relative error of 2.4% in isolation phase. The corresponding period in isolation phase was 1.2 seconds, which can be further prolonged by increasing the number of pressurized and depressurized cylinders according to Eqs. (45.3) and (45.5). Moreover, additional pressurized and depressurized cylinders do not increase the height of the 3DIVVS as they can be placed horizontally. Figure 45.6 shows the hysteretic curves of the prototype under two ground motions (El Centro and Wenchuan earthquakes) for the sake of brevity. Other results can be found in Chen et al. (2019). As the results show, the designed variable stiffness characteristic can be observed from the overall trend of the hysteretic curve. However, the area of hysteretic curves means that there exists energy dissipation in the prototype caused by damping force. The area of hysteretic curves mainly exists in normal phase. During this phase, only main cylinder and auxiliary cylinder were working. The oil pipe connecting the two cylinders was the only channel that the oil could flow and it led to a large current velocity. The damping force is in direct proportion to the square of the current velocity. Therefore, the large damping force caused the energy dissipation and changed the values of thresholds. The thresholds are designed based on hydrostatics, which does not consider the damping effect. When the current velocity of hydraulic oil in pipes is large enough, the damping force caused by the flow of oil in pipes cannot be neglected and its influence on thresholds must be considered. However, the damping force can be decreased by increasing the diameter of oil pipe, as found from the sinusoidal motion test.

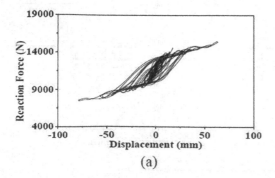
(a)

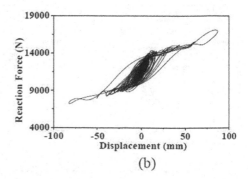
(b)

FIGURE 45.6   Hysteretic curve of the VS3DI under different ground motions: (a) EMC-150%; (b) Wen-300%.

45.4 CONCLUSIONS

The major findings and conclusions of this study are as follows:

1. By incorporating different hydraulic cylinders to work under different designed pressures, variable stiffness characteristics can be achieved for the vertical isolation device. Under self-weight and live load of the structures, the vertical stiffness of the 3DIVVS is larger than that in isolation phase to restrain the movements. As the vertical movements of the 3DIVVS increases, the 3DIVVS will be in isolation phase and its vertical stiffness is decreased. The vertical period of the base isolation system is prolonged because of the participation of the depressurized cylinder in isolation phase. In order to restrain large vertical displacement of vertical isolation device, the limit phase is designed to increase the vertical stiffness of the 3DIVVS and decrease the vertical period of the base isolation system.

2. The sinusoidal motion tests and shaking table tests results show that the vertical isolation device of the 3DIVVS can achieve vertical variable stiffness. However, the friction and damping force in the device can affect the hysteretic behavior of the device and increase the energy dissipation.

REFERENCES

Chen Z, Ding Y, Shi Y, Li Z. 2019. A vertical isolation device with variable stiffness for long-span spatial structures. *Soil Dynamics and Earthquake Engineering*; 123: 543–558.

Enokida R, Nagae T. 2016. Seismic Damage reduction of a structural system based on nontraditional sliding interfaces with graphite lubrication. *Journal of Earthquake Engineering*; 22(4): 666–686.

Fan F, Kong D, Sun M, Zhi X. 2016. Anti-seismic effect of lattice grid structure with friction pendulum bearings under the earthquake impact of various dimensions. *International Journal of Steel Structures*; 14(4): 77–84.

Furukawa S, Sato E, Shi Y, Becker TC. 2013. Full-scale shaking table test of a base-isolated medical facility subjected to vertical motions. *Earthquake Engineering and Structural Dynamics*, 42(11): 1931–1949.

Jia J, Ou J, Liu M, Zhang Z. 2012. Mechanical performance tests of a novel three-dimensional isolation bearing. *Journal of Civil, Architectural & Environmental Engineering*; 34(1): 29–34.

Jongwan H. 2016. Seismic analysis and parametric study of SDOF lead-rubber bearing (LRB) isolation systems with recentering shape memory alloy (SMA) bending bars. *Journal of Mechanical Science and Technology*; 30(7): 87–99.Takeuchi T, Xue S, Kato S. 2006. Recent development in passive control technologies for mental spatial structures. *Proceedings of IASS-ACPC Symposium*, Beijing, China, p. 18.

Warm GP, Ryan KL. 2012. A review of seismic isolation for buildings: historical development and research needs. *Buildings*; 2(3): 300–325.

Zhou Z, Wong J, Mahin S. 2016. Potentiality of using vertical and three-dimensional isolation systems in nuclear structures. *Nuclear Engineering and Technology*; 48: 1237–1251.

46 Detection of Crack in Euler-Bernoulli Beam Using Bayesian Inference

T. Wang
Southeast University

Mohammad Noori
California Polytechnic State University

Wael A. Altabey
Southeast University
Alexandria University

Z. Ying
Southeast University

R. Ghiasi
University of Sistan and Baluchestan

CONTENTS

46.1 INTRODUCTION

With an ever-increasing pace of urban development and increasing demand on existing infrastructure, the aging, degradation, damage, safety and sustainability of civil infrastructure have become important fields of research over the past two decades [1–6]. In most cases and types of damage and deterioration of structures and structural components, especially in concrete structures, development of cracks is a safety concern. Hence, early detection of the location and intensity of this common type of damage is an important problem. Subsequently, a large number of methods for crack parameter identification in structures have been proposed by researchers in order to monitor the presence and potential growth of cracks [7–12]. Given that vibration analysis is a common approach for the performance assessment of structures, fundamental dynamic characteristics such as mode shapes and natural frequencies of a structure, are viable metrics for assessing the deterioration of a structure, and numerous detection techniques have been introduced based on these inherent dynamic characteristics, such as vibration modes or natural frequencies [13–17].

As an important structural member, beams have been studied for over several decades. A series of theories and analytical methods have been developed to describe the static and dynamic deformation of beams, such as Euler-Bernoulli, Timoshenko and Reddy beam theories [18]. Cracks in a beam are usually modeled as massless rotational springs that connect the beam sections [19]. Based on this commonly used model, the closed-form solution is also given using generalized functions [20,21]. The analytical results of such models agree closely with finite element (FE) models and experimental results. This shows that the aforementioned closed-form solution is suitable for dynamic analysis of a cracked beam.

It should also be pointed out that the identification of damage is not always accurate because of the environmental effects and the contamination of the measurement data due to noise. The Bayesian model updating using vibration mode data is one of the most widely accepted approaches and has been used in practical engineering applications [22–27]. Combined with damage detection results, the Bayesian method can provide a comprehensive evaluation of structural behavior.

In this paper, a closed-form solution of the vibration modes of a cracked beam, in conjunction with the measured data for those modes, is used to obtain the parameters of cracks. By using an analytical solution the cracks are simulated to occur at any location along the beam and not only at nodal points, which is the case in FE analysis. Therefore, the proposed method avoids the error induced due to idealized FE models and demonstrates a better identification result. Besides, the uncertainty of identification is also taken into consideration by using Bayesian estimation. The identification result could be provided in the form of possible distribution functions.

46.2 FORMULATION FOR CRACK IDENTIFICATION

46.2.1 Closed-Form Solution of Euler-Bernoulli Cracked Beam Vibration Mode

As shown in Figure 46.1, an Euler-Bernoulli cracked beam with N cracks located along the x coordinate axes at locations x_1, x_2, ..., x_N is considered in this paper.

If we ignore the opening and closing of the crack which may happen due to cyclic loading, the vibration of cracked beam can be considered to be a linear vibration. Suppose that the cracks are equivalent to massless springs connecting the sub-beams. The flexural stiffness of the beam can be expressed as a uniform expression represented by the following equation as

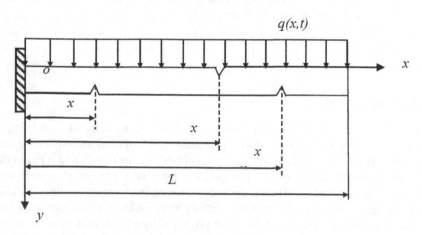

FIGURE 46.1 Euler-Bernoulli cracked beam.

$$\frac{1}{(EI)_e} = \frac{1}{(EI)_0} + \sum_{i=1}^{N} \frac{1}{K_i} \delta(x - x_i),$$
(46.1)

where $(EI)_e$ is the uniform flexural stiffness that varies with the coordinates of the beam section, $(EI)_0$ is the stiffness of un-cracked beam and δ is the Dirac delta function.

The equivalent spring of the ith crack K_i can be expressed as a function of crack's depth d/h and section height h as [28]

$$K_i = \frac{(EI)_0}{h} \frac{0.9[(d_i/h) - 1]^2}{(d_i/h)[2 - (d_i/h)]},$$
(46.2)

According to the flexural stiffness model in Eq. (46.1), the dynamic differential equation of the free vibration of Euler-Bernoulli cracked beam can be expressed as

$$\frac{\partial^2}{\partial x^2} \left[(EI)_e \frac{\partial^2}{\partial x^2} w(x,t) \right] + m \frac{\partial^2}{\partial t^2} w(x,t) = 0,$$
(46.3)

where $w(x,t)$ and m are the deflection function and linear mass density of the cracked beam.

For free vibration, the solution of Eq. (46.3) can be written as

$$w(x, t) = \varphi(x) y(t) = \varphi(x) e^{i\omega t},$$
(46.4)

Substituting Eq. (46.3) into Eq. (46.4), the differential equation of modal displacement can be written as

$$\frac{d^2}{dx^2} \left[(EI)_e \frac{d^2 \varnothing}{dx^2} \right] - m(\omega)^2 \varnothing(x) = 0,$$
(46.5)

Introducing a dimensionless parameter as

$$\xi = \frac{x}{L}, \ \phi(\xi) = \frac{\varphi(x)}{L}, \ \xi_i = \frac{x_i}{L}, \ K_i = \frac{K_i L}{(EI)_0} = \frac{L}{h} \cdot \frac{0.9[(d_i/h) - 1]^2}{(d_i/h)[2 - (d_i/h)]},$$
(46.6)

The governing equation of the vibration modes for cracked beam can be given as

$$\frac{d^2}{d\xi^2} \left[\overline{(EI)} \frac{d^2 \varnothing}{d\xi^2} \right] - \beta^4 \varnothing(\xi) = 0,$$
(46.7)

where parameter β is defined by the natural frequency of the cracked beam as

$$\beta^4 = m\omega^2 L^4 / (EI)_0,$$
(46.8)

The dimensionless stiffness $(\overline{EI})$ can be expressed as

$$\frac{1}{\overline{(EI)}} = 1 + \sum_{i=1}^{N} \frac{1}{K_i} \delta(\xi - \xi_i),$$
(46.9)

Solving Eq. (46.7), the exact closed-form solution of the dimensionless mode $\emptyset(\xi)$ can be obtained as

$$\emptyset(\xi) = C_1 \left[\frac{1}{2\beta} \sum_{i=1}^{N} \frac{\mu_i}{k_i} \Psi(\xi - \xi_i) + \sin(\beta\xi) \right] + C_2 \left[\frac{1}{2\beta} \sum_{i=1}^{N} \frac{v_i}{k_i} \Psi(\xi - \xi_i) + \cos(\beta\xi) \right]$$

$$+ C_3 \left[\frac{1}{2\beta} \sum_{i=1}^{N} \frac{\vartheta_i}{k_i} \Psi(\xi - \xi_i) + \sinh(\beta\xi) \right] + C_4 \left[\frac{1}{2\beta} \sum_{i=1}^{N} \frac{\eta_i}{k_i} \Psi(\xi - \xi_i) + \cosh(\beta\xi) \right] \qquad (46.10)$$

$$\Psi(\xi) = \left(\sinh(\beta\xi) + \sin(\beta\xi) \right) H(\xi) \qquad (46.11)$$

where $H(\xi)$ is the Heaviside function, and C_1, C_2, C_3, C_4 are undetermined constants.

If the boundary conditions are determined, the constant vector $\mathbf{C} = [C_1, C_2, C_3, C_4]$ is the solution of the linear equations of Eq. (46.12):

$$AC = 0, \qquad (46.12)$$

$\mathbf{C} = 0$ is a trivial solution of Eq. (46.12), but it means that the cracked beam stays in a static state. Therefore, the determinant of the coefficient matrix $\mathbf{A}$ has to be equal to zero for a non-trivial solution, as shown in Eq. (46.13):

$$G(\beta) = det\left[\mathbf{A}(\beta) \right] = 0, \qquad (46.13)$$

By solving Eq. (46.13), the frequency parameter β can be obtained.

46.2.2 BAYESIAN INFERENCE FOR CRACK DETECTION

Suppose that there are N_m measurement points along the cracked beam, and they measure M order modes corresponding to M measured natural frequency parameters $\tilde{\beta}_1 \ \tilde{\beta}_2 \cdots \tilde{\beta}_m$. The value of mth order vibration mode with measurement error in the N_mth measurement point $\tilde{\phi}_m(\xi_{cNm}, \tilde{\beta}_m)$ can be written as

$$\tilde{\phi}_m \left(\xi_{cNm}, \ \tilde{\beta}_m \right) = \phi_m \left(\xi_{cNm}, \ \tilde{\beta}_m \right) + \eta_{cNm, \, \phi_m}, \qquad (46.14)$$

$$\eta_{cNm, \, \phi_m} \sim N\left(0, \sigma_{cNm, \, \phi_m}^2 \right), \qquad (46.15)$$

where the exact value of vibration mode $\tilde{\phi}_m(\xi_{cNm}, \tilde{\beta}_m)$ is given by Eq. (46.8) and the measured error $\eta_{cNm, \, \phi_m}$ is supposed to follow the Gaussian distribution with a mean of zero and variance $\sigma_{cNm, \, \phi_m}^2$.

The Bayesian formulation defines the relationship between the prior Probability Distribution Function (PDF) and posterior PDF as it is presented by Eq. (46.16):

$$p\left(\mathbf{dh}, \Xi | \tilde{\Phi}, \tilde{\beta} \right) = cp\left(\tilde{\Phi}, \tilde{\beta} | \mathbf{dh}, \Xi \right) p\left(\mathbf{dh}, \Xi \right), \qquad (46.16)$$

where c is a normalized constant, and Ξ and $\mathbf{dh}$ are parameter vectors which include the crack coordinates and depths for each crack in the beam as they are described by Eqs. (46.17) and (46.18):

$$\mathbf{dh} = \left[\begin{array}{cccc} \dfrac{d_1}{h} & \dfrac{d_2}{h} & \cdots & \dfrac{d_n}{h} \end{array} \right]^T, \qquad (46.17)$$

$$\Xi = \begin{bmatrix} \xi_1 & \xi_2 & \cdots & \xi_n \end{bmatrix}^T, \tag{46.18}$$

$\tilde{\Phi}$ is the data matrix that includes all mode measurement data of the vibration modes and can be written as follows:

$$\tilde{\Phi} = \begin{bmatrix} \tilde{\phi}_1(\xi_{c1}) & \tilde{\phi}_2(\xi_{c1}) & \cdots & \tilde{\phi}_m(\xi_{c1}) \\ \tilde{\phi}_1(\xi_{c2}) & \tilde{\phi}_2(\xi_{c2}) & \cdots & \tilde{\phi}_m(\xi_{c2}) \\ \vdots & \vdots & \ddots & \vdots \\ \tilde{\phi}_1(\xi_{cNm}) & \tilde{\phi}_2(\xi_{cNm}) & \cdots & \tilde{\phi}_m(\xi_{cNm}) \end{bmatrix}, \tag{46.19}$$

$\tilde{\beta}$ is the vector of measured frequencies as

$$\tilde{\beta} = \begin{bmatrix} \tilde{\beta}_1 & \tilde{\beta}_2 & \cdots & \tilde{\beta}_n \end{bmatrix}, \tag{46.20}$$

The likelihood function $p(\mathbf{dh}, \Xi | \tilde{\Phi}, \tilde{\beta})$ can be expressed as Eq. (46.19) when Eq. (46.22) is presented in the form of the goodness-of-fit function J_g to be minimized. The goodness-of-fit function describes how well the proposed model fits the observation data. Minimization of J_g helps to select the best parameters.

$$p(\tilde{\Phi}, \tilde{\beta} | \mathbf{dh}, \Xi) = L(\mathbf{dh}, \Xi | \tilde{\Phi}, \tilde{\beta}) \propto \exp\left[-\frac{1}{2\sigma_e^2}\right] J_g(\mathbf{dh}^*, \Xi^*), \tag{46.21}$$

$$J_g(\mathbf{dh}, \Xi) = \sum_{i=1}^{m} w_m \sum_{j=1}^{Nm} \frac{\phi_i(\xi_{cj}, \mathbf{dh}, \Xi) - \tilde{\phi}_i(\xi_{cj})^2}{\phi_i(\xi_{cj}, \mathbf{dh}, \Xi)} + \sum_{i=1}^{m} w_m' \left(\frac{\beta_i(\mathbf{dh}, \Xi) - \beta_i}{\beta_i(\mathbf{dh}, \Xi)}\right)^2, \tag{46.22}$$

$$(\mathbf{dh}^*, \Xi^*) = \arg_{\mathbf{dh}, \Xi} \min J_g(\mathbf{dh}, \Xi), \tag{46.23}$$

$$\sigma_e^2 = \min J_g(\mathbf{dh}, \Xi) = J_g(\mathbf{dh}^*, \Xi^*), \tag{46.24}$$

where w_m $(1,2,\ldots m)$ and w' are the weights of mode-based identification part and frequency-based identification part and σ_e^2 is the variance parameter.

When there is no noise present, the exact answer cloud is calculated by solving equation $J_g(\mathbf{dh}, \Xi) = 0$, to obtain the identification results as $(\mathbf{dh}, \Xi)$. In that case, the PDF $p(\tilde{\Phi}, \tilde{\beta} | \mathbf{dh}, \Xi) = 1$, which means the identification result has been satisfactorily determined. However, equation $J_g(\mathbf{dh}, \Xi) = 0$ can be solved only when there are adequate number of sensors. Otherwise, the minimization of goodness-of-fit function is still necessary.

According to Bayesian theory, the final posterior PDF can be expressed as Eq. (46.25):

$$p(\Phi, \beta, \mathbf{dh}, \Xi | \tilde{\Phi}, \tilde{\beta}) \propto p(\tilde{\Phi}, \tilde{\beta} | \Phi, \beta, \mathbf{dh}, \Xi) p(\Phi, \beta | \mathbf{dh}, \Xi) p(\mathbf{dh}, \Xi)$$

$$\propto p(\tilde{\Phi}, \tilde{\beta} | \Phi, \beta) p(\Phi, \beta | \mathbf{dh}, \Xi) p(\mathbf{dh}, \Xi) \tag{46.25}$$

The prior PDF is usually given as uniform [29] or gamma distribution [9] depending on the damage form. Since the aim of this research is to identify the crack, we do not assume any prior knowledge about the crack. The probabilities of different crack locations and crack depths should be the same.

Suppose the prior PDFs of crack coordinate ξ_i and crack depth d/h are subjected to the uniform distributions U (0, 1) independently. Therefore, the prior PDF of crack parameters can be expressed as

$$p(\mathbf{dh}, \Xi) = 1, \tag{46.26}$$

The most probable values of the unknown crack parameters $(\mathbf{dh}, \Xi)$ can be found by maximizing the PDF shown in Eq. (46.25). The objective function can be redefined as

$$J(\Phi, \beta, \mathbf{dh}, \Xi) = \frac{1}{2}(\mathbf{dh} - \mathbf{dh}^\eta)^T \Sigma_{\mathbf{dh}}^{-1}(\mathbf{dh} - \mathbf{dh}^\eta) + \frac{1}{2}(\Xi - \Xi^\eta)^T \Sigma_{\Xi}^{-1}(\Xi - \Xi^\eta)$$

$$+ \left[-\frac{1}{2\sigma_e^2}\right]J_g(\mathbf{dh}, \Xi) + \frac{1}{2}\left[\begin{array}{c}\tilde{\Phi}-\Phi\\\tilde{\beta}-\beta\end{array}\right]^T \Sigma_{\Xi}^{-1}\left[\begin{array}{c}\tilde{\Phi}-\Phi\\\tilde{\beta}-\beta\end{array}\right] \tag{46.27}$$

Optimization of $J(\Phi, \beta, \mathbf{dh}, \Xi)$ can be described by the following steps:

1. Take the initial values as the nominal values, $(\mathbf{dh}, \Xi) = (\mathbf{dh}^\eta, \Xi^\eta)$ and the frequency as the measured values,
2. Update the estimates of vibration mode Φ^* by minimizing Eq. (46.27),
3. Update the estimates of frequency β^* by minimizing Eq. (46.27),
4. Update the estimates of crack parameters $(\mathbf{dh}^*, \Xi^*)$ by minimizing Eq. (46.27),
5. Iterate the previous steps 2–4 until the crack parameters in $(\mathbf{dh}^*, \Xi^*)$ satisfy some convergence criterion, resulting in the most probable values of crack parameters. The uncertainty can be estimated by Eq. (46.24).

The traditional Bayesian method mainly aims at updating the FE model, and the characteristic equation can be described as $\mathbf{K}\phi = \lambda\mathbf{M}\phi$, and the objective function can be optimized by a linear optimization method. However, since the determination of frequency requires solving non-linear Eq. (46.13), the objective function is difficult to minimize by using a linear optimization method. A series of intelligent optimization algorithms can be applied to minimize the goodness-of-fit function presented in Eq. (46.27), such as genetic algorithm (GA), artificial neural network (ANN) and particle swarm optimization (PSO).

In this research, PSO will be applied to optimize the function. Certainly, other optimization methods may also be useful in crack parameter identification. However, those are not considered in this paper.

The identification method of crack parameters consists of the following steps:

1. Estimate the possible number of cracks and obtain the length of parameter vectors of $\mathbf{dh}$ and Ξ.
2. Obtain the measurement data of vibration mode and frequency with errors and obtain the likelihood function defined by Eq. (46.22).
3. Minimize the likelihood function using smart algorithms and obtain the estimated crack parameter values for the present measurement $\mathbf{dh}^*$ and Ξ^* as well as posterior PDF in Eq. (46.25).

46.3 NUMERICAL EXAMPLES

46.3.1 CRACK IDENTIFICATION FOR A SINGLE-CRACK BEAM

As shown in Table 46.1, a series of Euler-Bernoulli beams with only one crack are considered. These beams are of different crack parameters and boundary conditions. The measurement data are

TABLE 46.1

Numerical Test Cases for Different Crack Parameters and Boundary Conditions

Case	Boundary Condition	Crack Coordinate ξ_1	Crack Depth d/h
1	Pinned-pinned (PP)	0.5	0.5
2	Clamped-Free ()	0.8	0.5

TABLE 46.2

Identification Results for Different Test Cases and SNR.

Boundary Condition	SNR (dB)	Crack Coordinate Identification Value $\hat{\xi}_1$	Crack Coordinate Identification Error $e_{\hat{\xi}}$ (%)	Crack depth Identification Value $\frac{\hat{d}}{h}$	Crack depth Identification Error $\frac{\hat{d}}{h}$ (%)
PP	30	0.50	0.00	0.50	0.00
	20	0.50	0.00	0.50	0.00
	10	0.52	4.00	0.48	4.00

simulated with different errors given in the form of different signal-to-noise ratios (SNR). For each cracked beam, SNRs of 10, 20 and 30 dB are simulated by coupled filed simulation (CFS).

There are five measurement points along the beam while the first three natural frequencies and vibration modes are measured. The weights in Eq. (46.22) are supposed to be the same and can be expressed as

$$w_m = \dot{w}_m = 1/6 (m = 1, 2, 3),\tag{46.28}$$

The model update number is nm = 10, which means the identification result is obtained after 10 independent mode and frequency measurements.

The standard PSO is used to solve the optimization problem while the number of particles is 100, the maximum iterations is 200, the inertia weight is 0.5, the learning factors c1 and c2 are 1.478 and the velocity of the particle is in the range of [0,1].

The result in Table 46.2 shows the effectiveness and effect of different boundary conditions and measurement errors. When the SNR is 30 or 20 dB, the identification errors of the crack depth and the coordinate for both pinned-pinned (PP) are 0.0%. That means the proposed method is accurate for low measurement error situations. However, when the SNR is 10 dB, although the errors exist, the results are still acceptable.

Figure 46.2 shows the PDFs for the identification of the crack's coordinate and depth for PP beam and clamped-free beam. As shown, the mean values of PDFs are close to those of the actual crack parameters. Therefore, the identification method is reliable even in a noisy measurement environment.

46.3.2 CRACK IDENTIFICATION FOR A MULTI-CRACK BEAM

To verify the effectiveness of the proposed method for a multi-crack beam, a multi-crack beam example (crack number n=2) is presented, similar to previous section. The measurement data are simulated with different errors given in the form of different SNRs. For each cracked beam, SNR values of 10, 20 and 30 dB are simulated by S. Table 46.3 shows the cracked beam with different crack parameters. There are five measurement points along the beam while the first three

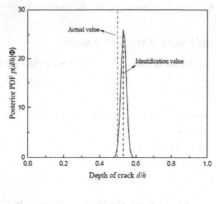 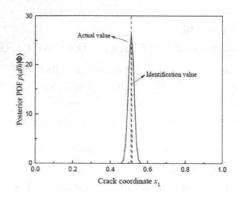

a)   PDFs of crack depth   b) PDFs of crack coordinate

FIGURE 46.2   Posterior PDFs of crack depth and coordinate for the PP beam when SNR = 10 dB.

TABLE 46.3
Numerical Test Cases for a Multi-Crack Beam with Different Crack Parameters

Boundary Condition	Crack Coordinates $[\xi_1, \xi_2, \xi_3]$	Crack Depth $\left[\dfrac{d_1}{h}, \dfrac{d_2}{h}, \dfrac{d_3}{h}\right]$
PP	[0.5, 0.8, 0]	[0.5, 0.6, 0]

TABLE 46.4
Identification Results for Different Test Cases and SNR

Boundary Condition	SNR (dB)	Crack Coordinate Identification Value $\left[\hat{\xi}_1, \hat{\xi}_2, \hat{\xi}_3\right]$	Crack Coordinate Identification Error $\left[e_{\hat{\xi}_1}, e_{\hat{\xi}_2}, e_{\hat{\xi}_3}\right]$	Crack Depth Identification Value $\left[\dfrac{\hat{d}_1}{h}, \dfrac{\hat{d}_2}{h}, \dfrac{\hat{d}_3}{h}\right]$	Crack Depth Identification Error $\left[e_{\hat{d}_1/h}, e_{\hat{d}_2/h}, e_{\hat{d}_3/h}\right]$
PP	30	[0.50, 0.80]	[0.00% 0.00%]	[0.50, 0.60]	[0.00% 0.00%]
	20	[0.50, 0.80]	[0.00% 0.00%]	[0.50, 0.60]	[0.00% 0.00%]
	10	[0.51, 0.83,]	[2.00% 3.75%]	[0.53, 0.62]	[6.00% 3.33%]

---

natural frequencies and vibration modes are measured. The weights are supposed to be the same as Eq. (46.27). The model update number is nm = 10. PSO is used for function optimization. The parameter selection of PSO is the same as the identification for a single crack.

Table 46.4 presents the identification results for the proposed test case and simulation errors. The identification results are relatively accurate when SNR is 30 and 20 dB. However, when SNR is 10 dB, the errors exist in both depth and coordinate identification. Besides, the identification of coordinate is more accurate overall than the depth identification. One possible reason is that the variation of the crack depth may have more influence on the vibration mode and frequency than the variation of the crack coordinate.

Figures 46.3 and 46.4 show the PDFs for the identifications of each crack's depth and coordinate.

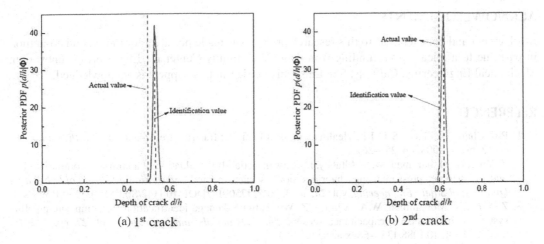

FIGURE 46.3   The posterior PDFs of crack depth for a PP beam with two cracks when SNR = 10 dB.

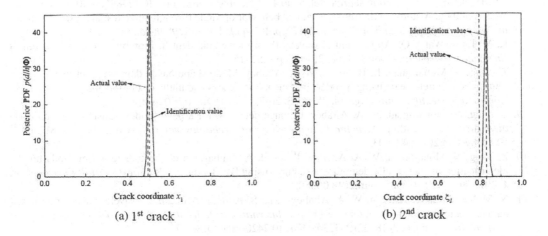

FIGURE 46.4   The posterior PDFs of crack coordinate for a PP beam with two cracks when SNR = 10 dB.

46.4   CONCLUSIONS

A new approach for crack identification in an Euler-Bernoulli beam based on closed-form solution of vibration modes is proposed by using Bayesian inference. The crack parameters (depth and coordinate) are used as unknown variables to derive the closed-form solution of the vibration modes. Based on the closed-form solution an intelligent algorithm and Bayesian inference are used to obtain the density function of crack parameters and to evaluate the uncertainty.

Based on the results presented in the paper the following conclusions can be drawn:

1. The proposed method performs well for both single and multiple crack cases. It has a reliable and acceptable robustness and can obtain the uncertainty of parameter identification.
2. In the case of a multi-crack beam, the proposed method is more accurate for the identification of the crack coordinate than the depth. Meanwhile, the errors and uncertainty increase by an increase in the number of cracks.

ACKNOWLEDGEMENTS

Involvement and contributions to this research project was made possible, for the second co-author, in part due to a release time funding from Donald E. Bently Center for Engineering Innovation, Mechanical Engineering, Cal Poly, San Luis Obispo. Herein this support is acknowledged.

REFERENCES

1. P. C. Chang, A. Flatau, S. C. Liu, Health monitoring of civil infrastructure. *Structural Health Monitoring*, vol. 2, no. 3, 2003, pp. 257–267.

2. Z. Ying, N. Mohammad, W.A. Altabey, L. Naiwei, Reliability evaluation of a laminate composite plate under distributed pressure using a hybrid response surface method. *International Journal of Reliability, Quality and Safety Engineering*, vol. 24, no. 3, 2017, 1750013, DOI: 10.1142/S0218539317500139.

3. Z. Ying, N. Mohammad, W.A. Altabey, Z. Wu, Fatigue damage identification for composite pipeline systems using electrical capacitance sensors. *Smart Materials and Structures*, vol. 27, no. 8, 2018, 085023. DOI: 10.1088/1361–665x/aacc99.

4. Z. Ying, N. Mohammad, W. A. Altabey, R. Ghiasi, Z. Wu, A fatigue damage model for FRP composite laminate systems based on stiffness reduction. *Structural Durability and Health Monitoring*, vol. 13, no. 1, 2019, 85–103, DOI: 10.32604/sdhm.2019.04695.

5. W. A. Altabey, An exact solution for mechanical behavior of BFRP Nano-thin films embedded in NEMS. *Advances in Nano Research*, vol. 5, no.4, 2017, pp. 337–357. doi: 10.12989/anr.2017.5.4.337.

6. W. A. Altabey; A study on thermo-mechanical behavior of MCD through bulge test analysis. *Advances in Computational Design*, vol. 2, no. 2, 2017, pp. 107–119. doi: 10.12989/acd.2017.2.2.107.

7. K. M. Liew, Wang, Q., Application of wavelet theory for crack identification in structures. *Journal of Engineering Mechanics*, vol. 124, no. 2, 1998, pp.152–157.

8. Z. Ying, N. Mohammad, B. B. Seyed, W. A. Altabey, Mode shape based damage identification for a reinforced concrete beam using wavelet coefficient differences and multi-resolution analysis. *Structural Control and Health Monitoring*, vol. 25, no. 3, 2017, 1–41. DOI: 10.1002/stc.2041.

9. Z. Ying, N. Mohammad, W. A. Altabey, Damage detection for a beam under transient excitation via three different algorithms. *Structural Engineering and Mechanics*, vol. 63, no. 6, 2017, 803–817, DOI: 10.12989/sem.2017.64.6.803.

10. Z. Ying, N. Mohammad, W. A. Altabey, T. Awad, A comparison of three different methods for the identification of hysterically degrading structures using BWBN model. *Frontiers in Built Environment*, 4, 2019, 80, DOI: 10.3389/fbuil.2018.00080.

11. N. Mohammad, W. Haifegn, W. A. Altabey, I. H. S. Ahmad, A modified wavelet energy rate based damage identification method for steel bridges. *International Journal of science & Technology, Scientia Iranica*, vol. 25, no. 6, 2018, 3210–3230, DOI: 10.24200/sci.2018.20736.

12. Z. Ying, N. Mohammad, W. A. Altabey, R. Ghiasi, Z. Wu, Deep learning-based damage, load and support identification for a composite pipeline by extracting modal macro strains from dynamic excitations. *Applied Sciences*, vol. 8, no. 12, 2018, 2564. doi: 10.3390/app8122564.

13. J. T. Kim, Y. S. Ryu, H. M. Cho, Damage identification in beam-type structures: frequency-based method vs mode-shape-based method. *Engineering Structures*, vol. 25, no. 1, 2003, 57–67.

14. W. A. Altabey, Free vibration of basalt fiber reinforced polymer (FRP) laminated variable thickness plates with intermediate elastic support using finite strip transition matrix (FSTM) method. *Vibroengineering*, vol. 19, no. 4, 2017, 2873–2885. DOI: 10.21595/jve.2017.18154.

15. W. A. Altabey, Prediction of natural frequency of basalt fiber reinforced polymer (FRP) laminated variable thickness plates with intermediate elastic support using artificial neural networks (ANNs) method. *Vibroengineering*, vol. 19, no. 5, 2017, 3668–3678. DOI: 10.21595/jve.2017.18209.

16. W. A. Altabey, High performance estimations of natural frequency of basalt FRP laminated plates with intermediate elastic support using response surfaces method. *Vibroengineering*, vol. 20, no. 2, 2018, 1099–1107. DOI: 10.21595/jve.2017.18456.

17. W. A. Al-Tabey, "Vibration analysis of laminated composite variable thickness plate using finite strip transition matrix technique," In *MATLAB verifications MATLAB- Particular for Engineer*, Kelly Bennett, ed, InTech USA, Charlotte, NC, vol. 21, 2014, 583–620, 980-953-307-1128-8. doi: 10.5772/57384.

18. J. N. Reddy, *An Introduction to Nonlinear Finite Element Analysis*. Oxford University Press, Oxford, 2004.

19. A. D. Dimarogonas, Vibration of cracked structures: A state of the art review. *Engineering Fracture Mechanics*, vol. 55, no. 5, 1996, 831–857.
20. S. Caddemi, I. Caliò, Exact closed-form solution for the vibration modes of the Euler–Bernoulli beam with multiple open cracks. *Journal of Sound & Vibration*, vol. 327, no. 3, 2009, 473–489.
21. S. Caddemi, A. Morassi, Multi-cracked Euler–Bernoulli beams: Mathematical modeling and exact solutions. *International Journal of Solids and Structures*, vol. 50, no. 6, 2013, 944–956.
22. F. L. Zhang, H. B. Xiong, W. X. Shi, Structural health monitoring of Shanghai Tower during different stages using a Bayesian approach. *Structural Control and Health Monitoring*, vol. 23, no. 11, 2016, 1366–1384.
23. S. C. Kuok, K. V. Yuen, Structural health monitoring of Canton Tower using Bayesian framework. *Smart Structures and Systems*, vol. 10, no. 4_5, 2012, 375–391.
24. W. A. Altabey, N. Mohammad, Detection of fatigue crack in basalt FRP laminate composite pipe using electrical potential change method. *Journal of Physics: Conference Series*, vol. 842, 2017, 012079. DOI: 10.1088/1742-6596/842/1/012079.
25. W. A. Altabey, Delamination evaluation on basalt FRP composite pipe by electrical potential change. *Advances in Aircraft and Spacecraft Science*, vol. 4, no. 5, 2017, 515–528. DOI: 10.12989/aas.2017.4.5.515.
26. W. A. Altabey, EPC method for delamination assessment of basalt FRP pipe: electrodes number effect. *Structural Monitoring and Maintenance*, vol. 4, no. 1, 2017, 69–84. DOI: 10.12989/smm.2017.4.1.069.
27. W. A. Altabey, N. Mohammad, Monitoring the water absorption in GFRE pipes via an electrical capacitance sensors. *Advances in Aircraft and Spacecraft Science*, vol. 5, no. 4, 2018, 411–434. DOI: 10.12989/aas.2018.5.4.499.
28. C. Bilello, *Theoretical and experimental investigation on damaged beams under moving systems*. Ph. D. Thesis, Universitadegli Studi di Palermo, Italy, 2001.
29. A. S. Sekhar, Crack identification in a rotor system: a model-based approach. *Journal of Sound and Vibration*, vol. 270, no. 4, 2004, 887–902.

47 Seismic Response of Counterweight-Roller-Rail Coupled System for High-Speed Traction Elevator

Wen Wang
Guangxi University of Science and Technology
Tongji University

Yan Jiang
Guangxi University of Science and Technology

CONTENTS

47.1 INTRODUCTION

As a vertical indispensable vehicle in high-rise buildings, the elevator is an important non-structural component of the building structure, and its seismic safety performance is gradually becoming a social concern. Especially for hospitals, schools, emergency command centers and other key buildings of disaster prevention, elevators are life safety passages and require continuous operation during earthquakes to ensure evacuation and post-disaster relief. Earthquake disasters show that the elevator system has poor seismic performance and suffer various degrees of damage during the earthquake (Suarez and Singh 2000, Du 2008). The most common type of damage is derailment, as shown in Figure 47.1. However, the seismic design of elevators in present codes is rather vague, and researches on the seismic behavior are limited in available literatures. At present, the research on the seismic performance of elevator systems is mainly focused on the counterweight-rail system. Yang et al. (1983) first investigated the response of the elevator guide rail-weight system under the base harmonic excitation using experiments and numerical simulation. Segal et al. (1996) developed dynamic models representing the structure and the elevator counterweight subsystems, with the nonlinear contact element between the counterweight and guide rail. Yao (2001) conducted experimental and numerical analysis of full-scale elevator counterweight to examine the mechanical behavior, damage types of elevator and to propose improvement measures. Wang et al. (2016) investigated the seismic performance of a function traction elevator as part of a full-scale five-story building shake table test, to study the acceleration amplifications of the elevator components subjected to dynamic excitations. Zhu et al. (2017) carried out a shaking table test on the full-scale counterweight-rail system, and performed numerical simulation of the test model to illustrate the behavior of the elevator counterweight system under earthquake action. These studies have

FIGURE 47.1   Counterweight derailment.

estimated responses of rail-counterweight system utilizing experimental and numerical methods based on the sliding shoes, but the contact mechanism is totally different from the roller guide-rail contact. Singh and Suarez (2002, 2004) first studied the linear and nonlinear seismic responses of the counterweight-rail system with roller guides, and calculated the horizontal seismic response of the counterweight at different positions. Zhu et al. (2013) established a coupled vibration model of the car body-car frame-rail, and the dynamic response of the elevator system under earthquake action was investigated.

All of the above studies investigated the seismic response when the elevator stopped at a floor, and the researches on the seismic performance of elevators under operating condition have not been deeply carried out. The author (Wang et al. 2018) has previously proposed a dynamic model of the coupled vibration of car-roller-rail for a traction elevator, which considered the nonlinear contacts among the elevator car, the roller guide shoe and the rail. In this paper, a coupled vibration model of the elevator counterweight-guide shoe-rail is established based on the previous model, which can study the seismic response of the elevator counterweight when the elevator is under operating condition.

47.2   DYNAMICAL MODEL

The model of elevator counterweight-guide shoe-rail system is shown in Figure 47.2. The counterweight is simulated by a rigid body with 2 degrees of freedom (DOF). The height of the counterweight is h_c, the mass is m_c, the moment of inertia around the axis is J_θ, the vertical distance from the upper guide shoe to its center of gravity is h_T and the vertical distance between the lower guide shoe and its center of gravity is h_B. Denote the horizontal displacement of the center of gravity as v_c, and the rotational angular displacement as θ_c. The horizontal displacement of the upper end of the counterweight is v_1^c, and that of the lower end is v_2^c. The quality of the four rollers is m_r, and their horizontal displacements are $v_{L,1}^r, v_{R,1}^r, v_{L,2}^r$ and $v_{R,2}^r$, respectively. The contact stiffness between the roller and the counterweight is k_{sr}, the damping is c_{sr}, the contact stiffness between the roller and the guide rail is k_{rr}; the contact stiffness between the retainer plate and the guide rail is k_{cr}. The guide rail is assumed to be a continuous Euler beam with elastic supports, the length of the single span is l_r, its total length is L, the bending stiffness is EI and the mass per unit length is ρ_r. The coordinates of the upper and lower rollers on the guide rail are, respectively, $s_1(t)$ and $s_2(t)$, and the spring stiffness of the jth rail bracket is $k_{b,j}$. Denoting the pre-tightening force between the roller and the guide rail as f_0, δ_0 is the pre-displacement between the roller and the guide rail, and d_0 is the initial clearance distance between the retainer plate and the guide rail.

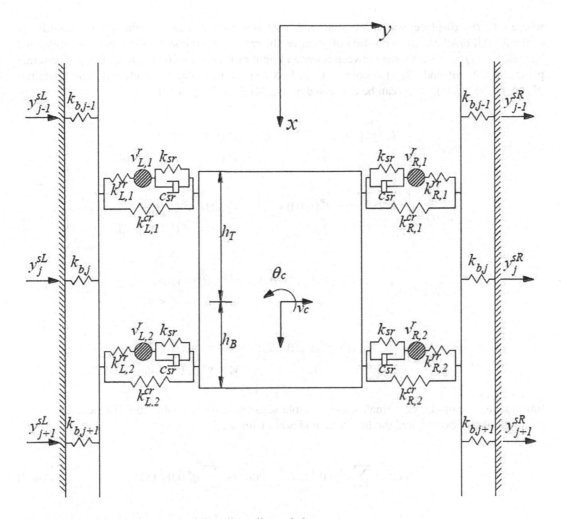

FIGURE 47.2   Model of counterweight-roller-rail coupled system.

The elevator guide rail is fixed to the wall of the elevator shaft in the building by brackets, and the vibration of the building is transmitted to the guide rail through rail brackets. The displacement of the jth bracket caused by the vibration of the building is $y_j^{sL}(t)$ and $y_j^{sR}(t)$. Equations of motion for the left and right rails can be expressed by the following partial differential equations:

$$
\mathrm{EI}\frac{\partial^4 y^L(x,t)}{\partial x^4} + \rho_r \frac{\partial^2 y^L(x,t)}{\partial t^2} + \sum_{j=0}^{N} k_{b,j}[y^L(x_j,t) - y_j^{sL}(t)]\delta(x - x_j)
$$

$$
= -\sum_{k=1}^{2} [f_{L,k}^{rr}(t) + f_{L,k}^{cr}(t)]\delta[x - s_k(t)]
$$

$$
\mathrm{EI}\frac{\partial^4 y^R(x,t)}{\partial x^4} + \rho_r \frac{\partial^2 y^R(x,t)}{\partial t^2} + \sum_{j=0}^{N} k_{b,j}[y^R(x_j,t) - y_j^{sR}(t)]\delta(x - x_j)
$$

$$
= \sum_{k=1}^{2} [f_{R,k}^{rr}(t) + f_{R,k}^{cr}(t)]\delta[x - s_k(t)]
$$

(47.1)

where $y^L(x, t)$ = displacement of left guide rail, $y^R(x, t)$ = displacement of right rail, x_j = coordinate of the jth rail bracket, $s_k(t)$ = position of roller on the rail, $f^{rr}_{L,k}(t)$ = contact force between roller and left rail, $f^{rr}_{R,k}(t)$ = contact force between roller and right rail, $f^{cr}_{L,k}(t)$ = contact force between retainer plate and left rail and $f^{cr}_{R,k}(t)$ = contact force between retainer plate and left rail. The quantities $f^{rr}_{L,k}(t), f^{rr}_{R,k}(t), f^{cr}_{L,k}(t), f^{cr}_{R,k}(t)$ can be expressed by Eq. (47.2) and Eq. (47.3).

$$f^{rr}_{L,k}(t) = \begin{cases} k_{rr}[y^L(s_k(t)) - v^r_{L,k}(t) + \delta_0] & y^L(s_k(t)) - v^r_{L,k}(t) > -\delta_0 \\ 0 & y^L(s_k(t)) - v^r_{L,k}(t) \le -\delta_0 \end{cases}$$

(47.2)

$$f^{rr}_{R,k}(t) = \begin{cases} k_{rr}[v^r_{R,k}(t) - y^R(s_k(t)) + \delta_0] & v^r_{R,k}(t) - y^R(s_k(t)) > -\delta_0 \\ 0 & v^r_{R,k}(t) - y^R(s_k(t)) \le -\delta_0 \end{cases}$$

$$f^{cr}_{L,k}(t) = \begin{cases} k_{cr}[y^L(s_k(t)) - v^c_k - d_0] & y^L(s_k(t)) - v^c_k > d_0 \\ 0 & y^L(s_k(t)) - v^c_k \le d_0 \end{cases}$$

(47.3)

$$f^{cr}_{R,k}(t) = \begin{cases} k_{cr}[v^c_k - y^R(s_k(t)) - d_0] & v^c_k - y^R(s_k(t)) > d_0 \\ 0 & v^c_k - y^R(s_k(t)) \le d_0 \end{cases}$$

Introducing the modal coordinates, the variable separation on the vibration displacement of the guide rail is performed, and the first N_rth modes are truncated.

$$y^L(x,t) = \sum_{n=1}^{N_r} q^L_n(t) Y^L_n(x) , \quad y^R(x,t) = \sum_{n=1}^{N_r} q^R_n(t) Y^R_n(x)$$

(47.4)

where $q^L_n(t)$, $q^R_n(t)$ = generalized coordinates and $Y^L_n(x)$, $Y^R_n(x)$ = model shape functions. Based on the orthogonality of the model functions, the vibration equations of the left and right guide rails can be discretized into the following equations:

$$\rho_r \ddot{q}^L_n(t) + 2\xi_n \omega_n \rho_r \dot{q}^L_n(t) + \rho_r \omega^2_n q^L_n(t) + \sum_{k=1}^{2}(k^{rr}_{L,k} + k^{cr}_{L,k}) \sum_{i=1}^{N_r} Y^L_n(s_k) Y^L_i(s_k) q^L_i(t) - \sum_{k=1}^{2} k^{rr}_{L,k} Y^L_n(s_k) v^r_{L,k}$$

$$= \sum_{k=1}^{2} k^{cr}_{L,k} Y^L_n(s_k) v^c_k + \sum_{j=0}^{N} Y^L_n(x_j) k_{b,j} y^{sL}_j(t) + \sum_{k=1}^{2}(-k^{rr}_{L,k}\delta_0 + k^{cr}_{L,k}d_0) Y^L_n(s_k)$$

$$\rho_r \ddot{q}^R_n(t) + 2\xi_n \omega_n \rho_r \dot{q}^R_n(t) + \rho_r \omega^2_n q^R_n(t) + \sum_{k=1}^{2}(k^{rr}_{R,k} + k^{cr}_{R,k}) \sum_{i=1}^{N_r} Y^R_n(s_k) Y^R_i(s_k) q^R_i(t) - \sum_{k=1}^{2} k^{rr}_{R,k} Y^R_n(s_k) v^r_{R,k}$$

$$= \sum_{k=1}^{2} k^{cr}_{R,k} Y^R_n(s_k) v^c_k + \sum_{j=0}^{N} Y^R_n(x_j) k_{b,j} y^{sR}_j(t) + \sum_{k=1}^{2}(k^{rr}_{R,k}\delta_0 - k^{cr}_{R,k}d_0) Y^R_n(s_k)$$

(47.5)

where $k_{L,k}^{rr}, k_{R,k}^{rr}, k_{L,k}^{cr}$ and $k_{R,k}^{cr}$ $(k=1,2)$ can be determined by

$$
\left\{
\begin{aligned}
k_{L,k}^{rr} &= \begin{cases} k_{rr} & y^L(s_k(t)) - v_{L,k}^r(t) > -\delta_0 \\ 0 & y^L(s_k(t)) - v_{L,k}^r(t) \le -\delta_0 \end{cases} & k_{L,k}^{cr} &= \begin{cases} k_{cr} & y^L(s_k(t)) - v_k^c > d_0 \\ 0 & y^L(s_k(t)) - v_k^c \le d_0 \end{cases} \\[2mm]
k_{R,k}^{rr} &= \begin{cases} k_{rr} & v_{R,k}^r(t) - y^R(s_k(t)) > -\delta_0 \\ 0 & v_{R,k}^r(t) - y^R(s_k(t)) \le -\delta_0 \end{cases} & k_{R,k}^{cr} &= \begin{cases} k_{cr} & v_k^c - y^R(s_k(t)) > d_0 \\ 0 & v_k^c - y^R(s_k(t)) \le d_0 \end{cases}
\end{aligned}
\right. \tag{47.6}
$$

The equation of motion of the left and right rollers is

$$
m_r \ddot{v}_{L,k}^r + c_{sr}(\dot{v}_{L,k}^r - \dot{v}_k^c) + k_{sr}(v_{L,k}^r - v_k^c) + k_{L,k}^{rr} \left[v_{L,k}^r - \sum_{i=1}^{N_r} Y_i^L(s_k) q_i^L(t) \right] = k_{L,k}^{rr}\delta_0 - f_0
$$

$$
m_r \ddot{v}_{R,k}^r + c_{sr}(\dot{v}_{R,k}^r - \dot{v}_k^c) + k_{sr}(v_{R,k}^r - v_k^c) + k_{R,k}^{rr} \left[v_{R,k}^r - \sum_{i=1}^{N_r} Y_i^R(s_k) q_i^R(t) \right] = f_0 - k_{R,k}^{rr}\delta_0 \tag{47.7}
$$

Displacement compatibility conditions can be written as

$$
v_1^c = v_c - h_T \theta_c, \qquad v_2^c = v_c + h_B \theta_c, \qquad h_c = h_1 + h_B \tag{47.8}
$$

Hence, the vibration equations of the counterweight are expressed as

$$
\frac{m_c h_B^2 + J_\theta}{h_c^2} \ddot{v}_1^c + \frac{m_c h_B h_T - J_\theta}{h_c^2} \ddot{v}_2^c + c_{sr}(2\dot{v}_1^c - \dot{v}_{L,1}^r - \dot{v}_{R,1}^r) + k_{sr}(2v_1^c - v_{L,1}^r - v_{R,1}^r)
$$

$$
+ k_{L,1}^{cr} \left[v_1^c - \sum_{i=1}^{N_r} Y_i^L(s_1) q_i^L(t) \right] + k_{R,1}^{cr} \left[v_1^c - \sum_{i=1}^{N_r} Y_i^R(s_1) q_i^R(t) \right] = (k_{R,1}^{cr} - k_{L,1}^{cr}) d_0 \tag{47.9}
$$

$$
\frac{m_c h_B h_T - J_\theta}{h_c^2} \ddot{v}_1^c + \frac{m_c h_T^2 + J_\theta}{h_c^2} \ddot{v}_2^c + c_{sr}(2\dot{v}_2^c - \dot{v}_{L,2}^r - \dot{v}_{R,2}^r) + k_{sr}(2v_2^c - v_{L,2}^r - v_{R,2}^r)
$$

$$
+ k_{L,2}^{cr} \left[v_2^c - \sum_{i=1}^{N_r} Y_i^L(s_2) q_i^L(t) \right] + k_{R,2}^{cr} \left[v_2^c - \sum_{i=1}^{N_r} Y_i^R(s_2) q_i^R(t) \right] = (k_{R,2}^{cr} - k_{L,2}^{cr}) d_0
$$

Combining Eqs. (47.5), (47.7) and (47.9), the coupled vibration equations of the counterweight-roller-rail system are rewritten in the matrix form.

$$
M\ddot{Y}(t) + C\dot{Y}(t) + K(t)Y(t) = F(t) \tag{47.10}
$$

where $Y = \left[q_1^L, \ldots q_{Nr}^L, v_{L,1}^r, v_{L,2}^r, v_1^c, v_2^c, v_{R,1}^r, v_{R,2}^r, q_1^R, \ldots q_{Nr}^R \right]^T$. Any suitable numerical procedures can be used to solve the equations of motion. In this study, the fourth order Runge-Kutta method is used to solve the numerical problem, and the numerical calculations are carried out using MATLAB®.

47.3   NUMERICAL CALCULATION

The building is assumed to be located in Shanghai. According to the current seismic design code (2016), the seismic fortification intensity is 7°, the design basic acceleration is 0.1 g, the site category is Class IV and the design feature period is 0.9 seconds. The elevator is installed in a 60-story high-rise building with 3 m high per floor. In order to compare the seismic response of the elevator under operation and docking conditions, it is assumed that the displacement excitation of the different rail supports caused by the vibration of the building under earthquake is the same, while the difference in response of different floors of the building is neglected. Assuming that the pre-tightening force f_0 between the roller and the rail is 120 N, the gap d_0 between the retainer plate and the guide rail is 5 mm. The following typical elevator parameters are selected for the analysis.

> Counterweight properties: m_c = 3,000 kg, l_c = 3.6 m, J_θ = 4,240 kgm², $h_B = h_T$ = 1.8 m.
> Roller: m_r = 12 kg, c_{sr} = 200 Ns/m.
> Guide rail: ρ_r = 13.54 kg/m, l_r =2.5 m, L =180 m, EI =2.1 × 10⁵ Nm².
> Stiffness coefficients: k_{rr} = 4 × 10⁵ N/m, k_{sr} = 4 × 10⁴ N/m, k_{cr} = 10⁷ N/m, k_b = 3 × 10⁶ N/m.

Considering the site conditions of the building, the design seismic grouping and the site characteristic period, the EI Centro wave (1979, EI Centro Array #6, Imperial Valley) is selected as the seismic input according to the relevant provisions of the Code for Seismic Design of Buildings (2016). The acceleration and displacement time history curves are shown in Figure 47.3. The accelerations corresponding to frequent, moderate and severe earthquakes in Shanghai are 0.035, 0.1 and 0.2 g, respectively. Since the counterweight derailment is the most common failure mode, the contact force and the disengagement displacement between the retainer plate and guide rail are selected as the evaluation indexes for assessing the seismic response of the elevator counterweight-guide shoe-rail system.

Figure 47.4 is the time histories of the contact force and disengagement displacement of the retainer plate and the guide rail in the 6–14 seconds when the elevator is under docking condition. In the figure, f_{c1} and f_{c2} are the contact forces between the upper retainer plates and the left- and right-side guide rails, respectively, f_{c3} and f_{c4} are the contact forces of the lower retainer plates and the left- and right-side guide rails. v_{c1} and v_{c2} are the relative displacements between the upper retainer plates and guide rails, v_{c3} and v_{c4} are relative displacements that the lower retainer plates disengaged from the left- and right-side guide rails. It can be seen from this figure that the retainer plates collided the guide rails during the vibration process, and the left and right contacts appeared alternately. Contact forces between the upper set of guide shoes and the guide rail are smaller than those of the lower set of guide shoes and the guide rail, but the results of the disengagements are reversed, which is caused by the upper set of guide shoes located in the guide rail span, while the lower set is located at the guide rail bracket. Figure 47.5 showed the seismic response time history of the counterweight-guide

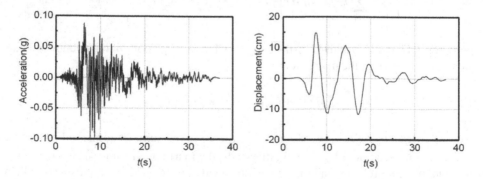

FIGURE 47.3   Acceleration and displacement time-history curves of El Centro.

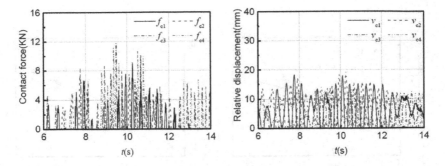

FIGURE 47.4 Seismic responses when elevator stops.

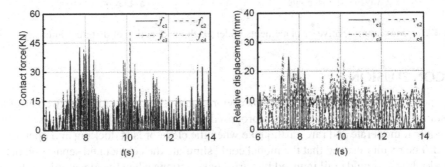

FIGURE 47.5 Seismic responses while running the elevator.

shoe-rail system when the elevator is under operation condition, and the maximum speed is 5.0 m/s. It can be seen from the figure that the maximum contact force between the retainer plate and the guide rail is about 3–4 times that when the elevator is stopped at a floor, which is mainly due to the change in the stiffness of the guide rail. Compared with the contact force, the amplification effect of the derailed displacement is not so obvious. The maximum contact force and disengagement displacement under different intensity earthquakes are compared in Figures 47.6 and 47.7. It can be concluded from Figure 47.6 that when the elevator is under operation condition, the maximum contact force of the system under the action of frequent earthquake is even larger than that caused by the rare earthquake when it is under docking condition, while the maximum disengagement displacement is close to the response under moderate earthquake as shown in Figure 47.7. It is further explained that the operation condition had a greater impact on the contact force of the elevator.

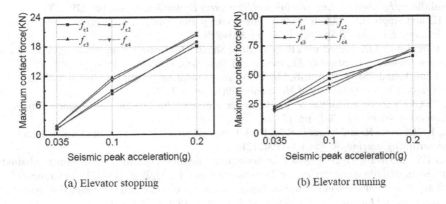

(a) Elevator stopping   (b) Elevator running

FIGURE 47.6 Maximum contact forces under earthquake action of different intensity.

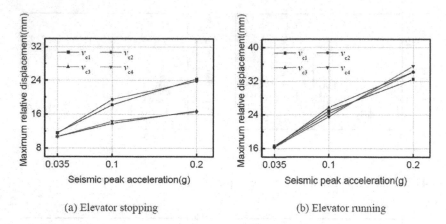

(a) Elevator stopping   (b) Elevator running

FIGURE 47.7   Maximum relative displacements under earthquake action of different intensity.

47.4 CONCLUSION

In this paper, a dynamic model of the coupled vibration of the elevator counterweight-guide shoe-rail is established based on the previously proposed model, and the seismic response of the system under frequent, moderate and rare earthquake when the elevator is at docking and operation state is examined. The results showed that the model could simulate the contact and separation between the roller guide shoe and guide rail induced by earthquake actions when the elevator is under operation condition. The influence of the parameter excitation due to the change of the guide-rail stiffness on the seismic response of the system could not be ignored, especially on the contact force. The calculation method proposed in this study can be utilized for the seismic design of elevators.

REFERENCES

Du P.H. 2008. From the Wenchuan seism: Statistics and analysis of elevator damage in Xi'an. *Elevator World*, 56(11), pp. 72–75.

GB50011-2010. 2016. *Code for seismic design of buildings*. Beijing: Building Industry Press of China. (In Chinese).

Segal F., Rutenberg A., and Levy R. 1996. Earthquake response of structure-elevator system. *Journal of Structural Engineering*, 122(6), pp. 607–616.

Singh M.P., and Suarez L.E. 2002. Seismic response of rail-counterweight systems in elevators. *Earthquake Engineering and Structural Dynamics*, 31(2), pp. 281–303.

Singh M.P., and Suarez L.E. 2004. Non-linear seismic response of the rail-counterweight system in elevators in buildings. *Earthquake Engineering and Structural Dynamics*, 33(33), pp. 249–270.

Suarez L.E., and Singh M.P. 2000. Review of earthquake performance, seismic codes, and dynamic analysis of elevators. *Earthquake Spectra*, 16(4), pp. 853–878.

Wang X., Hutchinson T.C., and Astroza R. et al. 2016. Shake table testing of an elevator system in a full-scale five-story building. *Earthquake Engineering & Structural Dynamics*, 46(3), pp. 391–407.

Wang W., Qian J., and Zhang A.L. 2018. Modelling and simulation of coupled vibration of car-roller-rail for an elevator. *Chinese Quarterly of Mechanics*, 39(1), pp. 107–116.

Yao G.C. 2001. Seismic performance of passenger elevators in Taiwan. *Earthquake Engineering and Engineering Seismology*, 3(2), pp. 17–26.

Yang T.Y., Kullegowda H. and Kapania R.K. 1983. Dynamic response analysis of elevator model. *Journal of Structural Engineering*, 109(5), pp. 1194–1210.

Zhu T., Lu F.Y. and Wang H. 2017. Study on horizontal vibration experiment and numerical simulation of counterweight-rail system in elevator. *Structural Engineers*, 33(2), p. 175–181. (In Chinese).

Zhu M., Zhang P. and Zhu C.M. et al. 2013. Seismic response of elevator and rail coupled system. *Earthquake Engineering and Engineering Vibration*, 33(4), p. 183–188. (In Chinese).

48 Modal Shape Estimation Based on a Parked Vehicle Induced Frequency Variation

W. Y. He and W. X. Ren
Hefei University of Technology

CONTENTS

48.1 INTRODUCTION

Mode shape estimation approaches can generally be categorized into two groups, namely, direct approaches with sensors installed on bridge and indirect approaches with sensors installed on vehicles (Yang et al. 2014). Though the former methods are extensively used in engineering practice, a few problems have been gradually discovered. For example, a huge number of sensors are required to obtain mode shapes with high spatial resolution, and a complicated procedure is required to mass normalize the mode shapes without excitation information (Bernal 2004). In view of this, the indirect approaches in which only one sensor is required to obtain mode shapes with high spatial resolution have received increasing attention. However, the unavoidable road surface roughness often leads to misleading results.

Based on the frequency variation caused by a parked vehicle, this paper aims to propose an approach to obtain high spatial resolution mode shapes with only one sensor and to overcome the disadvantage of high sensitivity to road surface roughness. First the relationship between the frequency variation induced by a parked vehicle and the amplitudes of the mode shapes is deduced. Then the mode shape is estimated by using the frequencies measured on a vehicle parked at different locations of the bridge progressively. Numerical example of a simply supported bridge is carried out to examine the performance of the proposed method.

48.2 ESTIMATION METHOD

A parked vehicle and the bridge will form a vehicle-bridge system, and the frequencies of the system and the bridge would be different (Chang et al. 2014). The frequency variation of the vehicle-bridge system and bridge can reflect the dynamic properties of the bridge. Figure 48.1 shows a parked vehicle as a simply supported bridge system, where (k_v) is large enough. The physical parameters of the bridge are length L, constant bending rigidity EI, constant mass density ρ, and constant cross section A. The vehicle is simplified as a sprung mass m_v.

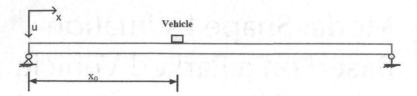

FIGURE 48.1   The parked vehicle-bridge system.

The motion equation of the vehicle-bridge system is (Yang et al. 2014)

$$\rho A\ddot{u}(x,t) + EIu''''(x,t) = [-m_v g - m_v \ddot{u}(x_0,t)]\delta(x - x_0) \tag{48.1}$$

where $u(x,t)$ is the vertical displacement of the bridge at the location x, t is time, and δ is the Dirac delta function.

The vertical displacement $(u(x,t))$ can be expressed as the superposition of mode shapes

$$u(x,t) = \sum_{i=1}^{\infty} q_i(t)\varphi_i(x) \tag{48.2}$$

where $\varphi_i(x)$ is the ith mass-normalized mode shape of the bridge and $q_i(t)$ is the corresponding modal coordinate.

Substituting Eq. (48.2) into Eq. (48.1)

$$\rho A\sum_{i=1}^{\infty} \ddot{q}_i(t)\varphi_i(x) + EI\sum_{i=1}^{\infty} q_i(t)\varphi_i''''(x) = \left[-m_v g - m_v\sum_{i=1}^{\infty} \ddot{q}_i(t)\varphi_i(x_0)\right]\delta(x - x_0) \tag{48.3}$$

Pre-multiplying the ith mode shape $(\varphi_i(x))$ to both sides of Eq. (48.3), computing the integral in the interval $[0,L]$, and applying the orthogonal condition and mass normalization character of the mode shapes, the following equation can be obtained.

$$\ddot{q}_i(t)[1 + m_v\varphi_i^2(x_0)] + w_i^2 q_i^2(t) = -m_v\varphi_i(x_0)\left[g + \sum_{n=1,n\neq i}^{\infty} \ddot{q}_i(t)\varphi_i(x_0)\right] \tag{48.4}$$

where w_i is the ith natural angular frequency of the bridge

$$w_i = \frac{i^2\pi^2}{L^2}\sqrt{\frac{EI}{\rho A}} \tag{48.5}$$

$w_i(x_0)$ is the ith natural angular frequency of the bridge-vehicle system when the vehicle is located at point x_0.

$$w_i(x_0) = \sqrt{\frac{w_i^2}{1 + m_v\varphi_i^2(x_0)}} \tag{48.6}$$

Then the relationship between frequency variation induced by a parked vehicle and the amplitudes of the mode shapes is

$$|\varphi_i(x_0)| = \sqrt{\frac{w_i^2 - w_i^2(x_0)}{m_v w_i^2(x_0)}} \tag{48.7}$$

Thus the mode shape estimation involves the natural angular frequency of the system with the vehicle parked at different locations and the natural angular frequency of the bridge. The process consists of the following steps: (1) Select the parking points according to the required the spatial resolution of the mode shape; (2) Determine the vehicle weight to make sure the frequency variation induced by the parked vehicle is visible; (3) Measure the frequencies of the bridge and the vehicle-bridge system under environment excitation; and (4) Estimate the mode shapes according to Eq. (48.7).

48.3   NUMERICAL EXAMPLE

A simply supported bridge with equal cross section is simulated to examine the feasibility of the proposed mode shape estimation approach. The parameters of the simply supported bridge are length $L = 20$ m, elastic modulus $E = 33.3$ GPa, cross-section $A = 1 \times 0.6$ m^2, and density $\rho = 2,500$ kg/m^3. The bridge is simulated via finite element method with 20 Euler beam elements.

The mass-normalized mode shape estimation approach is carried out through the four steps as presented in Section 2. 20 and equally spaced points are selected as parking points for the vehicle. More density points can be selected to obtain higher spatial resolution mode shapes when necessary. The weight of the vehicle is deemed as $m_v = 100$ kg. One wireless sensor is first installed on the bridge without the vehicle to measure the vibration responses of the bridge under environment excitation. Further the wireless sensor is moved on the vehicle to formulate the removable test equipment. Then removable test equipment is parked on the 19 parking points gradually. For each parking point, the wireless sensor records the dynamic responses of the vehicle-bridge system under environment excitation. The frequencies of the bridge and the vehicle-bridge system with the vehicle parked at different points are calculated via Fourier transform. Finally such frequencies are used to estimate the mode shape of the bridge according to Eq. (48.7).

Figure 48.2 plotted the estimated, reference mode shapes and their difference (deemed as "Error") of the bridge. Considering the fact that only low modes can be excited for girder bridges, only the first three mode shapes are estimated and compared. The results clearly indicate that the proposed approach can estimate the mass-normalized mode shapes for the simply supported bridge with high accuracy.

Compared to traditional mode shape estimation approaches, the proposed approach possesses several obvious advantages (He et al. 2018). (1) Only one wireless senor is required in obtaining high spatial resolution mode shapes. (2) It is insensitive to road surface roughness since the vibration test at each point is conducted independently instead of a continuous moving process. (3) The mode shapes are mass normalized, which is meaningful in structural finite element model updating and damage detection.

48.4   CONCLUSION

Estimating mode shape fast and accurately is essential in bridge damage detection and maintenance. This paper presented a mass-normalized mode shape estimation approach for bridges taking advantage of its bearing vehicle load property. The essential rule of frequency variation of a vehicle-bridge system with a vehicle parked at different locations was revealed via theoretical derivation. The relationship between the frequency variation induced by a parked vehicle and the amplitudes of the mode shapes was established accordingly. Then the procedures of the mass-normalized mode shape estimation approach were proposed. Finally a numerical example of a simulated simply supported beam indicated the proposed approach could estimate the mass-normalized mode shapes with high accuracy. Besides, it possessed several obvious advantages compared to traditional mode shape estimation methods.

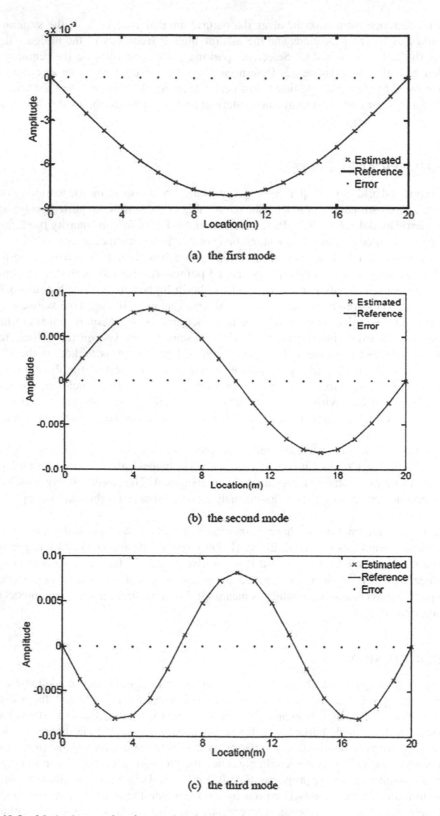

(a) the first mode

(b) the second mode

(c) the third mode

FIGURE 48.2 Mode shape estimation results of the simply supported bridge.

ACKNOWLEDGEMENTS

The authors are grateful for the financial support from the National Natural Science Foundation of China (No.51878234 and No.51778204), Fundamental Research Funds for the Central Universities (No. JZ2019HGPA0101 and No. PA2017GDQT0022).

REFERENCES

Bernal, D. 2004. Modal scaling from known mass perturbations. *J. Eng. Mech., ASCE*, **130**: 1083–1088.

Chang, K.C., Kim, C.W., and Borjigin, S. 2014.Variability in bridge frequency induced by a parked vehicle. *Smart Struct. Syst.*, **13**(5), 755–773.

He, W.Y., Ren, W.X., and Zuo, X.H. 2018. Mass normalized mode shape identification method for bridge structures using parking vehicle induced frequency change. *Struct. Control Health*, **25**, e2174.

Yang, Y.B., Li, Y.C., and Chang, K.C. 2014. Constructing the mode shapes of a bridge from a passing vehicle: a theoretical study. *Smart Struct. Syst.*, **13**(5), 797–819.

49 Wind Speed and Wind Direction Joint Distribution Model Based on Structural Health Monitoring

X. W. Ye and Y. Ding
Zhejiang University

CONTENTS

49.1 INTRODUCTION

With increase in bridge span, bridge structures become more sensitive to ambient excitations because of their slender and flexible characteristics. Among various types of structural responses caused by external dynamic excitations, wind-induced vibrations may be the primary effect on the serviceability and safety of bridge structures, especially for long-span cable-stayed and suspension bridges (Matsumoto et al. 1995, Fujino and Yoshida 2002). A typical example is the collapse of the Tacoma Narrows Bridge in 1940. Therefore, wind loads play a significant role in the design of long-span bridges, and their aerodynamic characteristics have been a focus of attention for structural safety evaluation. As wind characteristics depend on the geographical environment, terrain roughness and structural shape, it is essential to consider these site-dependent features to implement accurate wind-input simulation for bridge aerodynamics research (Scanlan and Gade 1997, Herb et al. 2007). In this regard, to facilitate the wind-resistant design and wind-induced fatigue assessment of long-span bridges, it becomes necessary to analyze the field wind characteristics, mainly involving average wind properties, e.g., average wind speed and direction, as well as the fluctuating wind characteristics, e.g., wind turbulence intensity, gust factor, turbulence integral scale and wind power spectral density (PSD).

As for the stochastic characterization of average wind features, three statistical analysis methods are commonly employed, i.e., the stationary random process method (Wen 1983), the maximum wind coefficient method (Simiu and Filliben 1981) and the joint probability distribution method (Cook 1982). Among them, the joint probability distribution has been a widely used statistical method to calculate the wind power and to evaluate the structural fatigue damage. Feng and Shen (2015) investigated the joint distribution of wind speed and direction for wind farm power calculation and layout optimization. Gu et al. (1999) established the joint probability density function (PDF) of wind speed and direction to estimate the fatigue life of steel girders of the cable-stayed Yangpu Bridge. Xu et al. (2009) assessed long-term wind-induced fatigue damage by integration of the joint distribution of wind speed and direction. A significant observation obtained from current research is that the estimated fatigue damage may have serious errors due to the inaccuracy of the derived continuous joint PDF of wind speed and direction (Alduse 2015).

One major factor which will affect the accuracy of the stochastic characterization of the average wind feature is the source of the wind data. Due to the difficulty in collecting complete wind data near the bridge site, previous investigators usually applied wind data measured by meteorological observatories near bridges. Simiu et al. (1985) exploited the wind data routinely collected from major weather stations in the United States and used the probability distribution to fit the largest yearly wind speed data in eight principal compass directions. Torres et al. (1999) utilized 10-minute mean wind data gathered from 11 meteorological stations distributed in Spain and estimated the Weibull parameters of 8 directional sectors. However, there are several disadvantageous aspects, e.g., the sampling frequency is set to be 10-minute or 1 day in the meteorological stations, the wind data are recorded with 16 orientation angles instead of random angles and the meteorological stations are usually far from the bridge site, and thus the wind data obtained from the meteorological stations cannot accurately reflect the wind characteristics at the bridge site. Recently, the structural health monitoring (SHM) technology has been rapidly developed in the civil engineering community, and this provides abundant valuable information for the evaluation of structural integrity, durability and reliability. As to bridge engineering, an SHM system usually contains several monitoring categories, and one of the major subsystems is the wind load and response monitoring system.

On the other hand, the joint PDFs of wind speed and direction are commonly employed in research on bridge wind engineering. Due to the complexity and inaccuracy of the discrete joint PDFs (Ge and Xiang 2002, Martin et al. 1999), research efforts have been devoted to the construction of continuous joint PDFs of wind speed and direction. For instance, Weber (1997) used an isotropic Gaussian model to describe the wind speed and direction by assuming that the variances of longitudinal and lateral components are the same. In consideration of different variances of these two components, Weber (1991) perceived the anisotropic Gaussian model. Qu and Shi (2010) applied the Farlie-Gumbel-Morgenstern (FGM) approach to construct the bivariate joint distribution to describe wind speed and air density simultaneously. Johnson and Wehrly (1978) used angular-linear (AL) distributions to model the joint distribution of bivariate random variables when one variable is directional and one is scalar. Erdem and Shi (2011) modeled seven different bivariate joint distributions by use of the AL, anisotropic lognormal and FGM methods, and then made a comparative study. Carta et al. (2008) established the AL distributions with specified marginal distributions and used normal-Weibull mixture distributions to build the marginal distribution of wind speed. However, the above-mentioned indirect modeling approaches usually build the joint distribution function by use of the marginal distributions of two variables, the consequences of which might lead to the deviation between the established joint distribution model and the measured data.

In this study, the stochastic characterization of wind field characteristics of an arch bridge is presented by analyzing the long-term wind monitoring data collected by the SHM system installed on the bridge. For the sake of describing the distribution features of the average wind, the genetic algorithm (GA)-based finite mixture modeling approach is proposed to construct the joint PDF of the average wind speed and direction. The Weibull distribution and the von Mises distribution are used to represent the distribution of the wind speed and direction, respectively. The GA-based parameter estimation method is employed to estimate the parameters of the mixed distribution models. The optimal models are selected by use of the Akaike's information criterion (AIC), Bayesian information criterion (BIC) and R^2 value. The results of the joint PDF of the wind speed and direction formulated by the proposed approach are compared with those obtained by use of the traditional AL distribution-based modeling approach.

49.2 LONG-TERM MONITORING OF WIND FIELD CHARACTERISTICS

49.2.1 DESCRIPTION OF THE INSTRUMENTED ARCH BRIDGE

The Jiubao Bridge, with an overall length of 1,855 m, is the first river-crossing arch bridge in China composed of a steel-concrete composite structure, which was opened to traffic on 6 July 2012. The bridge is located in Hangzhou, China and was built across the Qiantang River. The superstructure of the main bridge comprises a 3×210 m beam-arch composite structure and an 85 m composite box girder in the approaching bridge. The bridge deck is constructed with six traffic lanes and double-sided pavements, and the planned vehicle travel speed is set as 80 km/h.

The wind environment of the bridge site is complicated and dominated by both monsoon climate and typhoons. The climate condition around the Jiubao Bridge pertains to the humid subtropical with four distinct seasons. The monsoon circulation at the bridge site has a significant influence on the seasonal changes of local weather patterns. The southeast wind from the eastern sea in the summer and the north wind from northwest Siberia in the winter compose the primary wind loading acting on the Jiubao Bridge. Furthermore, in the summer, the bridge usually suffers from several typhoons. Figure 49.1 illustrates the wind rose diagrams in the summer and winter which are compatible with the monsoon climate.

49.2.2 MONITORING OF WIND FIELD

An SHM system was instrumented on the Jiubao Bridge to monitor the integrity, durability and reliability of the bridge during the in-service stage. The SHM system is mainly designed to monitor eight categories of structural physical quantities of the bridge, namely, deck geometry, temperature and humidity, wind loading, stress and strain, cable force, structural vibration, traffic loading, and support displacement. The sensory system consists of approximately 350 sensors and the condition of the bridge is continuously measured by these sensors. The sensory system includes anemometers, temperature and humidity sensors, a weigh-in-motion system, accelerations, seismic sensors, and cable force sensors.

In order to collect the wind field data near the bridge site during the construction and operation stages, two mechanical anemometers and one ultrasonic anemometer were installed on the Jiubao Bridge. One mechanical anemometer (ANE_L4) and one ultrasonic anemometer (UAN_L6) were installed on the upstream and downstream side of the bridge at the north of the main bridge, and the other mechanical anemometer (ANE_L35) was installed at the south of the main bridge. In order to minimize the effect of traffic on the accuracy of the measured wind field data, the anemometers were installed at approximately 6 m above the bridge deck. For the wind loading monitoring system, the wind direction angle 0° denotes north and 90° denotes east, rotating in a clockwise direction. The sampling frequency of the ultrasonic anemometer is set at 4 Hz and that of the mechanical anemometer is set at 0.1 Hz. The anemometer is able to detect a wind speed ranging from 0 to 60 m/s with an error within 0.01 m/s, and wind direction ranging from 0° to 360° (no dead angle) with an error within 0.1°.

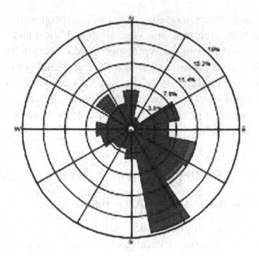

(a) Wind rose diagram of summer

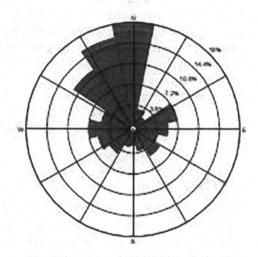

(b) Wind rose diagram of winter

FIGURE 49.1   Wind rose diagrams.

49.2.3   AVERAGE WIND SPEED AND DIRECTION

In the study, the wind data collected by the ultrasonic anemometer (UAN_L6) are selected to analyze the wind field characteristics, and the data from the mechanical anemometers are used to verify
the results. The recorded wind data include the wind speed time series $u(t)$ and the wind direction
time series $\phi(t)$, and two time series $u_x(t)$ and $u_y(t)$ can be calculated by the method of vector decomposition, in which x and y represent east and north directions, respectively. The basic time interval is
10 minutes in the specification, and 10-minute average horizontal wind speed U and wind direction
ϕ can be obtained by

$$U = \sqrt{\bar{u}_x^2 + \bar{u}_y^2}$$

(49.1)

$$\tan \phi = \frac{\bar{u}_x}{\bar{u}_y} \tag{49.2}$$

where $\bar{u}_x$ and $\bar{u}_y$ are the mean values of $u_x(t)$ and $u_y(t)$, respectively.

49.3 PROBABILISTIC MODELING METHODOLOGIES

49.3.1 AL DISTRIBUTION-BASED INDIRECT MODELING APPROACH

There are various indirect modeling approaches which construct the joint distribution by use of the marginal distributions of wind speed and direction, such as the FGM, AL, and anisotropic log-normal approaches. In this section, the AL distribution proposed by Johnson and Wehrly (1978) is chosen to establish the joint PDF of wind speed and direction. The joint PDF specified by the AL distribution is defined as

$$f_{v,\theta}(v,\theta) = 2\pi g(\xi) f_v(v) f_\theta(\theta) \tag{49.3}$$

where $f_v(v)$ represents the PDF of the wind speed, and $f_\theta(\theta)$ denotes the PDF of the wind direction, and $g(\xi)$ is the PDF of the circular variable ξ which can be expressed by

$$\xi = 2\pi \left[F_v(v) - F_\theta(\theta) \right] \tag{49.4}$$

In Eq. (49.4), the cumulative distribution functions of the wind speed and direction, $F_v(v)$ and $F_\theta(\theta)$, are given by

$$F_v(v) = \int_0^v f_v(v) dv \tag{49.5}$$

$$F_\theta(\theta) = \int_0^\theta f_\theta(\theta) d\theta \tag{49.6}$$

Due to the fact that the marginal probability distributions of wind speed and direction exhibit the characteristics of being multimodal and complex, the method of finite mixture distributions should be used to represent them. McLachlan et al. (2003) defined that the basic structure of finite mixture distributions for independent scalar or vector observations can be expressed by a weighted sum of the component distribution.

For the wind speed distribution function $f_v(v)$, the mixture of two-parameter Weibull distribution is chosen, which is expressed by

$$f_v(v) = \sum_{i=1}^N w_i \frac{\alpha_i}{\beta_i} \left(\frac{v}{\beta_i} \right)^{\alpha_i - 1} \exp \left[-\left(\frac{v}{\beta_i} \right)^{\alpha_i} \right] \tag{49.7}$$

where α_i $(i=1,\dots,N)$ represents the shape parameter, β_i $(i=1,\dots,N)$ is the location parameter, and w_i denotes the weight parameter which has a nonnegative quantity with a summation being ne, that is

$$\sum_{i=1}^N w_i = 1 \tag{49.8}$$

In order to construct the marginal PDF of the wind direction, $f_\theta(\theta)$, and the PDF of the circular variable, $g(\xi)$, a mixture of von Mises distribution (Carta et al. 2008) is employed which can be expressed by

$$f_\theta(\theta) = \sum_{i=1}^{N} \frac{w_i}{2\pi I_0(\kappa_i)} \exp\left[\kappa_i \cos(\theta - \mu_i)\right] \qquad (49.9)$$

where N is the number of the component distribution, κ_i ($i = 1,\ldots, N$) is the measure of the location, μ_i ($i = 1,\ldots, N$) represents the concentration parameter, and w_i ($i = 1,\ldots, N$) denotes the weight parameter. Additionally, in Eq. (49.9), $I_0(\kappa)$ is the modified Bessel function of the first kind and order zero, and can be calculated by

$$I_0(\kappa_i) = \frac{1}{\pi} \int_0^\pi e^{\kappa_i \cos(\theta)} \, d\theta \qquad (49.10)$$

After obtaining the marginal PDF of wind speed and direction, the circular ξ is calculated by Eq. (49.4). Then a mixture of von Mises distributions is used to model the PDF of circular ξ. Finally, the joint PDF of wind speed and direction can be constructed by Eq. (49.3) by the three univariate marginal distributions mentioned above.

In this section, the expectation maximization (EM) parameter estimation method is used to estimate the unknown parameters in each marginal PDF. However, the EM parameter estimation method typically converges to a local optimum, not the global optimum, and its convergence rate is not bound in general. Furthermore, the EM algorithm is arbitrarily poor in high dimensions and may cause an exponential number of local optima. Hence, it is necessary to develop an alternative technique for estimation of the parameters, especially in the high-dimensional setting.

49.3.2 GA-Based Finite Mixture Modeling Approach

49.3.2.1 Finite Mixture Distributions

The traditional AL distribution-based indirect modeling approach introduced above constructs the joint distribution usually by marginal distributions of two variables, so it is unavoidable that a difference exists between the joint distribution models and measured data. Due to the imprecision of the AL approach and the limitation of the EM algorithm in the high-dimensional setting, in the study, a GA-based finite mixture modeling approach is proposed to directly construct the joint PDF of wind speed and direction. The wind speed and direction are assumed to follow the finite mixture distribution with conditionally independent component distributions, and the joint PDF of the bivariate finite mixture distribution of wind speed and wind direction is defined by

$$f(\mathbf{x}; \mathbf{w}, \Theta) = \sum_{j=1}^{N} w_j f_v(v; \alpha_j, \beta_j) f_\theta(\theta; \mu_j, \kappa_j) \qquad (49.11)$$

where $f(\mathbf{x}; \mathbf{w}, \Theta)$ is the joint PDF of wind speed and direction, $f_v(v)$ is the PDF of wind speed, and $f_\theta(\theta)$ is the PDF of wind direction. $\mathbf{x}$ is the variable vector including wind speed v and wind direction θ. Θ is the parameters of the PDFs and w_j is the weight of each mixture component and satisfies the following:

$$\sum_{j=1}^{N} w_j = 1 \qquad (49.12)$$

In this finite mixture distribution function, we choose the Weibull distribution to represent the distribution of wind speed and use the von Mises distribution to represent the distribution of wind direction:

$$f(\mathbf{x};\mathbf{w},\Theta)=\sum_{j=1}^{N} w_j \frac{\beta_j}{\alpha_j}\left(\frac{v}{\alpha_j}\right)^{\beta_j-1}\exp\left[-\left(\frac{v}{\alpha_j}\right)^{\beta_j}\right]\cdot\frac{1}{2\pi I_0(\kappa_j)}\exp\left[\kappa_j cos\left(\theta-\mu_j\right)\right] \quad (49.13)$$

where parameters c and k are the shape and scale parameters of Weibull distribution, respectively. Parameters μ and κ are the location and concentration parameters of von Mises distribution, respectively. Because the proposed joint PDF is a bivariate distribution, we use the GA-based parameter estimation method to estimate the parameters of the proposed joint PDF. Using the GA-based parameter estimation method to estimate the parameters of a PDF is of increasing interest recently, since it can generate high-quality solutions for optimization and search problem in global scope, unlike the EM algorithm which is often subject to the problem of becoming stuck in local optima.

49.3.2.2 GA-Based Parameter Estimation

GA is proposed by Holland (1975), and is a widely used optimization algorithm inspired by Darwin's theory of evolution. In the GA, a group of candidate solutions of the optimization problem is evolved toward better solutions by the bio-inspired operators, i.e., mutation, crossover, and selection based on the fitness function. Thus, the establishment of the fitness function is the crucial part of GA.

Assuming that we have wind data $\mathbf{x}=[\mathbf{x}_1,\mathbf{x}_2,\mathbf{x}_3,\ldots,\mathbf{x}_n]^T$ which is a two-dimensional and n size dataset, $\mathbf{x}_i=[x_{i1},x_{i2}]$ as a two-dimensional vector including wind speed x_{i1} and wind direction x_{i2}, we select a section R_d $(d=1,2)$ which just contains all data for each dimension and then divide the section into n_d equal intervals with the width s_d, where n_d depicts the number of intervals for each dimension. For the two-dimensional wind data, the size of the vth interval I_v $(v=1,2,3,\ldots,n_1\times n_2)$ is $S_v=[s_{v1},s_{v2}]$ and the proportion of the vth interval is ξ_v, as $\xi_v=s_{v1}\times s_{v2}$. Using a histogram, the measure data are counted into these nonoverlapping and equally sized areas, and the quantity of observations N_v falling into the vth interval I_v is counted out. When $f(\mathbf{x};\mathbf{w},\Theta)$ is the continuous PDF, the probability that the observation $\mathbf{x}$ will fall inside the vth interval is given by

$$\iint_{I_v} f(\mathbf{x};\mathbf{w},\Theta)\,dx_1\,dx_2 \approx f(\mathbf{x}_v;\mathbf{w},\Theta)\xi_v \quad (49.14)$$

where Θ is the parameters of the joint PDF, I_v is the vth interval, ξ_v represents the proportion of vth interval, and $\mathbf{x}_v$ is the center point of the vth interval. The frequency p_v of observation data falling in the vth interval is given by

$$p_v=\frac{N_v}{N}\approx f(\mathbf{x}_v;\mathbf{w},\Theta)\xi_v \quad (49.15)$$

where N represents the total number of observation data and N_v is the number of observation data which fall into the vth interval.

The fitness function measures the quality of the produced parameters and tells the GA whether these parameters are optimal solutions. The closer the model and measured data distribution, the smaller the gap between frequency p_v and $\iint f(\mathbf{x};\mathbf{w},\Theta)dx_1dx_2$ is. So, we use the following function as the fitness function:

$$T=\frac{1}{\sum_{v=1}^{n_1\times n_2}\left(\dfrac{p_v-f(\mathbf{x}_v;\mathbf{w},\Theta)\xi_v}{f(\mathbf{x}_v;\mathbf{w},\Theta)\xi_v}\right)^2} \quad (49.16)$$

49.3.3 Selection of an Optimal Model

In this study, the AIC, BIC and R^2 values are selected to judge the fit performance in order to choose both the optimal number of components and evaluate the quality of joint PDF of wind speed and direction. The AIC is a measure of the quality of the statistical model (Akaike 1974), and is defined as

$$AIC = 2M - 2\ln(L) \tag{49.17}$$

where M is the number of unknown parameters in the finite mixture distribution, and L is the value of the maximum likelihood function of the mixed distribution. The preferred model is the one with the minimum AIC value.

The BIC is based on the likelihood function and is formally defined as

$$BIC = \ln(n)M - 2\ln(L) \tag{49.18}$$

where n is the total number of observations. The model with the lower BIC value is preferred.

The R^2 criterion represents the linear relationship between the estimated and observed frequencies, and is defined as

$$R^2 = 1 - \frac{SS_E}{SS_T} \tag{49.19}$$

where SS_T is the total sum of squares and SS_E is the sum of residuals. SS_T and SS_E are given by

$$SS_T = \sum (y_i - \bar{y})^2 \tag{49.20}$$

$$SS_E = \sum (y_i - f_i)^2 = \sum e_i^2 \tag{49.21}$$

where y_i is the probability of measured data falling in the ith interval, and f_i is the corresponding predicted PDF value. It is obvious that the value of R^2 is between 0 and 1, and the higher the R^2 value is, the better the fit performance is.

49.4 STOCHASTIC CHARACTERIZATION OF WIND SPEED AND DIRECTION

In this study, both the proposed GA-based finite mixture and the traditional AL distribution-based indirect modeling approaches are employed to construct the stochastic characterization of wind speed and direction. One-year wind monitoring data (from 1 September 2014 to 31 August 2015) recorded by the ultrasonic anemometer (UAN_L6) are extracted for comparative study. Figure 49.2 shows the values of AIC, BIC and R^2 with different numbers of components using the proposed approach. It can be seen from Figure 49.2 that the values of AIC, BIC and R^2 become stable when the number of components exceeds ten. This means that the formulated finite mixed model shows good performance with ten components. The joint distribution function of the wind speed and direction established based on the proposed approach is illustrated in Figure 49.3a. Figure 49.3b shows the histogram of the wind speed and direction. The estimated parameters of the established joint distribution model are listed in Table 49.1.

The MultiStart (MS) toolbox in MATLAB®, a gradient search method with randomized multiple starting points, is used to calculate the parameters of the joint PDF of wind speed and direction. In consideration of the increase of parameter number and the constraint of computation time, the number of multiple start points is set as ten thousand times the number of parameters and the

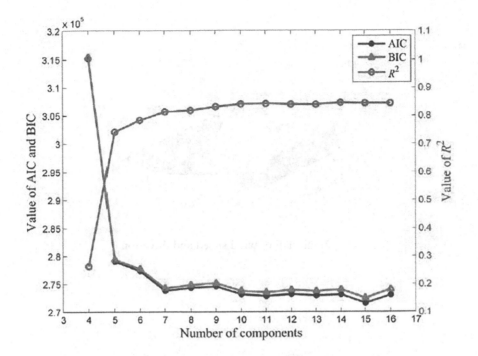

FIGURE 49.2   AIC, BIC and R^2 values calculated by the proposed approach.

distribution of start points adopts the uniform distribution. The Sequential Quadratic Programming (SQP) method is selected as the parameter estimation method to minimize the value of the fitness function by the fmincon function with the linear inequality constraint. Figure 49.4 shows the comparative analysis between MS and GA by use of the AIC value. It is seen from Figure 49.4 that the AIC value calculated by MS is gradually increased and cannot converge with the increase of the number of components, while GA exhibits good performance.

For the AL approach, the joint distribution model of the wind speed and direction is constructed based on the completion of the establishment of the distribution models of the wind speed, wind direction, and circular variable. Figure 49.5 shows the values of AIC, BIC and R^2 with different numbers of components for the probability distribution of the wind speed. It is seen from Figure 49.5 that the values of AIC, BIC and R^2 are stable with five components. Thus the mixture of five Weibull distributions is chosen as the wind speed distribution model, and the estimated parameters of the model are listed in Table 49.2. Likewise, as illustrated in Figure 49.6, the mixture of ten von Mises distributions is selected as the wind direction distribution model, and the estimated parameters of the model are listed in Table 49.3. After obtaining the distribution models of the wind speed and direction, the circular variable ξ can be derived for each group of the wind data. As shown in Figure 49.7, the values of AIC, BIC and R^2 are stable with two components, and thus the mixture of two von Mises distributions is determined as the distribution model of the circular variable, and the estimated parameters of the model are listed in Table 49.4. Figure 49.8 illustrates the obtained marginal distributions of the wind speed (Figure 49.8a), wind direction (Figure 49.8b), and circular variable (Figure 49.8c), the joint distribution of the wind speed and direction by use of the AL approach (Figure 49.8d), and the joint distribution of the wind speed and direction by the proposed approach (Figure 49.8e).

In addition, the stochastic properties of the wind monitoring data in different seasons are also characterized through establishing the distribution models by use of the proposed and AL approaches. Figures 49.9 and 49.10 show the formulated joint distribution of the wind speed and

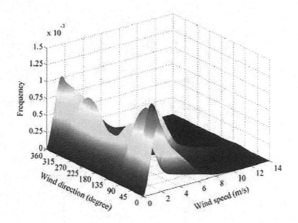

(a) Joint PDF of wind speed and direction

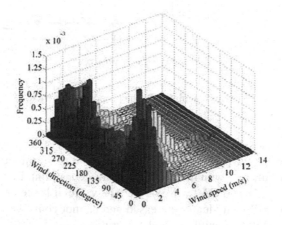

(b) Histogram of wind speed and direction

FIGURE 49.3   Joint distribution of wind speed and direction by the proposed approach.

TABLE 49.1
Estimated Parameters of Established Joint Distribution Model

Weight (ω)	Shape Parameter (α)	Location Parameter (β)	Concentration Parameter (μ)	Mean Direction (k)
0.101	6.588	1.948	2.100	11.846
0.195	3.847	1.482	0.327	0.063
0.154	3.441	2.109	1.322	11.439
0.141	2.265	2.966	4.669	1.616
0.040	5.974	3.653	5.120	6.149
0.065	5.241	4.831	2.250	10.443
0.078	1.331	3.166	3.905	1.401
0.100	3.852	2.360	6.065	11.908
0.073	1.745	2.922	0.398	9.526
0.053	3.294	3.355	2.730	10.260

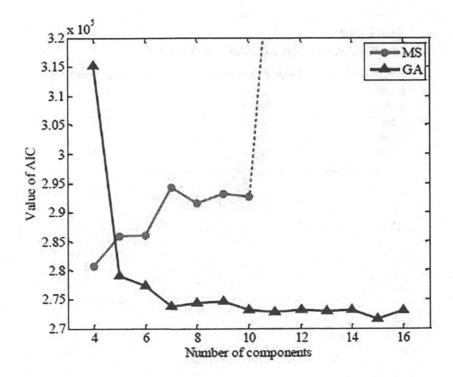

FIGURE 49.4   Comparative analysis between MS and GA.

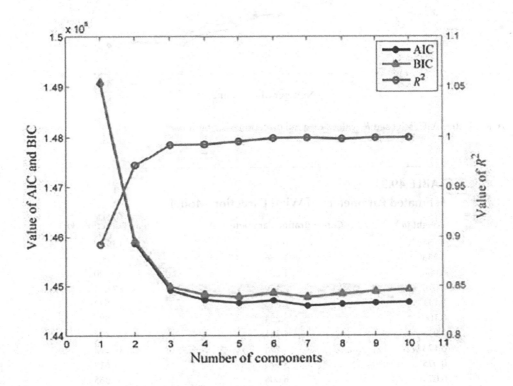

FIGURE 49.5   AIC, BIC and R^2 values of wind speed distribution.

TABLE 49.2

Estimated Parameters of Wind Speed Model

Weight (ω)	Shape Parameter (α)	Location Parameter (β)
0.200	4.035	2.703
0.200	1.428	2.819
0.201	2.379	3.285
0.199	4.865	1.946
0.200	3.526	2.121

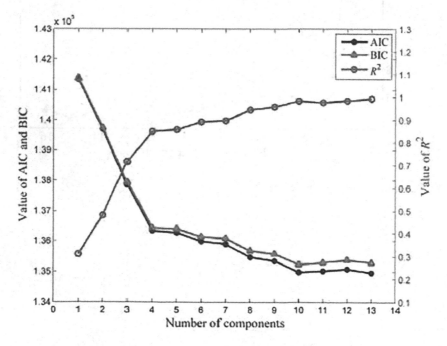

FIGURE 49.6   AIC, BIC and R^2 values of wind direction distribution.

TABLE 49.3

Estimated Parameters of Wind Direction Model

Weight (ω)	Concentration Parameter (μ)	Mean Direction (k)
0.136	0.165	13.790
0.238	1.466	4.062
0.058	1.263	95.490
0.070	2.120	87.818
0.111	2.540	14.522
0.100	3.254	4.583
0.015	4.281	335.821
0.123	4.595	10.272
0.075	5.472	21.618
0.074	6.006	62.365

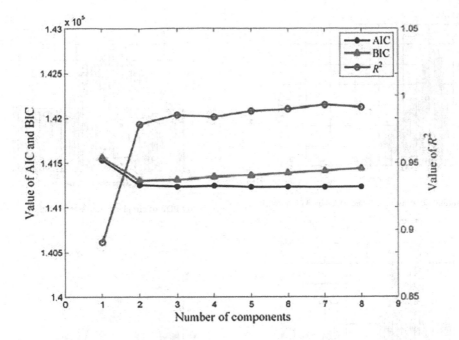

FIGURE 49.7   AIC, BIC and R^2 values of distribution of circular variable.

TABLE 49.4
Estimated Parameters of Circular Variable Model

Weight (ω)	Concentration Parameter (μ)	Mean Direction (k)
0.535	2.899	1.116
0.465	0.369	0.393

direction using these two methods. As listed in Tables 49.5–49.7, the AIC, BIC and R^2 values of the joint distribution of the wind speed and direction are calculated and used to compare the fitness performance of these two methods. As shown in Tables 49.5–49.7, it is obvious that the proposed GA-based finite mixture modeling approach has a better performance than the traditional AL distribution-based indirect modeling approach.

49.5   CONCLUSIONS

This paper addressed the stochastic characterization of wind field characteristics based on abundant wind monitoring data measured by the instrumented SHM system. To accurately describe the joint distribution features of the wind speed and direction, a GA-based finite mixture modeling approach was proposed to directly establish the joint PDF of the wind speed and direction. The Weibull and von Mises distributions were selected to represent the distribution of the wind speed and direction, respectively. The GA-based parameter estimation method was applied to estimate the parameters in the finite mixed models. The results of the proposed modeling approach were compared with those obtained by the traditional AL distribution-based indirect modeling approach in accordance with the AIC, BIC and R^2 criteria.

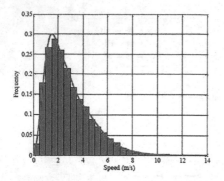

(a) Marginal distribution of wind speed by AL approach

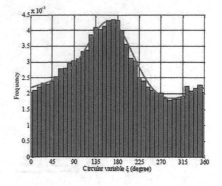

(c) PDF of circular variable by AL approach

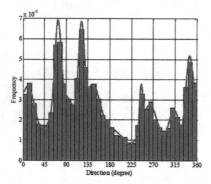

(b) Marginal distribution of wind direction by AL approach

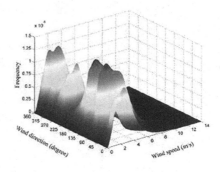

(d) Joint distribution of wind speed and direction by AL approach

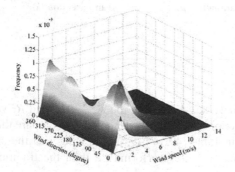

(e) Joint distribution of wind speed and direction by proposed approach

FIGURE 49.8   Joint distribution of wind speed and direction by AL and proposed approaches.

Through applying these two modeling approaches to describe the wind field characteristics of the arch Jiubao Bridge located in Hangzhou, China, the following conclusions can be drawn: (1) the long-term wind monitoring data are very helpful in stochastically representing the wind field characteristics around the bridge site; (2) the proposed GA-based finite mixture modeling approach can fulfill the task of directly modeling the wind speed and direction; (3) the GA-based parameter

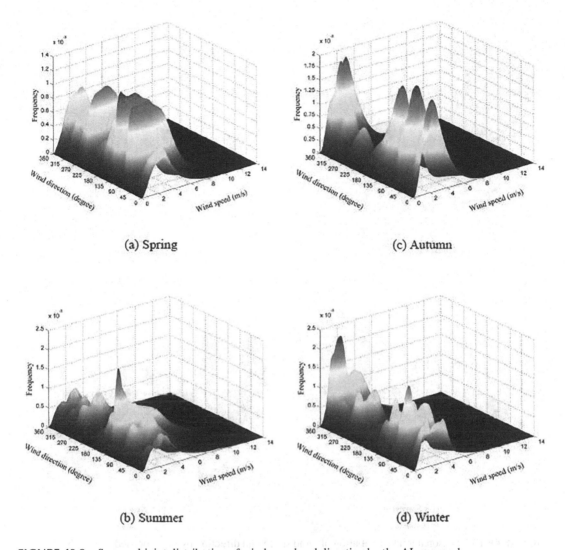

(a) Spring

(c) Autumn

(b) Summer

(d) Winter

FIGURE 49.9 Seasonal joint distribution of wind speed and direction by the AL approach.

estimation method has the ability of two-dimensional optimization and global search and can accurately estimate the parameters of the bivariate joint PDF; (4) the comparison between the proposed and AL modeling approaches based on 1-year wind monitoring data shows that the proposed modeling approach better represents the data; and (5) the results achieved from this study could facilitate the structural performance evaluation of long-span bridges under wind action and provide a useful input for wind-resistant design and wind-induced fatigue assessment.

ACKNOWLEDGEMENTS

The work described in this paper was jointly supported by the National Science Foundation of China (Grant No. 51778574), the Fundamental Research Funds for the Central Universities of China (Grant No. 2017QNA4024), and the Key Lab of Structures Dynamic Behavior and Control (Harbin Institute of Technology), Ministry of Education of the PRC.

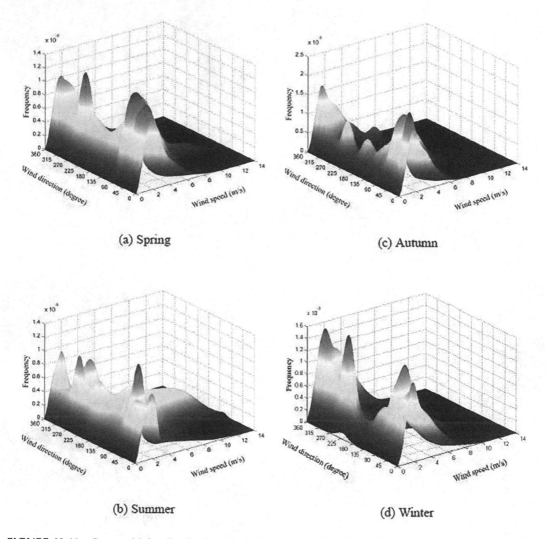

(a) Spring

(c) Autumn

(b) Summer

(d) Winter

FIGURE 49.10   Seasonal joint distribution of wind speed and direction by the proposed approach.

TABLE 49.5
AIC Values of Joint Distribution

	Indirect Modeling Approach	Direct Modeling Approach
One year	279145.935	273161.357
Spring	67302.670	66246.468
Summer	58916.987	56921.219
Autumn	74910.617	60246.843
Winter	75154.829	72252.439

TABLE 49.6
BIC Values of Joint Distribution

	Indirect Modeling Approach	Direct Modeling Approach
One year	279774.335	273789.861
Spring	67752.665	66696.463
Summer	59454.479	57458.711
Autumn	75524.650	60860.876
Winter	75768.152	72865.762

TABLE 49.7
R^2 Values of Joint Distribution

	Indirect Modeling Approach	Direct Modeling Approach
One year	0.681	0.842
Spring	0.607	0.856
Summer	0.513	0.776
Autumn	0.636	0.856
Winter	0.634	0.858

REFERENCES

Akaike, H. (1974). "A new look at the statistical model identification, automatic control." *IEEE Transactions on Automatic Control*, 19(6), 716–723.

Alduse, B.P., Jung, S., Vanli, O.A., and Kwon, S.D. (2015). "Effect of uncertainties in wind speed and direction on the fatigue damage of long-span bridges." *Engineering Structures*, 100, 468–478.

Carta, J.A., Ramirez, P., and Bueno, C. (2008). "A joint probability density function of wind speed and direction for wind energy analysis." *Energy Conversion and Management*, 49(6), 1309–1320.

Chen, B., Yang, Q.S., Wang, K., and Wang, L.A. (2013). "Full-scale measurements of wind effects and modal parameter identification of Yingxian wooden tower." *Wind and Structures*, 17(6), 609–627.

Cook, N.J. (1982). "Towards better estimation of extreme winds." *Journal of Wind Engineering and Industrial Aerodynamics*, 9(3), 295–323.

Erdem, E., and Shi, J. (2011). "Comparison of bivariate distribution construction approaches for analysing wind speed and direction data." *Wind Energy*, 14(1), 27–41.

Feng, J., and Shen, W.Z. (2015). "Modelling wind for wind farm layout optimization using joint distribution of wind speed and wind direction." *Energies*, 8(4), 3075–3092.

Fujino, Y., and Yoshida, Y. (2002). "Wind-induced vibration and control of Trans-Tokyo Bay crossing bridge." *Journal of Structural Engineering*, ASCE, 128(8), 1012–1025.

Ge Y.J., and Xiang, H.F. (2002). "Statistical study for mean wind velocity in Shanghai area." *Journal of Wind Engineering and Industrial Aerodynamics*, 90(12–15), 1585–1599.

Gu, M., Xu, Y.L., Chen, L.Z., and Xiang, H.F. (1999). "Fatigue life estimation of steel girder of Yangpu cable-stayed bridge due to buffeting." *Journal of Wind Engineering and Industrial Aerodynamics*, 80(3), 383–400.

Herb, J., Hoppmann, U., Heine, C., and Tielkes, T. (2007). "A new approach to estimate the wind speed probability distribution along a railway track based on international standards." *Journal of Wind Engineering and Industrial Aerodynamics*, 95(9), 1097–1113.

Holland, J.H. (1975). *"Adaptation in Natural and Artificial System."* The University of Michigan Press, Ann Arbor, MI, USA.

Johnson, R.A., and Wehrlyb, T.E. (1978). "Some angular-linear distributions and related regression models." *Journal of the American Statistical Association*, 73(363), 602–606.

Kaimal, J.C., Wyngaard, J., Izumi, Y., and Cote, O.R. (1972). "Spectral characteristics of surface-layer turbulence." *Quarterly Journal of the Royal Meteorological Society*, 98(417), 563–589.

Martin, M., Cremades, L.V., and Santabarbara, J.M. (1999). "Analysis and modelling of time series of surface wind speed and direction." *International Journal of Climatology*, 19(2), 197–209.

Matsumoto, M., Saitoh, T., Kitazawa, M., Shirato, H., and Nishizaki, T. (1995). "Response characteristics of rain-wind induced vibration of stay-cables of cable-stayed bridges." *Journal of Wind Engineering and Industrial Aerodynamics*, 57(2), 323–333.

McLachlan, G.J., Peel, D., and Bean, R.W. (2003). "Modelling high-dimensional data by mixtures of factor analyzers." *Computational Statistics and Data Analysis*, 41(3–4), 379–388.

Qu, X.L., and Shi, J. (2010). "Bivariate modeling of wind speed and air density distribution for long-term wind energy estimation." *International Journal of Green Energy*, 7, 21–37.

Scanlan, R.H., and Gade, R.H. (1977). "Motion of suspended bridge spans under gusty wind." *Journal of the Structural Division*, ASCE, 103(9), 1867–1883.

Simiu, E., and Filliben, J.J. (1981). "Wind direction effects on cladding and structural loads." *Engineering Structures*, 3(3), 181–186.

Simiu, E., Hendrickson, E.M., Nolan, W.A., Olkin, I., and Spiegelman, C.H. (1985). "Multivariate distributions of directional wind speeds." *Journal of Structural Engineering*, ASCE, 111(4), 939–943.

Torres, J.L., Garcia, A., Prieto, E., and de Francisco, A. (1999). "Characterization of wind speed data according to wind direction." *Solar Energy*, 66(1), 57–64.

Weber, R.O. (1991). "Estimator for the standard deviation of wind direction based on moments of the Cartesian components." *Journal of Applied Meteorology*, 30(9), 1341–1353.

Weber, R.O. (1997). "Estimators for the standard deviation of horizontal wind direction." *Journal of Applied Meteorology*, 36(10), 1403–1415.

Wen, Y.K. (1983). "Wind direction and structural reliability." *Journal of Structural Engineering*, ASCE, 109(4), 1028–1041.

Xu, Y.L., Liu, T.T., and Zhang, W.S. (2009). "Buffeting-induced fatigue damage assessment of a long suspension bridge." *International Journal of Fatigue*, 31(3), 575–586.

50 Application of a 3D Base Isolation and Overturn Resistance Device on the Large Height-Width Ratio Structure

X. Yan, S. Shi, and C. Zhang
Fuzhou University

H. Mao
Fujian University of Technology

L. Zhou
Fuzhou University

CONTENTS

50.1 INTRODUCTION

Recently, numerous methods have been used to control structural damage in earthquakes, including active, passive, and combined methods. In passive control, the dynamic characteristics of buildings are permanently changed to reduce the possibility of large internal forces generated in structural components. The isolation technique is a passive control method that filters out the high-frequency components of ground motion by setting the isolation layer between the base and the upper structure and reduces the seismic energy transferred to the upper structure to reduce the structure damage. The horizontal component of ground motion is believed to play a major role in the destruction of the general structure. Therefore, research and application of seismic isolation are mainly concentrated on the horizontal direction. The isolation structure generally refers to the horizontal base isolation structure. However, studies show that the structure responses under the three-dimensional (3D) ground motion are much more than those considering only one-dimensional ground motion. The vertical component tends to exacerbate the structure damage because of the multi-dimensional characteristics of ground motion (Sweedan 2009). The vertical component of ground motion also tends to exacerbate the structure damage for large height-width ratio structures. Therefore, a probabilistic seismic assessment of seismically isolated electrical transformers was performed by Kitayama et al. (2017) considering vertical isolation and vertical ground motion. Different vertical seismic isolation systems were also investigated by Milanchian et al. (2017) and Wakabayashi et al. (2009).

Considering the 3D base isolation for important buildings and infrastructure is very significant. Yan et al. (2010, 2017, 2018) experimentally studied the dynamic responses of the 3D isolated structure using the proposed 3D isolation devices. A 3D base isolation system adopting a negative stiffness device to control both the horizontal and vertical components of ground motion was presented by Cimellaro et al. (2018). The large height-width ratio isolation structure is at risk of overturning during an earthquake because of the large overturning moment. Therefore, Qi and Shang (2011) and Wang et al. (2007) studied the height-width ratio limited value, particularly for the base isolated structure. Shen et al. (2019) and Hu et al. (2017) developed tension-resistant devices for lead rubber bearings to increase the applicable height of the base isolation structure. Lu et al. (2016) investigated tension-resistant elastomeric isolation bearings, while Huang et al. (2006) studied Tension-resistant friction pendulum isolators. Using the base isolation principle, Kansas State University's stiffness decoupler for the base isolation of structures, which provide resistance to overturning, was investigated by Wipplinger (2004). The abovementioned studies focused on the single aspect of 3D isolation, height-width ratio limit, or tension-resistant bearing, respectively, but failed to comprehensively consider 3D isolation and overturn resistance in one system.

Therefore, the present study investigates a kind of 3D base isolation and overturn resistance device. The construction and design of the 3D base isolation and overturn resistance device are introduced. The numerical analysis of the large height-width ratio isolation structure is performed. The dynamic response of the 3D base isolation and overturn resistance structure is studied when the height-width ratio is gradually increased.

50.2   DESIGN OF THE 3D BASE ISOLATION AND OVERTURN RESISTANCE DEVICE

50.2.1   Design of the Horizontal Laminated Rubber Bearing

The 3D base isolation and overturn resistance device were composed of horizontal and vertical sub-devices. A tensile laminated rubber bearing was adopted as the horizontal sub-device, which had tensile strength, with the wire rope in the ordinary rubber bearing. Figure 50.1 shows the structural section. The ordinary rubber bearing bore pressure in most cases. Only when it produced a large horizontal shear deformation does the rubber bearing have tensile stress. The tension bearing capacity of the ordinary rubber bearing was assured by the connection between the steel plate and rubber. Although no obvious damage in the bearing was observed from the exterior when the bearing was under tension, it had many empty holes inside, which greatly influenced the bearing performance. After adding wire ropes to the ordinary laminated rubber bearing, the internal part of the bearing did not produce tensile stress in the case of a minor horizontal shear deformation. Meanwhile, the wire rope was in a state of relaxation; hence, it did not work and provide stiffness. In the case of a large horizontal shear deformation, the bearing may produce a large tensile stress, with the wire rope pulled tightly sharing most of the tension and ensuring the structure safety.

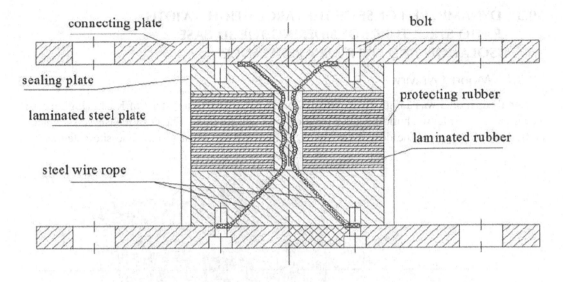

FIGURE 50.1   Section drawing of the tensile laminated rubber bearing.

50.2.2   Design of the Vertical Isolation Device

Figure 50.2 illustrates the adopted vertical isolation device. Its main component was a disk spring group, which made the vertical vibration isolation device have bearing, isolation, energy dissipation, and tensile capacity. The disk spring was orientated by a guide sleeve according to the literature (National standards of P.R.C 2005), and a 1 mm space existed between the guide sleeve and the spring.

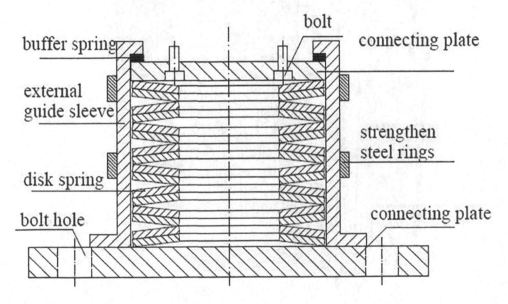

FIGURE 50.2   Section drawing of the vertical isolation device.

50.3 DYNAMIC RESPONSE OF THE LARGE HEIGHT-WIDTH RATIO STRUCTURE CONSIDERING THE 3D BASE ISOLATION AND OVERTURN RESISTANCE

50.3.1 MODEL OVERVIEW

The analysis model was a high-rise residential building in a certain district of Kashi (Figure 50.3). The high-rise residential building had a fortification intensity of 8°. The design basic seismic acceleration was 0.30 g, and the design earthquake group was the third group. The site category was

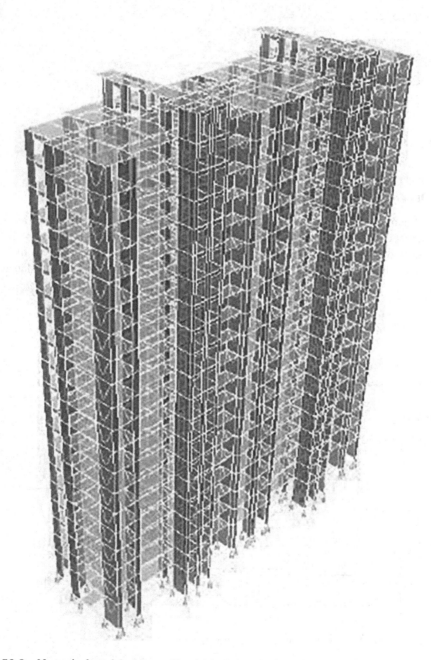

FIGURE 50.3 Numerical model of the residential buildings in Kashi.

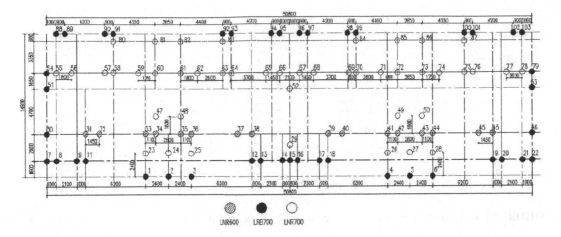

FIGURE 50.4 Layout of the isolation layer for the building.

Class II, and the characteristic period was 0.45 seconds. The height-width ratio was 5.16, indicating that the building has a large height-width ratio structure. The arrangement of the isolation layer was designed as shown in Figure 50.4. A 1/10 vertical stiffness of laminated rubber bearing (LRB) was adopted for the 3D isolation. In the ETABS, Isolator1 and Hook elements were connected in parallel to simulate the horizontal isolation and vertical tensile performances of the 3D base isolation and overturn resistance device. The Isolator1 element was used to simulate the laminated rubber bearing and the vertical isolation sub-device of the 3D base isolation and overturn resistance device. The Hook element was used to simulate the wire rope of the device that can only withstand tensile force and cannot withstand pressure. The Hook element provided rigidity to the system when the tensile deformation was greater than the set gap length of 100 mm. The Hook element did not provide rigidity to the system when the tensile deformation was less than or equal to the set gap length.

50.3.2 Analysis Result

50.3.2.1 Dynamic Analysis of the Non-Isolated, Horizontal Isolated, and 3D Base Isolation and Overturn Resistance Structures

A 3D excitation was applied to the model, that is, the excitation was input in the X, Y, and Z directions at a ratio of 1:0.85:0.65. A time-history analysis was performed using three acceleration time-history curves (i.e., El-Centro waves, Taft waves, and artificial waves). The results of the time-history analysis were compared with those of the mode decomposition reaction spectrum method, and larger values were obtained. Under the action of the design earthquake, the X direction inter-story displacement of the isolation layer of the common isolation bearing reached 113 mm; however, the X direction inter-story displacement of the isolation layer, where the 3D base isolation and overturn resistance device was located, was only 106 mm, which was smaller than that of the horizontal isolation bearing. The gap of the Hook element in ETABS was set to 100 mm, indicating that the Hook element played a role under the action of the earthquake.

Figure 50.5 shows the inter-story displacement of the non-isolated, horizontal isolated, and 3D base isolation and overturn resistance structures under the action of earthquake. The inter-story displacements of the horizontal isolated structure and the 3D base isolation and overturn resistance structure in the X direction were smaller than that of the non-isolated structure. The inter-story displacement of the 3D base isolation and overturn resistance structure was slightly smaller than that of the horizontal isolated structure. The inter-story displacement of the 3D base isolation and overturn resistance structure above the third floor in the Y direction was slightly larger than that of the

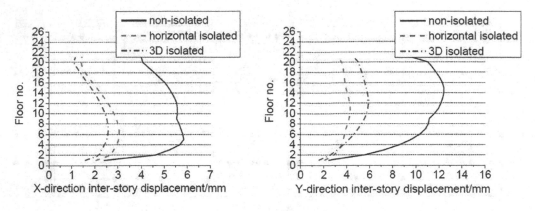

FIGURE 50.5   Inter-story displacement response of the three structures.

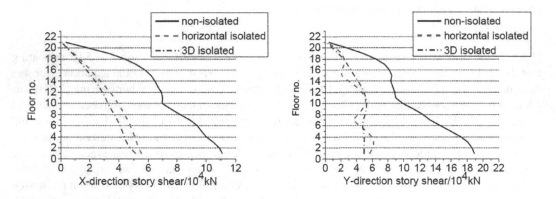

FIGURE 50.6   Story shear response of the three structures.

horizontal isolated structure, and the value below the third floor was smaller than in the horizontal isolated structure.

Figure 50.6 shows the story shear forces of the non-isolated, horizontal isolated, and 3D base isolation and overturn resistance structures under the action of earthquakes. The bottom shear force of the isolated structure was significantly smaller than that of the non-isolated structure in the X and Y directions. However, the story shear force in the Y direction fluctuated as the floor increased. According to the general information, no weak story formed, which met the vertical regularity requirements of the Chinese "Code for seismic design of buildings." The effect of the 3D base isolation was better than that of the horizontal isolation. Although the story shear force still had undulations, its distribution tended to be uniform, and the bottom shear force was slightly smaller than that of the horizontal isolated structure. Whether it was a horizontal isolated structure or a 3D base isolation and overturn resistance structure, the story shear force was smaller compared with that of the non-isolated structure, which was approximately half of the non-isolated structure.

50.3.2.2   Dynamic Response of the 3D Base Isolation and Overturn Resistance Structure Varying with the Height-Width Ratio

No major changes to the parameters and the superstructure of the 3D base isolation layer were needed to study the dynamic response of the 3D base isolation and overturn resistance structure with the increase of the structural height-width ratio. The height-width ratio was changed by adding floors. The dynamic analysis and the comparison of the 3D base isolation and overturn resistance structures with the height-width ratios of 5.16, 5.40, 5.64, 5.88, 6.12, 6.36, and 6.6 were performed.

Figure 50.7 shows the horizontal inter-story displacement of the 3D base isolation and overturn resistance structure in the X direction with the increase of the height-width ratio under the action of an earthquake. The Figure 50.8 shows that under the 16th floor, the inter-story displacement slightly decreased with the increase of the height-width ratio. The maximum inter-story displacement of the floor decreased most when the height-width ratio changed from 5.16 to 5.40, and then slightly decreased as the height-width ratio increased. The inter-story displacement fluctuated with the increase of the floor when the height-width ratio was 6.12 and 6.36. The fluctuation became more obvious when the height-width ratio was 6.36, and the maximum response appeared on the 10th and 24th floors. However, when the ratio increased to 6.60, this fluctuation tended to stabilize. Except for the structure with a height-width ratio of 6.36, the inter-story displacement of the top floor of other structures was smaller than that of the lower story. As shown in Figure 50.7, the lateral stiffness of the structure in the Y direction was smaller than that in the X direction; hence, the horizontal displacement in the structural Y direction was larger than that in the X direction. The horizontal inter-story displacement of the structure showed an increasing trend when the height-width ratio was increased from 5.16 to 5.88. The horizontal inter-story displacement was reduced when the ratio was increased to 6.12. The inter-story displacement tended to stabilize when the height-width ratio continued to increase until 6.60. Figure 50.8 shows that the structures with height-width ratios of 6.36 and 6.60 almost had the same inter-story displacement. The analysis of the inter-story displacement in two directions indicated that the 3D base isolation and overturn resistance structure had a height-width ratio in the changing process of displacement with the height-width ratio. With the height-width ratio, the displacement response of the structure was the largest, and the response curve most obviously fluctuated. Above and below this height-width ratio, the structural response tended to be stable. Therefore, the height-width ratio of the fluctuating section should be avoided in the actual structural design.

Figure 50.8 shows that the bottom shear of the structure in the X direction decreased as the height-width ratio increased. The story shear force in the X direction had the same growth trend when the height-width ratio was 5.16:5.4. The X direction story shear force of the structure with a height-width ratio of 5.4 was smaller than that of the structure with a height-width ratio of 5.16 below the 14th floor. The story shear force of the structure with a height-width ratio of 5.4 above the 14th floor was slightly larger than that of the structure with a height-width ratio of 5.16. The curve began to fluctuate as the floor increased when the height-width ratio reached 5.88. The degree of fluctuation reached the maximum when the height-width ratio reached 6.36. The fluctuations tended to stabilize after the height-width ratio reached 6.6. The curves of the structures with height-width ratios of 5.4, 5.64, 5.88, 6.12, and 6.36 intersected with the structure having a height-width ratio of 5.16 and 5.4 at the 13th floor, where their shear forces were smaller than those of the latter below the 13th floor and larger than those of the latter above the 13th floor. The structure with a height-width ratio of 6.6

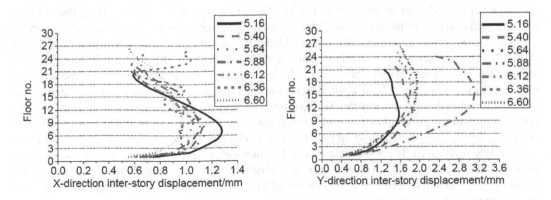

FIGURE 50.7 Inter-story displacement of the 3D isolated structure with the height-width ratio increase.

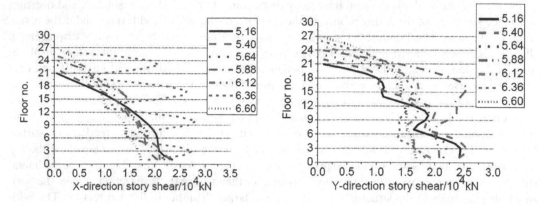

FIGURE 50.8 Story shear of the 3D isolated structure with the height-width ratio increase.

had the lowest bottom shear force and a slower growth of shear force. Its curve intersected with the structures with a height-width ratio of 5.16 and 5.4 at the 16th floor. As shown in Figure 50.8, the bottom shear force of the structure in the Y direction decreased as the height-width ratio increased. The story shear curves of the structures with a height-width ratio of 5.16 and 5.4 were substantially the same, but greatly fluctuated. The curve tended to stabilize when the height-width ratio was increased to 5.64. When increased to 5.88, the story shear force also greatly increased on the middle floor, but rapidly decreased with the increase of the floor in the upper floor. The curve shape was similar to that with the height-width ratio of 5.88 when the height-width ratio was increased to 6.12; however, the value was reduced to a level at a height-width ratio of 5.64. After which, the story shear force decreased with the increase of the height-width ratio, and the curve tended to stabilize. From the story shear curve in both directions, in one height-width ratio, the story shear of the structure was the maximum, and the shear force considerably fluctuated as the floor number increased. This height-width ratio in the X direction was larger than that in the Y direction. Therefore, the story shear force in two directions must be comprehensively considered, and a reasonable arrangement of the anti-lateral force members in the structure design must be made.

50.4   CONCLUSIONS

The construction and design of a 3D base isolation and overturn resistance device were introduced in this study.

The numerical simulation of the dynamic response of the large height-width ratio isolation structure was performed, particularly that of the large height-width ratio structure using the 3D base isolation and overturn resistance device. The results showed that the overturn resistance mechanism can work as required with reasonable settings. Under normal circumstances, the dynamic response of the 3D base isolation and overturn resistance structure in the X direction with a large height-width ratio was slightly larger than that of the horizontal isolated structure, while the dynamic response in the Y direction with a small height-width ratio was slightly smaller than that of the horizontal isolated structure. For the 3D base isolation and overturn resistance structure with the same architectural layout and structural arrangement, the dynamic response of the structure was greatly changed after the height-width ratio reached a critical value (i.e., when the height-width ratio was slightly increased). The most unfavorable height-width ratio of various types of the dynamic response of the 3D base isolation and overturn resistance structures was different. Moreover, the dynamic responses in different directions were different. The influence of the height-width ratio on the structural dynamic response should be comprehensively considered in the structure design. The existing structural scheme cannot be simply adopted when the total height of the 3D base isolation and overturn resistance structure changes; it needs to be re-analyzed and adjusted to ensure the structure safety and reliability.

Author Contributions: Conceptualization, X.Y. and S.S. Formal analysis, C.Z. Investigation, S.S. Project administration, X.Y. Supervision, C.Z. Visualization, H.M. Writing—original draft, S.S. Writing—review and editing, X.Y. and C.Z.

Funding: This research was funded by the National Natural Science Foundation of China, grant number 51578160, and the Fujian Science and Technology Plan Project, grant number 2018Y0057.

Conflicts of Interest: The authors declare no conflicts of interest.

REFERENCES

Cimellaro, G.P.; Domaneschi, M. and Warn, G. 2018. Three-dimensional base isolation using vertical negative stiffness devices. *Journal of Earthquake Engineering*, 1, pp. 1–29.

Hu, K.; Zhou, Y.; Jiang, L.; Chen, P. and Qu, G. A. 2017. Mechanical tension-resistant device for lead rubber bearings. *Engineering Structures*, 152, pp. 238–250.

Huang, S.; Islam, S. and Skokan, M. 2006. Essential facility design using tension-restraint friction pendulum isolators. *8th US National Conference on Earthquake Engineering*, San Francisco, CA, pp. 18–22.

Kitayama, S.; Lee, D.; Constantinou, M.C. and Kempner Jr, L. 2017. Probabilistic seismic assessment of seismically isolated electrical transformers considering vertical isolation and vertical ground motion. *Engineering Structures*, 152, pp. 888–900.

Lu, X.; Wang, D. and Wang, S. 2016. Investigation of the seismic response of high-rise buildings supported on tension-resistant elastomeric isolation bearings. *Earthquake Engineering & Structural Dynamics*, 45, pp. 2207–2228.

Milanchian, R.; Hosseini, M. and Nekooei, M. 2017. Vertical isolation of a structure based on different states of seismic performance. *Earthquakes and Structures*, 13, pp. 103–118.

National standards of P.R.C. 2005. GB/T1972-2005 Disk spring, China standard press. (in Chinese)

Qi, A. and Shang, H.J. 2011. Analysis on limit of height-width ratio of high-rise base-isolated structure. *Journal of Vibration and Shock*, 30, pp. 272–280. (in Chinese)

Shen, C.; Chen, Y.; Huang, X.; Tan, P. and Ma, Y.H. 2019. Numerical and experimental modeling of a new anti-tension equipment for high-rise isolation buildings. *Engineering Structures*, 179, pp. 129–138.

Sweedan, A.M.I. 2009. Equivalent mechanical model for seismic forces in combined tanks subjected to vertical earthquake excitation. *Thin Walled Structures*, 47, pp. 942–952.

Wakabayashi, N.; Ohmata, K. and Masuda, T. 2009. Vertical seismic isolation system using V-shaped link mechanism. *Journal of Japan Society of Mechanical Engineers*, 75, pp. 26–32.

Wang, T.Y.; Wang, H.D.; Zhang, Y.S. and Liu, W.G. 2007. Height-width ratio limited value for rubber bearing isolated structure computed by uniform design method. *Journal of Harbin Institute of Technology.*, 14, pp. 36–40.

Wipplinger, L.A. 2004. Dynamic testing of a masonry structure on a passive isolation system. *Journal of Architectural Engineering*, 10, pp. 15–21.

Yan, X.Y.; Zhang, Y.S.; Wang, H.D. and Wei, L.S. 2010. Shaking table test for the structure with three-dimensional base isolation and overturn resistance devices. *Engineering Mechanics*, 27, pp. 91–96. (in Chinese)

Yan, X.Y.; Qi, A.; Mao, H.M. and Xu, X. 2017. Multi-dimensional seismic response analysis of three-dimensional seismic-isolation mega-sub structure. *China Civil Engineering Journal*, 50, pp. 36–46. (in Chinese)

Yan, X.; Chen, W.; Shi, S. and Wang, X. 2018. Dynamic response of a combined isolation based mega-substructure under bidirectional near-fault ground motions. *Shock & Vibration*, pp. 1–15, 2018.

51 Non-Contact Video-Based Identification for Dynamic Behaviors of Beam Structures

Yu Cheng and Wen Xiong
Southeast University

CONTENTS

51.1 INTRODUCTION

The increasing of bridge operation time and the traffic flow will cause various degrees of damages to the bridge structure, which will seriously reduce the safety and durability of the structure. The dynamic performance parameters of the bridge structure, such as natural frequency, damping ratio, mode of vibration, dynamic impact number and dynamic response, are important indicators for the macro evaluation of the overall stiffness and operational performance of the bridge. It's also the key scale for the evaluation of the bridge's safety operational capability. The natural frequency and mode of vibration are the main parameters reflecting the dynamic characteristics of the structure, which are the fundamental for the dynamic analysis and damage identification under wind load, earthquake load and other dynamic loads. The identification of these dynamic characteristics mainly depends on the accurate measurement and high-frequency recording of the dynamic displacement of the structure.

The dynamic characteristic parameters of the bridge structure can be obtained through dynamic load experiments (such as hammer). According to the main modal characteristics of the structure in various modes within a certain frequency range, the actual vibration response of the structure inside and outside in this frequency under the action of various excitation sources can be estimated (Siringoringe & Fujino 2018). Dynamic load test generally needs to be combined with traditional contact measurement methods such as acceleration sensor and dynamic displacement meter to obtain the time-history curve of the bridge structure. However, this traditional method requires a large amount of manpower and material resources to install the sensor device; meanwhile, the sensor is susceptible to the influence of vehicle traffic and generates resonance. This sensing method makes the practical implementation more difficult in complex and harsh environments (Tian et al. 2019).

Digital image correlation (DIC) method (DICM) is a non-contact, full-field deformation optical measurement method based on digital image processing and numerical calculation (Pan et al. 2010). The 2D deformation of the surface of the object is measured by analyzing the images before and after the deformation. This method can be used to directly calculate the displacement and deformation of the object at various scales under the action of load. The bridge displacement measurement technology based on DICM has superior performance, which not only gives full play to the powerful data processing ability of computer but also realizes the non-destructive testing of the bridge through the non-contact of the measurement process (Busca et al. 2014).

Currently there are mainly two ways to study the dynamic displacement of the bridge method based on digital video technology, one kind is the speckle recognition method based on DIC. The displacement of each point on the surface of the measured object can be obtained by identifying the speckle changes of the structure. However, speckles need to be made manually, and the analysis speed is slow, which takes a long time in the bridge vibration experiment that requires a lot of data testing. The other kind is to install light emitting diode (LED) light targets on the structure under the condition of remote measurement to obtaining and calculating moving images. This method is suitable for the test environment with poor illumination, and can eliminate most of the interference in the identification process, but it is easy to leave smear on the image when the light is dim, thus affecting the accuracy of sampling (Ribeiro et al. 2014). To solve the problems of these two methods, a connected domain recognition method based on geometric feature extraction is proposed in this paper.

In this paper, digital video technology is used for testing and analyzing the hammer vibration of simply supported beams. During the vibration test, the structure images are obtained by smartphones and portable industrial cameras, and a zoom telescopic lens is used to obtain video images at different distances. The acquired video image is extracted from the region of interest, the connected domain is identified by the connected domain recognition algorithm based on geometric feature extraction proposed in this paper, and the target mark points are tracked to obtain the dynamic displacement curve when the structure vibrates, so as to calculate the modal parameters of the structure.

51.2 EXPERIMENTAL SYSTEM SETUP BASED ON DIGITAL VIDEO

51.2.1 MEASUREMENT SYSTEM

The measurement device used in this study is smartphone iPhone8, an industrial camera basler 1920–150uc, and the zoom lens is Walimex of 650–1,300 mm. The pixel of the smartphone is $4,032 \times 3,024$ and the video pixel is $1,920 \times 1,080$, the pixel size is 1.22 μm, and the frame rate is 60 fps. The industrial camera's video capture pixel is $1,920 \times 1,200$, and the size is 4.8 μm. The actual frame rate is affected by the bandwidth of the computer and the exposure time during shooting, so the frame rate used in this experiment is between 49 and 112 fps. The theoretical value of the first-order natural vibration frequency of the simply supported beam model used in this paper is 6.25 Hz. According to the Shannon sampling theorem, to recover the analog signal without distortion the sampling frequency should be greater than 2 times the highest frequency in the signal spectrum ($F_s > 2F_{max}$). To identify the first-order natural frequency of the simply supported beam model, the sampling frequency should be no less than 12.5 fps. The sampling frequency of the accelerometer used in the experiment was set to 128 fps, and the sampling frequency of the camera was set from 49 to 122 fps, both of which meet the above requirements.

51.2.2 TARGET RECOGNITION METHOD

There are two commonly used methods of dynamic digital image processing in structural vibration test: mark recognition and speckle recognition. The mark point recognition method has a fast recognition speed and does not need to make speckle manually. The speckle analysis method can measure the shape, deformation,* and motion of the whole field, but it needs to make the speckle manually, At the same time, the analysis speed is slow, and in the face of the huge amount of bridge vibration test, this method takes a long time. From the point of view of computational economy, the time-history curve of structure is obtained by using the mark point identification method which takes less time. The currently used marks are in the form of circular mark, diagonal mark, LED light cursor target, and other forms derived therefrom. Circular and diagonal marking calculations are convenient and accurate, but needs good quality of illumination during the test, and it is easy to produce errors in the case of uneven illumination; LED light targets are suitable for test environments with poor lighting. In the recognition process, most of the interference can be eliminated, but it is easy to leave a smear on the image, affecting the accuracy of the sampling. Considering the reduction in exposure can seriously affect the camera's frame rate in poor illumination, this paper selects the ordinary circular mark and the diagonal mark to carry out the experiment in the daytime when the illumination is good.

For circular marks, the use of connected domain identification and shape determination is a very efficient identification method. Visually, points that are connected to each other from an area, and points that are not connected form different areas. A collection of all points that are connected to each other is called a connected area. For a circular mark point, the white circular area enclosed by the ring belongs to the connected domain of high-efficiency closure, and the use of the connected domain for recognition will result in efficient and accurate results.

At present, there are many connected domain recognition algorithms. The one with high efficiency is to use a scan to traverse the image, mark the target pixels that appear, and record the continuous pixel runs in each row (or column). Since the same connected domain may produce different tag values, a complete set of equivalent tag pairs will be generated after scanning. The sparse matrix and the Dulmage-Mendelsohn decomposition algorithm are used to eliminate the equivalence tag pair, and then the original image is re-marked through the equivalence pair.

Since only a small portion of the entire image is the area of interest, this area is also known as the region of interest (ROI). Selecting the ROI for further processing and identification can highly reduce the processing time and improve the recognition accuracy; therefore, only the connected domain identification is performed on the ROI area during the processing. After the connected domain identification of ROI, other impurities connected domain will inevitably be generated. In order to remove the impurity connected domain, a filter was set up according to the geometric features of circular mark points to filter the identified connected domain.

There are many algorithms that can be used for the filtering of geometric features of circular marks. Among them, the more robust one is to use shape feature extraction to identify the circularity of the circular marker points (Figure 51.1).

Cameras, lenses and support systems

FIGURE 51.1   Measurement system based on digital video technology.

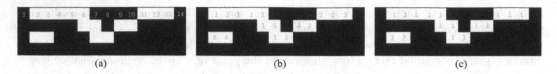

(a)   (b)   (c)

FIGURE 51.2   Connected domain tag diagram. (a) An image of 14 pixels in length. (b) The results of four connected identifications. (c) The results of octal connectivity recognition.

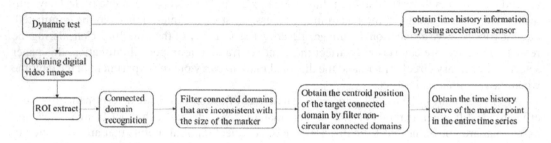

FIGURE 51.3   Identification flow chart.

The definition of circularity is given by Eq. (51.1)

$$C = \frac{P^2}{4\pi A} \text{ or } C = \frac{A}{A'} \tag{51.1}$$

where P is the graphics circumference; A is the graphics area; and A' is the Figure 51.2 minimum circumscribed circle area. The flow chart of target recognition above is shown in Figure 51.3.

   In current mature algorithms, although a commonly used algorithm, speeded up robust features (SURF), feature point recognition can identify circular and rectangular landmarks with obvious graphic features, it cannot identify specific targets independently. When the background is messy, the recognition results will contain a large number of non-target points, greatly increasing the workload of result data processing. Meanwhile, since the commercial software GOM also has the function of identifying the point, the result of the recognition of the points is likely to be misaligned or even missing some points when there is a great deal of mark points, the inconspicuous contrast of the image and the large inter-frame. In this paper, the connected domain recognition algorithm based on geometric feature extraction was used to identify and track circular and rectangular diagonal points. Meanwhile, SURF feature and commercial GOM point recognition algorithms were used to compare the results and verify the accuracy and practicability of this method.

51.3   EXPERIMENTAL VERIFICATION BASED ON SIMPLY SUPPORTED STEEL BEAM

51.3.1   INTRODUCTION OF EXPERIMENTAL MODEL

In order to verify the validity and correctness of the proposed structural parameter identification method based on digital video, an indoor experiment on the vibration of simply supported beams was carried out. The experimental settings are shown in Figure 51.4. The experimental model is a simply supported steel beam with a length of 1.4 m, which is excited by hammering, and the multiple acceleration sensors are used to obtain the structural response of the model under the action of hammer; a checkerboard mark is pasted on the surface of the accelerometer, and a plurality of circular mark points is pasted at the edge of the corresponding steel beam to identify the dynamic

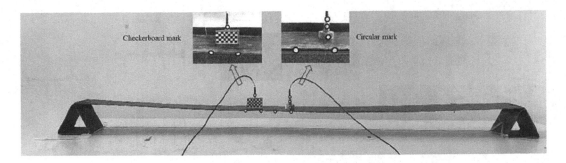

FIGURE 51.4 A schematic diagram of experimental model.

displacement of the model during the experiment. The size of each grid is 2.5 mm×2.5 mm, the outer diameter of circular mark points is 10 mm, and the inner diameter is 5 mm. Mark points and experimental model are shown in Figure 51.4.

The experiment uses a smartphone to acquire dynamic displacement video, and the shooting distance is about 1 m. Multiple hammer experiments were carried out, and multiple sets of data were obtained by adding a fulcrum to the model to change the frequency of the model. Since the acceleration sensor placed in the experiment can only measure the vertical acceleration, and the vibration of the simply supported beam is also vertical, only vertical displacement and acceleration are considered in the result processing.

51.3.2 CLOSE-RANGE EXPERIMENT IN THE LABORATORY

The images captured by video were initially split into a subset of the ROI, and then the connected domain recognition algorithm was used to calculate the centroid position of each marker point. The circular mark points occupy 16 pixels in the image, each square of the checkerboard has a size of 4 pixels, the outer diameter of the circular mark point is 10 mm, and the size of each square of the checkerboard is 2.5 mm × 2.5 mm. By setting the scaling factor $s_f = 0.625$ mm/pixel, the actual vertical displacement of the mark points between each frame can be determined. The physical displacement can be obtained by multiplying the pixel displacement by the scaling factor. Setting the tracking displacement to d_{t_i}, dividing the change in displacement by the time interval $\Delta t = t_{i+1} - t_i$ to get the corresponding instantaneous speed v_{t_i}, represented as $v_{t_i} = d_{t_i} / \Delta t$.

On this basis, the acceleration time history is obtained by using the numerical differential method of the four-point difference algorithm and the velocity time history. The mathematical formula can be expressed as

$$a_{t_i} = \left[v_{t_{i-2}} - 8v_{t_{i-1}} + 8v_{t_{i+1}} - v_{t_{i+2}} \right] / 12\Delta t \tag{51.2}$$

where $a_{t_i} =$ the acceleration and $v_{t_i} =$ velocity at time t_i.

The acceleration calculated by the connected domain recognition method is in good agreement with the measured result, and the correlation coefficient of two kinds of data in the stable vibration attenuation stage is 0.9447, which meets the requirements of engineering accuracy (Figure 51.5).

To verify the precision of the connected domain identification method, the SURF feature points recognition algorithm and commercial software GOM were used respectively to identify circle mark point, and the results were compared with the connected domain recognition method, as shown in Figure 51.6. The results of these methods are basically consistent, and the results of connected domain recognition were closest to the acceleration sensor, while the results of GOM and SURF feature point recognition were slightly different from the sensor results (Table 51.1).

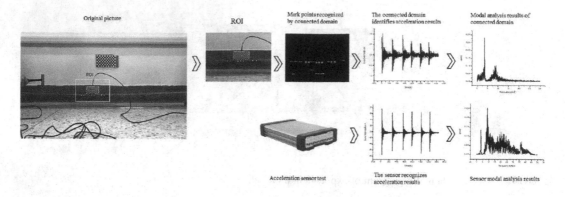

FIGURE 51.5   Model analysis of simply supported beams using connected domain recognition algorithm.

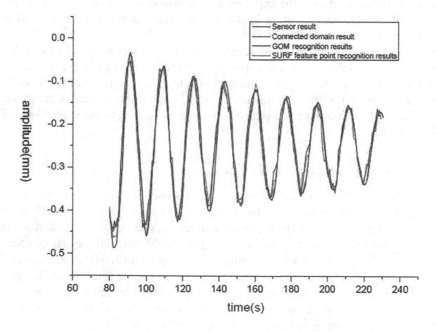

FIGURE 51.6   Comparison of results obtained by different methods over a period of time.

TABLE 51.1

The First-Order Frequency Measured by Different Algorithms (Hz)

	Frequency of the 1st Order			Frequency of the 2nd Order		
	1st Experiment	2nd Experiment	3rd Experiment	1st Experiment	2nd Experiment	3rd Experiment
Sensor results	6.348	6.372	5.029	12.085	11.963	11.206
Connected domain results	6.355	6.384	5.036	12.002	11.953	11.426
Deviation (%)	−0.110	−0.188	−0.139	0.687	0.084	−1.963
GOM results	6.357	6.387	5.039	12.129	12.129	11.426
Deviation (%)	−0.142	−0.235	−0.199	−0.364	−1.388	−1.963
SURF results	6.357	6.387	5.039	12.002	12.129	11.426
Deviation (%)	−0.142	−0.235	−0.199	0.687	−1.388	−1.963

51.3.3 LONG-DISTANCE EXPERIMENT IN THE LABORATORY

In order to verify the effectiveness and robustness of this method at a long distance, industrial camera and telephoto lens are used to acquire dynamic displacement in an empty indoor site. The shooting distance is about 30 m, and the shooting focal length is 700 mm. Experiments are carried out when the illumination is weak, medium, and strong to simulate the recognition effect under different weather conditions.

In the case of weak illumination, in order to ensure the accuracy of recognition, an external light source is applied while adjusting the exposure of the camera to meet the requirements of sufficient contrast between the target and the background during shooting. At this intensity of exposure, the frame rate used in this experiment was 67 fps. The position and support of the intermediate support point of the simply supported beam are changed several times to obtain different natural vibration frequencies. Under medium intensity illumination, the camera frame rate is 112 fps. The position support of the intermediate support point of the simply supported beam is also changed several times to obtain different natural vibration frequencies. In the case of strong illumination, in order to investigate whether the sampling frequency of the camera will affect the recognition of the natural frequency, the camera exposure and sampling frequency are changed several times, and the sampling frequency is 112 fps, 99 fps, 66 fps, and 49 fps, respectively. The images and processed data sampled during the experiment are shown in Figure 51.7.

The time-history curve obtained was subjected to modal analysis using a commercial modal and dynamic analysis system DASP. Under weaker light and medium illumination, the deviation of the first-order natural vibration frequency and the measured value obtained by various video image processing methods are less than 1%, which is in good agreement with the measured value. The error of the second-order natural vibration frequency and the measured value is slightly larger, but it is also within the range of 1.5%, which satisfies the requirements of engineering precision. However, due to the difference in sampling frequency between weak and medium light, the image quality and sampling frequency obtained under medium light are higher than that it was under weak light, so the data obtained are more accurate. Therefore, the recognition result under medium light is better than that under weak illumination in general. Under weak light conditions, the use of commercial software GOM and SURF feature point recognition algorithms frequently produces dislocation and failure to recognize some mark points due to the poor image quality, which affects the recognition efficiency.

Under the condition of strong light, with the decrease of the sampling frequency, video image processing methods of the first- and second-order natural frequencies and the error of the measured values increased gradually, especially when the sampling frequency is 49 fps. When using commercial software GOM and SURF feature points recognition algorithm to identify the fundamental frequency of the results, the error is more than 4%, which because when the exposure time is too long and the sampling frequency is too low, the phenomenon of ghosting was generated in the video images, affecting the judgment precision of the dynamic displacement.

Combined with the above experimental data, it can be seen that the connected domain recognition based on geometric features can not only identify the mark points with specific geometric features more efficiently and accurately but also avoid the omission and dislocation of points in the recognition processing. Then, the other feature points not of interest are removed to save the workload. Meanwhile, the ultra-long zoom lens was used in the long-distance experiment, which caused a slight shake of the lens itself under the disturbance of indoor air. Therefore, the accuracy of the results of the indoor long-distance experiment is slightly lower than that of the results taken by the smartphone at close range. To get more accurate results in remote outdoor experiments, it is necessary to maintain the stability of the air around the lens during the shooting process and to adopt a more stable camera support system (Tables 51.2–51.4).

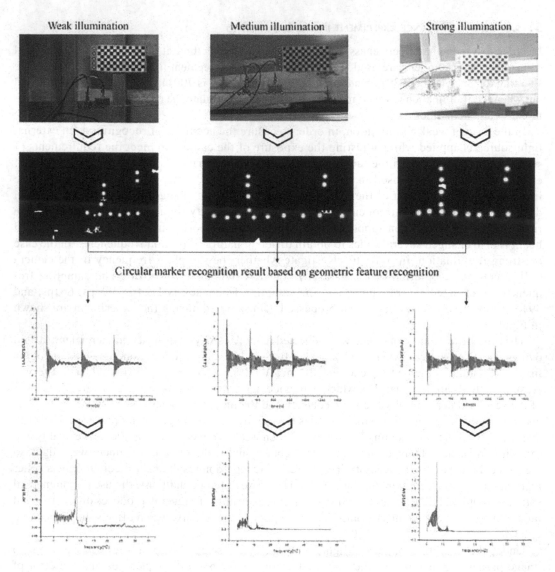

FIGURE 51.7 Schematic diagram of experimental results of different illuminations and sampling frequencies.

TABLE 51.2
Frequency Measured by Different Algorithms in the First Experiment (Hz)

	Frequency of the 1st Order		Frequency of the 2nd Order	
	1st Experiment	2nd Experiment	1st Experiment	2nd Experiment
Sensor results	6.344	8.344	12.834	15.25
Connected domain results	6.347	8.375	12.793	15.303
Deviation (%)	−0.047	−0.372	0.319	−0.348
GOM results	6.349	8.375	12.661	15.403
Deviation (%)	−0.079	−0.375	1.348	−1.003
SURF results	6.351	8.375	12.709	15.377
Deviation (%)	−0.110	−0.375	0.974	−0.833

TABLE 51.3

Frequency Measured by Different Algorithms in the Second Experiment (Hz)

	Frequency of the 1st Order			Frequency of the 2nd Order		
	1st Experiment	2nd Experiment	3rd Experiment	1st Experiment	2nd Experiment	3rd Experiment
Sensor results	6.188	6.625	7.875	12	15.75	15.875
Connected domain results	6.180	6.617	7.875	11.977	15.859	15.805
Deviation (%)	0.126	0.943	0	0.192	−0.692	0.441
GOM results	6.178	6.508	7.875	11.867	15.859	15.805
Deviation (%)	0.126	1.769	0	1.108	−0.692	0.441
SURF results	6.180	6.563	7.875	11.866	15.859	15.805
Deviation (%)	0.126	0.490	0	1.117	−0.692	0.441

TABLE 51.4

Frequency Measured by Different Algorithms in the Third Experiment (Hz)

	Frequency of the 1st Order				Frequency of the 2nd Order			
Sampling frequency (fps)	112	99	66	49	112	99	66	49
Sensor results	6.125	6.125	6.125	6.188	12.125	12.125	12.125	12.125
Connected domain results	6.125	6.123	6.091	6.125	12.104	12.093	12.179	12.253
Deviation (%)	0	0.032	0.558	1.010	0.173	0.264	−0.445	−1.056
GOM results	6.125	6.123	6.091	5.934	12.092	12.144	12.193	12.278
Deviation (%)	0	0.032	0.558	4.104	0.272	−0.157	−0.561	−1.262
SURF results	6.125	6.123	6.091	5.934	12.106	12.093	12.193	12.273
Deviation (%)	0	0.032	0.558	4.104	0.157	0.264	−0.561	−1.221

51.4 CONCLUSION

In this paper, a connected domain identification algorithm based on geometric feature extraction is proposed to identify the mark points in the bridge dynamic test. This method has the characteristics of fast and accurate detection, which can filter all the non-target points in the data acquisition stage and track the mark points with specific geometry accurately. At the same time, this algorithm has fast operation speed and high efficiency, which is suitable for video detection method of bridge dynamic test with large amounts of data. The displacement time-history curve of simply supported steel beam model is tested by this algorithm.

According to the displacement time-history curve of the model in the test, the acceleration time-history curve of the model is obtained, and then the spectrum of the structure is calculated. Compared with the results of the traditional contact speed sensor, the accuracy of the method is verified under different experimental conditions. When the indoor air is relatively static, the error of the first- and second-order natural frequency and the measured value obtained by the video image processing method are less than 1.5%, which is in good agreement with the measured value. In the actual application at outdoor, some objective conditions and environmental factors will also affect the measurement results and accuracy of the machine vision displacement measurement system, such as air turbulence, atmospheric refraction, environmental temperature, camera support deformation, ground subsidence, and earth curvature, which needs further research.

REFERENCES

G. Busca, A. Cigada, P. Mazzoleni and E. Zappa. 2014. Vibration monitoring of multiple bridge points by means of a unique vision-based measuring system. *Experimental Mechanics*, 54, pp. 255–271.

B. Pan, D-F. Wu and Y. Xia. 2010. Study of speckle pattern quality assessment used in digital image correlation. *Journal of Experimental Mechanics*, 25, pp. 120–129.

D. Ribeiro, R. Calcada, J. Ferreira and T. Martins. 2014. Non-contact measurement of the dynamic displacement of railway bridges using an advanced video-based system. *Engineering Structures*, 75, pp. 164–180.

D. M. Siringoringe and Y. Fujino. 2018. Seismic response of a suspension bridge: Insights from long-term full-scale seismic monitoring system. *Struct Control Health Monitoring*, pp. 2511–2536.

Y. Tian, J. Zhang and S. Yu. 2019. Rapid impact testing and system identification of footbridges using particle image velocimetry. *Computer-Aided Civil and Infrastructure Engineering*, 34, pp. 130–145.

52 Bridge Damage Localization Using Moving Embedded Principal Component Analysis with a Single Sensor

Z. F. Shen and Y. K. Xie
Jinan University

Z. H. Nie
Jinan University
The Key Laboratory of Disaster Forecast and Control in Engineering

CONTENTS

52.1 INTRODUCTION

Structural health monitoring (SHM) is considered as the important role for the structural operational status. It includes four parts[1]: (1) the whole monitoring to test if the structural damage situation has changed; (2) the partial detection to localize structural damage; (3) the quantification of structural damage severity and (4) the estimation of residual life of structure. Especially, the partial detection is highly concerned among these four parts.

With the development of science and technology, numerous approaches of structural damage detection are proposed and applied in SHM. Generally, structural vibration integrated analysis is one of the popular approach in SHM, which includes structural vibration, system identification, signal acquisition and analysis and other interdisciplinary theories. Principal component analysis (PCA), one of the data-driven method, is able to mine raw measured responses by tracking changes in signals without accurate analytical model and remove environmental impact from the vibration characteristics, which is suitable for the complexity of structure and the uncertainty of environmental excitation.[2–4]

However, there are some issues of current engineering implementation: (1) superfluous sensors installed on structure; (2) massive data during monitoring and analyzing; etc. Therefore PCA exposes some problems as the other techniques of structure detection. On one hand, the application of standard PCA for SHM is in great demand of sensors, which causes massive data and high costs. On the other hand, PCA can only analyze the whole measured responses at a time so that damage features may drown in numerous data. To extract the damage features more exactly, a method named moving PCA (MPCA) was put forward. In MPCA, the portion of tiny damage characteristics analyzed in a moving window was improved compared to that analyzed in the whole dataset.[5,6] In spite of that, the great demand of sensors and massive data is still not solved.

To deal with the above issues, this paper proposes a new method named moving embedded component analysis (MEPCA) for damage localization using only a single sensor. This method uses a moving window to cut out the time series of the single response, and the windowed data are embedded to be a multiple matrix which is proceeded by the standard PCA. As the window moves with time, the eigenvalue in the window is considered to be the local damage index that can locate the damage.

52.2 METHODOLOGY

The PCA is used to transform original data in high-dimensional into low-dimensional data while retaining the most information of the data. Consider a matrix $\mathbf{X}_{m \times n}$ representing the dataset of m sampling points measured by n sensors, the steps of PCA are listed as follows:

First, the covariance matrix $\mathbf{C}$ needed to be calculated is shown as

$$\mathbf{C} = \frac{1}{m}\mathbf{X}^T\mathbf{X}. \tag{52.1}$$

Second, the eigenvectors and eigenvalues of the covariance matrix $\mathbf{C}$ can be calculated as

$$\mathbf{C}\mathbf{V} = \mathbf{\Lambda}\mathbf{V}, \tag{52.2}$$

where each column of $\mathbf{V}$ is the eigenvector of covariance matrix $\mathbf{C}$ and $\mathbf{\Lambda}$ is a diagonal matrix where each diagonal element λ_n is the eigenvalue of the covariance matrix $\mathbf{C}$, respectively. And then, the principal component is calculated as

$$G^{(n)} = \mathbf{X}\mathbf{V}^{(n)}, \tag{52.3}$$

where $G^{(n)}$ is the n th component that represent the n th projection of the original dataset and $\mathbf{V}^{(n)}$ is the n th column of eigenvectors. Finally, to ascertain the several principal components without the loss of the most information, cumulative contribution ratios (CCRs) are defined as

$$\mathrm{CCR}_i = \frac{\lambda_i}{\sum\limits_i^n \lambda_i}, \quad i = 1,2,\ldots, n. \tag{52.4}$$

The component number is selected considering the CCR_i is larger than a threshold of 90% for the first i principal components, which means the considered components are able to represent most information from the original dataset.

To reduce the number of sensors and keep the effect of damage detection, a lot of methods were put forward by transforming time domain into space domain to form a multi-dimensional phase space.[7,8] Among these methods, the reconstructed phase space using time delay to form a multi-dimensional matrix based on the responses of a single sensor[9] is defined as

$$\mathbf{D}(n) = \left[d(n), \; d(n + T_{\mathrm{au}}),\ldots, d(n + q \cdot T_{\mathrm{au}}) \right], \tag{52.5}$$

where $\mathbf{D}(n)$ is the reconstructed multi-dimensional matrix, $d(n)$ is the measured response, T_{au} is the delay time, q is the number of reconstructed dimensions and $d(n + T_{\mathrm{au}})$ is the delayed series, respectively. In this paper, the T_{au} is set to be 1 and the q is set to be 5.

In order to locate the damage, a moving window is used in MEPCA, and the length of window l is obtained as

$$l = k\frac{f_s}{f_1},\tag{52.6}$$

where f_1 is the fundamental frequency and f_s is the sampling frequency, respectively. In this paper, k is set to be 2 and the f_1 can be identified by the Fourier spectrum of the response.

Based on the number of the reconstructed dimensions q and the defined moving window, the analyzed dataset B in the t th window is as

$$B_t = \begin{pmatrix} b_t(1) & b_t(2) & \cdots & b_t(q) \\ b_t(2) & b_t(3) & \cdots & b_t(q+1) \\ \vdots & \vdots & \ddots & \vdots \\ b_t(l-q+1) & b_t(l-q+2) & \cdots & b_t(l) \end{pmatrix}.\tag{52.7}$$

By discomposing the dataset B_t, the score S, eigenvectors $\mathbf{V}$ and the matrix of eigenvalue Λ could be calculated as

$$\text{PCA}(B_t) = [S, \mathbf{V}, \Lambda].\tag{52.8}$$

In this paper, the first eigenvalue λ_1 is used in the following section for the first principal component containing the most information of the origin data compared to other principal components. To figure out the change of λ_1 in damage location more accurately, the curvature of the eigenvalue CE is computed as

$$\text{CE}_t = \left(\lambda_{(t+1),1} - 2\lambda_{t,1} + \lambda_{(t-1),1}\right)/2\Delta t,\tag{52.9}$$

where Δt is the time interval of sampling and is set to be 1 in this paper.

52.3 NUMERICAL SIMULATION

To illustrate MEPCA, a beam bridge subjected to a moving mass is simulated by the software ANSYS. The beam bridge model is shown in Figure 52.1. In this bridge, the density is 7,850 kg/m³, the Yang's modulus is 210 GPa and the Poisson's ratio is 0.3, respectively. In this simulation, the moving mass is 400 kg with the velocity of 0.2 m/s. The bridge is divided into 800 elements along the length direction, while it is divided into ten elements along the height direction of the bridge. Two sensors are installed at the locations of 0.5l and 0.875l, respectively, to test the vibration of the bridge in order to access the effect of the method using the sensor at different positions. The sampling frequency is 200 Hz. The damage was created at 0.4l by deleting elements. The damage depths in scenarios 1 and 2 are 0.02 m and 0.04 m, respectively.

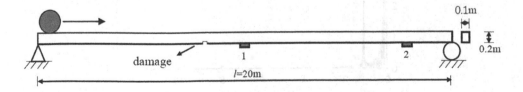

FIGURE 52.1 Simple beam model.

Based on the theory mentioned in Section 52.2, the steps to carry out MEPCA are as follows:

First, the displacement response is shown in Figure 52.2. The fundamental frequency is obtained by the FFT spectrum, shown in Figure 52.3. As shown, the fundamental frequency is 1.135 Hz. Second, with Eq. (52.6), the length of the window l is determined to be 352. Third, with Eqs. (52.4) and (52.8), the first CCR_1, shown in Figure 52.4, is nearly 100%. Because the first principal component contains the main information of the original responses, the status of the bridge can be monitored by the first eigenvalue λ_1.

Using MEPCA, first eigenvalue λ_1 is shown in Figure 52.5, in which the curves have no significant change at the damage location. Hence, the curvature of the first eigenvalue CE is computed using Eq. (52.9), as shown in Figure 52.6. It is obvious that there are sharp changes of CE

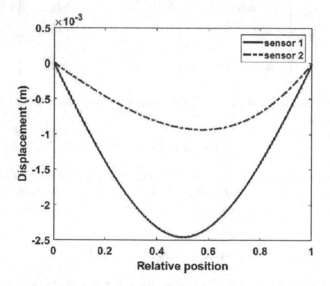

FIGURE 52.2   The displacement responses of sensors of scenario 1.

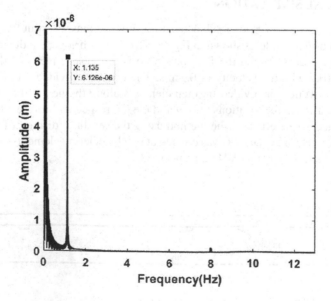

FIGURE 52.3   The Fourier spectrum of the structure of scenario 1.

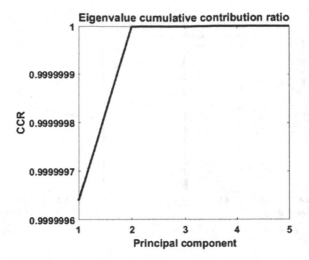

FIGURE 52.4 The CCR of the structure of scenario 1.

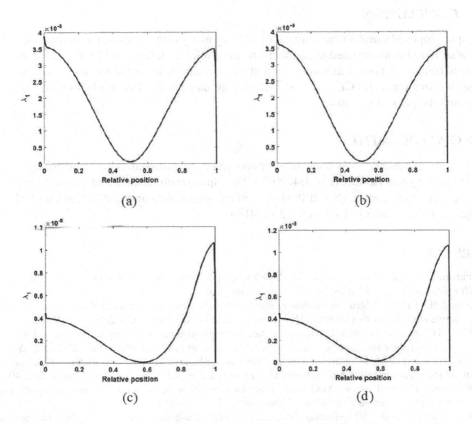

FIGURE 52.5 The first eigenvalue λ_1: (a) λ_1 of sensor No. 1 in scenario 1; (b) λ_1 of sensor No. 1 in scenario 2; (c) λ_1 of sensor No. 2 in scenario 1; (d) λ_1 of sensor No. 2 in scenario 2.

in the damage location of each damage scenario using different sensors. These results indicate that the damage index CE can well locate the damage of the beam bridge. It is worth to note that this method is a baseline-free data-driven approach which is more suitable for the application of practical engineering.

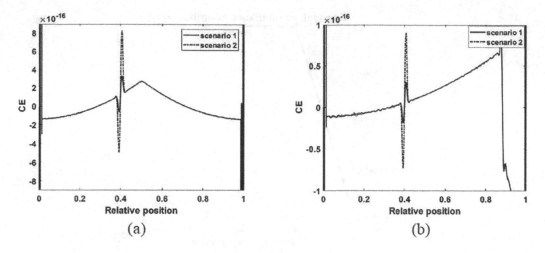

FIGURE 52.6   The curvature of the first eigenvalue CE: CE of sensor (a) No. 1 and (b) No. 2.

52.4 CONCLUSION

This paper proposed a new method named MEPCA by using a single sensor for damage localization. From what have been discussed above, it can be concluded as follows: (1) Embedded dimension is feasible to transform a one-dimensional dataset into a multi-dimensional one for multi-dimensional analysis by time delay; (2) CE is a good indicator for damage localization by observing the sharp change and the peak of the curve.

ACKNOWLEDGEMENTS

The authors acknowledge the financial supports from the National Natural Science Foundation of China under grant number (No. 11402098), The Guangzhou Science and Technology Planning Project under grant number (No. 201804010498) and Innovation and Cultivation Fund of Central Colleges and Universities of China (No. 21617413).

REFERENCES

1. Farrar, C.R. et al. 1994. *Dynamic Characterization and Damage Detection in the I-40 Bridge over the Rio Grande[R]*. Los Alamos: Los Alamos National Laboratory.
2. Nie, Z.H. et al. 2013. Structural damage detection based on the reconstructed phase space for reinforced concrete slab: Experimental study[J]. *Journal of Sound and Vibration*, 332(4): 1061–1078.
3. Gui, G.Q. et al. 2017. Data-driven support vector machine with optimization techniques for structural health monitoring and damage detection[J]. *KSCE Journal of Civil Engineering*, 21(2): 523–534.
4. Lin, Y.Z. et al. 2016. Data-driven damage detection for beam-like structures under moving loads using quasi-static responses[C]. *International Conference on Fuzzy System and Data Mining*, 281: 403–411.
5. Posenato, D. et al. 2008. Model-free data interpretation for continuous monitoring of complex structures[J]. *Advanced Engineering Informatics*, 22(1): 135–144.
6. Cavadas, F. et al. 2013. Damage detection using data-driven methods applied to moving-load responses[J]. *Mechanical Systems and Signal Processing*, 39(1–2): 409–425.
7. Zhu, X.Q. and Law, S.S. 2006, Wavelet-based crack identification of bridge beam from operational deflection time history[J]. *International Journal of Solids and Structures*, 43(7–8): 2299–2317.
8. Zhang, W.W. et al. 2017. Damage detection in bridge structures under moving loads with phase trajectory change of multi-type vibration measurements[J]. *Mechanical Systems and Signal Processing*, 87: 410–425.
9. Nie, Z.H. et al. 2017. Reconstructed phase space-based damage detection using a single sensor for beam-like structure subjected to a moving mass[J]. *Shock and Vibration*, 2017: 1–20.

53 Damage Detection of RC Bridges Based on Comprehensive Information of Influence Lines

Zhiwei Chen
Xiamen University

Weibiao Yang
CSCEC Strait Construction and Development Co., Ltd

Jun Li
Curtin University

CONTENTS

53.1 INTRODUCTION

The number of highway bridges in China has exceeded 779,000 by the end of 2015, and 90% of them are short- and medium-span bridges which accounts for the vast majority. Among these bridges, the ordinary reinforced concrete (RC) and prestressed concrete are the most widely used build bridges. Due to the combined actions of the ordinary (or overweight) traffic loading and corrosive environmental effect, cracks are prone to initiate in the tensile regions of RC bridges at the early stage and then gradually propagate. To this end, detecting the abnormality of RC bridges at an early stage and taking the protective measures in advance are effective ways. Therefore, an efficient damage detection method sensitive to early damage is essential for the maintenance of RC bridges.

In the past few decades, civil engineers and researchers have studied several types of methods for condition and damage assessment of bridge structures. However, the relation between such signs of damage and the corresponding "condition" of the structure is often very difficult to establish (Aktan and Catbas 2002). Based on the fundamental idea that damage-induced changes in the physical properties (mass, damping, and stiffness) will cause detectable changes in the modal properties (notably frequencies, mode shapes, and modal damping), the vibration-based damage detection techniques underwent a rapid development in the civil engineering society (Doebling et al. 1998; Fan and Qiao 2011; Li and Hao 2016). A well-known family of them is based on structural dynamic characteristics and their derivatives, such as frequencies, mode shapes, damping ratios, and strain mode shapes, whereas their representative derivatives include frequency response function, modal strain energy, energy transfer ratio, and flexibility matrix. Although these methods demonstrated varying degrees of success in previous studies, detecting early damage of bridge structures remains a big challenge. One of the main obstacles is that structural dynamic characteristics are either insensitive to early damage or too sensitive to measurement noises and changes in the operational environment conditions, such as temperature.

Several studies on influence line (IL)-based damage detection method have been carried out for applying to bridge structures. Aktan and Catbas (2002) have demonstrated that unit ILs (UIL) of strain or displacement could be a promising structural condition index. Furthermore, integrating with video images and sensor data, UIL was identified and utilized to detect damage with an experimental setup of a four-span steel bridge structure (Zaurin and Catbas 2011). Recently, the methodology was applied to detect and locate common damage scenarios on a steel bascule bridge (Zaurin et al. 2016). Chen et al. (2015) also applied the stress IL-based method to detect the local damage of a long suspension steel bridge, and its efficacy was successfully validated by case studies. Other IL-related studies were included in some recent literatures, such as Choi et al. (2004), Sun et al. (2015), and Ding et al. (2017). Although above studies have proved that IL is a promising damage index for bridge structures and applied the methods to steel bridges, few studies involve RC bridges that are widely used in practice. Compared with steel bridges, RC bridge has much more complicated modes of internal force distribution and structural failure, and thus damage-induced variations on IL indices of RC bridge deserve a careful and in-depth investigation.

This paper first introduces key issues in establishing a systematic methodology framework for damage detection based on bridge ILs. A damage index is first defined for bridge damage detection, and both displacement and strain ILs are utilized to detect the occurrence and location of damages on bridges. To improve the efficiency of damage localization, information fusion technique that synthesizes ILs of multiple locations is used for damage detection. Finally, the feasibility of the systematic method from IL identification to damage detection is verified through experimental tests of a three-span continuous RC beam.

53.2 DAMAGE DETECTION BASED ON COMPREHENSIVE INFORMATION OF MULTIPLE ILs

53.2.1 DAMAGE INDEX BASED ON IL CHANGE

After a certain service period, an RC bridge may have local damages, such as cracks, spalls, chemical deterioration, and corrosion. ILs can provide a suitable comparison of bridge response under the same load, a moving unit force, between the current and original states of the bridge. The bridge deflection or strain responses of a healthy bridge to a unit vertical force are taken as the baseline ILs. When a bridge is subject to several severe local damages, ILs may exhibit apparent changes that can be detected through a comparison with baseline ILs. Therefore, the IL change can be regarded as a damage index as follows:

$$\Omega(x) = \Phi(x) - \Phi_{BL}(x) \tag{53.1}$$

where $\Phi(x)$ and $\Phi_{BL}(x)$ are the newly obtained and baseline ILs, respectively, both of which are functions of the abscissa x of unit force in the longitudinal direction. If the bridge does not suffer any damages or the location of damage is far away from the measurement location of ILs, the damage index $\Omega(x)$ of ILs should be minimal and negligible. Otherwise, the magnitude of $\Omega(x)$ may increase when a unit force moves on the bridge. Member stiffness loss in statically determinate structures leads to a change in displacement IL, but no change in strain IL takes place unless a strain gauge is directly installed at the damage location. Fortunately, most bridges are statically indeterminate. In this study, both displacement and strain ILs are adopted to detect the occurrence and location of damages on RC bridges.

53.2.2 Damage Localization Based on Information Fusion of Multiple ILs

Multiple ILs are identified from sensors installed on different locations of a bridge, and thus it is necessary to propose a damage localization approach which could synthesize ILs of multiple locations through information fusion technique. By assuming n sensors installed in the concerned region of bridge and m candidate locations of interest $x_j (j = 1, 2, \cdots, m)$ for damage localization, IL information from multiple sensors are available, thus the mass function matrix of ILs can be defined as

$$m = \begin{bmatrix} m_1(x_1) & m_1(x_2) & \cdots & m_1(x_m) \\ m_2(x_1) & m_2(x_2) & \cdots & m_2(x_m) \\ \cdots & \cdots & m_i(x_j) & \cdots \\ m_n(x_1) & m_n(x_2) & \cdots & m_n(x_m) \end{bmatrix} \tag{53.2}$$

in which, the mass function coefficient $m_i(x_j)$ denotes the degree of belief regarding the damage at jth location according to the IL change from the ith sensor, and can be calculated by the equation as follows:

$$m_i(x_j) = \frac{|\Omega_i(x_j)|}{\sum_{j=1}^{m} |\Omega_i(x_j)|} \tag{53.3}$$

A greater value of $|\Omega_i(x_j)|$ implies a higher probability of damage occurrence at the location x_j, and thus is assigned with a greater mass function coefficient. The evidence from multiple information sources (i.e., sensors) can be combined using Dempster's combination rule (Basir and Yuan 2007). By fusing information of n sensors, the joint mass function coefficient corresponding to the jth candidate location can be computed by

$$m(x_j) = \prod_{i=1}^{n} m_i(x_j) \bigg/ \sum_{j=1}^{m} \left(\prod_{i=1}^{n} m_i(x_j) \right) \tag{53.4}$$

When the ith sensor location is distant from the candidate location x_j, the evidence $m_i(x_j)$ may not be reliable even though a great value is detected; when a sensor is mounted very close to the concerned location x_j, the measurement is more sensitive to the possible damage, and the provided evidence information should be treated more credible than those from other distant sensors. To account for this distance effect, a weighted joint mass function coefficient is proposed to take into account the relative credibility of each strain sensor in the information fusion process:

$$m(x_j) = \prod_{i=1}^{n} w_{ij} m_i(x_j) \bigg/ \sum_{j=1}^{m} \left(\prod_{i=1}^{n} w_{ij} m_i(x_j) \right) \tag{53.5}$$

where w_{ij} is the weighting factor determined based on the distance D_{ij} between the ith sensor and the jth location. Given that totally n sensors are involved in information fusion, the relationship between w_{ij} and D_{ij} is established:

$$w_{ij} = \frac{1/D_{ij}}{\sum_{i=1}^{n}(1/D_{ij})} \tag{53.6}$$

in which

$$D_{ij} = |x_i - x_j| \tag{53.7}$$

where x_i is the location of the ith sensor and x_j stands for the jth location. By introducing the weighting factor into calculation, information from sensor at a small distance can be assigned with a great weighting factor; whereas the distance being very far, a minimal weighting factor should be assigned and the contribution of the corresponding sensor should be nearly ignored. To avoid infinite or extremely large weighting factor in the data fusion, $D_{ij} = 0.1m$ is taken when the sensor is very close to the concerned damage location (i.e., $D_{ij} < 0.1m$).

By integrating IL-based damage detection and information fusion technologies, the weighted joint mass function coefficient is defined after a comprehensively consideration of multiple ILs, and a greater value of the coefficient (between 0 and 1) indicates a larger probability of damage. Therefore, it could be a promising method for determining local damage locations in a more accurate and reliable way.

53.3 EXPERIMENTAL VERIFICATION ON A THREE-SPAN CONTINUOUS RC BEAM

53.3.1 EXPERIMENTAL SETUP

As a simplification of RC bridges in the real world, a three-span continuous RC beam is used for verifying the feasibility and performance of the proposed approach for applications to damage detection in RC bridges. A T-shaped beam is specially designed for the convenience of a steel-made loaded vehicle moving on it.

The three-span continuous RC beam was fabricated in the laboratory, and a pulling force generated by an electric motor was applied to pull a steel-made loaded vehicle to cross the bridge, as shown in Figure 53.1. Many steel plates are superimposed on the vehicle to ensure a reasonable vehicle bridge mass ratio and make the deflections or strains of beam large enough for a precise measurement.

To identify the quasi-static strain and deflection ILs, five strain gauges and five displacement meters were installed on the middle span of the beam. For a better comparison between the dynamic and static properties of the beam, eleven accelerometers are installed to identify the first few modal frequencies. The layout of the sensor installing on the beam is displayed in Figure 53.2.

53.3.2 VERIFICATION OF IL IDENTIFICATION METHOD

To verify the proposed approach of quasi-static IL identification based on multiple monitoring data, the moving vehicle testing and static single-point loading testing are respectively performed to determine the baseline and identified ILs.

The strain or deflection time histories are required to construct response vector for IL identification. The strain and deflection time histories induced by a loaded vehicle moving from one end of the beam to the other were measured by strain gauges and displacement meters installed in the

FIGURE 53.1 Loaded vehicle moving on a continuous RC beam.

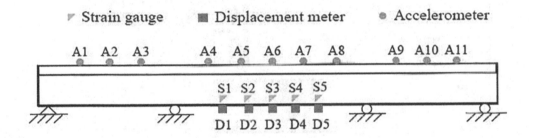

FIGURE 53.2 Layout of sensors installed on the beam.

middle span of this three-span continuous RC beam. The dynamic strain and deflection time histories measured by sensors installing at the mid-span are shown in Figure 53.4a and b. It can be found from both figures that a peak value occurs when a vehicle approaches the sensor location at the mid-span. In addition, obvious fluctuations are found from the measured strain and deflection time histories, and this phenomenon mainly attributes to the vehicle-bridge dynamic interaction effect and measurement noises. Although the vehicle weight has been increased to 1 ton and responses at the mid-span are larger than other locations, the response peaks of time histories recorded at the mid-span (shown in Figure 53.3) are still very small, which indicates that the RC beam has a high rigidity. For the same level of measurement noise, using small amplitude of response leads to a decrease of signal-to-noise ratio which makes the precise IL identification more challenging.

The identified strain and deflection ILs at the mid-span of beam are shown in Figure 53.4a and b, respectively. The horizontal axis represents the location of a unit force (1 kN) acting on the beam, and the vertical axis indicates the strain or displacement caused by a unit force at the corresponding location. In addition to the IL identified from moving vehicle testing, the baseline IL is identified from static single-point loading testing, in which a strip-shaped vertical loading is applied to the beam at each time and then successively moves in the longitudinal direction with a small interval. The baseline ILs are also shown in Figure 53.4 for the comparison. It can be observed from Figure 53.4a that the identified strain IL can perfectly fit with the baseline IL not only for its shape but also for its peak value. The identified result of deflection IL is not as good as strain IL, and the relative change ratio between the identified and baseline strain ILs is 5.6%, but it is still acceptable

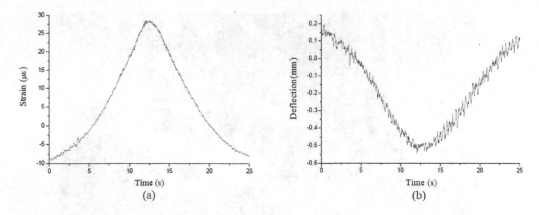

FIGURE 53.3 Moving vehicle induced dynamic response time history. (a) Strain time history. (b) Deflection time history.

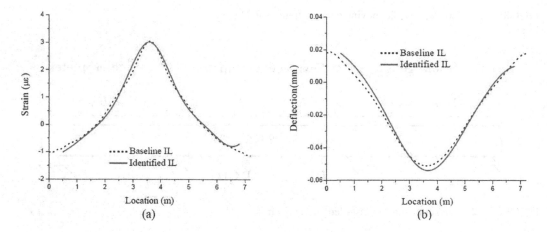

FIGURE 53.4 Comparison of baseline and identified ILs. (a) Strain IL. (b) Deflection IL.

for application to damage detection. The results demonstrate that the proposed IL identification method has a high accuracy and thus has great potentials in engineering applications.

53.3.3 VERIFICATION OF DAMAGE DETECTION METHOD

To verify the efficacy and performance of applying the proposed damage detection approach for RC bridges, a minor damage was generated by applying external vertical loadings on the mid-span of the laboratory beam through an electro hydraulic servo loading system. The system stopped loading when two tiny cracks appeared at the bottom of the loading region as shown in Figure 53.7b. Cracks are located in the sections at distances of 3.51 and 3.82 m from the left end of the beam. After unloading, the cracks will close and become so inconspicuous that made them rather difficult to detect by naked eyes (Figure 53.5).

In order to make comparisons with modal frequencies of the continuous beam before and after damage, dynamic tests have been carried out on the beam. The excitation is provided by an impact hammer applied at the mid-span and one-fourth span of the beam. Eleven force-balance accelerometers are installed on locations as shown in Figure 53.2 and are used to measure the dynamic responses. The frequency domain decomposition method is adopted to identify the modal frequencies of the intact and damage states. The frequencies of the first seven modes are identified, and the

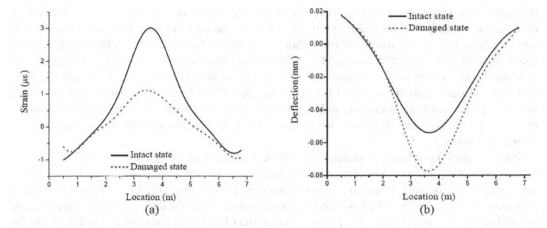

FIGURE 53.5   ILs in the intact and damaged states. (a) Strain IL. (b) Deflection IL.

changing rate index (CRI) between the intact and damage state for each mode is calculated. It can be found that the largest CRI is only 8.3% and the second largest one is less than 5%.This phenomenon indicates that modal frequency change indices are not sensitive to the minor damage at the early stage, and the frequency change induced by the temperature variation may be even greater than that caused by the minor damage, as reported by previous studies.

A similar procedure that is used to identify IL in the intact state is also applied to identify IL in the damaged state. Figure 53.8a and b show the identified strain and deflection ILs in the mid-span of the beam under the intact and damaged states. Significant differences can be observed in the ILs from these two states, and the largest difference appears at the peak value of ILs, which is just located in the damage region. Compared with the CRI values of modal frequency, CRI values of IL peak are found to be much higher.

For a comprehensive study of understanding the change in ILs induced by the occurrence of minor damage, ILs before and after damage are compared among different monitoring locations. The monitoring data collected by strain gauges and displacement meters installed at five critical locations (as shown in Figure 53.2) are used to identify strain and deflection ILs. Using the baseline and the newly obtained ILs, IL change defined as a damage index is calculated by Eq. (53.1) and is shown in Figure 53.6. Taking the deflection IL change curves for example, some uniform observations are obtained by comparing five curves in Figure 53.6b, which include that all curves have apparent fluctuations, and the locations of their peaks are close to each other. The obvious

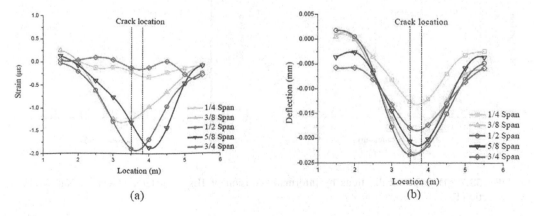

FIGURE 53.6   Changes in ILs at different monitoring locations. (a) Strain ILs. (b) Deflection IL.

fluctuation can be regarded as an indicator to detect the occurrence of damage, and the location of apparent peaks can be used to roughly determine the damage location. Peak values of the five curves fall in the main damage region without exception, which indicates deflection IL change could be an excellent index for damage localization. Compared with deflection IL, only the change of strain IL at the location of 1/2 span in Figure 53.6a can be used to accurately localize damage based on its peak location, and strain ILs at the location of one-fourth span and three-fourth span are almost impossible for accurately judging the existence of damage. This phenomenon could be explained by the fact that strain/stress is sensitive to local damage, but it is limited to a small region around the damage.

Without prior knowledge of damage, it will be very difficult to make a reliable decision on damage detection because conflicts are likely to exist in the findings from single ILs of different locations. Thus, the information fusion of multiple ILs proposed before is applied to localize damage of the RC beam. Based on the IL change before and after damage, each mass function coefficient is calculated by Eq. (53.3) to construct a mass function matrix defined in Eq. (53.2). By fusing information of multiple sensors, the weighted joint mass function coefficients for all the candidate locations are computed by Eq. (53.5). 127 candidate locations of interest at an interval of 0.05 m are selected from the experimental beam. Damage localizations by information fusion of different numbers of sensors are subsequently examined and compared. The results are shown in Figure 53.7, with the number of sensors measured for strain and deflection IL information fusion varying from one, three to five. The joint mass function value at the damage location around the mid-span increases along with the number of sensors. By contrast, after fusing information of three or five sensors, the joint mass function around the damage location can be easily distinguished from the other non-damage locations, and the mass function at the locations faraway from damage is reduced to near zero.

Furthermore, the efficiency of damage localization is validated by considering the information fusion with different types of sensors. In this study, the joint mass function values are calculated by information fusion of only strain ILs, only deflection ILs, and all ILs, and then the three curves are demonstrated in Figure 53.8 for comparison. The identification results indicate that different types of sensors adopted for information fusion are beneficial to damage localization with a certain extent. In conclusion, the information fusion technique can effectively strengthen the consistent information (such as damage-induced structural change) and eliminate non-consistent information (such as "noise" effect) from multiple sensors installed around the damage.

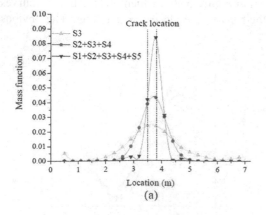 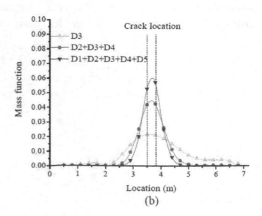

FIGURE 53.7   Damage localizations by information fusion of ILs at different locations. (a) Strain IL. (b) Deflection IL.

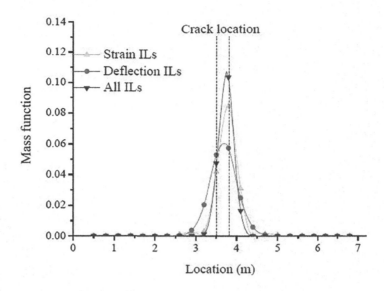

FIGURE 53.8 Damage localizations by information fusion of all ILs from different sensors.

ACKNOWLEDGMENTS

The authors wish to acknowledge the financial supports from the National Natural Science Foundation of China (NSFC-51778550), the Natural Science Foundation of Fujian Province of China (No. 201701101), and the Principal Fund of Xiamen University (No. 20720180060). Any opinions and concluding remarks presented in this paper are entirely those of the authors.

REFERENCES

Aktan, A.E. and Catbas, F.N. (2002), "Condition and damage assessment: issues and some promising indices", *Journal of Structural Engineering-ASCE*, **128**(8), pp. 1026–1036.
Basir, O. and Yuan, X.H. (2007), "Engine fault diagnosis based on multi-sensor information fusion using Dempster–Shafer evidence theory", *Information Fusion*, **8**(4), pp. 379–386.
Chen, Z.W., Zhu, S.Y., Xu, Y.L., Li, Q. and Cai, Q.L. (2015), "Damage detection in long suspension bridges using stress influence lines", *Journal of Bridge Engineering-ASCE*, **20**(3), p. 05014013.
Choi, I.Y., Lee, S.J., Choi, E. and Cho, H.N. (2004), "Development of elastic damage load theorem for damage detection in a statically determine beam", *Computer and structures*, **82**(29–30), pp. 2483–2492.
Ding, Y.L., Zhao, H.W. and Li, A.Q. (2017), "Temperature effects on strain influence lines and dynamic load factors in a steel-truss arch railway bridge using adaptive FIR filtering", *Journal of Performance of Constructed Facilities-ASCE*, doi: 10.1061/(ASCE)CF.1943-5509.0001026.
Doebling, S.W., Farrar, C.R. and Prime, M.B. (1998), "A summary review of vibration-based damage identification methods", *Shock and Vibration Digest*, **30**(2), pp. 91–105.
Fan, W. and Qiao, P. (2011), "Vibration-based damage identification methods: a review and comparative study", *Structural Health Monitoring*, **10**(1), pp. 83–111.
Li, J. and Hao, H. (2016), "A review of recent research advances on structural health monitoring in Western Australia", *Structural Monitoring and Maintenance*, **3**(1), pp. 33–49.
Sun, S.W., Sun, L.M. and Chen, L. (2015), "Damage detection based on structural response induced by traffic load: methodology and application", *International Journal of Structural Stability and Dynamics*, **16**(4), p. 1640026.
Zaurin, R. and Catbas, F.N. (2011), "Structural health monitoring using video stream, influence lines, and statistical analysis", *Structural Health Monitoring*, **10**(3), pp. 309–332.
Zaurin, R., Khuc, T. and Catbas, F.N. (2016), "Hybrid sensor-camera monitoring for damage detection: case study of a real bridge", *Journal of Bridge Engineering-ASCE*, **21**(6), p. 05016002.

54 Hybrid DIC-Meshless Method for Evaluating Strain Field around the Crack

Zhu Zhihui, Luo Sihui, Wang Fan,
Feng Qianshuo, and Jiang Lizhong
Central South University

CONTENTS

54.1 INTRODUCTION

Digital Image Correlation (DIC) technology, also known as digital speckle correlation, was proposed by Peters and Ranson (1982) of the University of South Carolina in the United States and Yamaguchi (1981) of Japan in the early 1980s. Sutton et al. (1986) made a series of research and development. The method can achieve non-contact and full-field strain measurement, with simple experimental equipment and operation, and has low requirements on the measurement environment (Dildar et al. 2019). It has been widely used in the field of solid mechanics, material properties and experimental research of fracture mechanics (Fawad et al. 2012; Javad et al. 2017).

DIC technology is now widely used in the measurement of displacement field, which is an optical measurement method by comparing two high-definition camera images (Zhong and Quan 2018). According to the DIC measurement technique, the appropriate subset value is used for the analysis of the speckle quality. At this time, the displacement field is often not obtained around the crack and at the sample boundary (Molteno and Becker 2015). When using small subsets for analysis, the finite element based DIC strain field is not accurate. Therefore, the existing DIC technology is difficult to obtain an accurate crack tip strain field. In order to obtain accurate strain results around the crack, many problems have been raised to solve this problem. Réthoré et al. (2008) proposes a DIC algorithm based on enriched finite element subsets. Poissant and Barthelat (2010) provide corrections to the DIC algorithm to represent crack regions, but these methods are quite complex and are not widely used. Chen et al. (2018, 2019) proposed an expansion procedure to construct the full-field displacement and strain information by using either a model or a non-model approach. Barhli et al. (2017) proposed a solution combining DIC test results with finite element theory models, but it needs to re-model the crack part and the local encryption mesh, which is more troublesome.

Therefore, a hybrid DIC-meshless method is proposed to evaluate the strain field around the crack. A coordinate position and a displacement measurement result of a field node in a full field range in a target image are obtained based on the DIC technique, wherein the crack tip portion is analyzed by a smaller subset and the node is refined. The meshless model is based on the nodes of the DIC measurement. The displacement field obtained by the DIC test is added as a boundary condition to the meshless model, and the stress and strain at the crack tip are calculated based on the Element-free Galerkin (EFG) method. Based on DIC technology, a quasi-static cyclic static tensile test is performed on a CT specimen with Q235 material to validate the method experimentally.

54.2   HYBRID DIC-MESHLESS METHOD

54.2.1   DIC Method

In DIC, two digital images before and after the deformation are addressed. By searching the deformed image for the subset with the highest correlation of the original image subset, the surface displacement and strain of the test piece are calculated by comparing the gray scale (Barhli et al. 2017). The digital image before the deformation is generally referred to as the "reference image," and the deformed digital image is referred to as the "target image." Its basic principle is shown in Figure 54.1. A rectangular reference image sub-region of size $(2m+1)$ pixel $\times (2m+1)$ pixel centered on the point (x, y) to be sought is taken in the reference image, and a deformed image is taken in the deformed image. In the sub-region, the deformed image sub-region and the reference image sub-region have a certain correlation, which can be expressed by correlation coefficients. According to a search method, a deformed image sub-region centered on (x', y') is found. The correlation reaches a maximum value, and the reference image sub-region corresponds to the deformed image sub-image after deformation. The displacement information of the point can be obtained according to the coordinate difference between the two points (Gonzáles et al. 2017).

The correlation function C used is the sum of squared errors of the pixel values in the reference image sub-region and the target image sub-region. The smaller the sum, the higher the similarity.

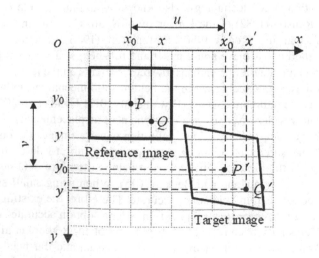

FIGURE 54.1   Principle of the DIC method.

$$C(x,y,u,v) = \sum_{i,j=-n/2}^{n/2} \left(\left(I(x+i,y+j) - I^*(x+u+i,y+v+j) \right) \right)^2 \tag{54.1}$$

where x, y are the pixel coordinates in the reference image and u, v are the actual coordinate values. I and I^* represent the gray values of the reference and deformed images, respectively.

54.2.2 EFG METHOD

In the meshless method, the solution $u(x)$ of the structure can be represented by the local approximate solution $u^h(x)$,

$$u(x) \approx u^h(x) = \sum_{i=1}^{m} p_i(x)a_i(x) = p^T(x)a(x) \tag{54.2}$$

where $p_i(x)$ is the basis function, m is the number of the term of $p_i(x)$ and $a_i(x)$ is a specific coefficient matrix, which is related to the spatial coordinate x. The function is solved based on the moving least squares method (MLS), by introducing the weight function $\omega(x - x_I)$ for each discrete node in the meshless region.

$$u^h(x) = \sum_{i=1}^{n} N_I^k(x)u_I = N^k u \tag{54.3}$$

where N^k is a shape function of the EFG method, which can be evaluated according to references (Mehdi et al. 2016).

Considering the particularity of the crack tip, when calculating the strain, the node search for the shape function should be searched by the diffraction method. The distance between the crack tip and the node around the crack to the integration point is

$$s(x) = \left(\frac{s_1 + s_2(x)}{s_0(x)} \right)^\lambda s_0(x) \tag{54.4}$$

among them, $s_0(x) = \|x - x_I\|$, $s_1(x) = \|x_c - x_I\|$ and $s_2(x) = \|x - x_c\|$. The definition of each distance is as shown in Figure 54.2 and λ is a parameter, which is generally 1 or 2. The diffraction method eliminates discontinuities in the domain and is more suitable for non-convex boundaries.

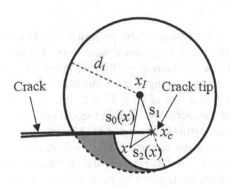

FIGURE 54.2 Diffraction method.

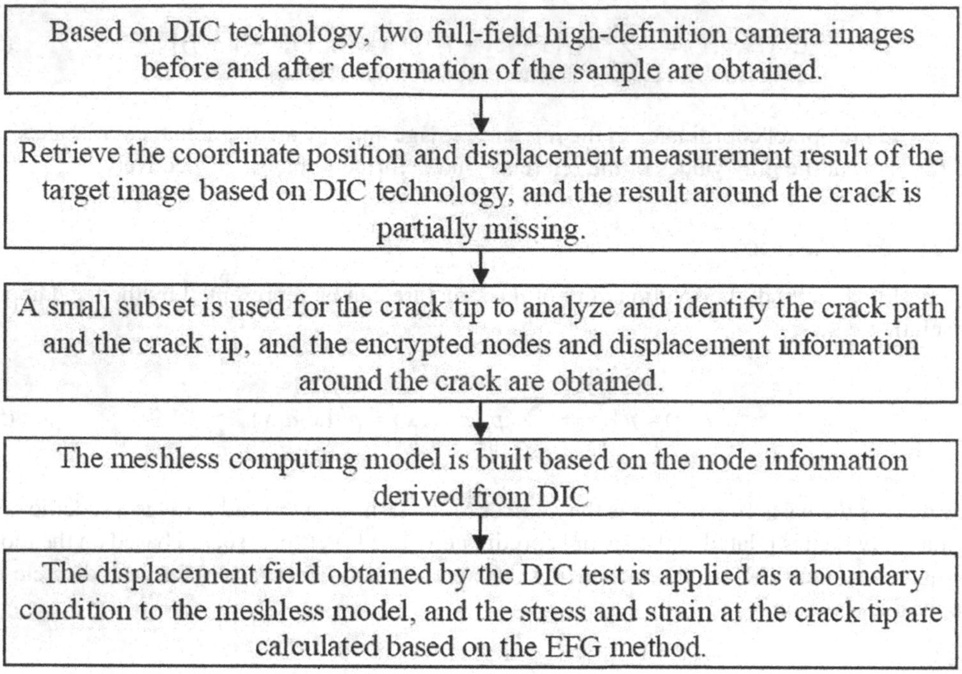

Based on DIC technology, two full-field high-definition camera images before and after deformation of the sample are obtained.

↓

Retrieve the coordinate position and displacement measurement result of the target image based on DIC technology, and the result around the crack is partially missing.

↓

A small subset is used for the crack tip to analyze and identify the crack path and the crack tip, and the encrypted nodes and displacement information around the crack are obtained.

↓

The meshless computing model is built based on the node information derived from DIC

↓

The displacement field obtained by the DIC test is applied as a boundary condition to the meshless model, and the stress and strain at the crack tip are calculated based on the EFG method.

FIGURE 54.3   The process of the hybrid DIC-meshless method.

54.2.3 Hybrid DIC-Meshless Method

The process of hybrid DIC-meshless method can be organized as follows: DIC retrieves the full-field information by using different subsets, developing the meshless calculation model and obtaining the full-field displacement field. EFG method is used to calculate the stress-strain field. The field nodes required for the meshless model can be derived from the DIC analysis. The nodes are distributed according to the distance delta × delta in the full-field range, and the node of the crack tip portion is refined appropriately. The DIC will automatically identify the crack tip position and derive the crack edge position. Finally, the meshless model is formed by combining the full-field area field node information and the crack tip area field node information. Based on the EFG method, combined with the displacement field, the strain field and the stress field are calculated (Figure 54.3).

54.3 QUASI-STATIC CYCLE TEST OF CT SPECIMENS

54.3.1 Specimen

To study the accuracy and effectiveness of the proposed hybrid DIC-meshless method, the fracture properties of the Q235qE steel CT specimen were studied. The mechanical properties of steel Q235qE were as follows: elastic modulus $E = 210\,\text{GPa}$, Poisson's ratio $\nu = 0.33$, yield strength of the material $\sigma_s = 235\,\text{MPa}$, tensile strength $\sigma_t = 390\,\text{MPa}$, density $\rho = 7.85$ g/cm^3 (Guangyong et al. 2018). The CT sample had a size of $75 \times 72 \times 5\,\text{mm}$, as shown in Figure 54.4. The fatigue pre-cracking of the specimen was performed before the test so that the total crack length was about 35 mm, the error should be kept within the range of $\pm 0.25\,\text{mm}$, which can be used to study the stress intensity factor (SIF) variation of the material under load. According to the reference (Zhangyan et al. 2002), the strain gauge was used to measure the strain near the crack. Therefore, the strain gauge G_1 was instrumented at the corresponding position of the crack. The orientation of the strain gauge was $\theta = 54.27°$, $r = 12.5\,\text{mm}$ and $\phi = 68.01°$.

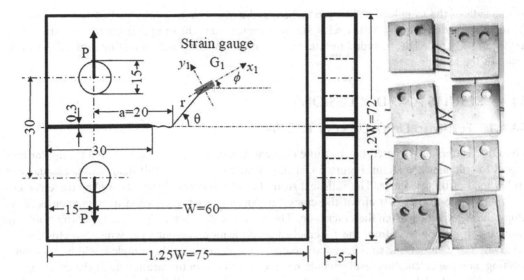

FIGURE 54.4 The CT specimen.

54.3.2 EXPERIMENTAL SETUP

The test set up was shown in Figure 54.5. The fatigue and tensile loads were applied by the 25T-MTS fatigue tester. The specimens were clamped on the fatigue tester by U-shaped clamps. The test results of the strain gauges were collected by the HBM QuantumX comprehensive data acquisition system. The two cameras of the non-contact full-field strain measurement system were mounted at a distance of 1.2 m from the surface of the sample, and the camera height was the same. The angle between the cameras was about 40°. The corresponding lighting equipment and image acquisition computer were also shown in the figure. The fatigue pre-cracking was conducted with a load ratio $R = 0.1$ (maximum/minimum load) and a frequency $f = 10\,\text{Hz}$ harmonic load to obtain fatigue cracks with a length of $5 \pm 0.25\,\text{mm}$. In the formal test, a non-reflective black speckle pattern was applied

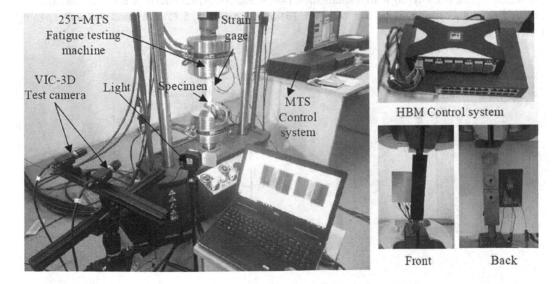

FIGURE 54.5 The test setup.

to one side of the sample after being polished, and a strain gauge was attached to the other side. Loaded in a series of quasi-static cycles, stepwise increased to the sample fracture, each stage of the load was 1 kN. The DIC recorded the state at the time of loading and unloading as the deformation result of the loading.

54.4   RESULTS AND DISCUSSIONS

54.4.1   FULL-FIELD DISPLACEMENT OF DIC TEST

The DIC technology is based on the finite element theory and aims to establish a triangular mesh model to calculate the strain. Figure 54.6a shows the results of the full-field strain test under the 10 kN load of the specimen. The full-field strain has a symmetric distribution. And there is a concentrated three-direction strain at the crack tip. An obvious compression phenomenon is shown along the X-axis away from the crack side. The plate doesn't have a significant deformation at the crack tip. It can be seen from the ε_{yy} cracked amplification contour that, when searching with a relatively large subset, the strain around the crack and the edge of the sample could be completely missing, and the calculation results would have some effects on the strain field of the crack tip.

The DIC calculation indicates that when subset = 59, the full-field average uncertainty interval reaches a minimum of only 0.01 pixels. To validate the test results, Figure 54.6b shows the displacement-load curve at the Point G1 from either strain gauge or DIC. A very good overlay is shown between the two curves. If ignoring the system error caused by the DIC instrumentation, the maximum error of the measurement between the DIC and the strain gauge is 8.33%. When the load is greater than 13 kN, a nonlinear deformation occurs at Point G1. Thus, when subset = 59, the strain measured by the DIC is proved to be accurate by the data from the strain gauge.

54.4.2   MESHLESS MODEL OF A CT SAMPLE

A meshless model of CT specimens is developed based on the theory in Section 54.2.3, as shown in Figure 54.7. The DIC can derive evenly distributed node information by setting the range and spacing. First, DIC uses subset = 59 to calculate and uniformly distribute the field nodes according to the node spacing of delta × delta = 2 mm × 2 mm in the whole space. The transition area ABCD node range is $x \in (-1.5, 15)$ mm, $y \in (-11.5, 11.5)$ mm, and the node spacing is delta/2 = 1.0 mm. The transition region is analyzed with a small subset = 19, and the ABCD derived nodes of the inserted cracked-edge encryption region are $x \in (3.75, 15)$ mm, $y \in (-5.75, 5.75)$ mm, and the node spacing is delta/4 = 0.5 mm.

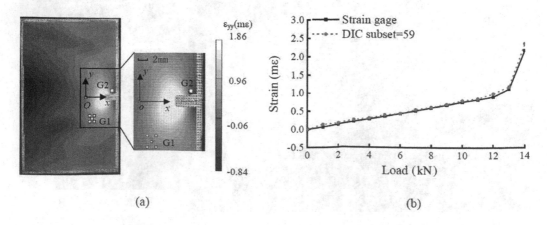

(a)   (b)

FIGURE 54.6   DIC test ε_{yy} strain test results. (a) Full-field strain map. (b) strain at Point G1 versus load.

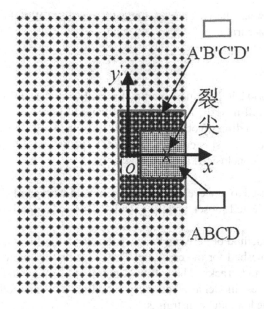

FIGURE 54.7   Meshless computing model.

54.4.3   COMPARISON OF STRAINS IN DIFFERENT METHODS

Based on the EFG method, the full-field strain distribution is calculated. Figure 54.8a shows the Y-direction strain contour of the DIC-meshless calculation on the P51 sample under 10 kN load. It can be seen from Figure 54.8a that the crack strike is roughly parallel to the X-axis, so the strain at the crack tip is not symmetric about the X-axis. The crack tip coordinate is (9.75, −0.25). Since the DIC calculation loses the high-stress gradient part of the crack tip, the strain range of the DIC test is only (−0.84, 1.86) $m\varepsilon$ (as shown in Figure 54.6). Comparably, the DIC-meshless method can have a strain range of (−1.85, 7.53) $m\varepsilon$. The hybrid DIC-meshless method can obtain the detailed information of the crack tip and crack and thus can better characterize the stress-strain distribution of the crack tip. To validate the strain calculated by the proposed method, Figure 54.8b shows the strain curve at Point G2 with different loads, obtained either by hybrid DIC-meshless method or DIC. Even if the scattered spots of the crack tip portion of the meshless model are composed of two parts of the DIC calculation results (subset = 59 and 19), the strain of the crack tip is still effective.

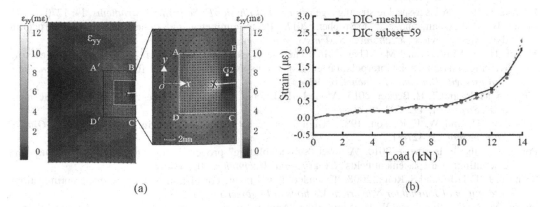

(a)   (b)

FIGURE 54.8 ε_{yy} DIC-meshless calculation for P51 sample. (a) Full-field strain map. (b) strain at Point G2 versus load.

The good agreement between the two curves indicates that the full-field strain calculated by the DIC-meshless method is accurate.

54.5 CONCLUSION

This paper proposes a hybrid DIC-meshless method for analyzing the strain field at the crack tip. The DIC is used to obtain the full-field displacement and node information and then develop a meshless model. The EFG method can directly obtain full-field stress and strain. Based on DIC technology, a quasi-static cyclic static tensile test is conducted on a CT specimen with Q235 material to validate the proposed method. The conclusions are as follows:

1. DIC technology is used to retrieve the displacement field with good stability, and it is stable with the different selected subset, and the calculated strain is greatly affected by the size of the subset.
2. Based on DIC, the method of deriving nodes to develop a meshless model does not need to be remodeled and meshed for the crack tip. The modeling is simple for structures having irregular geometry and cracks. The full-field measured displacement from DIC is introduced into the meshless model as a boundary condition. This method is easy to implement without knowing the load and constraints.
3. The hybrid DIC-meshless method can obtain the accurate full-field strain including the crack tip, which is a great challenge for the DIC. The proposed method can be used for fracture and fatigue properties of cracked structures in the future.

REFERENCES

Barhli, S. M., M. Mostafavi, A. F. Cinar, D. Hollis, and T. J. Marrow. 2017. 'J-integral calculation by finite element processing of measured full-field surface displacements', *Experimental Mechanics*, 57: 997–1009.
Dildar, D. E, E. Tsangouri and K. Spiessens. 2019. 'Digital image correlation (DIC) on fresh cement mortar to quantify settlement and shrinkage', *Archives of Civil and Mechanical Engineering*, 19: 205–14.
Fawad, T., M. Z. Siddiqui, and N. Naz. 2012. 'Practical Application of DIC in Fatigue and Fracture Toughness Testing'.
Gonzáles, G. L. G., J. A. O. González, and J. T. P. Castro. 2017. 'A J-integral approach using digital image correlation for evaluating stress intensity factors in fatigue cracks with closure effects', *Theoretical and Applied Fracture Mechanics*, 90: 14–21.
Guangyong, S., L. Xinglong, and Z. Gang. 2018. 'On fracture characteristics of adhesive joints with dissimilar materials – An experimental study using digital image correlation (DIC) technique', *Composite Structures*, 201: 1056–75.
Ichirou, Y. 1981. 'A laser-speckle strain gauge', *Journal of Physics E: Scientific Instruments*, 14: 1270.
Javad, B., P. Peyman, and N. Christopher. 2017. 'Photogrammetry and optical methods in structural dynamics–a review', *Mechanical Systems Signal Processing*, 86: 17–34.
Mehdi, D., A. Mostafa, and M. Akbar. 2016. 'The use of element free Galerkin method based on moving Kriging and radial point interpolation techniques for solving some types of Turing models', *Engineering Analysis with Boundary Elements*, 62: 93–111.
Molteno, M. R. and T. H. Becker. 2015. 'Mode I–III decomposition of the J-integral from DIC displacement data', *Strain*, 51: 492–503.
Peters, W. H. and W. F. Ranson. 1982. 'Digital image techniques in experimental mechanics', *Optical Engineering*, 21: 427–31.
Poissant, J and F. Barthelat. 2010. 'A novel "subset splitting" procedure for digital image correlation on discontinuous displacement fields', *Experimental Mechanics*, 50: 353–64.
Réthoré, J., F. Hild, and S. Roux. 2008. 'Extended digital image correlation with crack shape optimization', *International Journal for Numerical Methods in Engineering*, 73: 248–72.
Sutton, M. A., C. Mingqi, and W. H. Peters. 1986. 'Application of an optimized digital correlation method to planar deformation analysis', *Image and Vision Computing*, 4: 143–50.

Yuanchang, C., J. Dagny, and A. Peter. 2018. 'Underwater dynamic response at limited points expanded to full-field strain response', 140: 051016.

Yuanchang, C., L. Patrick, and A. Peter. 2019. 'Non-model based expansion from limited points to an augmented set of points using Chebyshev polynomials', *Experimental Techniques*: 1–23.

Zhangyan, Z., L. Yunbing, and S. Guozheng. 2002. 'Experiment measuring fracture toughness of Q235 steel by J integral', *Journal of Wuhan University of Technology*, 24: 111–12.

Zhong, F and C. Quan. 2018. 'Efficient digital image correlation using gradient orientation', *Optics and Lasers Technology*, 106: 417–26.

55 Identification of Moving Forces in the Case of Unknown Bridge Structural Parameters Using Truncated Generalized Singular Value Decomposition Algorithm

Zhuhong Ouyang and Chiu Jen Ku
Shantou University

CONTENTS

55.1 INTRODUCTION

During the actual operation of the bridge, various types of vehicles pass through the bridge at different speeds. In addition, the bridge deck itself is not smooth, which makes the interaction between the bridge and vehicles very complex. Compared with the static wheel load, the dynamic wheel load may increase pavement damage 2–4 times, which can lead to a reduction in the service life of the bridge structure (Cebon 1987). Due to the interaction between vehicle and bridge it is difficult to directly measure the interaction force. However, it would be beneficial if we can use the measured response data of bridges and indirectly identify the interaction force.

In recent decades, the problem of vehicle and bridge interaction force identification has been widely studied. O' Connor and Chan (1988) estimated the equivalent static load by using the response data measured by the bridge as well as their dynamic variation with time. The time domain method (TDM) based on mode superposition was proposed by Law and Chan (1997). Its result shows that using acceleration response or using both acceleration and bending moment response to identify moving force is effective. Yu and Chan (2007) reviewed four developed moving force identification methods and found that TDM and frequency-time domain method were relatively accurate, and using singular value decomposition method can improve the identification accuracy. However, due to the ill-posed nature of the inverse problem, it is still found that the identification results of the moving force are sensitive to noise and exhibit large fluctuations (Yu et al. 2016). Subsequently, many researchers used regularization method to solve the ill-posed of the inverse problem, and at the same time reduce the impact of noise on the identification result. Zhu and Law (2006) combined TDM and regularization method to identify the moving force on multi-span continuous beams. Zhu and Law (2016) carried out a literature review on solving the inverse problem of a bridge-vehicle system. Numerical examples and experimental studies showed that compared with the least-square method, the regularization method could better estimate the entire time history of force, especially at the beginning and end of the time duration. The TDM is more accurate than the frequency domain method in solving the ill-posed problem of moving force identification. Based on least-squares regularization, an updated static component technology was proposed to identify the dynamic loads of multi-axle on a multi-span continuous bridge (Asnachinda et al. 2008). González and Rowle (2008) used first-order Tikhonov regularization and dynamic programming methods to identify the moving force of an elaborate three-dimensional vehicle and orthotropic bridge interaction system from strain response. Based on the TDM, Chen and Chan (2017) adopted truncated generalized singular value decomposition method (TGSVD) to improve the ill-posed of moving force identification, which significantly improved the identification accuracy, and the algorithm was not sensitive to noise.

However, the above papers of moving force identification are all based on the known bridge structure parameters, such as the bending stiffness or damping ratio. In reality, it is usually impossible to obtain accurate bridge structure parameters, which limits the application of these methods. So, the best approach is to identify the structural parameters and the moving force simultaneously when the structural parameters are unknown. Li and Chen (Li and Chen 2003, Chen and Li 2004) proposed an iterative identification procedure based on least-square technique to identify the input excitation and structural parameters using output-only measured response, and proved analytically that when the system parameters and excitation are unknown, the iterative solution can converge. Zhu and Law (2007) developed an iterative method to identify the interaction force and bridge structural parameters from the measured response of the bridge with a full sensor placement. Lu and Law (2007) proposed an iterative gradient-based model updating method based on sensitivity of structural response to identify system parameters and excitation forces of a bridge structure. To use limited measured response data to identify structural parameters and external forces simultaneously, Sun and Feng et al. (Sun et al. 2015, Feng et al. 2015) proposed a numerical algorithm and applied it to shear-type building, truss bridge and vehicle bridge models, respectively. The estimation of unknown axle loads is incorporated in the framework of an iterative parameter optimization process, wherein the objective is to minimize the error between the measured and predicted system responses, which ultimately is solved by the damped Gauss-Newton method. The part of moving force identification is solved by Bayesian inference regularization method to solve the ill-posed problem.

Based on the TDM, this paper proposes a new method to simultaneously identify the structural parameters of a bridge and the moving forces on it from the acceleration response based on TGSVD algorithm. The simultaneous identification is an iterative optimization process, in which the objective is to minimize the difference between the measured and estimated acceleration responses. The TGSVD algorithm is incorporated in this process to solve the ill-posed least-square problem for the estimation of unknown moving forces. In the process of the simultaneous identification, the structural parameters and the moving force can be constantly updated, so as to realize the identification of the moving force when the parameters of the bridge structure are unknown. In this paper, the numerical analysis of a single-span simply supported beam bridge is carried out to verify the correctness and effectiveness of the algorithm.

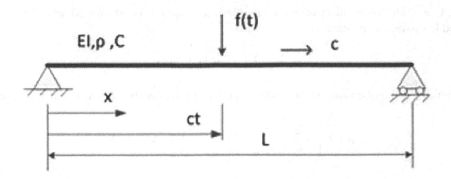

FIGURE 55.1 Moving force $f(t)$ on a simply supported beam.

55.2   BACKGROUND OF THEORY

The bridge-vehicle system is modeled as a time-varying force moving on a simply supported beam, as shown in Figure 55.1. The bridge is simplified as a Bernoulli-Euler beam which is of constant cross section, having a linear, viscous proportional damping with small deflections. And the effects of shear deformation and rotary inertia are not taken into account. The length of the bridge is L and constant mass per unit length is ρ. The viscous damping parameter of the bridge is C and the flexural stiffness of the bridge is EI. The force moves from left to right at a constant speed c, then the motion equation in terms of the modal coordinate can be written as

$$\ddot{q}_n(t) + 2\xi_n \omega_n \dot{q}_n(t) + \omega_n^2 q_n(t) = \frac{2}{\rho L} p_n(t) \tag{55.1}$$

where $\omega_n = \left(n^2\pi^2/L^2\right)\sqrt{EI/\rho}$ is the nth modal frequency; $\xi_n = C/(2\rho\omega_n)$ is the nth modal damping ratio and $p_n(t) = f(t)\sin(n\pi ct/L)$ is the nth modal force.

Equation (55.1) can be solved in time domain by the modal superposition and convolution integral. The acceleration $\ddot{v}(x,t)$ of the beam at point x and time t can be obtained as Law and Chan (1997)

$$\ddot{v}(x,t) = \sum_{n=1}^{\infty} \frac{2}{\rho L} \sin\frac{n\pi x}{L} \left[p_n(t) + \int_0^t \ddot{h}_n(t-\tau)p_n(\tau)d\tau \right] \tag{55.2}$$

where $\omega_n' = \omega_n\sqrt{1-\xi_n^2}$, $\ddot{h}_n(t) = \frac{1}{\omega_n'} e^{-\xi_n\omega_n t}\left\{\left[\left(\xi_n\omega_n\right)^2 - \omega_n'^2\right]\sin\omega_n' t + \left(-2\xi_n\omega_n\omega_n'\right)\cos\omega_n' t\right\}$.

By discretizing Eq. (55.2) and writing it in matrix form, the identification equation of the moving force can be obtained:

$$\ddot{\mathbf{v}} = \mathbf{A}\cdot\mathbf{f} \tag{55.3}$$

where $\ddot{\mathbf{v}} \in \mathbf{R}^{N\times 1}$, $\mathbf{A} \in \mathbf{R}^{N\times(N_B-1)}$, $\mathbf{f} \in \mathbf{R}^{(N_B-1)\times 1}$, N, $N_B = L/(c\Delta t)$ is respectively the number of sample points of acceleration response and moving load and Δt is the sampling interval.

55.3   TGSVD FOR MOVING FORCE IDENTIFICATION

By the derivation of the above TDM, the moving force identification problem is transformed into solving a linear equation $\mathbf{Ax} = \mathbf{b}$. Adopting regularization method is an efficient way to solve the ill-posed problems of moving force identification (MFI) and ill-conditioning of the $\mathbf{A}$ matrix.

Based on Tikhonov regularization, the regularized matrix $\mathbf{L}$ is introduced, and the solution of ill-posed least-squares problem is

$$\mathbf{x}_\lambda = \left(\mathbf{A}^T\mathbf{A} + \lambda^2\mathbf{L}^T\mathbf{L}\right)^{-1}\mathbf{A}^T\mathbf{b} \tag{55.4}$$

The generalized singular value decomposition (GSVD) of the matrix pair $(\mathbf{A}, \mathbf{L})$ can be expressed as

$$\mathbf{A} = \mathbf{U}\begin{pmatrix} \Sigma & 0 \\ 0 & \mathbf{I}_{n-p} \end{pmatrix}\mathbf{X}^{-1}, \quad \mathbf{L} = \mathbf{V}\begin{pmatrix} \mathbf{M}, & 0 \end{pmatrix}\mathbf{X}^{-1} \tag{55.5}$$

where $\mathbf{A} \in \mathbf{R}^{m\times n}$, $\mathbf{L} \in \mathbf{R}^{p\times n}$, $\mathbf{U} \in \mathbf{R}^{m\times n}$ and $\mathbf{V} \in \mathbf{R}^{p\times p}$ are orthonormal matrices, $\mathbf{X} \in \mathbf{R}^{n\times n}$ is a nonsingular matrix, $\Sigma = \mathrm{diag}\left(\sigma_1, \sigma_2, \ldots, \sigma_p\right)$ and $\mathbf{M} = \mathrm{diag}\left(\mu_1, \mu_2, \ldots, \mu_p\right)$ are diagonal matrices and satisfy $0 \leq \sigma_1 \leq \cdots \leq \sigma_p \leq 1$, $1 \geq \mu_1 \geq \cdots \geq \mu_p \geq 0$, $\sigma_i^2 + \mu_i^2 = 1$ $(i = 1, 2, \cdots, p)$. The generalized singular values of the matrix pair $(\mathbf{A}, \mathbf{L})$ are defined as $\gamma_i = \sigma_i/\mu_i$ $(i = 1, 2, \ldots, p)$.

Substituting Eq. (55.5) into Eq. (55.4) to obtain the regularization solution:

$$\mathbf{x}_\lambda = \mathbf{X}\begin{bmatrix} \mathbf{F}_\lambda\Sigma^+ & 0 \\ 0 & \mathbf{I}_{n-p} \end{bmatrix}\mathbf{U}^T\mathbf{b} = \sum_{i=1}^p f_i\frac{\mathbf{u}_i^T\mathbf{b}}{\sigma_i}\mathbf{x}_i + \sum_{i=p+1}^n \mathbf{u}_i^T\mathbf{b}\mathbf{x}_i \tag{55.6}$$

where $\mathbf{F}_\lambda = \mathrm{diag}\left(f_1, f_2, \ldots, f_p\right)$ are diagonal matrices, $f_i = \gamma_i^2/\left(\gamma_i^2 + \lambda^2\right)$. Σ^+ is the pseudo inverse of Σ.

Same as truncated singular value decomposition (TSVD), the solution of TGSVD $\mathbf{X}_k$ is introduced by substituting for $\mathbf{F}_\lambda$ in Eq. (55.6) a diagonal matrix with k unit elements corresponding to the k largest σ_i in Σ and otherwise 0, thus simply neglecting the contributions corresponding to the p - k smallest σ_i (Hansen 1989).

$$\mathbf{x}_k = \mathbf{X}\begin{bmatrix} \hat{\Sigma}_k^+ & 0 \\ 0 & \mathbf{I}_{n-p} \end{bmatrix}\mathbf{U}^T\mathbf{b} = \sum_{i=p-k+1}^p \frac{\mathbf{u}_i^T\mathbf{b}}{\sigma_i}\mathbf{x}_i + \sum_{i=p+1}^n \mathbf{u}_i^T\mathbf{b}\mathbf{x}_i \tag{55.7}$$

where $\hat{\Sigma}_k^+ = \mathrm{diag}\left(0, \ldots, 0, \sigma_{p-k+1}^{-1}, \ldots, \sigma_p^{-1}\right)$. k is the truncated parameter.

Besides, in this paper, the regularized matrix $\mathbf{L}$ we use is

$$\mathbf{L}_1 = \begin{pmatrix} -1 & 1 & & & \\ & -1 & 1 & & \\ & & \ddots & \ddots & \\ & & & -1 & 1 \end{pmatrix}_{(n-1)\times n} \tag{55.8}$$

55.4   STRUCTURAL PARAMETER IDENTIFICATION

The parameter identification of bridge structure is considered as an iterative process of parameter optimization. The objective is to find the optimal parameter estimation vector to minimize the difference between the real measured acceleration response and the estimated acceleration response. Assume $F(t_k)$ is the true moving force input of the bridge structure system at the time instant t_k where $k = 0, 1, 2, \ldots, n_t$, and n_t denotes the total number of sampling points. $a(t)$ is the true acceleration response of the bridge structure at the time instant t_k. Then the relationship between the estimated moving force and acceleration response can be expressed as

$$\hat{\mathbf{F}}(t_k) \xrightarrow{\hat{\theta}} \hat{\mathbf{a}}(t_k) \tag{55.9}$$

where $\hat{\theta} = (\theta_1, \theta_2, \ldots, \theta_n)$ is the estimated structure parameter vector where n is the number of unknown bridge structure parameters. $\hat{\mathbf{F}}(t_k)$ is the estimated moving force vector and $\hat{\mathbf{a}}(t_k)$ is the estimated acceleration response vector.

Therefore, the objective function can be defined as

$$\Pi(\theta) = \sum_{k=1}^{n_t} \mathbf{r}_k^T(\theta)\mathbf{r}_k(\theta), \quad \mathbf{r}_k(\theta) = \hat{\mathbf{a}}(t_k) - \mathbf{a}(t_k) = \left\{ \begin{array}{c} \hat{a}_1(t_k) - a_1(t_k) \\ \hat{a}_2(t_k) - a_2(t_k) \\ \vdots \\ \hat{a}_{n_s}(t_k) - a_{n_s}(t_k) \end{array} \right\} \tag{55.10}$$

where $\mathbf{r}_k(\theta)$ is the residue vector where n_s is the number of measure points.

Herein, structural parameter identification has been transformed into solving a nonlinear optimization problem, which will be solved by damped Gauss-Newton method.

55.5   THE SIMULTANEOUS IDENTIFICATION ALGORITHM

Incorporating the TGSVD method into the iterative process of parameter optimization, the identified structure parameter and moving force in this process will be constantly updated, until the objective function satisfies the corresponding convergence condition. The identification process assumes that the unit length mass of the beam ρ and the speed of moving force c is known, the stiffness EI and damping ratio of the structure are unknown, and the moving force is also unknown.

The process of simultaneous identification solved by damped Gauss-Newton method is described as follows:

a. Given the measured acceleration response as input.

b. Assume initial value of structure parameter vector $\hat{\theta}^{(0)}$ and damping factor $h^{(0)}$.

c. The system matrix $\hat{\mathbf{A}}$ is constructed according to the structural parameter vector $\hat{\theta}^{(0)}$, and the estimated moving force $\hat{\mathbf{F}}$ is calculated by the TGSVD method (Eq. (55.7)).

d. Calculate the predicted system acceleration response: $\hat{\mathbf{a}} = \hat{\mathbf{A}}\hat{\mathbf{F}}$.

e. Calculate the objective function: $\Pi(\hat{\theta}^{(0)})$.

f. Calculate the Jacobian matrix of residual function:

$$[\mathbf{J}_r]_{ij} = \frac{\partial \mathbf{r}_i(\hat{\theta}^{(0)})}{\partial \theta_j} \approx \frac{\mathbf{r}_i(\theta_1, \ldots, \theta_j + \delta\theta_j, \ldots \theta_{n_\theta}) - \mathbf{r}_i(\theta_1, \ldots, \theta_j - \delta\theta_j, \ldots \theta_{n_\theta})}{2\delta\theta_j}$$

g. Update structure parameter vector: $\hat{\theta}^{(1)} = \hat{\theta}^{(0)} + \left[\mathbf{J}_r^T\mathbf{J}_r + h^{(0)}\text{diag}(\mathbf{J}_r^T\mathbf{J}_r)\right]^{-1}\mathbf{J}_r^T\mathbf{r}_i(\hat{\theta}^{(0)})$

h. Update damping factor using the adaptive tracking approach: $h^{(0)} \to h^{(1)}$ (Marquardt 1963).

i. Use $\hat{\theta}^{(1)}$ and $h^{(1)}$ replace $\hat{\theta}^{(0)}$ and $h^{(0)}$ in step (b), and repeat steps (c)-(i) until the following convergence condition is met: $\left|\Pi(\hat{\theta}^{(\text{iter})}) - \Pi(\hat{\theta}^{(\text{iter}-1)})\right| \Big/ \left|\Pi(\hat{\theta}^{(\text{iter})})\right| < \varepsilon$, where the superscript iter stands for the current calculation step, ε is the allowable error and generally is 1×10^{-6}.

55.6   NUMERICAL SIMULATION

55.6.1   SIMULATION PARAMETERS

To check on the correctness and effectiveness of the proposed simultaneous identification method, identification of the two-point forces is simulated:

$$f_1(t) = 20,000[1 + 0.1\ \sin(10\pi t) + 0.05\ \sin(40\pi t)]\ N$$

$$f_2(t) = 20,000[1 - 0.1\ \sin(10\pi t) + 0.05\ \sin(50\pi t)]\ N$$

The moving speed of moving forces are 40 m/s, and the distance between two moving forces is 4 m. The parameters of the beam are as follows: $L = 40$ m, $EI = 1.274916 \times 10^{11}$ N·m^2, $\rho = 1.2 \times 10^4$ kg/m. The first three natural frequencies for the simply supported beam are 3.2, 12.8 and 28.8 Hz, and the first three damping ratios are 2%. The sampling frequency is 400 Hz. In the process of simultaneous identification, the initial value of the assumed structural parameters is $EI = 1.019933 \times 10^{11}$ N·m^2, $\xi = 1.6\%$ (0.8 of the true EI and ξ, respectively).

The acceleration responses at 1/2 span, 1/4 span, 3/4 span, 1/5 span, 2/5 span, 3/5 span and 4/5 span of the beam are used. Three combinations of measurement points are adopted. Two measurement points case S2: $L/4$ and $L/2$ of the beam. Three measurement points case S3: $L/4$, $L/2$ and $3/4L$ of the beam. Four measurement points case S4: $L/5$, $2/5L$, $3/5L$ and $4/5L$ of the beam.

According to the above assumed bridge structure parameters and moving force, the real acceleration response of the bridge is calculated, and then the noise is added to the calculated acceleration response to obtain the simulated measured acceleration response:

$$\ddot{v} = \ddot{v}_{calculated} + E_p \times \sigma(\ddot{v}_{calculated}) \times \mathbf{N}_{oise} \tag{55.11}$$

where E_p denotes error level, $\sigma(\ddot{v}_{calculated})$ is the standard deviation of the calculated acceleration response, $\mathbf{N}_{oise}$ is a standard normal distribution vector with zero mean value and unit standard deviation.

The percentage error between the true moving load and the identified load are calculated by the following equation:

$$Error = \frac{\|\mathbf{f}_{identified} - \mathbf{f}_{true}\|}{\|\mathbf{f}_{true}\|} \times 100\% \tag{55.12}$$

55.6.2   RESULT

The cases of identifying two-point forces with the unknown flexural stiffness EI and damping ratio ξ moving on a simply supported beam are studied. Besides, the effects of the measurement points, the initial estimates of identified bridge structural parameters and the measurement noise on the identification accuracy of the moving forces are respectively investigated. The simulation results are shown in Tables 55.1 and 55.2 and Figures 55.2 and 55.3.

When $E_p = 0$, namely no noise is embedded in the measured acceleration response, the identification results of the moving force and structure parameter are accurate under various situations, this means that the proposed method is correct. When the $E_p = 5\%$ or 10%, the identification results of the moving force and structure parameter show different degrees of error, but the error is controlled in the range of 5%. With the increase of noise level, the increase of error is not obvious, which indicates that the algorithm is not sensitive to noise.

As shown in Table 55.1, with the increase of number of acceleration measurement points, the identification accuracy of moving forces and structural parameters is improved to some extent. The identified results from cases of 3 and 4 measurement points are better than results

TABLE 55.1

Percentage Error of Two Moving Forces Identification

Measurement Points Case	Noise Level (%)	0	5	10
S2	Force F_1 (%)	0	2.97	3.07
	Force F_2 (%)	0	3.73	4.03
	EI (%)	0	0.07	0.27
	ξ (%)	0	0.90	4.43
S3	Force F_1 (%)	0	2.30	2.62
	Force F_2 (%)	0	2.85	3.10
	EI (%)	0	0.44	1.05
	ξ (%)	0	0.42	0.25
S4	Force F_1 (%)	0	2.52	2.62
	Force F_2 (%)	0	3.07	3.57
	EI (%)	0	0.23	0.49
	ξ (%)	0	0.53	0.60

TABLE 55.2

Percentage Error of Two Moving Forces Identification (4 Measurement Points Case: S4)

Initial Estimate	Noise Level (%)	0	5	10
0.6	Force F_1 (%)	0	2.71	2.65
	Force F_2 (%)	0	3.15	3.58
	EI (%)	0	0.21	0.56
	ξ (%)	0	0.72	0.56
1.2	Force F_1 (%)	0	2.52	2.66
	Force F_2 (%)	0	3.20	3.63
	EI (%)	0	0.45	0.38
	ξ (%)	0	0.45	0.77
1.4	Force F_1 (%)	0	2.52	2.66
	Force F_2 (%)	0	3.20	3.63
	EI (%)	0	0.45	0.38
	ξ (%)	0	0.45	0.77

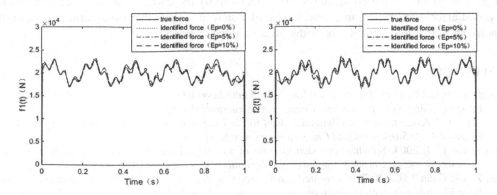

FIGURE 55.2 The result of two moving forces identification with the structure stiffness EI and damping ratio ξ unknown.

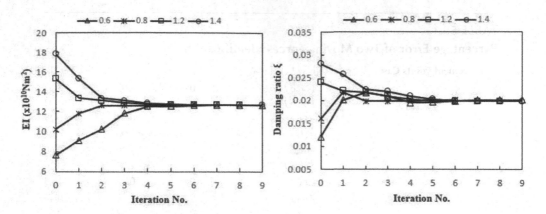

FIGURE 55.3 The evolution process of the identified flexural stiffness and damping ratio for case S4 with 10% noise level.

from 2 measurement points. The identified moving forces from 3 measurement points have the smallest error. Figure 55.2 shows that the identified moving force histories from 3 measurement points agree quite well with the true forces, with only small discrepancies at the entry and exit moment of the moving forces. Generally, the identification accuracy is better if more measurement points are adopted.

In addition, the sensitivity of the identification results to different initial estimates of identified bridge structural parameters (0.6, 0.8, 1.2 and 1.4 of the true EI and ξ, respectively) are studied, shown in Tables 55.1 and 55.2. It can be seen that identified results from different initial estimates of bridge structural parameters are similar. Figure 55.3 shows the evolution process of the identified flexural stiffness and damping ratio for case S4 with 10% noise level. It is seen that all the parameters efficiently converge to the true values. Therefore, the algorithm is not sensitive to the initial estimates of bridge structural parameters.

55.7   CONCLUSION

In this paper, a method based on TGSVD is proposed to identify moving forces in the case of unknown bridge structural parameters from the acceleration response. The TGSVD is added to the framework of structural parameter iterative optimization to realize the identification of moving forces under the condition of unknown bridge structural parameters. A numerical simulation of a single-span simply supported beam is carried out to verify the correctness and effectiveness of the proposed algorithm. At the same time, the simulation results show that the algorithm is insensitive to noise and the initial estimates of bridge structural parameters.

REFERENCES

Asnachinda P. and Pinkaew T. and Laman J.A. 2008. Multiple vehicle axle load identification from continuous bridge bending moment response. *Engineering Structures*, **30**, pp. 2800–2817.

Cebon D. 1987. Assessment of the dynamic wheel forces generated by heavy road vehicles. *Symposium on Heavy Vehicle Suspension and Characteristics*, Canberra, Australia.

Chen J. and Li J. 2004. Simultaneous identification of structural parameters and input time history from output-only measurements. *Computational Mechanics*, **33**(5), pp. 365–374.

Chen Z. and Chan T.H.T. 2017. A truncated generalized singular value decomposition algorithm for moving force identification with ill-posed problems. *Journal of Sound and Vibration*, **401**, pp. 297–310.

Feng D.M. and Sun H. and Feng M.Q. 2015. Simultaneous identification of bridge structural parameters and vehicle loads. *Computers and Structures*, **157**, pp. 76–88.

González A. and Rowle C. 2008. A general solution to the identification of moving vehicle forces on a bridge. *International Journal for Numerical Methods in Engineering*, **75**(3), pp. 335–354.

Hansen P.C. 1989. Regularization, GSVD and truncated GSVD. *BIT Numerical Mathematics*, **29**(3), pp. 491–504.

Law S.S. and Chan T.H.T. 1997. Moving force identification: a time domain method. *Journal of Sound and Vibration*, 201(1), pp. 1–22.

Li J. and Chen J. 2003. A statistical average algorithm for the dynamic compound inverse problem. *Computational Mechanics*, 30(2), pp. 88–95.

Lu Z.R. and Law S.S. 2007. Identification of system parameters and input force from output only. *Mechanical Systems and Signal Processing*, 21, pp. 2099–2111.

Marquardt D.W. 1963. An algorithm for least-squares estimation of nonlinear parameters. *Journal of the Society for Industrial and Applied Mathematics*, 11(2), pp. 431–441.

O'Connor C. and Chan T.H.T. 1988. Dynamic wheel loads from bridge strains. *Journal of Structural Engineering, ASCE*, 114(8), pp. 1703–1723.

Sun H., Feng D., Liu Y. and Feng M.Q. 2015. Statistical regularization for identification of structural parameters and external loadings using state space models. *Computer-Aided Civil and Infrastructure Engineering*, 30(11), pp. 843–858.

Yu L. and Chan T.H.T. 2007. Recent research on identification of moving loads on bridges. *Journal of Sound and Vibration*, 305(1–2), pp. 3–21.

Yu Y., Cai C.S. and Deng L. 2016. State-of-the-art review on bridge weigh-in-motion technology. *Advances in Structural Engineering*, 19(9), pp. 1514–1530.

Zhu X.Q. and Law S.S. 2006. Moving load identification on multi-span continuous bridges with elastic bearings. *Mechanical Systems and Signal Processing*, 20, pp. 1759–1782.

Zhu X.Q. and Law S.S. 2007. Damage detection in simply supported concrete bridge structure under moving vehicular loads. *Journal of Vibration and Acoustics*, 129(1), pp. 58–65.

Zhu X.Q. and Law S.S. 2016. Recent developments in inverse problems of vehicle–bridge interaction dynamics. *Journal of Civil Structural Health Monitoring*, 6, pp. 107–128.

56 Estimation of Repeated Slip Surface in Cut Slope Stability Analysis

Zulkifl Ahmed, Shuhong Wang, and Dong Furui
Northeastern University

CONTENTS

56.1 INTRODUCTION

The promotion of China Pakistan Economic Corridor (CPEC) strategy has led to a large number of engineering construction and exploitation in Pakistan. The stability of such large-scale engineering activities beside road directly related to the progress of the project, economy and the safety of the workers (Azimi et al. 2018). The site is located near the country's much active seismic west zone (Jackson and McKenzie 1984). The seismic effect on rock slope stability usually have two characteristics such as triggered and accumulated effects (Pantelidis 2009). The former performance is largely due to the increase of water pressure and surface failure, which was triggered by dynamic forces. The latter performance is mostly for a serious state of rock slope uncertainty moment, sand liquefaction and weak deposit thixotropic unstiffening which is also triggered by seismic loading. Wang et al. (2012) analyzed the effect of earthquake mechanism on slope failure process, and Qi et al. (2007) have made a great contribution to dig out the shaking instability aspects: one is inertial force and the other is water pressure.

Sliding surface provides essential information and has widely been used to judge the stability of slopes in two-dimension (2D) (Cheng and Lau 2014). In pseudo-static analysis or static analysis, the slip surface (SS) is the surface that produces the minimum FS. The smallest value of safety factor (SF) has been calculated by studying circular arc with cohesion (*c*) over internal friction (*φ*) ratio (Hajiazizi and Tavana 2013). The method of judgment of the sliding surface is directly connected to the method of determining the minimum SF (Baker 1980). To determine minimum SF for a sliding surface, limit equilibrium technique may allow a precise and accurate evaluation method during large-scale stability investigation. It is more important to provide a model and technique for the estimation of repeated failure surfaces of a three-dimensional weathered rock slope that is much similar to actual sliding surfaces of a rock slope under static and dynamic loading conditions.

This paper estimates the repeated SSs with slice method for a three-dimensional weathered rock (heterogeneous) cut slope beside road. The SF and stability of slope are calculated with strength reduction (SR) method (SRM) within the framework of finite difference (FD) software FLAC^{3D} (Fast Lagrangian Analysis of Continua) under both seismic and static conditions.

56.2 METHODOLOGY

Initially, Fellenius (1936) located the critical failure surface of a slope. In the current study slice method and an FD computer program FLAC^{3D} are used to demonstrate the technique in which conventional FD stress analysis data are used to generate the critical SSs (CSSs) and corresponding minimum SF. FLAC^{3D} is an unambiguous FD computer program simulating rock construction behavior that experiences flexible movement at their yield limit (Itasca 2000). It must be eminent that the SRM based on FLAC^{3D} code has been mostly used for the slope in which the failure surface is circular and the inclination is regular. The reason is that the weak or strong rock layers usually control the circular failure within the rock mass. But, this numerical method cannot be used to locate the circular failure surfaces in blocky rock masses.

56.2.1 Study Area and Geology

Fort Munro hilly area is located near Dera Ghazi Khan City in Pakistan and is selected for a case study. A slope has a height of 70 m, a length of 190 m and an angle of 29°, as shown in Figure 56.1. The geology of the slope simply comprised a sandstone material with shallow surface loose deposits (cohesionless boulder) with a friction angle of 25° (Figure 56.1). The density of rock material was 2,530 (kg/m³), cohesion was 0.0161 (MPa) and compressive strength was 0.71 (MPa). The coordinates (x and y) of slope boundary were measured with total station as presented in Table 56.1. Pore-water pressure (Ui) was measured with a piezometer during field investigation, and the other important properties (i.e., c and φ) were determined from the laboratory.

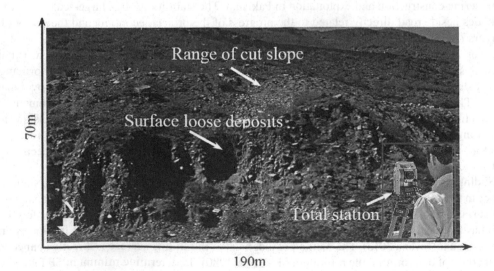

FIGURE 56.1 General view and topographical conditions of the selected slope.

TABLE 56.1

Coordinates of Stratum Boundary Line

No.	X (m)	Y (m)
1	0.000	414.000
2	5.000	414.700
3	11.090	416.851
4	12.243	418.189
5	13.397	419.435
6	14.458	420.404
7	16.995	422.018
8	19.902	423.587
9	22.163	426.078

56.2.2 SF CALCULATION

In coordinate direction, force components which are positive and act on the cut surface are described with a sequence of horizontal (P_f) and vertical forces (P_r) acting at the middle (x_i) of the slice (Sengupta and Upadhyay 2009). The factual points of exploitation were usually defined with the moments (M_y) which is relative to x_i forces (Baker 1980). The analysis is carried out in terms of boundary water pressure and total unit weight. The external loads (P_f, P_r and M) on the cut slope surface are considered as effect of water. Mathematically this is taken as

$$\text{Factor of Safety, FOS} = \frac{\Sigma S}{\Sigma T} \tag{56.1}$$

$$T_i = \frac{K_i}{F}\left[C_i + (N_i - U_i)\psi_i\right] \tag{56.2}$$

$$K_i = \left\{0 \text{ if } (N_i - U_i) < S_i, K_i = \left\{1 \text{ if } (N_i - U_i) > S_i \right.\right. \tag{56.3}$$

$$0 \leq S_i \leq \frac{C_i}{\psi_i} \tag{56.4}$$

$$P_r = F_{sp} \cdot \Sigma T - \left[\Sigma\left\{(N_i - U_i)\cdot \tan\varphi\right\} + \Sigma(C \cdot l)\right] \tag{56.5}$$

$$N = W \cdot \cos\theta + QN \tag{56.6}$$

$$U = u \cdot b \cdot \cos\theta \tag{56.7}$$

$$T = W \cdot \sin\theta + QT \tag{56.8}$$

$$\Sigma(N - U)\tan\varphi + C \cdot \Sigma l = FS \cdot \Sigma T \tag{56.9}$$

where ΣS: skid resistance, ΣT: sliding force, C_i: the force due to effective cohesion, U_i: the force due to pore water pressure, ψ_i: tan φ_i, φ_i: internal friction angle, K_i: cut-off function, S_i: tensile strength, SF: Safety factor, F_{sp}: planned SF, P_r: required prevention force, N: normal force by gravity of slice, T: tangential force by gravity of slice, b: Slice width, l: sliding surface length of slice, W: slice weight, θ: sliding surface inclination angle, u: unit pore water pressure, Q_N: vertical load component force, Q_T: vertical load component force. From the above equations system, the upper limit is consistent with linear Mohr-Coulomb criterion and the lower limit corresponds to zero tensile strength.

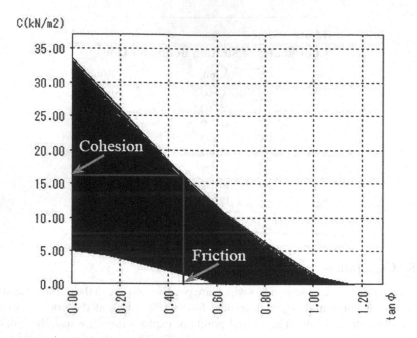

FIGURE 56.2 C-tan φ relationship diagram.

56.2.3 ESTIMATION OF REPEATED SSs

To estimate the repeated SSs and corresponding forces a C-tan φ relationship diagram is first created, (Figure 56.2), to back calculate unknown material constants (c and φ) from multiple sliding surfaces. Two curves showing the C/φ relationships can be plotted by using Eq. (56.9). Set the hypothetical SF of the said slope intended for repeated calculation to be one. The following lists the strata whose soil constants are (1) sandy soil: wet weight $\gamma_t = 17.00$ (kN/m³), friction angle $\varphi(°)$ and cohesion force c (kN/m²) are determined from reverse calculation by using density of water γ_w as 9.80 (kN/m³) and (2) weathered rock: wet weight $\gamma_t = 21.00$ (kN/m³), friction angle $\varphi = 35.00$ $\varphi(°)$ and cohesion $c = 9,999$ (kN/m²). The unknown constants (c and φ) for sandy soil were then determined from C-tan φ relational expression diagram (Figure 56.2). With reference to Figure 56.2, the value of c and tan φ is noted as $c = 16.09$ kN/m² and tan $\varphi = 0.466$ ($\varphi = 25.0°$). The repeated SSs and corresponding SF are then estimated by using the above equation system.

56.3　NUMERICAL SIMULATION

56.3.1 NUMERICAL MODEL OF SLOPE

To compute SF, displacement and seismic stability of a weathered rock slope, a slope model was created in FLAC³ᴰ. In the model, Mohr-Coulomb criterion and elastoplastic constitutive model are selected. The boundary conditions are applied according to slope geometry to generate a real model in the FD computer program. The right and left boundaries were non-slip boundaries, and the bottom boundary was fixed (Figure 56.3). In all the vertical boundaries, the horizontal movement is constrained. The appropriate grid size (0.5 m×0.5 m) is selected. The size and shape of the FD elements identical as that of the rock material parameters were assigned in two phases, Table 56.2, by using FISH language in FLAC³ᴰ as (Itasca 2000). FD code can be used to display the weak zones of a slope (Melentijevic et al. 2017), and also the failure surface of a weathered rock slope was situated in the area with a high displacement value.

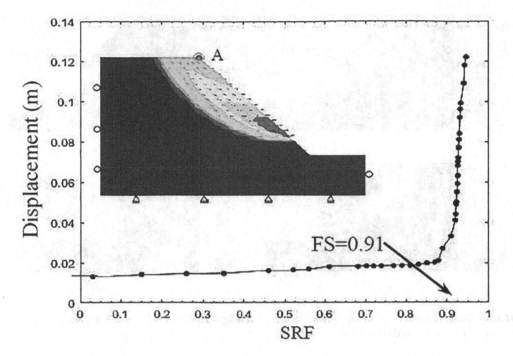

FIGURE 56.3 Change in displacement with SRF.

TABLE 56.2
Geological Features of the Case Slope

Orientation	Unit Weight γ_t (kg/m³)	Cohesion (MPa)	Friction Angle (°)	Reduction Angle (°)	Bulk Modulus (MPa)	Shear Modulus (MPa)	Strength (MPa)
Initial phase	2,000	/	/	/	3,000	1,000	/
Solution phase	2,530	0.0161	25.00	25.00	3050	1330	0.710

56.3.2 FAILURE CRITERION AND SF

The failure criterion and SF are evaluated in this subsection by using SR technique (SRM). Strength reduction factor reduces the internal friction (φ) and cohesion (c) as follows:

$$cf = \frac{c}{F} \tag{56.10}$$

$$\varphi f = \tan^{-1}\left(\frac{\tan\varphi}{F}\right) \tag{56.11}$$

where: cf and φf are reduced cohesion and internal friction, respectively. Subsequently, by using various trial values of f a series of simulations are accomplished until slope failure happens. The trial value of f that causes slope failure is reflected as the SF of the slope (Xiao et al. 2016). For slope stability investigation, the failure condition is based on the relation of safety reduction factor (SRF) values with displacements (Figure 56.3) at monitoring point A.

A relationship curve between horizontal displacement and SRF is shown in Figure 56.3. Inside a large number of interfaces an abrupt change is observed in the displacement, and the critical value

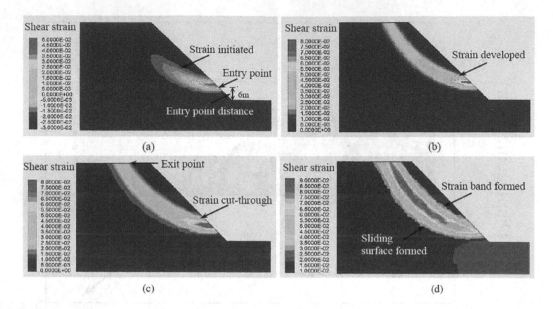

FIGURE 56.4 Contour plot of an equivalent shear strain by FLAC³ᴰ. (a) $T = 0.337$ seconds, (b) $T = 0.349$ seconds, (c) $T = 0.362$ seconds and (d) $T = 0.376$ seconds.

of SRF is considered as the SF of the slope. Conferring to Figure 56.3, the value of SRF is 0.91 for the rock cut slope.

56.3.3 DEVELOPMENT OF REPEATED SSs

The development of repeated SS of rock cut slope, including SS entry and exit points (i.e., the points at which shear strain starts to penetrate in slope and cuts through the crest of slope are entry and exit points of SS, respectively), and entry point distance (i.e., the vertical distance from slope toe to strain penetration point), depicted by shear strain at different times (Figure 56.4). The slid of rock cut slope starts at $T = 0.337$ seconds (Figure 56.4a). As observed, with strain localization color band the failure of the rock cut slope is depicted (Figure 56.4), which propagates above 6 m from the toe of slope to the crest of slope. The accumulated shear strain region started to form at $T = 0.349$ seconds as shown in Figure 56.4b. Next, it penetrates the entire slope at $T = 0.362$ (Figure 56.4c), and finally, a SS is formed at $T = 0.376$ as shown in Figure 56.4d. The geometry of the repeated SS is observed to be circular and a surface failure can happen. The reason is that a weak zone is found in between the toe and top of the slope, and the repeated SS is always situated around the slope surface.

56.4 RESULTS AND DISCUSSIONS

The results of SF, maximum displacements, maximum shear strain and seismic wave propagation at both conditions (static and seismic) of the slope are analyzed by using an FLAC³ᴰ software. Numerical results of SF, displacement and shear strain rate are illustrated in Table 56.3. The seismic impact on cut slope stability was more than static impact (Table 56.3). Under both conditions the value of SF is less than one, which indicates that the slope is found to be unstable. In Figure 56.5, the contours of maximum displacement, for the slope, can be plotted under static condition. For static conditions maximum displacement 13.33 m (10^{-3}) was observed at the cut-face of slope (Figure 56.5). Also, the contour of maximum shear strain is plotted under seismic condition (Figure 56.6). The value of maximum shear strain was noted to be 9×10^{-3}. The maximum concentration of shear strain is 10 m above from the toe of the cut slope (Figure 56.6), which is different from static conditions (Figure 56.4).

TABLE 56.3
Results from FLAC³ᴰ Analyses

Conditions	SF	Maximum Displacement (mm)	Maximum Shear Strain (10^{-3})
Static	0.91	5.33	2.33
Seismic	0.68	13.33	6.00

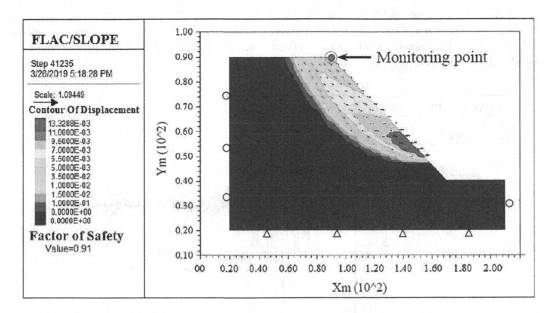

FIGURE 56.5 Contour of displacement calculated by FLAC³ᴰ.

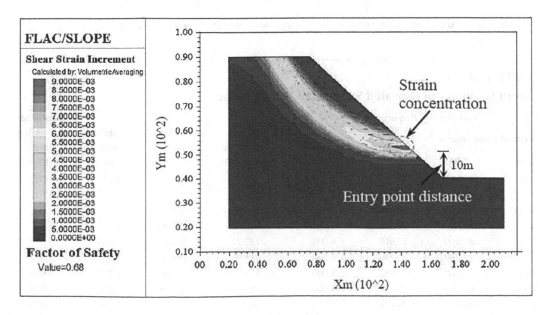

FIGURE 56.6 Safety coefficient, an increment of shear strain.

The input of the dynamic loading is a record of the seismic waves, as shown in Figure 56.7. Figure 56.7 is an acceleration-time history for an earthquake. All energy released during 4 seconds of the highest acceleration and the SF was 0.68 (Figure 56.7). The seismic wave duration may be related to the cumulative effect as a permanent loss of strength, liquefaction of rock material and deformation of slope. The horizontal and vertical stresses of cut slope were increased after the earthquake. Also in a very small period of time (4 seconds) the vertical variables should be increased and later reached a plateau.

Also, the coordinates, maximum depth and corresponding SF and other significant information for slope reinforcement design are evaluated by using the above equation system in an excel spreadsheet. As the depth of SS increases the value of the corresponding SF increases as presented in Table 56.4. The sizes of the sliding surfaces were similar, and each sliding surface was chosen as the critical one. The important information related to repeated sliding surface and the corresponding SF are summarized in Table 56.4.

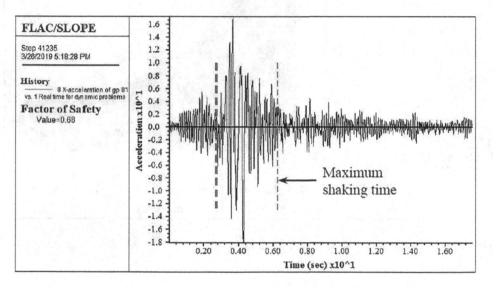

FIGURE 56.7 Acceleration time history for a seismic wave and the time of strong shaking.

TABLE 56.4
Result Summary of Repeated SS by C-tan φ Calculation

Circular Arc Requirements				Skid Resistance	Sliding Force	Safety Factor	Required Prevention Force
Central Coordinates		Radius	Max. Depth				
X (m)	Y (m)	r (m)	(m)	S (kN/m)	T (kN/m)	SF	P_r (kN/m)
1.00	444.00	27.036	4.500	791.11	769.08	1.028	1.028
−9.00	455.00	41.598	4.500	824.99	799.03	1.032	1.032
−6.00	454.00	39.149	5.000	850.03	823.13	1.032	1.032
−7.00	453.00	38.788	4.500	811.02	785.30	1.032	1.032
3.00	440.00	22.610	4.000	737.77	714.43	1.032	1.032
−8.00	454.00	40.193	4.500	817.93	792.33	1.032	1.032
−1.00	448.00	31.482	5.000	836.23	809.51	1.033	1.033
−3.00	445.00	29.901	3.500	718.48	694.96	1.033	1.033
−3.00	444.00	29.129	3.500	706.43	683.05	1.034	1.034

56.5 CONCLUSION

In the current study Fellenius modified slice method is used first to estimate the repeated SSs and the corresponding SF of a highly weathered rock beside road in Pakistan. Then SR technique is used to analyze the stability factor of a rock cut slope within the framework of $FLAC^{3D}$ under static and seismic conditions. The result showed that theoretically failures can occur where the shear forces encouraging failure are equal to the shear resistance in the ground. The failure mechanism can be a circular slope failure, and repeated SS have entry and exist points located above the toe at the crest of the slope. The cut slope is globally unstable under static and seismic conditions with inadequate SFs of slope less than one. The results in the current research can provide the foundation for reinforcement treatment of the Fort Munro cut slope.

ACKNOWLEDGEMENT

This work was conducted with supports from the National Natural Science Foundation of China (Grant Nos. 51474050 and U1602232), the Fundamental Research Funds for the Central Universities (N170108029), Doctoral Scientific Research Foundation of Liaoning Province (Grant No. 20170540304; 20170520341) and the Research and Development Project of China Construction Stock Technology (CSCEC-2016-Z-20-8) to Dr Shuhong Wang.

REFERENCES

Azimi, S.R., Nikraz, H. and Yazdani-Chamzini, A. 2018. Landslide risk assessment by using a new combination model based on a fuzzy inference system method. *KSCE Journal of Civil Engineering*, pp. 1–9.

Baker, R. 1980. Determination of the critical slip surface in slope stability computations. *International Journal for Numerical and Analytical Methods in Geomechanics*, **4**, pp. 333–359.

Cheng, Y. and Lau, C. 2014. *Slope Stability Analysis and Stabilization: New Methods and Insight*. CRC Press. Boca Raton, FL.

Fellenius, W. 1936. Calculation of stability of earth dam. *Transactions on 2nd Congress Large Dams*, Washington, DC, 1936, pp. 445–462.

Hajiazizi, M. and Tavana, H. 2013. Determining three-dimensional non-spherical critical slip surface in earth slopes using an optimization method. *Engineering Geology*, **153**, pp. 114–124.

Itasca, F. 2000. *Fast Lagrangian Analysis of Continua*. Itasca Consulting Group Inc. Minneapolis, Minn.

Jackson, J. and McKenzie, D. 1984. Active tectonics of the Alpine—Himalayan Belt between western Turkey and Pakistan. *Geophysical Journal International*, **77**, pp. 185–264.

Melentijevic, S., Serrano, A., Olalla, C. and Galindo, R. 2017. Incorporation of non-associative flow rules into rock slope stability analysis. *International Journal of Rock Mechanics and Mining Sciences*, **96**, pp. 47–57.

Pantelidis, L. 2009. Rock slope stability assessment through rock mass classification systems. *International Journal of Rock Mechanics and Mining Sciences*, **46**, pp. 315–325.

Qi, S., Wu, F., Yan, F. and Liu, C. 2007. *Rock Slope Dynamic Response Analysis*. Science Press. Beijing.

Sengupta, A. and Upadhyay, A. 2009. Locating the critical failure surface in a slope stability analysis by genetic algorithm. *Applied Soft Computing*, **9**, pp. 387–392.

Wang, H., Chen, Z.X. and Zhang, D.M. 2012. Rock slope stability analysis based on FLAC3D numerical simulation. Applied Mechanics and Materials. Trans Tech Publications, pp. 375–379.

Xiao, J., Gong, W., Martin II, J.R., Shen, M. and Luo, Z. 2016. Probabilistic seismic stability analysis of slope at a given site in a specified exposure time. *Engineering Geology*, **212**, pp. 53–62.

Index

Note: **Bold** page numbers refer to tables and *italic* page numbers refer to figures.

Printed in the United States
By Bookmasters